Lecture Notes in Control and Information Sciences 279

Editors: M. Thoma · M. Morari

Springer

Berlin
Heidelberg
New York
Barcelona
Hong Kong
London
Milan
Paris
Tokyo

S. Engell, G. Frehse, and E. Schnieder (Eds.)

Modelling, Analysis, and Design of Hybrid Systems

With 248 Figures

Springer

Editors

Prof. Dr. Sebastian Engell
Dipl.-Ing. Goran Frehse
Universität Dortmund
Fachbereich Chemietechnik
Lehrstuhl für Anlagensteuerungstechnik
Emil-Figge-Str. 70
44227 Dortmund
Germany

Prof. Dr.-Ing. Eckehard Schnieder
Technische Universität Braunschweig
Institut für Regelungstechnik und Automatisierungstechnik
Langer Kamp 8
38106 Braunschweig
Germany

Cataloging-in-Publication Data applied for
Die Deutsche Bibliothek – CIP-Einheitsaufnahme
Modelling, analysis, and design of hybrid systems / S. Engell ... (ed.). -
Berlin ; Heidelberg ; New York ; Barcelona ; Hong Kong ; London ; Milan ;
Paris ; Tokyo : Springer, 2002

(Lecture notes in control and information sciences ; 279)
(Engineering online library)
ISBN 3-540-43812-2

ISBN 3-540-43812-2 Springer-Verlag Berlin Heidelberg New York

Springer-Verlag Berlin Heidelberg New York
a member of BertelsmannSpringer Science + Business Media GmbH

http://www.springer.de

Typesetting: Digital data supplied by author. Data-conversion by PTP-Berlin, Stefan Sossna e.K.
Cover-Design: design & production GmbH, Heidelberg
Printed on acid-free paper SPIN 10883612 62/3020Rw - 5 4 3 2 1 0

Preface

In 1995, the Deutsche Forschungsgemeinschaft (DFG), the largest public research funding organization in Germany, decided to launch a priority program (Schwerpunktprogramm in German) called KONDISK– *Dynamics and Control of Systems with Mixed Continuous and Discrete Dynamics*. Such a priority program is usually sponsored for six years and supports about twenty scientists at a time, in engineering and computer science mostly young researchers working for a doctoral degree. There is a yearly competition across all disciplines of arts and sciences for the funding of such programs, and the group of proposers was the happy winner of a slot in that year. The program started in 1996 after an open call for proposals; the successful projects were presented and re-evaluated periodically, and new projects could be submitted simultaneously. During the course of the focused research program, 25 different projects were funded in 19 participating university institutes, some of the projects were collaborative efforts of two groups with different backgrounds, mostly one from engineering and one from computer science.

There were two main motivations for establishing KONDISK. The first was the fact that technical systems nowadays are composed of physical components with (mostly) continuous dynamics and computerized control systems where the reaction to discrete events plays a major role, implemented in *Programmable Logic Controllers* (PLCs), *Distributed Control Systems* (DCSs) or real-time computer systems. These two elements interact closely, and the resulting behavior can be surprisingly complex even for very simple systems, as demonstrated for the filling of three or more tanks by a switched server (Chase et al., 1993, Engell et al., 1997). Such complex behavior can neither be analyzed nor synthesized by methods that are based on either purely continuous or purely discrete systems theory. Despite the lack of theoretical tools or even powerful simulation environments for systems with mixed continuous and discrete dynamics, such systems have been engineered successfully on a trial-and-error basis, applying a combination of "divide and conquer" and "separation of concerns". The price that has to be paid, however, is extensive testing, frequent iterations in the design process, and the lack of guarantees for safety and performance properties.

The second important factor in the creation of the priority program KONDISKwas the growing awareness of the need for a more comprehensive approach to hybrid systems both in the computer science and the control engineering communities, and the fact that important foundations had been laid in both camps – and sometimes across their borders as well. It was one of the key ideas in the call for proposals that the interaction of scientists from computer science and control engineering should be stimulated, and this resulted in several interdisciplinary projects. These projects not only led to interesting and novel results but generally created a deeper understanding of the complementary theories and issues on both sides. Some of these cooperative projects can be recognized from the list of authors of the contributions in this volume.

A good German tradition in Engineering Science is that the results of academic research should somehow be applied to reality, if not in industry then at least in laboratory experiments. Thus several practical examples of systems with mixed continuous-discrete dynamics were studied in the projects of KONDISK, e.g.

- a conveyor belt, p. 26,
- an annealing furnace, p. 29,
- a wire stretching plant, p. 43,
- a membranous filtration process, p. 63,
- a batch evaporator, pp. 99, 212,
- a combined heating and material handling process, pp. 116, 201, 291, 302.
- a two-tank-system, pp. 167, 187, 297,
- a distillation column, p. 260,
- a titration plant, p. 280,
- a diesel engine, p. 288
- a multi-arm transportation task, p. 324,
- an underactuated two degree-of-freedom robot arm, p. 327,
- a chemical reactor, p. 349,
- an aircraft elevator system, p. 369,
- a three-tank-system, p. 409,
- and a multi-fingered robotic hand, p. 437.

Benchmark problems were formulated, most prominently the batch evaporator control problem, which provided the focus for a special issue of the European Journal of Control (Kowalewski et al., 2001b), and the combined heating and material handling process (Nixdorf and Lunze, 2000b).

This volume summarizes the results of KONDISK. Its structure follows the general scheme of most books on controller design: it is divided into the chapters Modeling, Simulation, Analysis, Controller Synthesis, and finally Applications. Of course, several contributions cover more than one of these topics, but an effort was made to arrange the papers according to their main focus.

Modeling

In continuous system theory, there is one single underlying modeling framework that provides a compact and powerful description of physical dynamic systems and serves as a starting point of most pieces of theoretical work: systems of differential and algebraic equations. In mixed continuous-discrete systems, the diversity of the available descriptions of reactive discrete event systems and the various choices to restrict the dynamic behavior of the continuous part for the sake of decidability and efficiency of the analysis result in a large number of possible combinations of discrete and continuous formalisms, each one with its own advantages and often tailored to specific application domains. This leaves the choice of the modeling framework open for discussion. Rather than trying to find out which paradigm might possibly be applicable to all potential problems, it pays off to analyze which one is particularly suited for the problem at hand. Following this line of thinking, the chapter on modeling presents several different proposals for the modeling of mixed dynamic systems.

The chapter starts with a somewhat provocative statement by Lunze that challenges hybrid systems theory by putting forward that only those systems are truly hybrid that contain jumps of the continuous state whereas other forms of discontinuities can be dealt with in either the discrete or the continuous domain. In the second article by Drath et al, discrete-event Petri Nets are enhanced with continuous components such as firing speeds for the transitions or continuous attributes of the tokens. A hierarchy is introduced by an object-oriented encapsulation of subnets. The contribution by Bender et al. presents a hybrid extension to the *real time object-oriented methodology* (ROOM) for the model based development of hybrid systems, and discusses its application in two tools for simulation and test case generation.

The fourth article by Münnemann et al. defines encapsulations and templates for function blocks in a manner similar to ROOM. That way, standardized components, such as generic control units, can be reused. Their behavior is formally specified by condition/event systems. Many problems in hybrid systems can be dealt with efficiently on the basis of a simpler, sometimes even purely discrete model. This model must match the behavior of the original model in the sense of an over-approximation, or abstraction. The article by Lunze and Raisch examines the properties of discrete abstractions of continuous systems, and discusses the implications of the resulting non-deterministism of the discrete-event models.

Simulation

Compared to the models used for analysis and controller synthesis, much more complex models can be handled efficiently in simulation. Currently, simulation is the only available tool that can cope with nonlinear dynamics interacting with complex discrete event dynamic systems. The description of a large hybrid system must be represented in a manner that combines (re-)usability with efficient simulation. On the algorithmic level, there is ongoing research how to integrate structural changes and discrete event handling with the established methods for solving differential equations or DAE-systems.

Remehle et al. introduce a software environment for the integration of complex hierarchical discrete-event models in MODELICA, a powerful object-oriented language for continuous systems. Graphical editors and translators for the discrete parts of the overall system support various formalisms, modularity and hierarchy. The overall model can be solved efficiently using the preprocessing and event-handling capabilities of MODELICA-based simulators.

The paper by Pawletta et al. presents a hierarchical modeling approach that supports time-varying structures of coupled systems. A combination of modular and monolithic simulation techniques avoids the overhead that is otherwise necessary for the coordination of the subsystems in strictly modular simulators. In the contribution by Nordwig, the software engineering concepts of restricted genericity and structural dynamics are applied to the modeling of hybrid systems. Based on the object-oriented specification language ZimOO, the graphical tool zooed is presented.

Analysis and Verification

The survey paper by Kowalewski gives an introduction to the formal analysis of hybrid systems. It highlights different directions from which hybrid models and their analysis are approached in computer science and engineering. Fundamental problems arising from the combination of discrete and continuous dynamics are discussed, and the following articles in this chapter are related to the different basic approaches.

The contribution by Nenninger et al. presents the so-called net-state-model, a combination of a continuous (ODE) system and a Petri Net, and discusses the reachability analysis for a class of hybrid systems. A control design scheme for hybrid systems with piecewise affine dynamics is introduced that is based on left eigenvector assignment. The analysis of fluid stochastic Petri Nets is discussed in the paper by Wolter et al. An improved numerical solution algorithm based on discretization is proposed for nets with two continuous dimensions. An example illustrates how performance metrics can be obtained from such models.

Simon et al. describe a method to deal with time critical problems in the field of automatic control of manufacturing systems. They determine values of the parameters of timestamp Petri Nets which prevent the net from getting blocked because of timing conditions by solving a linear optimization problem. Finally, the use of formal methods in the analysis and control of hybrids systems is reviewed in the paper by Huuck et al. They introduce some formal concepts and models, and present a compositional approach to the verification of hybrid automata based on the assumption/commitment paradigm.

Controller Synthesis

Analysis and verification are ex post activities which require that a controller has been designed beforehand. In analogy to synthesis procedures for continuous controllers, which have been the ultimate goal of control theory for decades, the obvious alternative is to come up with synthesis procedures for hybrid systems so that the desired properties are satisfied by design and no verification step is necessary. Moor and Raisch use the abstraction of a continuous system with discrete external signals, as presented in the chapter on modeling by Lunze and Raisch, to synthesize a discrete controller. They show that the temporal evolution of the quantization cells can be conveniently over-approximated if the dynamics of the system is monotone.

In the case of a fault in a system, it may be necessary to change not only the parameters but also the structure of the controller. The paper by Lunze and Steffen presents a method where the faulty system is modeled by a timed stochastic automaton. First, a discrete controller takes the system to an equilibrium state by choosing new actuators, sensors or set-points. Then a linear controller is designed to stabilize the system around the new equilibrium.

Wegele et al. present an iterative scheme for the optimal control design for hybrid systems. Each iteration of the overall optimization consists of an automatic controller design and then testing the controlled system for the violation of given constraints. A violation results in an additional penalty term in the cost function associated with the controller. The optimization algorithm can modify the controller parameters as well as choose a different design method. A procedure to detect and prevent undesired transitions of a discrete control of a continuous system is proposed by Müller et al. The system is modeled by Place/Transition nets that are fully deterministic. Undesired transitions are detected in the condensation of the evolution graph and excluded by modifications of the Petri Net and the firing conditions.

Two papers are concerned with the optimization of hybrid systems. Buss et al. present an approach for the computation of optimal trajectories of nonlinear hybrid systems. The continuous subproblem, including resets and switchings at fixed points in time, is solved by direct collocation. For the remaining problem of determining the times when discrete transitions occur, two alternative methods are proposed: finding suboptimal solutions on a grid and solving a mixed integer program. The method is applied to examples from robotics and a scalable hybrid travelling salesman problem. The paper by Stursberg et al. describes a linear mixed-integer discrete-time approximation of a hybrid system for the calculation of optimal continuous and discrete inputs for a linear cost function. In contrast to other approaches, a disjunctive formulation is used in the transformation to a mixed-integer linear program, and the combinatorial explosion is reduced by using a moving horizon and variable time steps.

Applications

The first paper by Decknatel et al. deals with the performance analysis of moving and fixed-block train protection systems. An extension of Colored Petri Nets is used where the tokens contain the current value of an attribute and parameters describing the continuous behavior of that value. Performance analysis is carried out using the tool Design/CPN. In the contribution by Mosterman et al., a complex object-oriented model of aircraft dynamics, hydraulic actuators, and continuous controllers is combined with a complex redundancy management system and the overall system is simulated. The general approach is the one described by Remehle et al. in the chapter on modeling with extensions to handle structural changes in DAE systems.

Manz and Göhner present an online monitoring method based on a combination of qualitative and dynamic models. An online state space reduction and an online analysis for failure detection and hazard predicton are carried out, based on the qualitative model. The approach is applied to a three-tank-system. Models of traffic flow of varying detail are combined in the paper by Czogalla et al. in order to obtain a model that provides sufficient accuracy as well as acceptable computational performance. The approach decomposes the overall model and uses more abstract submodels where certain effects of the refined model can be neglected

The final contribution by Schlegl et al. presents a hybrid controller for a robotic hand. It involves a hybrid planning scheme for grasping and re-grasping, impedance control algorithms based on sensor information, and a formal compensation method for discrete contact state errors. The resulting performance is illustrated by dynamical simulations and experiments.

Summarizing the whole impact created by the research of this priority program, a substantial progress concerning different aspects can be observed. One indication is the apparent and increasing international visibility of the research on this topic in Germany, quantified by the number of publications presented on national and international conferences, e.g. EKA, WODES, ADPM, ECC, etc., and journals. The volume's extensive bibliography witnesses the broad publishing activities. In addition, having reached their academic merits relying on this work, the young researchers equipped with comprehensive knowledge of and theoretical experience in handling complex facts have been employed in strategic positions within leading industrial companies, mainly of the automotive branch. Moreover, some researchers decided to start up an entrepreneurship by founding a company of their own.

Regarding the scientific and first practical results – in comparison with evolution in biology – some of the early approaches to tackle the methodological challenges of continuous-discrete systems have been observed to survive the selection by efficiency and effort. Hence, the long-term evidence of this priority program on hybrid systems research will only become apparent after several years, maybe a decade. It may, however, be assumed that the promising research has theoretically opened the inherent potential of the advanced approaches to solve more complex applications on the one hand, and to exhaust the given boundaries of existing systems on the other.

Acknowledgment

In the name of all participants in the research program KONDISK, the editors would like thank the DFG for the financial support and particularly Dr. Klaus Wefelmeier for his efforts. Both were essential to the success of this program.

Dortmund, Braunschweig,
April 2002

Sebastian Engell
Goran Frehse
Eckehard Schnieder

Contents

Part I

Modeling

What Is a Hybrid System?

Jan Lunze

Institute of Automation and Computer Control
Ruhr-University Bochum
Universitätsstraße 150
D-44780 Bochum
phone: +49 234 32 24071
http://www.ruhr-uni-bochum.de/atp

Abstract. Hybrid systems have become a major research topic in Control Engineering and other disciplines. Many different models have been proposed for describing them. However, the question what a hybrid system is has remained a matter of debate. This paper argues that state jumps are the basic hybrid phenomenon that cannot be represented and analysed by methods elaborated either in continuous or in discrete systems theory. Hence, a system has to be considered as a hybrid system if both the continuous movement and the state jumps are important for the control task to be solved.

1 Introduction

During the last decade, hybrid dynamical systems have become a major research topic. The conference proceedings like (Alur et al., 1996, Antsaklis et al., 1999, Antsaklis et al., 1995, Antsaklis et al., 1997, Grossman et al., 1993, Lynch and Krogh, 2000), the special journal issues (Antsaklis and Nerode, 1998a, DEDS'98, 1998) and (Automatica 35(3), 1999) and this book describe the different research directions and results obtained. The hybrid nature of such systems has attracted the interest of mathematicians, control engineers and computer scientists. The methods applied and the results obtained are as diverse as the backgrounds of these researchers. No common definition of a hybrid system is available.

A major argument in the hybrid system literature says that a given dynamical system should be considered a hybrid system if (and only if) it is impossible to deal with it either as a purely continuous–variable system or as a purely discrete–event system without ignoring important phenomena that result from the combination of continuous and discrete movements of this system. Here and in the rest of the paper, the terms *continuous* and *discrete* are used with respect to the range of the signals and not with respect to the time over which the signals are defined.

This argument does not clarify what a hybrid system is. Most of the theoretical papers start with a *given* hybrid system and do not consider whether and why hybrid systems theory has to be applied to the system under investigation. Likewise, application papers use hybrid models and analysis tools but do not elaborate the main reasons why the system had to be dealt with as a hybrid system. Most of the chapters of this book adopt the same position and investigate different kinds of hybrid systems.

S. Engell, G. Frehse, E. Schnieder (Eds.): Modelling, Analysis and Design of Hybrid Systems, LNCIS 279, pp. 3–14, 2002.

This situation is due to the fact that the theories of continuous and discrete systems have been elaborated completely separately until recently. Hybrid systems pose the problem of bridging the gap between both theories. This has been done until now not only by considering a combination of continuous and discrete subsystems but also by investigating different extensions of either continuous or discrete systems. Timed or hybrid Petri Nets and timed automata generalise the untimed models that are typically used in discrete systems theory whereas switched systems extend continuous systems by discrete phenomena, to mention some example systems investigated in this book.

From a theoretical point of view the definition of the field of hybrid systems by a collection of different kinds of systems or models is not satisfactory. It is important to know why the new class of hybrid systems has to be defined, because this clarifies the *novelty* of dynamical phenomena to be investigated and the *necessity* of developing new concepts, methods and tools. This chapter gives an answer to the question of what a hybrid system is. It argues that a new notion should only be defined if the class of systems under investigation does not fall within the framework of the already existing theory. The characterisation of hybrid systems given here is in some contradiction with definitions used in literature. It should initiate a thorough discussion of this new notion.

The problem of defining a hybrid system is rather involved because it is not the characteristics of a given dynamical system alone that distinguishes continuous from discrete and hybrid systems. For example, a tank system is usually considered to be continuous if a level controller is to be designed but it is tackled as a discrete system if it is analysed as a part of a batch process, in which only the discrete state of a full or an empty tank are distinguished. The intentions of modelling, analysis or control have a considerable influence on the fact whether a system has to be considered as a hybrid system or not.

Hybrid dynamical systems have existed for long, but before the appearance of the currently developing theory on hybrid systems, these systems have either been considered as purely continuous or as purely discrete. The reason why they became a hot topic in research is given by the fact that many modern technological processes cannot be analysed and controlled by investigating only the continuous or only the discrete movements. As the theories of continuous or discrete systems have made contradictory basic assumptions, which have to be satisfied by the systems in order to make their representation in the framework of the respective theory possible, the consideration of both continuous and discrete phenomena makes it necessary to develop a new theory. Roughly speaking, if the main assumptions of both theories are not satisfied, a system has to be dealt with as a hybrid system.

Definition 1. A *hybrid system* is a dynamical system that cannot be represented and analysed with sufficient precision either by the methods of the continuous systems theory or by the methods of the discrete systems theory.

Remarks. *Continuous systems theory* assumes that the system under consideration can be described by some differential equation

$$\dot{\boldsymbol{x}} = \boldsymbol{f}(\boldsymbol{x}(t), \boldsymbol{u}(t), t), \quad \boldsymbol{x}(0) = \boldsymbol{x}_0 \tag{1}$$

$$\boldsymbol{y}(t) = \boldsymbol{g}(\boldsymbol{x}(t), \boldsymbol{u}(t), t) \tag{2}$$

where $\boldsymbol{x} \in \mathbb{R}^n$ is the state vector, $\boldsymbol{u} \in \mathbb{R}^m$ the input vector and $\boldsymbol{y} \in \mathbb{R}^r$ the output vector. $\boldsymbol{x}_0$ denotes the initial state. More generally, (1) can be replaced by a set of difference and algebraic equations, which then is called a differential–algebraic system (DAE system).

The key assumption of continuous systems theory concerns the fact that the functions $\boldsymbol{f}$ and $\boldsymbol{g}$ satisfy a Lipschitz condition. With respect to the state $\boldsymbol{x}$ this smoothness assumption means for the function $\boldsymbol{f}$ that a constant L has to exist for which the inequality

$$||\boldsymbol{f}(\boldsymbol{x}, \boldsymbol{u}, t) - \boldsymbol{f}(\hat{\boldsymbol{x}}, \boldsymbol{u}, t)|| \leq L \cdot ||\boldsymbol{x} - \hat{\boldsymbol{x}}|| \tag{3}$$

holds for all $\boldsymbol{x}$, $\hat{\boldsymbol{x}}$, $\boldsymbol{u}$ and t, where $||\cdot||$ symbolises a vector norm. A similar condition should be satisfied with respect to $\boldsymbol{u}$. Under this assumption, uniqueness and existence results can be derived for the solution of the differential equation (1). Furthermore, many analysis methods assume the property (3).

Discrete systems theory considers systems whose signals have discrete range. That is, all signals are assumed to be binary–valued or have values from a finite or infinite discrete value set. Due to this assumption, the system jumps from one discrete state to another but the continuous movements of the system cannot be described. The abrupt state changes are called events.

These basic assumptions of the continuous and discrete systems theories are, on the one hand, contradictory, but on the other hand complementary. A system can be either continuous or discrete, because it either moves Lipschitz–continuously from one real–valued state to another or it jumps among different discrete states. In many situations it is satisfactory to consider either the continuous movement or the discrete jumps of the system.

2 Hybrid Phenomena

Hybrid phenomena are state transitions that cannot be represented or analysed by the methods developed in continuous *or* discrete systems theory. That is, these phenomena do not satisfy the Lipschitz condition (3) and cannot be represented as a sequence of discrete state values. Hence, neither a purely continuous nor a purely discrete representation is appropriate for the task at hand.

Hypothesis 1 *Consider a dynamical system subject to some continuous input* $\boldsymbol{u}$. *The basic hybrid phenomenon is a combination of continuous state changes and abrupt state jumps.*

Remarks. Figure 1 shows a state trajectory $\boldsymbol{x}(t)$ of an autonomous first–order system which includes a state jump at time t_k. The state jump is an autonomous jump which occurs after the state has reached a threshold $\boldsymbol{x}_s$. This trajectory cannot occur as the solution of models that are developed either in continuous or in discrete systems theory. On the one hand, the Lipschitz condition (3) is not satisfied for $\boldsymbol{x} = x_s$ (and $\boldsymbol{u} = 0$) and, hence, the system cannot be dealt with by the methods of continuous systems theory. On the other hand, the continuous movement of the system cannot be represented by models with discrete–valued state, because such a model could only describe the discrete state jump between $\boldsymbol{x}(t_k - 0) = \boldsymbol{x}_s = 3$ and $\boldsymbol{x}(t_k + 0) = 1$ but not the system behaviour for time $t \neq t_k$.

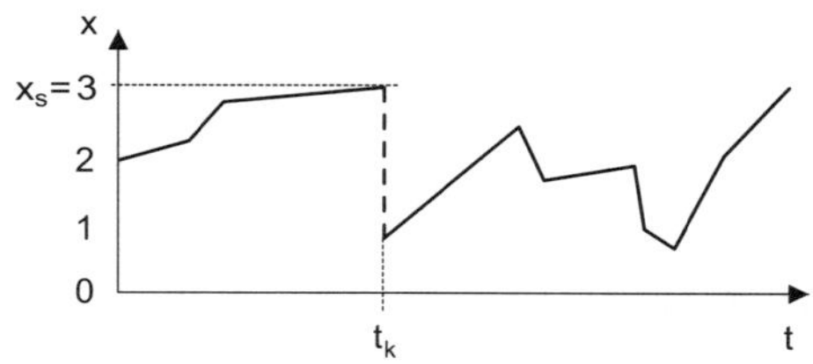

Fig. 1. Hybrid trajectory

In the literature on hybrid systems four phenomena are called hybrid (Branicky, 1995), (Branicky, 1996): autonomous and controlled jumps and autonomous and controlled switches. The following remarks show how these phenomena relate to Hypothesis 1. The discussion concentrates on the state evolution $\boldsymbol{x}(t)$, but extensions to systems with outputs $\boldsymbol{y}(t)$ are mentioned later.

Autonomous state jumps. The first situation concerns the fact that the state of a system may jump after it has reached a threshold $\boldsymbol{x}_s$ or, more generally, the border of a subset $\mathcal{X}$ of the continuous state space. Hence, the derivative $\dot{\boldsymbol{x}}$ exceeds all bounds but has the form of a Dirac impulse. A simple example is the first–order system

$$\dot{x} = -x(t) + \delta(x(t) - x_s), \quad x(0) = x_0 \tag{4}$$

for which the function

$$f(x) = -x + \delta(x - x_s)$$

violates the Lipschitz condition (3) for $x = x_s$. If both the continuous movement between the jumps and the state jumps have to be considered when solving a given modelling, analysis or control task, the system has to be considered as hybrid.

The jump occurs at time t_k for which the condition

$$x(t_k) = x_s \tag{5}$$

is satisfied. As the jump has no duration, the time instances $t_k - 0$ before the k-th jump and $t_k + 0$ after the jump are often distinguished. The point here is that the time

t_k (or $t_k - 0$) is *implicitly* described by the system equation (4). It depends upon the movement of the system and, hence, upon the initial state x_0. If the system trajectory is to be determined, the time t_k has also to be determined. For the system (4) an explicit representation can be obtained

$$t_k = -\ln \frac{x_s}{x_0} \quad \text{if } |x_0| > |x_s| \text{ and } x_0 x_s > 0, \tag{6}$$

but for higher–order linear systems and for most nonlinear systems, such an explicit formula cannot be found.

State jumps occur in particular, if the dynamics of a DAE system change when the state reaches a hypersurface in the state space. Then the state vector may even change its length. Details about such systems are described in (Verghese et al., 1981).

Controlled state jumps. The state of a system may change discontinuously if the continuous input reaches a given bound $\boldsymbol{u}_s$. Then the vector field $\boldsymbol{f}(\boldsymbol{x}, \boldsymbol{u}, t)$ violates the Lipschitz condition for $\boldsymbol{u} = \boldsymbol{u}_s$. A simple example is the first–order system

$$\dot{x} = -x(t) + \delta(u(t) - u_s), \quad x(0) = x_0 \tag{7}$$

whose state jumps if the input reaches the threshold u_s. The occurrence time of the jump depends upon the input and, more generally, may also depend upon the state, as in the system

$$\dot{x} = -x(t) + \delta(x(t) - u(t) - u_s), \quad x(0) = x_0. \tag{8}$$

Like autonomous state jumps, controlled jumps occur if the combined vector $(\boldsymbol{x}' \;\; \boldsymbol{u}')'$ reaches a threshold (or a hyperplane). Contrary to autonomous jumps, controlled jumps can be forced or prevented to occur by appropriately choosing the input $\boldsymbol{u}$. For example, for the control $u(t) = x(t)$ the state of the system (8) never jumps.

These controlled jumps have to be distinguished from jumps that occur due to input jumps or impulses. In systems theory, Dirac impulses are considered as a possible input to the system. Then a state jump occurs due to the infinite magnitude of the input which implies an infinite magnitude of the vector field $\boldsymbol{f}$. Such jumps do not represent a hybrid phenomenon. Therefore, in Hypothesis 1 the input $\boldsymbol{u}$ has been restricted to be continuous. Then, state jumps are a result of the dynamical properties of the system. In summary, state jumps that are brought about by continuous input signals represent a hybrid phenomenon.

Autonomous switching. An abrupt change of the vector field $\boldsymbol{f}$ if the state $\boldsymbol{x}$ reaches a given bound is called switching. Formally, the system can be represented by two or more different vector fields $\boldsymbol{f}_q$ together with conditions that describe the validity of these vector fields, for example by

$$\dot{\boldsymbol{x}} = \boldsymbol{f}(\boldsymbol{x}) \tag{9}$$

with

$$f = \begin{cases} f_1(x) & \text{for} \quad h(x) \leq 0 \\ f_2(x) & \text{for} \quad h(x) > 0. \end{cases} \tag{10}$$

If the system is currently described by the vector field f_1 and the state reaches the border $h(x) = 0$ of the region of validity of this vector field, the vector field switches to f_2 which is valid until the border described by $h(x) = 0$ is reached from the other side.

The notion of switching has been introduced in order to denote the fact that the system is governed by two (or more) different differential equations, which can be analysed separately by well known results from continuous systems theory but whose common analysis poses new problems. If the vector fields are linear and the system is described by

$$\begin{aligned} \dot{x} &= A_1 x \quad \text{for} \quad h(x) \leq 0 \\ \dot{x} &= A_2 x \quad \text{for} \quad h(x) > 0, \end{aligned}$$

the separate analysis of the two models is very easy but the analysis of the system as a whole is difficult. It is known, for example, that the system may be unstable even if the two matrices A_1 and A_2 are stable.

This consideration shows that the notion of switching has been introduced mainly due to the modelling phenomenon that the vector field f cannot be described by a unique analytical expression. This, however, does not imply that the system is hybrid in the sense defined above. A Lipschitz condition may be satisfied even if the vector field switches. If x_s denotes a state on the border between the two regions of validity of the vector fields f_1 and f_2, the condition

$$\|f(x_s) - f(\hat{x})\| \leq L \cdot \|x_s - \hat{x}\| \tag{11}$$

can be satisfied for some constant L for all $\hat{x}$ and x_s satisfying the relation $h(x_s) = 0$. Hence, switching does not imply state jumps and, therefore, does not describe a hybrid phenomenon. If the vector field f in (10) satisfies the condition (11), the system can be dealt with by continuous systems theory.

If the Lipschitz condition is not satisfied, because for some state x_s the relation

$$\lim_{\varepsilon \to 0} f_1(x_s - \varepsilon) \neq f_2(x_s)$$

holds, the vector field includes some "step" and represents a piecewise continuous function. Then the trajectory $x(t)$ is not differentiable but piecewise differentiable. Even in this case, the trajectory $x(t)$ does not include any state jumps. The system is continuous although its analysis is more involved due to the properties of the vector field f. However, the system is not hybrid. This consequence is in contrast with the fact that switching systems are currently considered as an important subject of hybrid systems theory.

At this point of the argument it should be stressed that Hypothesis 1 has been formulated with the aim to investigate the necessity of introducing the new notion of hybrid systems. It does not question the fact that switching systems pose a lot of unsolved theoretical problems, which even represent a focus of the current literature of systems theory.

Controlled switching. The same arguments apply to systems with controlled switching where the vector field also changes abruptly in response to an input command $\boldsymbol{u}$. Here, the notion of switching is used for systems with piecewise constant input, where for a given time interval the input is fixed to some value $\bar{\boldsymbol{u}}$ and, hence, the vector field is fixed to

$$\boldsymbol{f}(\boldsymbol{x}, \bar{\boldsymbol{u}}) = \bar{\boldsymbol{f}}(\boldsymbol{x}).$$

This kind of switching is nothing else than a change in the vector field due to a given input. As the autonomous system can be analysed more easily than a system with (arbitrary) input, the system with piecewise constant input is considered as an autonomous system with switching dynamics, which does not imply that the system exhibits hybrid phenomena.

Switched linear systems occur for this reason. If the input to the system

$$\dot{\boldsymbol{x}} = \boldsymbol{A}\boldsymbol{x}(t) + \boldsymbol{B}\boldsymbol{u}(t)$$

is piecewise constant and can assume only the discrete values $\boldsymbol{u}_i$ $(i = 1, ..., q)$, the system is governed by the affine differential equation

$$\dot{\boldsymbol{x}} = \boldsymbol{A}\boldsymbol{x}(t) + \boldsymbol{B}\boldsymbol{u}_i = \boldsymbol{A}\boldsymbol{x}(t) + \boldsymbol{b}_i$$

as long as the input does not change. The system is *considered* as a switching system, although it is merely a linear system with piecewise constant input.

Similarly to systems with autonomous switching, it is an interesting and useful way to analyse the vector field that drives the system between the switchings separately and to combine the results to get an analysis result for the overall system. However, from the viewpoint adopted here it becomes clear that this way of analysis does not make the system hybrid.

Extension to systems with outputs. For systems where the output $\boldsymbol{y}$ is generated by some function $\boldsymbol{g}$ as described by (2) the considerations above have to be extended. Then, the Lipschitz condition has to be imposed on $\boldsymbol{g}$ as well. Depending on whether both $\boldsymbol{f}$ and $\boldsymbol{g}$ satisfy Lipschitz conditions or not, the system may have different kinds of state or output jumps. As shown in (Verghese et al., 1981), systems may have a continuous output $\boldsymbol{y}$ although the state $\boldsymbol{x}$ jumps. On the other hand, quantised systems (Lunze, 2000) are systems with continuous state evolution but jumping output.

Note also that the jumps considered here concern the state or the output of the system for continuous input. Output jumps may occur even if the function $\boldsymbol{f}$ and the input $\boldsymbol{u}$ remain finite. An example for this is the linear system

$$\dot{\boldsymbol{x}} = \boldsymbol{A}\boldsymbol{x} + \boldsymbol{B}\boldsymbol{u}$$
$$\boldsymbol{y} = \boldsymbol{C}\boldsymbol{x} + \boldsymbol{D}\boldsymbol{u}$$

with $\boldsymbol{D} \neq \boldsymbol{O}$. This system simply transfers a jump in the input directly to the output. Hence, jumps of the output do not necessarily indicate a hybrid system behaviour. It has to be analysed whether the jumps originate from the input or from the system dynamics. Only in the latter case, the system is hybrid.

3 Representation of Hybrid Systems by Differential Equations

Time is continuously changing. Therefore, the most natural description of any dynamical system includes signals that are defined over the real time axis $\mathbb{R}^+$. As the hybrid systems are considered here with a continuous time axis, the question arises whether they can be described by differential equations. The answer is in the affirmative.

Hypothesis 2 *A hybrid dynamical system can be described by the differential equation*

$$\dot{\boldsymbol{x}} = \boldsymbol{f}(\boldsymbol{x}, \boldsymbol{u}, t) \tag{12}$$

where the function $\boldsymbol{f}$ *includes a Lipschitz–continuous part* $\tilde{\boldsymbol{f}}(\boldsymbol{x}, \boldsymbol{u}, t)$ *and Dirac impulses:*

$$\boldsymbol{f}(\boldsymbol{x}, \boldsymbol{u}, t) = \tilde{\boldsymbol{f}}(\boldsymbol{x}, \boldsymbol{u}, t) + \sum_{i=1}^{n_x} \boldsymbol{f}_i\, \delta(g_i(\boldsymbol{x}(t), \boldsymbol{u}(t), t)) \tag{13}$$

Remarks. A state jump occurs at time t_k if the state derivative $\dot{\boldsymbol{x}}$ includes a Dirac impulse

$$\boldsymbol{f}_i\, \delta(g_i(\boldsymbol{x}(t), \boldsymbol{u}(t), t)) = \delta(t - t_k).$$

This impulse occurs whenever the state $\boldsymbol{x}$ and the input $\boldsymbol{u}$ reach at time t_k the hyperplane described by $g_i(\boldsymbol{x}, \boldsymbol{u}, t) = 0$. Note that the time t_k is implicitly defined by the movement of the system, the input and the definition of the hyperplane.

For example, the system

$$\dot{x} = -x(t) + \delta(x(t) - x_s), \quad x(0) = x_0 \tag{14}$$

has a continuous trajectory until the state assumes the value x_s at time t_k. In this case the system jumps to the new state

$$x(t_k + 0) = x(t_k - 0) + 1.$$

Hypothesis 2 and this example point to the fact that a hybrid system can be represented by a differential equation and that the hybrid nature of a system does not imply that the system can only be described by a combination of differential equations and automata, which is often argued in hybrid systems theory (cf. Sect. 4). However, they also show that hybrid systems are nonlinear systems, whose vector fields do not satisfy the smoothness assumption made in nonlinear systems theory. The existence of Dirac impulses in the representation of the vector field is a consequence of Hypothesis 1.

It should be mentioned that the arbiter, which has been used in (Branicky, 1995) (p. 110) as an example system that cannot be represented by any differential equation, *can* be described by a differential equation of the form (13). The argument given in (Branicky, 1995) is only true with respect to Lipschitz–continuous vector fields. If the restriction concerning the continuity of the vector field is relaxed, a differential equation also exists for the arbiter.

4 Decomposition of Hybrid Systems

Hybrid systems theory has tackled the question of which structure a hybrid system may be appropriately represented. Many authors use the decomposition of the hybrid system into a continuous and a discrete subsystem. This decomposition is reasonable, because this definition makes it possible to use, at least in part, the models and analysis methods that have been elaborated in the two corresponding theories (Fig. 2). The continuous input $\boldsymbol{u}$ and the continuous output $\boldsymbol{y}$ are associated with the continuous–variable subsystem whereas the discrete–valued input v and output w concern the discrete–event subsystem. Both subsystems have to be connected through interfaces that transform continuous signals into discrete and vice versa. These interfaces are called quantiser and injector. Note that all signals represented in Fig. 2 by some arrow are defined over the continuous time axis.

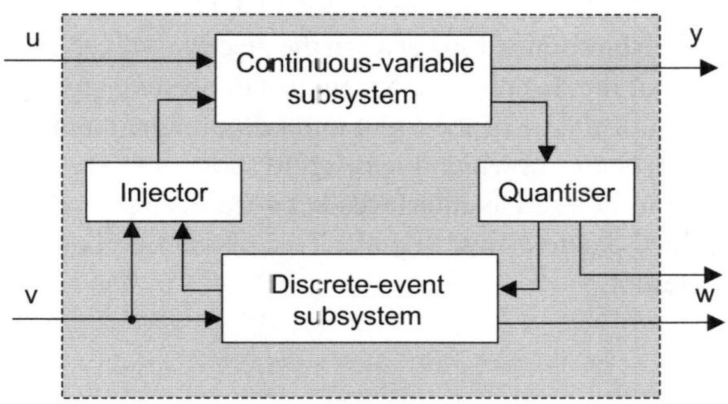

Fig. 2. Decomposed hybrid system

Different models have been proposed in literature in order to satisfy two aims:

- The model should be capable of describing dynamical systems that exhibit hybrid phenomena.
- It should be possible to analyse the model by rigorous methods.

These aims are contradictory. Whereas the first aim necessitates that the model should be as general as possible, the second aim can be satisfied only if the model is as specific as possible.

Whereas in (13) the continuous and the discrete movement of the state are described in a common model, the decomposition depicted in Fig. 2 describes the system by two different mathematical models. As this decomposition should take advantage of the methods of discrete systems theory, untimed discrete-event models are often used. This, however, brings about a new theoretical difficulty which is to be explained below.

The discrete subsystem, like the continuous subsystem, changes its state over the continuous time t and can be described by a differential equation. The state z jumps among different state values if the vector field includes impulses like

$$\dot{z} = \delta(t - t_k)$$
$$\dot{z} = \delta(z - v).$$

The first equation describes a discrete system in which a state jump occurs after a certain time t_k is elapsed. The second equation represents a system whose state jump occurs at the time instant at which the input v assumes the discrete value z.

Such a description is called a *timed* model in discrete–event systems theory. Many models proposed in hybrid systems theory assume that the discrete model is untimed. Typically, (untimed) automata or Petri Nets are used. The time at which a state transition occurs is given by the discrete input v which generally consists of discrete control inputs to the system and of quantised continuous-variable signals (cf. Fig. 2). The quantiser is used not only to determine the discrete value of the input v but also to determine the time t_k at which the untimed model changes its state.

In a more precise representation, the connection between the discrete-valued continuous-time input v and output w of the discrete subsystem has to use two signals: One that describes the values $v(t_k)$ or $w(t_k)$ of the input and output signal and the second which describes the time instant t_k at which the discrete subsystem changes its state (Fig. 3). Two blocks have to be used as interfaces between discrete-valued continuous-time and discrete-valued discrete-time signals. This distinction is often ignored in the hybrid system literature.

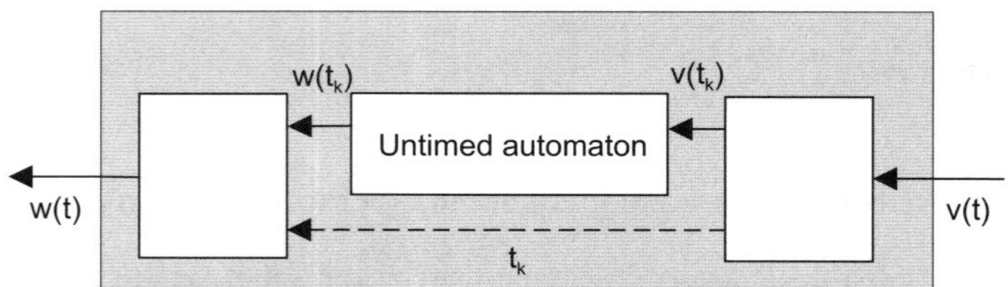

Fig. 3. Quantisation with discrete signal and clock signal

The representation of a hybrid system as a composition of a continuous and a discrete subsystem has several advantages:

- The representation shows the hybrid nature of the system explicitly.
- The methods available for continuous and discrete systems can be applied to the separate subsystems (although the results obtained for the isolated subsystems are not valid for the overall hybrid system.)

For example, automata theory can be applied to the discrete subsystems whereas results from continuous systems theory (controllability analysis, stability analysis etc.) can be applied to the continuous subsystem.

However, the representation of a system in the form depicted in Fig. 2 does not imply that the system is hybrid. Switched linear systems may be represented in this form where the quantiser determines the region in which the state resides and the discrete subsystem switches to a new model "number" after the boundary of a region is reached. As the discussion above has shown, such a system can be represented as a nonlinear system with a piecewise continuous vector field.

5 Conclusions

Hybrid systems has emerged as a new field of research. To see the novelty of this field, it has to be precisely defined what a hybrid system is. Only with this definition it becomes clear which new questions have to be answered.

This paper shows that state jumps within a continuous movement is the basic and probably the only hybrid phenomenon. Hence only those modelling, analysis and control tools that deal with this particular phenomenon are specific for hybrid systems.

This arguments can be used to evaluate the growing number of papers entitled hybrid systems and to elaborate interesting research topics that bring forward the main ideas and intentions of hybrid systems. For example, analysis methods for switched linear systems are certainly an interesting research topic but from the viewpoint adopted in this chapter this class of systems does not exhibit the main problems encountered in hybrid systems. Likewise timed Petri Nets or timed automata are interesting extensions of classical discrete-event systems, but they should be combined in future research with continuous state or output variables to contribute to the main ideas of hybrid systems theory.

This chapter concentrated on the question of what hybrid systems are. It did not concern the problem whether in a specific situation a hybrid system should be really tackled by the methods developed in hybrid systems theory. For example, methods for the discrete-event representation of quantised or hybrid systems surveyed in the chapter on "Discrete models for hybrid systems" concern the question of how the hybrid nature of the system can be neglected in order to simplify the analysis or control tasks. The resulting discrete-event model ignores the continuous movement of the system and, hence, cannot describe the hybrid system precisely. On the other hand, it may be interesting to analyse continuous systems by using methods developed in hybrid systems. The example of switching systems, which can be represented by coupled discrete and continuous subsystems, has already been mentioned. Here, the application of a hybrid representation scheme should not be confused with the hybrid nature of the system under investigation.

Finally, the question of how to decide from an input-output viewpoint whether a given system is hybrid, is still open. A continuity index has been proposed in (Lichtenberg and Kamau, 2001) to distinguish continuous systems from hybrid systems. This method is, however, only the first step towards a thorough analysis of this important identification problem.

Description of Hybrid Systems by Modified Petri Nets

Rainer Drath

ABB AG in Heidelberg, Germany

Abstract. This contribution focuses on the modeling of hybrid systems by means of modified Petri Nets. The main goal of this approach is to get a formal description language for hybrid systems, which combines the advantages of a graphical description with the possibility of a transparent visualization, simulation and documentation. For this, several enhancements are proposed. The first enhancement combines the classical discrete Petri Net approach and the concept of continuous Petri Nets. The resulting Petri Net class is called *Hybrid Dynamical Nets* (HDN). In the second enhancement, the aspect of the system complexity is covered by introducing object oriented concepts like encapsulation and information hiding. The resulting *Hybrid Object Nets* (HON) combine the advantages of HDN with those of the object-oriented paradigm. The third proposed enhancement is a combination of the HDN approach with the concept of colored Petri Nets in order to increase the flexibility of the HDN. The resulting class of nets is called *Attributed Hybrid Dynamical Nets* (AHDN). The proposed concepts are explained with several examples.

1 Introduction and Background

The topic of hybrid systems includes several aspects which are about to be investigated. Beside the modeling aspect, also the aspects of analysis and verification (Henzinger et al., 1995, Nenninger and Krebs, 1998), controller synthesis and the aspect of system identification and diagnosis (Rakoto-Ravalontsalama and Aguilar-Martin, 1998) are in the focus of interest.

In fact, the modeling of hybrid systems is the basic condition for all further investigations and requires the availability of a well defined modeling language[1]. Therefore, the main goal of this contribution is to introduce a modeling language for hybrid systems, which combines the advantages of a graphical description with the possibility of a transparent visualization, simulation and documentation. Since the analysis of hybrid systems is still a general problem and under investigation, e.g. (Henzinger et al., 1995, Branicky, 1995, Nenninger and Krebs, 1998), this topic is not considered here.

The modeling language presented here is based on the powerful concept of Petri Nets (Petri, 1962). Petri Nets were originally developed as a method for describing and analyzing discrete systems. They allow the modeling of causal coherences. For this, two node types are defined: transitions and places. Directed arcs connect the nodes and describe the structure of the system.

[1] This contribution is confined to an informal description of the proposed nets. A formal definition of the syntax and semantic is given in (Drath, 1999).

S. Engell, G. Frehse, E. Schnieder (Eds.): Modelling, Analysis and Design of Hybrid Systems, LNCIS 279, pp. 15–36, 2002.

In the following sections, several enhancements of Petri Nets are explained in order to model hybrid behavior. The first enhancement combines the classical discrete Petri Net approach with the concept of continuous Petri Nets. The resulting Petri Net class is *Hybrid Dynamical Nets* (HDN). In the second enhancement, the aspect of the complexity of the system is covered by introducing object oriented concepts like encapsulation and information hiding. The resulting *Hybrid Object Nets* (HON) combine the advantages of HDN with those of the object-oriented paradigm. The third enhancement is a combination of the HDN approach with the concept of colored Petri Nets in order to increase the flexibility of the HDN. The resulting class of nets is called *Attributed Hybrid Dynamical Nets* (AHDN). The proposed concepts will be explained by with several examples.

2 Hybrid Dynamical Nets

2.1 Background

The modeling of hybrid systems needs a combination of description methods for discrete systems and for continuous systems. The classical timed Petri Net approach with its discrete state space is well suited for the field of discrete systems but not for continuous systems. For the field of continuous systems, the continuous Petri Net approach after (David and Alla, 1987) is useful because it offers a continuous state space.

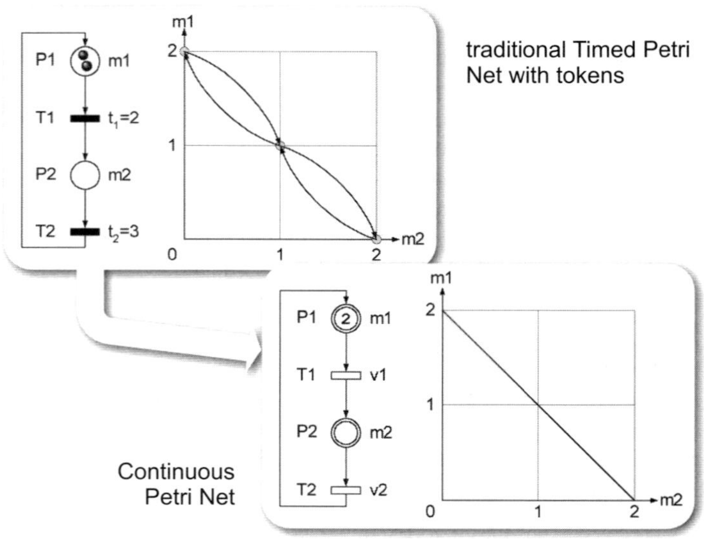

Fig. 1. Conventional Petri Nets and continuous Petri Nets

The main idea of the continuous Petri Net approach is explained in Fig. 1. The state space of the upper traditional net is consists of only three discrete states. The two initial tokens in place $P1$ can distribute themselves due to the times of the transition as represented here. The model below shows the extension with *continuous places* and *transitions*. Here, a token is not any more an individual, but a real quantity of token fragments. The continuous transitions move the token fragments with a velocity of flow from the place before into the place thereafter. The state space becomes infinitely in this way and this opens up the possibility of modeling continuous dynamics. The combination of a discrete and a continuous state space is a main condition for the modeling of hybrid systems. The main idea of HDN is: If we could describe the behavior of continuous systems with continuous Petri Nets, and if we could combine these models with the discrete world of common timed Petri Nets[2], we would be able to model the complex behavior of hybrid dynamical systems with *one* graphical description language.

The *Hybrid Petri Nets* (HPN) in (David and Alla, 1998) are not sufficient to model hybrid systems because they do not cover the possibility of modeling the dynamic behavior of continuous systems. Since continuous Petri Nets allow the modeling of real values, they can be used to model continuous state variables only. Therefore, the following enhancements are suggested:

- In HDN, the firing speed of continuous transitions can be given as a function of token quantities. This opens up the possibility of modeling the behavior of continuous dynamics. The values of this function can become positive as well as negative.
- In HDN, the token quantity of continuous places can take as well positive as negative values. This is needed to model positive as well as negative continuous system variables, whereas HPN only allow positive values.

Because of the differences between the origin Petri Nets and this new approach, it is no longer called Petri Net but *Hybrid Dynamical Net* (HDN). The proposed solution allows the modeling of the dependencies of all system parts with only *one* description language. The following conventional discrete and new continuous net elements are included:

The arcs shown below are used for the main relations between the different transition and places.

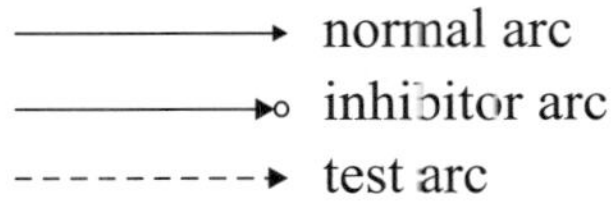

Fig. 2. Graphical notation of the arcs

[2] The time concept that is used here assigns a time value to each of the discrete transitions by using the strong firing rule, see (Drath, 1999). The considered time is continuous and only needs to be quantized by using a digital simulation environment.

Table 1. Graphical notation of the net element

Name mi [c]	discrete place capacity C	Name mj	continuous place token quantity m_j
Name t_s	discrete transition firing time t_s	Name v	continuous transition firing speed function v

2.2 Combination of Net Elements in HDN

The combination of discrete and continuous subnets allow the modeling of hybrid systems with one integrated description language. Table 2 shows all possible combinations of continuous and discrete net elements and briefly explains the usability, the validity and the firing rules of the valid combinations.

2.3 Modeling of Continuous Dynamics with HDN

Typical Continuous Basic Element. Usually, a continuous system is described by its input, output and by its system behavior. With HDN, the input and output variables are each described with a continuous place. The transition $T1$ is always active, the system behavior is described with the firing speed function v = f(u,y), see Fig. 3. The test arc between $P1$ and $T1$ does not allow token flow. This enables to model feedbackless subsystems because the token quantity of $P1$ is not influenced.

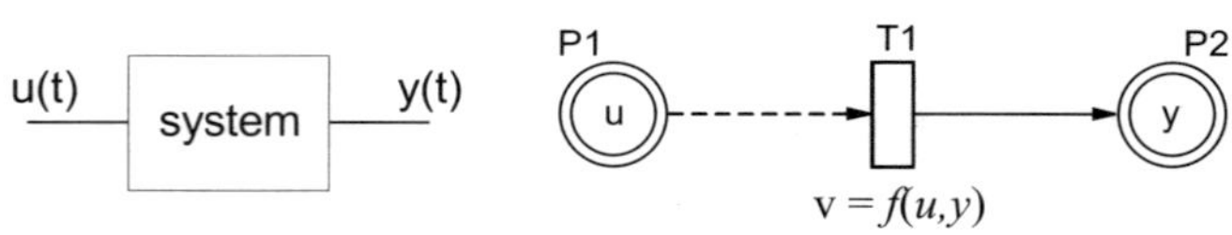

Fig. 3. Continuous basic element

If a single input arc is directed to $P2$, we get (1).

$$v = \frac{dy(t)}{dt} = \dot{y}(t) \tag{1}$$

For a general place P_j with the marking $m_j$and i input arcs we get (2).

$$\dot{m}_j(t) = \frac{dm_j(t)}{dt} = \sum_i v_i \tag{2}$$

This corresponds to the node theorem. Continuous input and output transitions supply their part to increase or to decrease m_j. We can model different basic elements in this way. Even non-linear coupled subsystems can be described since v may be a non-linear function.

Table 2. Combination of net elements

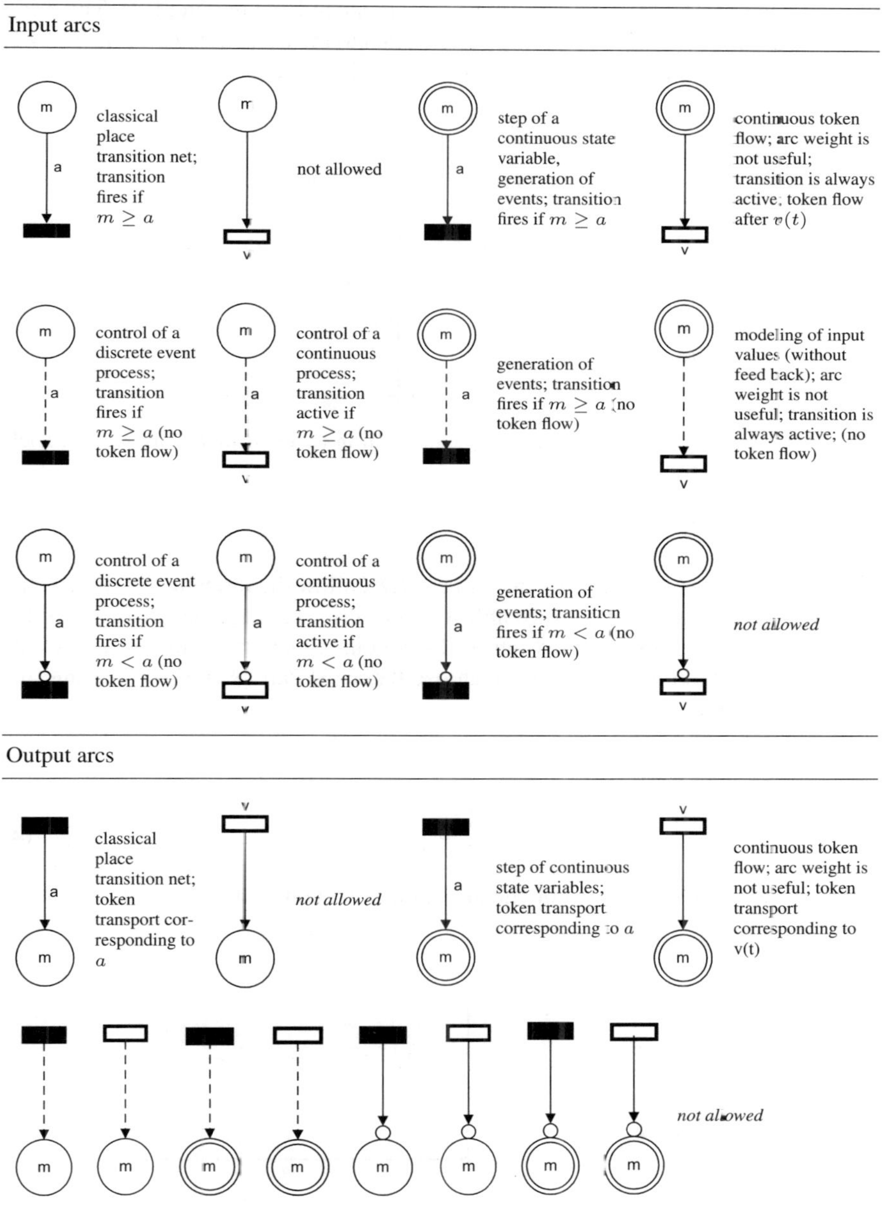

m token quantity

a arc weight (function)

v fire speed function

First Order System. The first order system shown in Fig. 4a is a basic continuous element whose HDN model in Fig. 4b has one input and one output place. The firing speed function $v = 1/T_1(K_P \cdot u - y)$ is assigned to the transition $T1$ (with $T_1 = 1$, $K_P = 10$). Figure 4c shows the step response of the system.

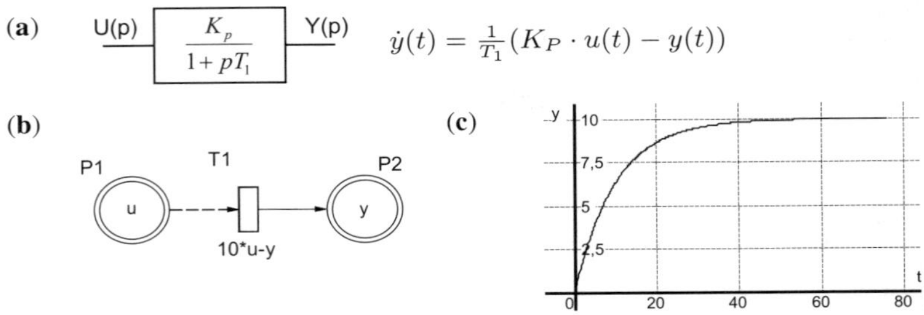

Fig. 4. First order system (**a**) first order system given in terms of time constants (**b**) HDN model of the system (**c**) simulation results

Oscillator (Second Order System). Since HDN allow the modeling of differential equations of first order it is also possible to model differential equations of nth order. For this, they have to be formed into n differential equations of first order. So the well-known differential equation for an oscillator (3) can be transformed into the equation system (4).

$$y + \frac{2D}{\omega_0}\dot{y} + \frac{1}{\omega_0^2}\ddot{y} = K_P \cdot u \tag{3}$$

$$\begin{bmatrix} \dot{x}_1 \\ \dot{x}_2 \end{bmatrix} = \begin{bmatrix} 0 & 1 \\ -\omega_0^2 & -2D\omega_0 \end{bmatrix} \cdot \begin{bmatrix} x_1 \\ x_2 \end{bmatrix} + \begin{bmatrix} 0 \\ \omega_0^2 \cdot K_P \end{bmatrix} \cdot u \tag{4}$$

with $y = x_1$ and $x_2 = \dot{x}_1$. For the simulation we consider $D = 0.25$, $\omega_0 = 1$, $K_P = 1$.

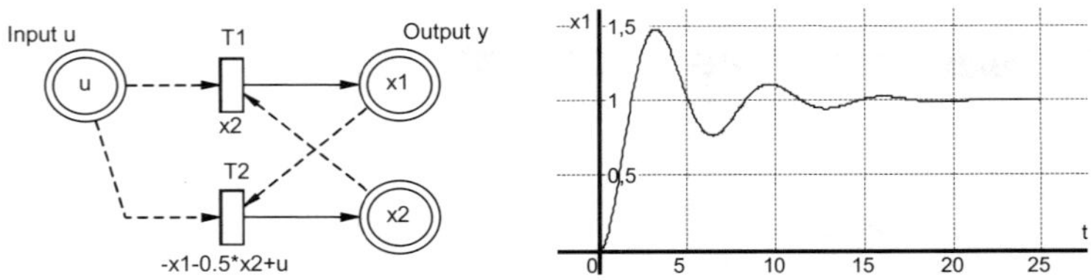

Fig. 5. Continuous second order system (Oscillator)

Since the output of a subsystem may be the input of another subsystem, several of the basic elements can be combined in order to model complex continuous behavior.

Modeling of Linear Continuous Systems of nth Order. The following general approach shows how to model differential equations of nth order with HDN. This is possible, since they can be transformed into n differential equations of first order. The modeling of the differential equation system occurs by combination of several basic first order elements like shown in Fig. 4.

$$\begin{bmatrix} \dot{x}_1 \\ \dot{x}_2 \\ \dots \\ \dot{x}_n \end{bmatrix} = \begin{bmatrix} a_{11} & a_{12} & \dots & a_{1n} \\ a_{21} & a_{22} & \dots & a_{2n} \\ \dots & \dots & \dots & \dots \\ a_{n1} & a_{n2} & \dots & a_{nn} \end{bmatrix} \cdot \begin{bmatrix} x_1 \\ x_2 \\ \dots \\ x_n \end{bmatrix} + \begin{bmatrix} b_{11} & b_{12} & \dots & b_{1n} \\ b_{21} & b_{22} & \dots & b_{2n} \\ \dots & \dots & \dots & \dots \\ b_{n1} & b_{n2} & \dots & b_{nn} \end{bmatrix} \cdot \begin{bmatrix} u_1 \\ u_2 \\ \dots \\ u_n \end{bmatrix} \quad (5)$$

Considering the state equation of the state space model (5), we use each one of the basic elements for every row of the equation system. The coupled in- and outputs of the system will be represented by arcs. This results in the net shown in Fig. 6. The firing speed functions, assigned to the continuous transitions, can be gained line by line from the equation system (5). The initial values of the system are described by the initial values of the variables $x_1, \dots, x_n$ and $u_1, \dots, u_m$.

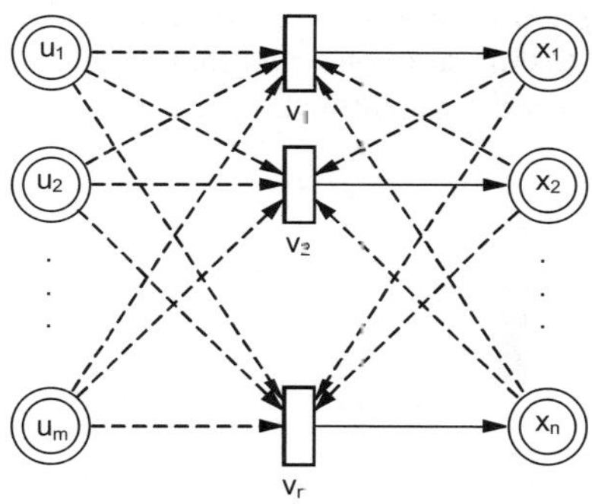

$$v_1=a_{11}x_1+a_{12}x_2+\dots+a_{1n}x_n+b_{11}u_1+b_{12}u_2+\dots+b_{1m}u_m$$
$$v_2=a_{21}x_1+a_{22}x_2+\dots+a_{2n}x_n+b_{21}u_1+b_{22}u_2+\dots+b_{2m}u_m$$
$$\dots$$
$$v_n=a_{n1}x_1+a_{n2}x_2+\dots+a_{nn}x_n+b_{n1}u_1+b_{n2}u_2+\dots+b_{nm}u_m$$

Fig. 6. HDN of a differential equation system of nth order

3 Modeling of Hybrid Dynamics with HDN

The main property of hybrid systems is the interaction between discrete and continuous subsystems. This interaction is also the main reason that classical methods of description are not sufficient in order to model such systems. In HDN, the following three essential interactions between continuous and discrete subnets are defined:

Discrete Control of Continuous Processes. The discrete control of continuous processes is described by discrete control places. They can influence the activity of continuous transitions in accordance with the firing rule, see (Drath, 1999). Figure 7 represents this by means of the discrete place $P3$. If $P3$ is marked, the transition $T1$ is activated, otherwise it is deactivated.

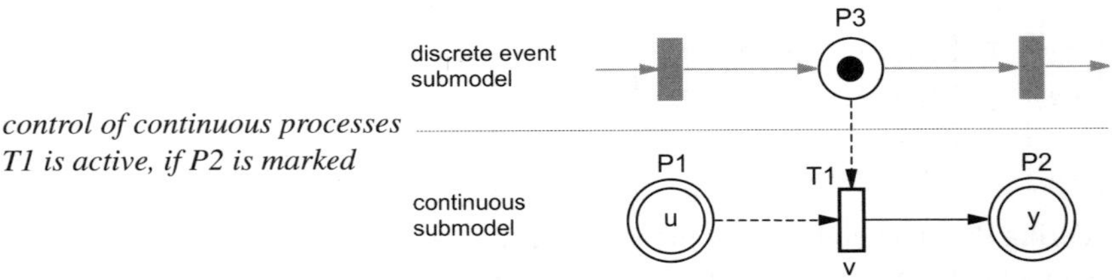

Fig. 7. Discrete control of a continuous process

Generation of Step Functions. In HDN, the generation of step functions is realized by the firing of discrete transitions where the token value of a continuous place is modified timelessly. This is illustrated in Fig. 8 by the transition $T2$. If $T2$ fires, the value u changes immediately.

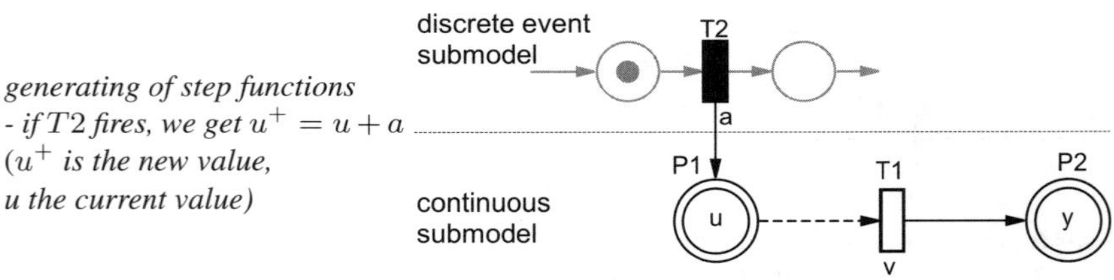

Fig. 8. Generation of step functions

Event Generation from Continuous State Variables. In HDN, the event generation from continuous state variables is described by the firing of discrete transitions. Figure 9 shows the two ways that are available with HDN. The transition $T1$ becomes activated, if $x1 > a$; $T2$ on the other hand is activated if $x2 < b$. The use of a test arc or the inhibitor arc respectively allows an feedbackless event generation.

Since both discrete and continuous subsystems can interact with each other, we become able to model hybrid dynamical systems in this way. This is demonstrated in (Drath, 1999) by a number of examples.

event derivation from continuous systems
- *T1 fires, if* $x1 \geq a$
- *T2 fires, if* $x2 < b$

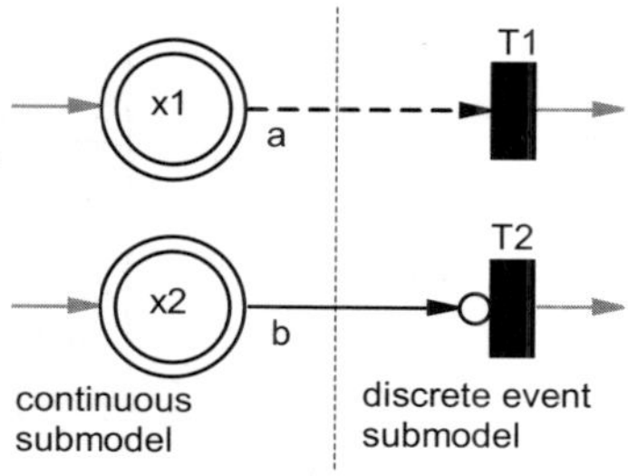

Fig. 9. Generation of events

3.1 Example for Hybrid Modeling with HDN

The following example consists of a continuous system of 3rd order which is coupled with a discrete event subsystem that cyclically generates step functions of the input values u_1 and u_2.

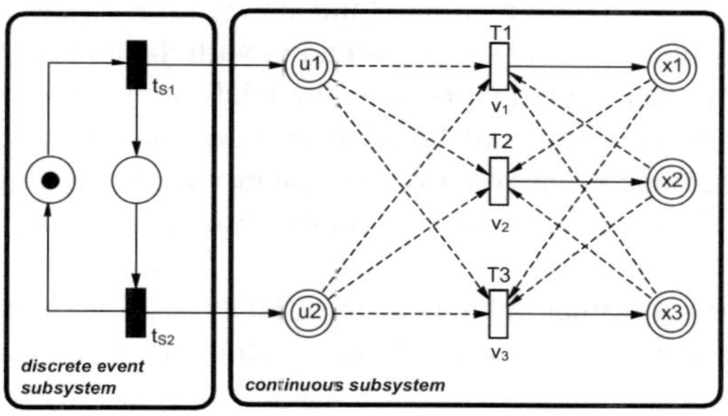

$$
\begin{aligned}
v_1 &= -0.05x_1 + 0.1x_2 - 0.1u_2 \\
v_2 &= -0.361x_2 + 0.361x_3 \\
v_3 &= -200x_1 - 10x_3 + 10u_1 \\
t_{S1} &= 20 \\
t_{S2} &= 20
\end{aligned}
$$

Fig. 10. Example for hybrid modeling with HDN

The discrete token in the discrete event subnet alternately activates the discrete transitions t_{S1} and t_{S2} which changes u_1 and u_2 in a discontinuous way. The simulation results show the behavior of the input and output values of the continuous subsystem.

This example demonstrates that it is possible to model combined discrete-continuous with HDN in a vivid way. It is obvious that the simple discrete subnet can easily be replaced by a more complex Petri Net that models extensive and real logic control functions. The combination of non-trivial logic functions with non-trivial continuous systems leads to real complex hybrid system behavior.

3.2 Summary

The main focus of this section was to demonstrate the capabilities of the HDN approach for the modeling of continuous system behavior and the interaction between

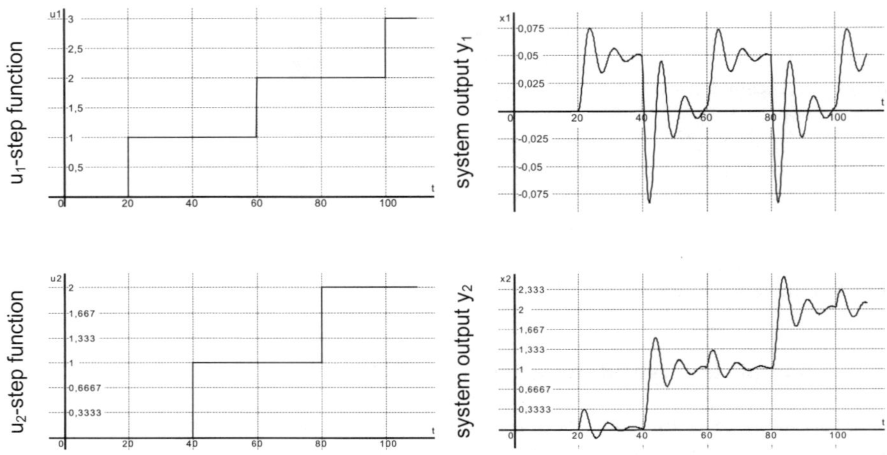

Fig. 11. Simulation results

both types of systems. The possibility of modeling linear (and also nonlinear) differential equations of nth order is a valuable result. In combination with the powerful discrete modeling capabilities of classical timed Petri Nets, the HDN are suited for the modeling of hybrid systems within one integrated modeling approach. This also includes the possibility of modeling concurrent systems and synchronization (in the discrete system parts) and of modeling parallel processes (in the discrete and continuous system part).

The following section introduces a further enhancement of HDN that improves the capabilities of modeling large systems with the HDN approach by means of the object oriented paradigm.

4 Hybrid Object Nets (HON)

One of the obvious problems regarding to the modeling of hybrid systems is that HDN – but also the known mathematical, textual and graphical methods – rise into fundamental problems in treating larger real systems due to the system complexity. The resulting nets become large, difficult to understand and difficult to modify. In order to solve these handling problems, this section applies the object oriented paradigm to HDN. Similar approaches are known for discrete Petri Nets (Bastide, 1995) but not for hybrid systems.

4.1 Object-Oriented Concepts

The object-oriented paradigm offers special qualities to encapsulate and reuse subsystems. The main purpose of the proposed approach is to encapsulate subnets within object frames. Objects can interact with each other using defined interfaces. This results in a new method to describe hybrid systems with reduced effort: *Hybrid Object*

Nets (HON). This has also been demonstrated in (Drath, 1999) with a number of examples.

Subsystems are described by classes. Classes are templates, which describe the general properties of objects. Classes are used to create objects, which are called *instances* of these classes. If an object is instantiated from a class, it gets all attributes and operations (methods) defined in this class.

A further advantage of the object paradigm is that classes can be inherited in order to create new classes. A new class, inherited by a parent class, inherits the whole object description. Afterwards the new class can be refined. Reuse, the most important advantage of the object-oriented paradigm, is given in this way. Since the subsystems can be encapsulated and interact over public interfaces, the models are more easily understandable.

4.2 General Properties of the HON

HON includes concepts for attributes, methods, interfaces, encapsulation, inheritance, abstraction, data exchange and reusing. *Attributes* are represented by places and their contained token quantities. *Methods* are given in the form of the net structure. *Information hiding* is realized by encapsulation of the detailed information of the net structure, and by publishing selected places using an interface. *Abstraction* is the step from a concrete net structure to a class: it is realized by filling the objects into a class hierarchy. *Instantiation* is the step from a class to a concrete object, a so-called instance of the class. *Inheritance* is the step from a class to a child class. If a new class is inherited from a class, it inherits the whole net structure including the interface. *Data exchange* is given by the token flow between the objects. Discrete tokens can model method calls and discrete system states; continuous tokens model continuous variables. *Reusing*, the most important quality of object orientation, is given by inheriting or instantiating classes. Derived objects can be refined; places, transitions, arcs and objects can be added, but no inherited element can be deleted.

To construct a new class, a suited subsystem must be modeled using the HDN. The places, which shall be published, and the parent class have to be defined. Afterwards the net can be fitted into a class hierarchy.

To generate an object, it has to be instantiated from a class. In HDN, every object has a hierarchical structure, which contains three layers, see Fig. 12, extending the two layer concept in (DS97). In the supreme layer the object frame is presented, which encapsulates the inner net structure of the object, and which allows the communication with the environment. In the underlying second layer, the net, inherited by the class, is enclosed in an object frame. In this layer, further net elements and objects can be added in order to modify the behavior of the object and form new subclasses. In the lowest layer, the net, inherited by the parent class, is represented. It can not be modified here.

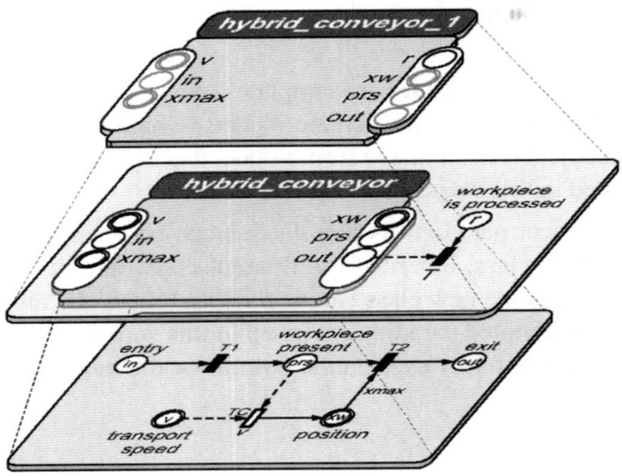

Fig. 12. Basic idea of HON: Encapsulation of subnets

4.3 Application Example

The following example from (Drath et al., 1999) explains the advantages of the HON. The considered example consists of a conveyor belt which transports workpieces. This subsystem may be part of a hybrid plant that processes passing workpieces. The movement of the parts occurs in positive direction of the x-axis of the system (see Fig. 13). The position of the workpiece is denoted as xw. The length of the range of the conveyor belt is denoted as $xwmax$.

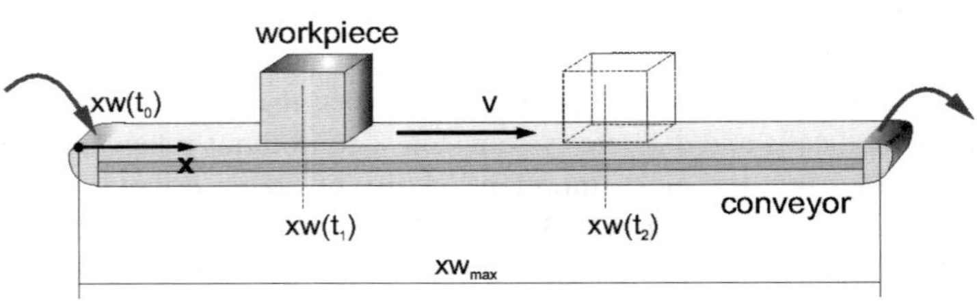

Fig. 13. Principal Sketch of the conveyor belt necessary parameters

The conveyor belt can be described by the net shown in Fig. 14a. If a workpiece reaches the conveyor belt, the place *entry* will be marked ($in = 1$). Afterwards the transition $T1$ fires a token into the place *workpiece present* ($prs = 1$). The token remains here during the entire stay time of the workpiece. This enables the continuous transition *TB* which increases the position value xw according to its firing speed v. After reaching the value $xmax$, the discrete Transition $T2$ is enabled and transports the token to the place *exit* and resets the value xw. Figure 14b shows the encapsulated version of the net model which is characterized by the object frame, the headline and a set of interface places which allow the interaction with the environment.

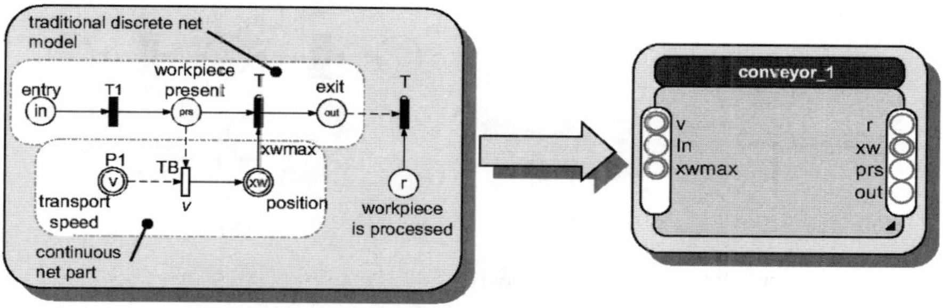

Fig. 14. HDN model and HON model for the conveyor belt

4.4 Summary

The HON concept enhances the HDN approach by means of object oriented concepts. The advantage of the HON approach is the ability to encapsulate complex system behavior within objects. Each object encapsulates its own state variables and its behavior. By composing several of such objects, the HON approach allows the modeling of systems whose size could not sufficiently be handled by a flat net approach. If the system structure is modeled close to the real system, the resulting nets are easy to design and to understand.

The following section suggests another enhancement of the HDN approach that improves the flexibility in modeling hybrid systems.

5 Attributed Hybrid Dynamical Nets

5.1 Motivation

In HDN, continuous state variables are modeled by continuous places whereas discrete state variables are modeled by discrete places. The separation of both variable types is advantageous for lots of hybrid systems. But in special cases this separation leads to modeling problems. This section demonstrates, that a combination of HDN with the known concept of the token attributes allows a more effective modeling in order to create smaller nets with reduced modeling effort. This is demonstrated by several examples of application.

5.2 Example of Application: Conveyor Belt

Figure 15a shows again the conveyor belt and the corresponding HDN in Fig. 15b. The HDN model is small and understandable. But it fails if the positions of more than one workpiece have to be described at the same time. Discrete tokens are well suited for the modeling of several entities, but the marking of continuous places is only suited for *one* continuous control variable. This is a result of the separation of both pieces of information.

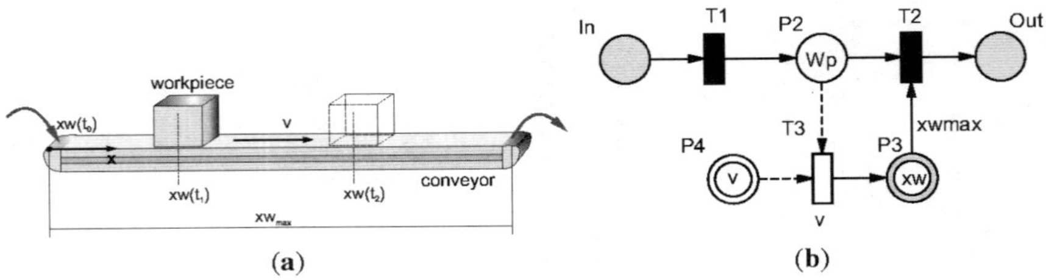

Fig. 15. Conveyor belt (**a**) and hybrid model (**b**)

In order to describe the position of n workpieces on the same conveyor belt with the HDN approach, the net would have to be duplicated n times as shown in Fig. 16. Arriving tokens are distributed between the sub nets, the individual positions are described by the token value of the several continuous places. The net works correctly but is inflexible and can become large according to the maximum number of workpieces n.

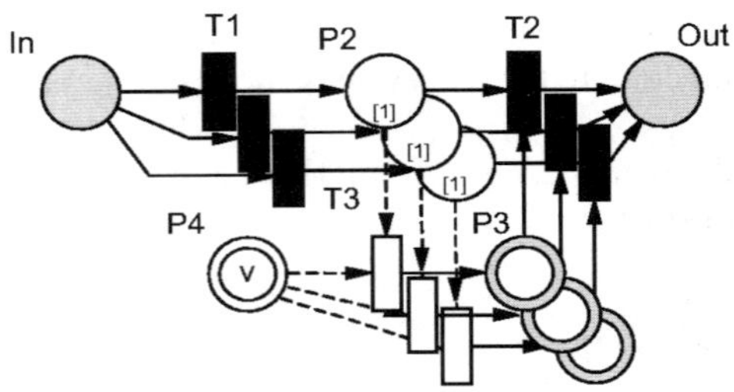

Fig. 16. HDN for a hybrid description of the transport of several workpieces on a conveyor belt

In the net shown in Fig. 15b, the discrete information *workpiece present* and *workpiece position* are separated into two places of the net. The unity of both pieces of information is given by its context but forbids the description of several workpieces on the same conveyor belt. To avoid this separation, both pieces of information can be assigned to one discrete token. This is done by the attribute concept proposed in (Drath and Schwuchow, 1997).

The main idea is that discrete tokens can be assigned to a set of information which describe the properties of the workpiece which may be continuous or discrete, e.g. the position of the workpiece and its color (see Fig. 17).

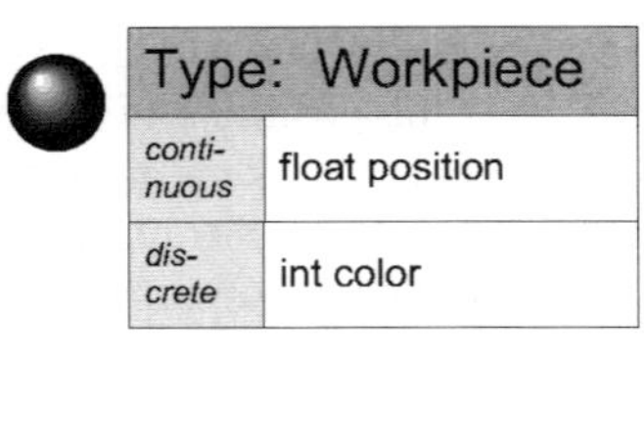

Fig. 17. Hybrid attributing of a discrete token

Figure 18a again represents the HDN model for the conveyor belt and points out both pieces of information which are assigned to the workpiece: the discrete information *Wp* (workpiece present) and the continuous information *xw* (workpiece position). The use of token attributes simplifies the information handling and can be used to simplify the HDN model: Fig. 15b illustrates the net modeled with AHDN. The variable *xw* is now assigned to the discrete token in $P2$. This leads to a smaller and more flexible net which can handle several workpieces on the conveyor belt.

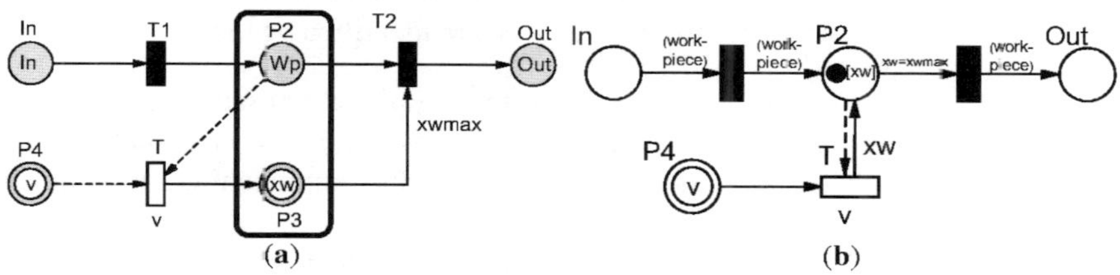

Fig. 18. Simplification of the HDN model (**a**) by using hybrid token attributes (**b**)

5.3 Example of Application: Continuous Heating Process

Figure 19a shows a HDN that models a workpiece which is heated up. The continuous transition *TK* changes the temperature *T* of the workpiece according to the assigned firing speed function and is controlled by the discrete control place. The HDN is easy and understandable – but works only for a single workpiece. If this model had to be enhanced in order to model n heated workpieces, the net would have to be duplicated n times.

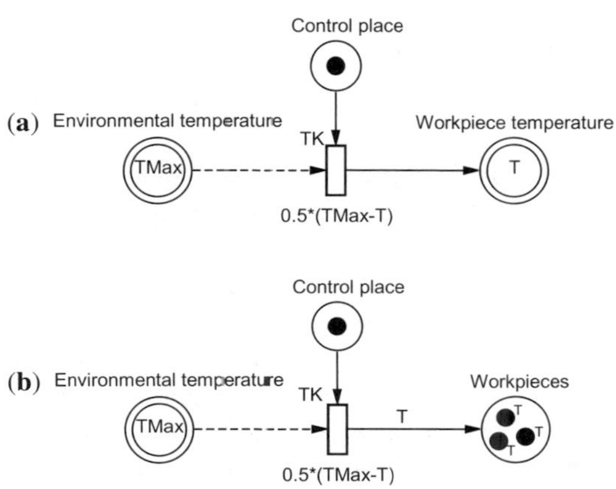

Fig. 19. (**a**) HDN for a heating process of a workpiece (**b**) AHDN for a heating process for several workpieces with each its own individual temperature T_i

Using AHDN, this succeeds in a much more simple way as illustrated in Fig. 19b. Each workpiece is interpreted as an object and described by one discrete token which is assigned to its individual attribute T. All tokens are placed inside the place *Workpieces*. The continuous transition *TK* increases the temperature T_i of each individual workpiece. The advantage of this model is obvious: the net model can hold an arbitrary number of workpieces without changing the net structure. For this, the following rules are valid:

- The firing speed function is applied to each individual token.
- The attributes that are about to be influenced are identified by the arc expression.

The concept of the token attributes is not new, the traditional attribute concepts (e.g. in (Hubert et al., 1991, Jensen, 1992)) also propose to assign individual data to each token. But there are significant differences between this approach and the known ones.

In the known attribute concept, the manipulation of the token attributes occurs during the firing of a transition. But this also means, that the manipulation of the attributes requires that the token is first taken from the place and is replaced after the firing operation has been executed. In contrast to that concept, here the tokens don't have to leave their places but are manipulated by the continuous activity of the continuous transitions without token flow.

Furthermore, in contrast to discrete approaches as in (Drath and Schwuchow, 1997), the continuous control variables are not changed in a time discrete way but in a continuous way. Continuous HDN are indeed continuous models whereas the modeling of continuous processes by means of discrete Petri Nets are in principle time discrete, the time discrete algorithm is translated into the net structure. The simulation of HDN can be done by several continuous simulation algorithms. The stepwise simulation process is only necessary if a digital computer is used.

5.4 Comparison of HDN and AHDN

Table 3 gives an overview of the net elements that are used in HDN and AHDN respectively.

Modeling of Hybrid Dynamics with AHDN. In HDN, three essential interactions between continuous and discrete subnets are defined in order to model hybrid system behavior. In the following, it is demonstrated how these interactions can be realized with AHDN.

In HDN, the *discrete control of continuous processes* is described by discrete control places since they can influence the activity of continuous transitions in accordance with the firing rule (see (Drath, 1999)). Figure 20a represents this by means of the discrete place $P3$. With the AHDN shown in Fig. 20b, this occurs in a similar manner. $P1$ and $P2$ are replaced by discrete places. The activity of the continuous transition $T1$ is also controlled by the discrete control place $P3$.

Table 3. Comparison of the net elements used in HDN and AHDN respectively

HDN		attributed HDN	
[C]	discrete place capacity C	[C]	discrete place capacity C
m	continuous place without capacity limitation	m	continuous place without capacity limitation
t_s	discrete transition with firing time	action firing time	discrete transition with action code and firing time
v	continuous transition with firing speed	v	Continuous transition with firing speed
●	discrete token	● (Attribute)	attributed discrete token (several attributes possible)
a → b →o c - - →	arcs with arc weight	expression → expression →o expression - - →	arcs with expression the expression of input arcs are conditions the expressions for output arcs are a list of attributes

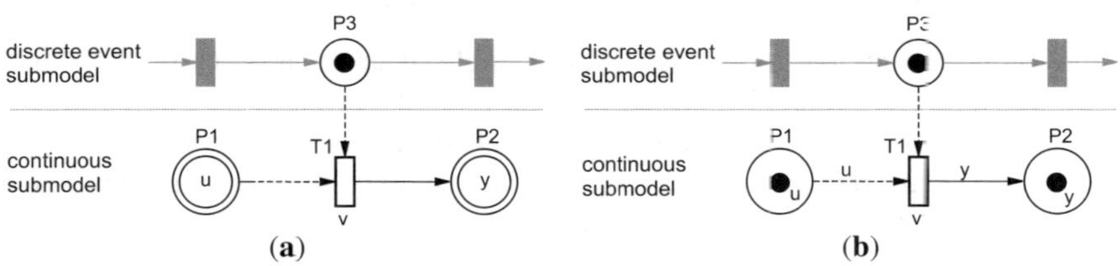

Fig. 20. Discrete control of a continuous process

In HDN, the *Generation of step functions* is realized by the firing of discrete transitions where the token value of a continuous place is modified timelessly. This is illustrated in Fig. 21a by the transition $T2$. With AHDN, the discrete event change of the token value occurs as it is known for other higher Petri Nets, e.g. in (Drath and Schwuchow, 1997): between $T2$ and $P1$ there must exist an arc backward from $P1$ to $T2$. The action code $u := u + a$ is assigned to $T1$, cf. Fig. 21b, and is executed during the firing of $P1$.

In HDN, the *event generation from continuous state variables* is described by the firing of discrete transitions. Figure 22a shows the two possibilities that are available with HDN. The transition T1 becomes activated, if $x1 > a$; $T2$ on the other hand is activated if $x2 < b$. The use of a test arc or the inhibitor arc respectively allows a feedbackless event generation. In AHDN, this occurs equally, see Fig. 22b: $T1$ is activated if one of the token attributes $x1 \geq a$; $T2$ on the other hand is activated if one of the token attributes $x2 < b$.

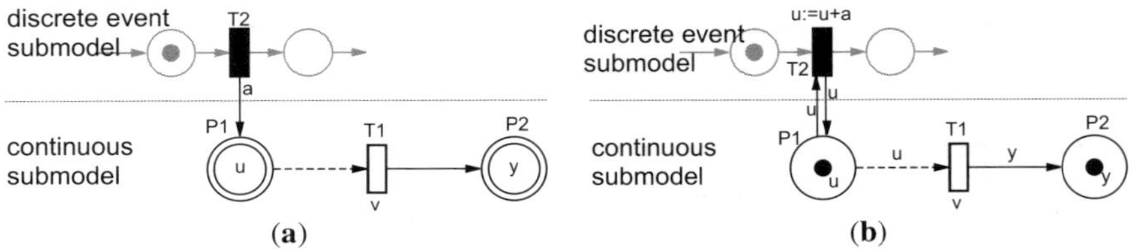

Fig. 21. Generation of Step functions

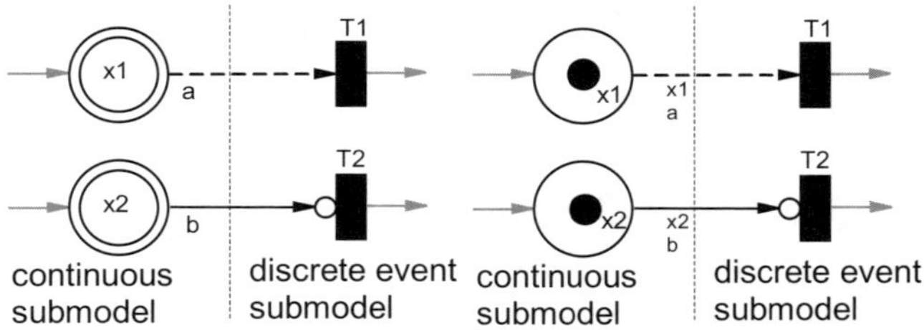

Fig. 22. Generation of events

5.5 Example of Application: Hybrid Heating Process (Annealing Furnace)

This section presents the modeling of an example of application as in (Fahrland, 1970). The presented model is based on a simplified description of this example in (Wieting, 1998) and is characterized by a high structural variability in the continuous part. The discrete part includes non-trivial logical processing functions which cause hybrid system behavior in interaction with the continuous subsystem.

Overview. Inside an annealing furnace, steel blocks are heated up in a number of chambers before they can be rolled in the next processing step. An incoming steel block is inserted into the furnace if at least one of the chambers is free. The temperature h of a steel block changes under the influence of the furnace according to the differential equation:

$$\dot{h} = (ho - h) \cdot c. \tag{6}$$

The temperature of the furnace is ho, c is an individual temperature coefficient for every steel block. The temperature of the furnace ho is heated up by a heater with the maximum temperature $Tmax$ according to:

$$\dot{ho} = (Tmax - ho) \cdot c. \tag{7}$$

The structural variability in the continuous subsystem of the annealing furnace results from the heating process of the remaining steel blocks that is influenced by adding/removing a “cold” or “hot” steel block. While adding a steel block it is supposed that the temperature of the furnace *ho* is reduced in a discrete way by a value that is calculated from the difference between the furnace temperature *ho* and the temperature of the steel block h divided by the number of blocks. The reduced furnace temperature increases the heating time of the remaining blocks.

Initial Conditions and Assumptions.

- $c = 0.2$ for all steel blocks, $Tmax = 2600$.
- At the beginning the furnace is empty.
- The starting temperature in the furnace is 1800.
- The temperature of the incoming steel blocks is 400.

Rule Basis.

- The steel blocks are heated up until one of them reaches a temperature of 2200.
- If the temperature of one of the steel blocks reaches 2200, it has to be removed as well as all further blocks with a temperature of at least 2000.
- If no chamber is free when a steel block arrives, it is stored in an input buffer in front of the furnace. The temperature of the blocks remains there constantly at 400.

Model of the Steel Blocks. AHDN allow to model the steel blocks by discrete tokens which are assigned to the attribute h, see Fig. 23. Each steel block has its individual temperature in this way, which will change during the heating process.

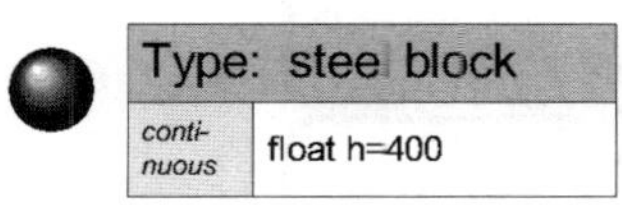

Fig. 23. Representation of a steel block by an attributed token

Hybrid Model of a Furnace with Several Chambers. Unlike the HDN approach, in which every chamber would need its own subnet, the furnace can be represented here by a single discrete place that includes the attributed tokens. Simply the number of the chambers has to be determined by the capacity of the discrete place *Furnace*.

The net in Fig. 24 shows the model for the furnace that contains a number of steel blocks as well as the complete discrete-event rule base. $T2$ can switch if for one of the attributes is valid: $h \geq 2200$. The corresponding token will be removed and will move to $P1$ and then to $P3$. If further steel blocks with the temperature of $h \geq 2000$ are available, they are diverted by $T4$ and moved to $P3$.

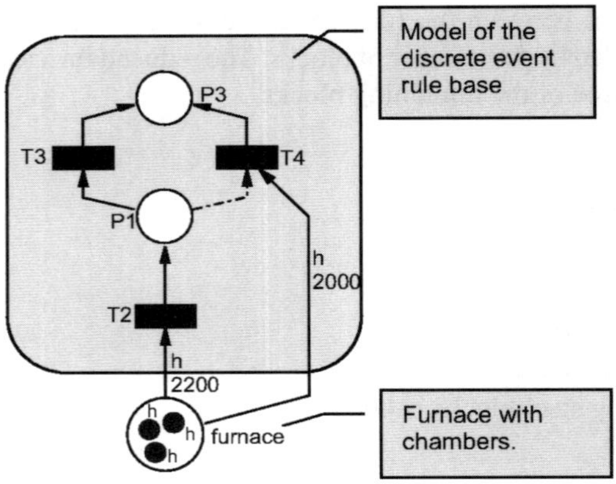

Fig. 24. Representation of the annealing furnace with included steel blocks

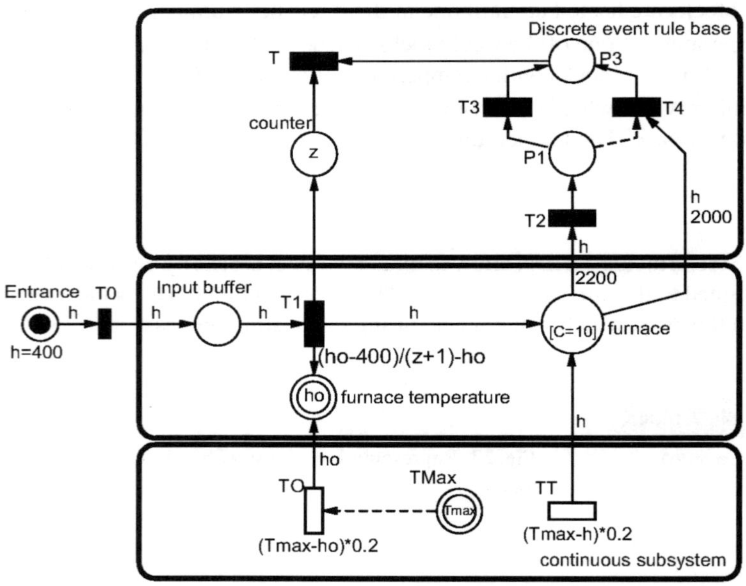

Fig. 25. Hybrid model of the entire annealing furnace

Hybrid Model of the Entire Annealing Furnace Plant. Starting with a model of the furnace with n chambers, the delivery and departure of steel blocks as well as the continuous behavior of the steel blocks inside the furnace can be modeled easily.

The net represented in Fig. 25 describes the entire plant and can be divided into 3 subnets – the already described discrete event subnet for the modeling of the discrete event rule base, a continuous subnet for the modeling of the continuous system behavior as well as the hybrid subnet that combines the two continuous and discrete event subnets with each other.

The function of the net is vivid and starts with the marking of the place *Entrance*. The token is attributed and has the initial temperature of 400. The transition $T0$ fires and transports the token into the place *Input buffer*. The transition $T1$ can only fire if the place *furnace* offers at least one free chamber (capacity concept), otherwise the token remains in the place *Input buffer*. If $T1$ fires, this increments the counter value z and moves the token into the place *furnace*. Furthermore the furnace temperature *ho* is reduced by a discrete value.

The always active continuous transitions *TO* and *TT* describe the heating process of the furnace and the steel blocks according to their fire speed function. *TO* increases the furnace temperature. *TT* increases the temperature of the individual steel blocks proportionally to the temperature difference between the current furnace and steel block temperature. The discrete subnet checks whether one of the steel blocks achieve the temperature of $h = 2200$. In this case, the appropriate token is removed and is filed into $P1$. By means of $T4$ it is determined whether further tokens reach the temperature of 2000 – the corresponding tokens are removed. They will be collected in the place $P3$ which enables the discrete transition T and decrements the counter value z.

During the modeling with AHDN it becomes clear that attributed tokens allow a more flexible modeling than HDN. The use of AHDN is especially advantageous for processes in which identical continuous processes are applied to several objects, for example workpieces which pass a manufacturing plant and both discrete and continuous state variables are changing. The furnace model can be expanded simply by modifying the capacity of the place *furnace* without changing the net structure: With the HDN, a separate net would have to be constructed for each chamber.

6 Conclusion

The main goal of this contribution was to introduce a modeling language for hybrid systems which combines the advantages of a graphical description with the possibility of a transparent simulation, visualization and documentation. For this, the continuous Petri Net approach from (David and Alla, 1987) has been enhanced in such a way that it allows modeling continuous system behavior. In combination with the traditional timed Petri Nets, the resulting Hybrid Dynamical Nets (HDN) allow the modeling of coupled discrete-event and continuous process dynamics within one integrated description language. This also includes the possibility of modeling concurrent systems and synchronization (in the discrete system parts) and of modeling parallel processes (in the discrete and continuous system part). In order to simplify the application of HDN in modeling larger systems, they were combined with object concepts. The advantage of the resulting Hybrid Object Nets (HON) is the ability to encapsulate complex system behavior within objects. Each object encapsulates its own state variables and its behavior. By composing several of such objects, the HON approach allows the modeling of systems whose size could not be handled sufficiently by a flat net approach.

The modeling of hybrid systems with HDN and HON is based on the separation of continuous and discrete state variables. This separation is advantageous in many cases, but sometimes this leads into modeling problems. Therefore, the HDN approach is combined with the known concept of token attributes. The resulting Attributed Hybrid Dynamical Nets (AHDN) allow a more effective modeling in order to create smaller nets with reduced modeling effort.

By the way: The applicability of the HDN and HON is not only limited to the field of technical systems. In several contributions (Matsuno and Doi, 2000, Matsuno et al., 2000, Matsuno et al., 2001, Matsuno and Miyano, 2000) the Hybrid Dynamical Nets are successfully used in modeling biological systems which are characterized by interacting discrete event and continuous subsystems. The considered *gene regulatory networks*, which are researched by the human genome center of the university of Tokyo, have been successfully modeled with HDN and HON. One of the main experiences of using the HDN/HON approach is that the graphical notation of the nets indeed represents the expectations of the biologists which are usually not familiar with differential equations. HON and HDN are available with the tool Visual Object Net++ (Visual Object Net ++, 2000).

Acknowledgment. This research is supported by the DFG (Deutsche Forschungsgemeinschaft, German research association) as a part of the investigation project "Analysis and synthesis of mixed continuous and discrete technical systems" (KONDISK) with the subject "Analysis and synthesis of hybrid sub processes in flexible manufacturing systems – examinations to an object oriented systems engineering".

Model Based Development of Hybrid Systems: Specification, Simulation, Test Case Generation*

Klaus Bender[1], Manfred Broy[2], István Péter[1], Alexander Pretschner[2], and Thomas Stauner[3]

[1] Lehrstuhl für Informationstechnik im Maschinenwesen
Technische Universität München, Boltzmannstr. 15, 85748 Garching, Germany
[2] Software&Systems Engineering, Institut für Informatik
Technische Universität München, Arcisstr. 21, 80290 München, Germany
[3] BMW Car IT, Petuelring 116, 80809 München, Germany

Abstract. This paper gives an overview of our approach to the development of discrete-continuous systems in a general model based setting. This includes formalized description techniques, CASE support for modeling and simulation, and test harness as well as test case generation. HyROOM is presented, a formally founded notation for the integration of continuous activities into MaSiEd, a CASE tool prototype based on the ROOM methodology. In addition, an approach to the automated generation of test cases for discrete and also discretized hybrid systems specified within a second CASE tool, AutoFocus, is presented.

1 Introduction

The development of hybrid systems, which operate mixed discrete and continuous data streams, is an interdisciplinary task. Engineers from different disciplines are involved in their designs. On a conceptual level, the artifacts in question are operational abstractions of aspects such as functionality, structure, logical and technical (deployment) architecture, data, communication, scheduling, fault tolerance, and quality-of-service related issues.

Integration, one key aspect of model based development, is needed: it is desirable (1) for integrating these not entirely orthogonal aspects, concerning (2) the process and its different created artifacts over time, and (3) for different levels of abstractions. While not in general true, graphical description techniques in the domain of hybrid systems turn out to ease communication between engineers from different disciplines. The descriptions are representations of *models* that form the essence of the system under development during its stages of increasing precision that eventually lead to possibly generated production code.

It is difficult to envision model based development without machine support. Complex systems require sophisticated management and design techniques for consistent models and their relationship. Simulation and code generation facilities are

* This work was supported with funds of the Deutsche Forschunsgemeinschaft under reference numbers Be 1055/7-1–7-3 and Br 887/9 within the priority programs *KONDISK* and *Design and design methodology of embedded systems*.

S. Engell, G. Frehse, E. Schnieder (Eds.): Modelling, Analysis and Design of Hybrid Systems, LNCIS 279, pp. 37–51, 2002.

desirable in integrated model based development. Requirements tracing is impossible without tool support. Test case generation should be automated and therefore a feature of an integrated tool.

Just as for safety critical discrete systems, it is desirable to apply a high degree of mathematical rigor in the development of safety critical hybrid systems, provided it does not lay too much burden on the engineer, and it is simple. Formalism itself certainly does not solve any problems. Applications that require a (transparent) formalization include semantics-preserving design steps like refactoring, refinements from continuous to discrete time (Stauner, 2002, Stauner, 2001), and testing (Pretschner et al., 2001). Consequently, the two tools presented in this paper have been given a formal semantics.

This paper gives an overview of our activities in IMMA (Integrated Mathematical Machine Modeling), a project within the DFG priority program KONDISK. We cover model based development in general, modeling and description techniques, semantics, tool support, and test case generation. The nature of an overview article implies a high degree of abstraction. Technical details have been previously published: (Stauner et al., 2001, Péter et al., 2000) define the formal semantics and present the case study. For refinements of HyCharts, e.g., the transition from continuous to discrete time, see (Stauner, 2002, Stauner, 2001); automatic test case generation is treated in, e.g., (Lötzbeyer and Pretschner, 2000, Pretschner, 2001, Pretschner et al., 2001, Pretschner et al., 2000, Schätz and Pretschner, 2002).

The paper kicks off with a brief overview of model based development and its relationship with approaches like HW/SW codesign or Simultaneous Engineering (Bender and Kaiser, 1995). Being aware of the fact that model based development with integration along the dimensions of time (process) and content (product) is a rather ambitious undertaking, we then present partial approaches to implementing this paradigm. Two CASE tools are presented, MaSiEd and AutoFocus. MaSiEd basically integrates the Real time Object Oriented Modeling methodology (ROOM) description techniques with continuous activities as specified by Matlab block diagrams. The simplicity of the execution semantics of AutoFocus, on the other hand, is the basis for effectively and efficiently applying verification and validation techniques like model checking, theorem proving, or test case generation.

Following a glimpse of related work, we summarize the basic ideas behind model based development in Sect. 2. MaSiEd is described along the lines of a wire stretching plant (Sect. 3). With the application of generating test harnesses, this also includes the automated translation of hybrid scenarios into hybrid state machines. In Sect. 4, a compositional and incremental approach to the automated generation of test sequences for hybrid systems in AutoFocus is sketched. Sect. 5 concludes.

1.1 Related Work

(Mosterman, 1999) contains an overview of simulation packages for hybrid systems. The reason for presenting yet another description technique for hybrid systems is that in popular tools like Matrix_X or Matlab/Simulink/Stateflow systems components are either discrete or continuous, but not both. Often these packages offer convenient, sometimes application specific, graphical description techniques, but, with the exception of Charon (Alur et al., 2000a), a formal semantics is usually not defined for them. There are also simulation tools with a strong formal background, see (Fábián et al., 1998). Their focus is not on visual specification.

A central issue of our work is the research for a convenient modeling methodology for hybrid systems which is suitable for practice and can be put on a formal basis, see (Stauner, 2001) for a detailed overview. Therefore, most of the work cited above is complementary to our approach, either dealing with the modeling and simulation of hybrid systems, or with formal models for them. A notable exception is the work in the context of (Friesen et al., 1998a) where UML's class diagrams are extended for hybrid systems and coupled with Z specifications.

There is a large body of literature on testing labeled transition systems, see (Schätz and Pretschner, 2002) for an overview. Lack of space prohibits a description of all the available frameworks, tools and techniques, e.g., Lurette (Raymond et al., 1998) or TorX (Vries et al., 2000). The main difference with our approach is that we do not explicitly construct labeled transition systems but rather work on composed finite state machines that describe the behavior, which enables us (1) to compute with sets of values by means of symbolic execution, and (2) to easily incorporate heuristic search strategies.

2 Model Based Development

Even though the notion of model based development was coined a decade or so ago, the search for clear definitions of this concept results in a hard time. We give a brief overview of our understanding of this idea (Schätz and Pretschner, 2002).

2.1 Models

Mastering the development of complex systems requires the use of suitably chosen abstractions for describing the essence of the system under development. This essence may differ for the points of view an engineer takes: it may be concerned with the above mentioned aspects of structuring the system, or with documentation, code generation, or analysis. This necessitates projections of integrated models.

For a particular purpose, abstractions discard details that are not relevant. Since they are simplifications, the artifacts under development become manageable. Clearly, for development, simplifications cannot go too far – remember that complexity is an essential rather than an accidental property of software. Embedding models, or rather code that is generated from them, into their target context (legacy systems, operating systems, sensors and actuators, different technical deployment architectures) obviously requires suitable concretizations. By now, we are only able to cope with them in an ad-hoc manner. Automatization of this task is the subject of current work since we consider bridging the gap between the modeling and implementation levels to be the key challenge in model based development. However, models as simulations of actual programmable logic controllers (PLCs) allow for simultaneous engineering (Bender and Kaiser, 1995) of hardware and control systems which was one of the driving forces behind the development of the MaSiEd tool.

Undoubtedly, the development from Assembler to higher-level programming languages like Ada or C has caused an enormous increase in productivity. The essence of this transition lies in abstractions of control, data and program structures. In terms of control flow, constructs for procedures (no explicit call stack), repetition, sequence, alternative, and, more recently, exceptions have been incorporated into such higher languages. Structured data types with dedicated access mechanisms exempt engineers from treating data on a memory cell level. Many languages are equipped with abstractions for inter-process communication – just consider monitors or Linda as an implementation of the communication paradigm in tuple spaces. Concepts like modules allow for structuring and mastering larger projects. Abstractions are ubiquitous: some object-oriented and declarative language implementations provide automatic garbage collectors, and window toolkit APIs like Swing are readily available. Java's comprehensive libraries and the buzzword of componentware are further developments in this direction. Well understood, this list is far from being complete.

The vision of model based development is to take these ideas a step further. Necessarily domain-specific essential entities and their relationship are encoded in the (syntactical) meta model. In the case of embedded systems, these may be components, ports, connectors, etc. Concepts for describing behavior (functions, statecharts, Mealy machines, Petri Nets) are also part of the meta model. For some application domains, e.g., time triggered bus architectures, synchronous Mealy machines may turn out to be a good choice. For others, like dedicated smart card operating systems with a focus on cryptography, Petri Nets with their possibility of implicitly encoding command interleavings, may be a better choice.

The kind of properties, refinements, and semantics needed to describe a system are encoded in the system model. Meta and system model together form the product model. As stated above, in model based development, the product-oriented point of view has to be complemented by a process-centered perspective. Interrelated with the product model, the process model defines the different incremental development steps (add functionality, perform a refinement in the mathematical sense, etc.). This also includes coping with variants and versions of a system under development (Schätz and Pretschner, 2002).

2.2 Process

The systematic use of models does not prescribe any particular process. In fact, processes like the Rational Unified Process or Cleanroom operate with models as the basic entities. Languages/methodologies like the Spark or Ravenscar subsets of Ada encourage the use of abstract design but, like the RUP and Cleanroom, do not emphasize the dependency on a particular domain.[1] In fact, model based development should not be seen as the philosopher's stone for every single problem. It might turn

[1] Unless safety critical software is considered a domain – Spark allows explicit code annotations for verification purposes.

out that it is beneficiary only in dedicated parts of an agile family of processes. We are concentrating on iterative processes (grow, not build software (Pretschner et al., 2001)) with executable artifacts right from the beginning. Briefly, due to the possibility of frequently checking back with a customer, the key advantage of this kind of process is intellectual control over the process. Increments occur along the three aforementioned dimensions of level of abstraction, development over time – versions, variants, configurations, elaboration of aspects like function, data, etc. – and projections for the purpose of analysis or generation. Due to the complexity of the involved systems, CASE support for development, requirements tracing, validation, and ensuring integrity is mandatory for model based processes.

In principle, the ideas of model based development also carry over to code-centered processes like Extreme Programming since the essence of a model is clearly independent of its representation. However, languages used in model based tools like AutoFocus deliberately restrict the power of general-purpose languages, as do Spark and Ravenscar (tasking). The reason is that this facilitates design steps that are correct by definition or that can be validated by machines (the generation of proof obligations would be a first step), which is in general impossible for full-fledged languages like Ada or C^{++}. As a side benefit, using more abstract model-based, possibly graphical, notations renders systems development language-independent.

3 MaSiEd (Machine Simulator/Editor)

MaSiEd is a CASE tool for modeling, simulating and analyzing the I/O behavior of general discrete, continuous, and hybrid systems. It has been tailored to the needs of field bus based manufacturing systems with the aim of testing the associated PLC software. The possibility to create virtual machine models of manufacturing plants is a prerequisite for PLC tests.

3.1 Modeling Discrete Systems

The I/O behavior of modern manufacturing systems can be characterized as a mainly event driven discrete behavior (with incorporated continuous behaviors; the focus, however, is on discrete systems which decreases the adequacy of tools such as Matrix_X that focus mainly on continuous parts). The MaSiEd CASE tool enables one to model reactive systems using the *real time object oriented modeling methodology* (ROOM, now a substantial part of the UML-RT), see (Selic et al., 1994). ROOM's emphasis is on the seamless use of models from the requirement/high-level design phase down to the low-level design and testing stages. The primary concepts of the ROOM modeling language are actors, protocols, ports, bindings, and ROOMcharts, and they are used to model architectures consisting of hierarchies of communicating concurrent components.

An actor is a concurrent active object that hides its implementation from other actors in its environment. Fig. 2, left, shows an architecture diagram where actors are depicted as boxes. The behavior of actors is specified by a variant of the statechart formalism called ROOMcharts. ROOMcharts basically are extended state machines with hierarchic states, but unlike statecharts without parallel composition of states: parallel composition is defined using architecture diagrams like in Fig. 2, left. This formalism can model asynchronous event driven real-time systems.

3.2 Modeling Continuous and Hybrid Systems

Even though the I/O behavior of most modern manufacturing systems can be mainly characterized as an event driven discrete behavior, there are, in addition, parts that have to be modeled in a continuous/hybrid manner.

The primary concepts added to ROOM in order to obtain the hybrid ROOM (HyROOM) modeling language are block diagrams, stores, and state activities. These concepts can be used to model hierarchies of communicating concurrent hybrid components. In order to support the modeling task of continuous subsystems we adopted the block diagram notation (Fig. 2, bottom right) as used in control theory. The block diagram notation is a widely used formalism for modeling, simulating, and analyzing dynamic systems. Block diagrams basically represent sets of differential equations. Note that block diagrams are, among other things, a means for architectural specifications of continuous systems.

For modeling hybrid systems, we extended ROOMcharts with the concept of continuous activities. Figure 2, right, shows such an extended automaton. An ad hoc way of enabling control-loop behavior modeling is to specify a state's activity in the form of block diagrams. Variables assigned to connectors in the block diagram associated to the activity can be evaluated in the transition conditions belonging to the respective state. Numerical algorithms associated with the block diagram stop execution upon exiting from the state. Different actors in a model may be multi rate and thus updated at different rates.

The newly introduced concept of a *store* enables the transfer of real valued message data from state machines to block diagrams. The last message arriving in a store can serve as input to a block diagram. Stores may be connected to other actors with input for continuous or hybrid behavior or analog outputs to external hardware.

3.3 Modeling and Simulation Infrastructure

MaSiEd provides a user-friendly graphical design interface where hierarchical block diagrams and ROOM models with inheritance can be edited in the same environment. Inheritance on both the structural and behavioral levels provides a basis for reuse. In the same modeling environment, it is also possible to capture the system requirements using HySCs (hybrid Sequence Charts, e.g., Fig. 1, right) and later to use the captured requirements for validating the model.

MaSiEd includes an incremental model compiler to translate HyROOM models into C^{++} source code programs that are then compiled to run on a ROOM virtual machine. A DDE (dynamic data exchange) interface to Matlab/Simulink enables the use of an automatic C program segment generation based on Matlab Real-Time Workshop and the evaluation of continuous models in early stages of the development.

The C code corresponding to the block diagrams translated by the Matlab Real-Time Workshop and the C^{++} code generated from the rest of the model are combined automatically. The generated model-specific code is linked with pre-compiled run time system libraries (MicroRTS, developed by ObjecTime Ltd.). Once the model compiled, it can be downloaded from the developing environment to a target computer running the VxWorks or RTLinux real-time operating system.

3.4 Example: Wire Stretching Plant

We chose to include a sketch of this industrial case study, previously described in (Stauner et al., 2001, Péter et al., 2000), in order to show how different description elements – architecture diagrams, extended state machines, continuous block diagrams, and hybrid Sequence Charts – can be connected within the MaSiEd tool. The system's purpose is to wind wire of different thicknesses on reels. The case study was done in order to test the discrete process control; the actual PLC has been connected to MaSiEd for this purpose. The system's structure is as follows. The environment produces wire that enters the system at a variable speed. This wire has to be wound up on a reel. The turning reel's velocity has to be almost equal to the incoming wire's velocity in order to guarantee a homogeneously wound wire. Its velocity is controlled by a device between reel and environment, called the *dancer*, that consists of a set of pulleys the wire runs over (Fig. 1, left).

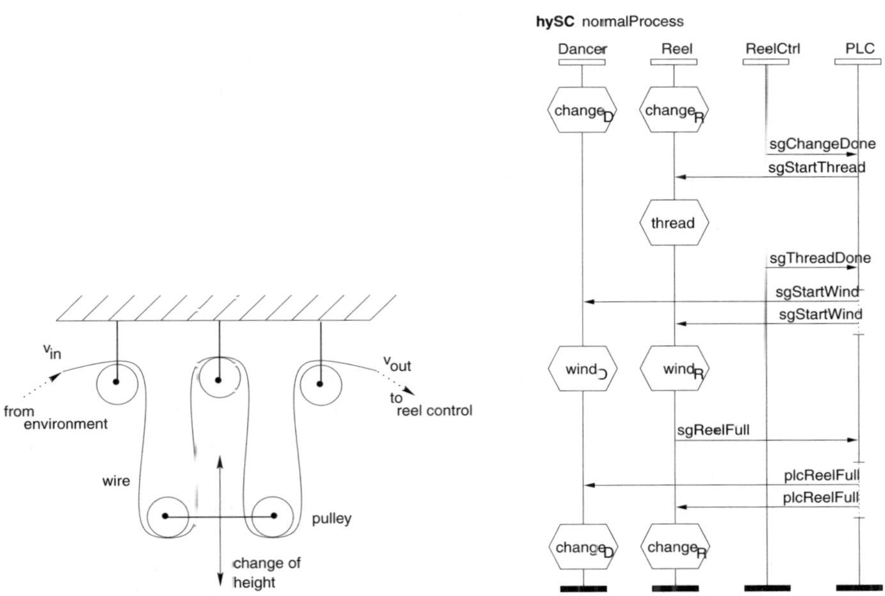

Fig. 1. Dancer (left), HySC: normal operation (right)

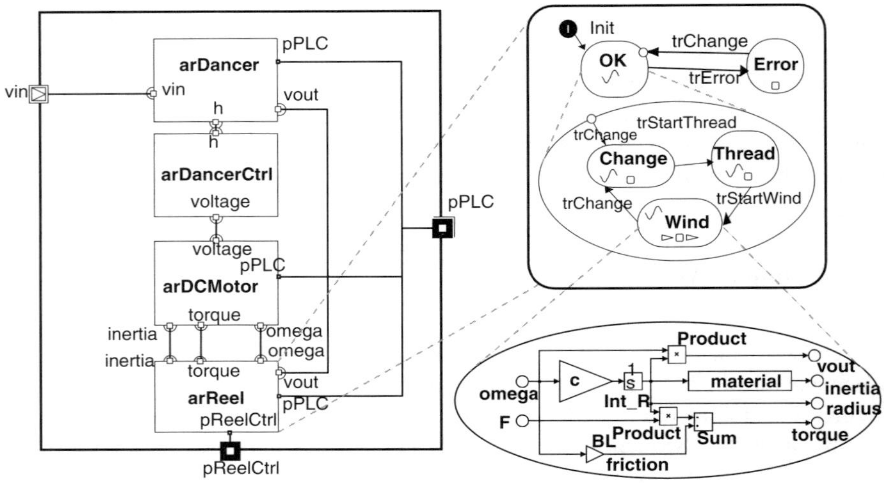

Fig. 2. Hybrid subsystem's architecture and reel's behavior

Not all of the pulleys are fixed so that the wire's velocity is dependent on the vertical position of the loose pulleys in this device. Once a reel is totally wound up it has to exit the system. This is achieved by a table that brings a new (empty) reel in position after the full one has been put on a belt. This is a complex, mostly discrete process that involves moving the table, fixing the new wheel on the motor's axis, cutting the wire, and making the new reel turn. There are two main conveyor belts involved in the system, one for empty, and one for wound up reels. This part of the system is omitted here for brevity's sake. In addition to hydraulic aggregates that guarantee the fixed position of a (turning) reel on the axis of the associated motor – the motor that interacts with the dancer via a controller for the turning speed – the last main component of this system is the PLC part with roughly 180 I/O ports. The MaSiEd model consists of roughly 100 discrete actors, 20 block diagrams, and about 10 hybrid actors.

3.5 Hybrid Subsystem

The hybrid subsystem that consists of the dancer, the DC motor for driving the reel, and the controller connecting the DC motor with the dancer is used for demonstrating the different description techniques. Its basic structure is depicted in Fig. 2, left, where continuous ports are marked with a semi-circle around a box. The systems input is the wire's continuously changing input velocity, v_{in}. The system communicates discretely with the PLC via port *pPLC*, and with the reel control via port *pReelCtrl*. The reel control takes care of exchanging a full reel in the system by an empty one. Figure 1, right, contains a *Hybrid Sequence Chart* (HySC), see (Grosu et al., 2000), depicting a typical use case for this system. HySCs are a variant of UML's Sequence Charts (Rational UML, 1997) and use the standard *Message Sequence Charts* (MSCs) notation, see (ITU, 1999). Unlike MSCs, HySCs employ a synchronous time model. They use the MSC condition boxes (depicted as hexagons) to refer to the (qualitative) state of one or more components. Dotted parts of an axis indicate that the associated signals occur simultaneously. In Fig. 1, right, a use case for the normal operation mode is specified: First, an empty reel has to be inserted in the system (states *change*). Once the change is done, the *threading* process starts; the wire is put onto the new reel, and it is cut from the old one. If this process successfully completes, the actual *winding* process is initiated; compared with the *change* state, its main characteristic is a relatively high velocity of the reel. When the reel is full, the PLC re-initiates the process of *changing* the reel by moving the full one out of the system and bringing an empty one in position. For the sake of brevity, we omit the predicates that describe the states as well as the differential equations describing the different continuous behaviors. Figure 2, bottom right, exemplifies the use of block diagrams. The DC motor is a standard PID controlled motor with its own controller. Its inputs are a voltage (which is proportional to the PID controlled dancer's height) that directly controls the motor's angular velocity as well as the reel's torque and inertia. We omit the (standard) details for brevity's sake. The *motor* component consists of just two states, *on*, and *off*. The third hybrid component of interest is the reel itself. Given the wire's input velocity, it keeps track of the reel's inertia, its torque, and its continuously growing radius (wire is being wound up; e.g., Fig. 2, bottom right). When state *change* is reached, actor *reel* is reset: the reel's radius is set to zero When the new reel has been fixed to the motor, the wire then needs to be threaded in state *thread*.

3.6 From HySCs to State Machines

HySCs also support test case implementation in MaSiEd. Traditional testing of simulation models is manual, time consuming and error prone. In order to facilitate model testing, MaSiEd supports the automatic generation of complete unit and integration test harnesses directly from HySC specifications. In contrast to the generation of test cases that we describe in the next chapter, scenarios have to be fully described.

The algorithm used in MaSiEd for the automated synthesis of complete unit and integration harnesses directly from HySC test case specifications is based on the maximum progress algorithm (Leue et al., 1998). Test case specifications in form of HySC are analyzed with respect to their software architectural content, including structure and behavior, and are represented in terms of HyROOM. Every concurrent instance (axis) in the HySC specification is represented by exactly one concurrent HyROOM actor. The motivation for the maximum progress algorithm is to determine maximum progress transitions in the HySC specification and to map these onto HyROOM behavior descriptions. This means that synthesized HyROOMchart transitions can span events originating from more than one HySC. Since we do not use a hierarchical state machine structure, the synthesized ROOMcharts will be flat. One HyROOMchart per instance in the HySC specification is generated.

3.7 Semantics

MaSiEd has been given a precise formal semantics (Stauner et al., 2001) on the grounds of HyCharts (Grosu et al., 1998). Roughly speaking, HyCharts provide means of specifying both the structure (HyACharts) and the behavior (HySCharts, DiCharts) of hybrid systems. The semantics is given by stream processing functions (Broy, 2001): (infinite) input trajectories are mapped to output trajectories. While developed independently, it turns out that there is a natural semantic mapping from HyROOM into HyCharts. We omit any technicalities for the sake of brevity; the semantics is defined in (Stauner et al., 2001); applications like program transformations are treated in (Stauner, 2002, Stauner, 2001).

4 Model Based Testing with AutoFocus

Formal methods like model checking and theorem proving are concerned with properties of a *model* that provides an abstraction. Proving or approximating properties of the actual *implementation* is the mandatory second step. Model based testing includes generating test cases from models and executing them. These test cases are used for testing different iterations (and/or projections) of the current stage of the product. Besides disambiguating requirements, the aim is to reach a valid model of a system. Generating test cases is thus a part of the requirements capture as well as implementation or design activities.

Models are used for hardware-in-the-loop simulations, for generating production code, or for validating existing systems. In the latter case, the idea is to perform conformance tests of a system with its model. This may require suitable concretizations of the respective test cases. Clearly, an automatic assignment of verdicts for functional test cases only makes sense if the same model is not used for generating production code and test cases. Otherwise, the system would be tested against itself (in this case, test cases may not be suitable for establishing functional conformance, but they might help in verifying environmental assumptions or the correctness of code generators). Using models for verification is a natural choice if for organizational reasons, quality assurance and implementation departments are to be separated, if efficient code generators for a particular target language do not exist, or if the system contains large legacy parts.

Currently, we are unable to generate test cases from MaSiEd models. The main reason is the use of C^{++} as transition annotation language which in general eludes automated formal analysis. This is why we implemented a test sequence generator for the CASE tool AutoFocus (Huber et al., 1997) which uses a functional language for guard specifications instead. The remainder of this section briefly describes AutoFocus, and explains how test case generation with Constraint Logic Programming (CLP) is performed. Note that what we do here is different from the generation of test harnesses from HySCs as described above where full discrete control and continuous signal information is to be provided. The technique described here aims at computing this information.

4.1 AutoFocus

Similar to MaSiEd, the main description elements of AutoFocus are concerned with structure, behavior, data, and interaction specifications as encoded in the meta model. Hierarchic system structure diagrams depict components (actors, capsules). They encapsulate data and behavior, and they thus provide a means of functionally decomposing a system. Bottom level components are assigned a behavior in terms of a Mealy-like state machine. Transitions consist of statements that read input channels, of a guard for establishing whether or not a transition may fire, assignments that update local variables, and of statements that compute outputs.

Guards and assignments are specified in a Haskell-like functional language. Components communicate over typed channels. The rationale for using a functional language for typing is that in embedded systems, data modeling with elaborate constructs like class diagrams is rarely necessary. Simple sum and product types turn out to be sufficient.

Similar to clock-synchronous hardware circuits, all components perform their computations simultaneously: they read values from their input channels, compute updates for local variables and output channels, and write these updates so that at the next clock tick, the values are available. This results in a time-synchronous communication scheme with buffer size one – staying with the analogy of clocked hardware, each channel contains an implicit latch, or shift register, respectively. The rationale behind choosing this admittedly restricted semantics is that it is exactly this simplicity that allows AutoFocus models to be formally analyzed, e.g., model checked, or used for test case generation. By using recursive list types, it is also possible to implement asynchronous communication. This semantics is inherently discrete. Continuous system parts are coped with by discretization (Pretschner et al., 2000). Matlab block diagrams are automatically translated.

4.2 Test Case Generation

It turns out that the simple clock-synchronous semantics is naturally encoded by Horn clauses with axiomatizations of natural numbers or reals (Lötzbeyer and Pretschner, 2000). The resulting CLP code may be used for simulation by giving inputs to the system for each step, similar to what is done with other simulation code generators as well. It is also possible to partially specify inputs, outputs, or constraints over them – for instance, a maximum number of signals to occur, or temporal dependencies – without specifying their exact timing. By enumerating all traces of a bounded length, the LP engine then computes those traces that satisfy the constraints imposed on inputs, outputs, states, transitions, or local variables.

Conceptually, the generation of test sequences is hence achieved by formalizing the test purpose by means of existential specifications of the kind "given a set of constraints, make the system reach state q_1, q_2, etc." where each q_i specifies a desired constrained value of the variables for control states, data states, inputs, or outputs. The resulting I/O traces are the test sequences we are interested in. These existential test case specifications are sufficient for covering use cases from requirements capture activities or finding test sequences that satisfy a given coverage criterion. These can be reduced to a set of test case specifications, each of which makes the system reach a certain state or condition.

Computationally, this would be too simple to work efficiently. In fact, our approach is akin to bounded explicit model checking or other state space exploration techniques. State space explosion is the commonly accepted hindrance of these approaches for acceptance in the industrial practice. We use dedicated heuristic A*-like search algorithms in order to find those q_i we are interested in (Pretschner, 2001). Furthermore, our system allows for explicitly specifying environmental and efficiency constraints for manually pruning the search tree. In terms of continuous or hybrid subsystems, environmental and efficiency constraints may include gradients of the respective curves, or restrict certain values to given intervals. This kind of constraints is taken care of by predefined constraint solvers connected to typical available CLP systems.

This not only reflects the need for manual intervention; experiences with industrial partners have shown that test engineers are in fact capable of identifying those parts of a system that may be sliced away. Constraints are used for taking care of temporal dependencies, numerical properties, excluded or enforced occurrences of certain signals. Furthermore, they allow to compute with and efficiently store sets of states (Pretschner, 2001).

Test case specifications may also include restrictions of the search space. They are provided directly as constraints with temporal operators, as sequence diagrams, or as finite state machines. In this latter case, the test case specification often is a combination of a partial environment model and the formalized test purpose. When testing protocols, for instance, the test case specification, given as an automaton, specifies certain typical runs or threat scenarios. It is also possible to define transition probabilities. As in the case of general models, the essence of a test case specification clearly is independent of its representation, be it a formula, a sequence diagram, or a state machine.

Specifications do not contain only existential properties. Universal properties like invariance, safety, or liveness are also specified. Since testing is, by definition, a finite activity, these properties cannot be tested exhaustively. We thus approximate the universal property by a set of existential properties. Justified by the success of limit testing in the setting of testing transformative systems, we compute traces that come as close as possible to a state that violates the invariance. This is done on the grounds of the same A* like heuristics used for finding particular elements in the state space (Pretschner, 2001).

4.3 Process: Procedure, Regression Tests, Compositionality

Test sequence generation proceeds as follows. The automatically translated AutoFocus model is conjoined with the (existential) test case specification, environmental and efficiency constraints. The resulting test sequences are used for debugging the model itself. This is done by (manually) comparing every I/O sequence to what one would have expected – at this stage, there usually is no formal operational specification to compare with. Instead, the model itself is the executable specification.

In an incremental setting, models are developed iteratively. For the sake of brevity, we only consider increments that add functionality to a model. If feedback from the customer suggests changes in increment $\mathcal{I}_n$, it becomes a modified part of increment $\mathcal{I}_{n+1}$. $\mathcal{I}_n$ might also remain unchanged in $\mathcal{I}_{n+1}$. For each $\mathcal{I}_j$, we consider functional and structural[2] test case specifications to be given by the engineer. The test case specifications are then used for computing actual test sequences. These traces can be computed separately for each of the increments. Validity of the traces has to be checked manually. It is, however, possible, to use test sequences for increments $\mathcal{I}_j$, $\mathcal{T}(\mathcal{I}_j)$, with $j \leq n$ for regression testing increments $\mathcal{I}_k$ for $k > n$. We simply feed the test sequences $\bigcup_{j=1}^{n} \mathcal{T}(\mathcal{I}_j)$ into $\mathcal{I}_{n+1}$, and are hence able to automatically assign verdicts to these tests. $\mathcal{I}_{n+1}$ is checked for conformance with $\mathcal{I}_j$ for $j \leq n$.

These verdicts have to be taken with caution. The problem is that adding functionality may actually restrict the behavior of a system; false negatives are the result. This is, for instance, the case if a timer that periodically emits a timeout is composed to a system $\mathcal{I}_n$. The test sequences for $\mathcal{I}_n$ may consist of traces that respond to timeouts that occur erratically.

By inverting the above idea, we get a compositional approach to generating test cases. Consider a system $\mathcal{I}_{n+1}$ consisting of increment $\mathcal{I}_n$ that is composed with a component k such that there is a channel between the two in each direction. It is then possible to use $\mathcal{T}(\mathcal{I}_n)$ for generating test cases for k and for $\mathcal{I}_{n+1}$. We can use the outputs of $\mathcal{T}(\mathcal{I}_n)$ as a driver for k, and thus get new test sequences for k, and, consequently, for $\mathcal{I}_{n+1}$. Conversely, we can use the inputs of $\mathcal{T}(k)$ as putative outputs of $\mathcal{I}_n$. Remember that using CLP allows us to partially specify outputs and make the system compute those fully instantiated I/O traces that eventually result in the specified output. Ignoring the problem of running into the same problem as with regression testing, we directly get new test sequences for $\mathcal{I}_n$, and for $\mathcal{I}_{n+1}$.

4.4 Example

We do not give the AutoFocus diagrams of our case study here since, apart from block diagrams, they are almost identical to the MaSiEd specification. Neither do we provide any actual test case specifications or computed test sequences for this system since this would require a rather deep level of technicality. We do, however, give some informal test purposes that readily translate into formalized test case specifications and that we have used for test sequence generation. Among others, test purposes include the following. For each of the discrete PLC, environment, and other components coverage on states, transitions, or guards is a test case specification. Reflecting the composition of components, these unit test sequences are combined in order to derive new test sequences for the connected components, as described above.

[2] When adequately modeled, structural coverage criteria like state coverage may well be considered as functional tests. This is because each control state encodes a certain functional unit.

Furthermore, for the dancer, there are HySCs from the requirements capture activities. We easily translate these into automata and use them as test case specification such the diagram depicted in Fig. 1, right. As a last example, in terms of universal properties, we compute a test suite for the property *whenever state Error is reached, we can escape from it*. Clearly, many more test case specifications are conceivable. For the sake of brevity, we omit the discussion of assessing the quality of a test suite.

5 Conclusion

Major advances in software and systems engineering seem to be bound to the use of abstractions as the key metaphor. Artifacts at increasing levels of abstraction enable intellectual control over highly complex systems. Integrated tool support, ranging from specification, implementation, verification to requirements tracing and documentation is desirable for an efficient workflow.

We have presented our approach to model based development which relies on suitably chosen abstractions for the essential constructs in a particular domain. Tool support for modeling, simulation, code generation, and test case generation for two CASE tools, MaSiEd and AutoFocus, has been presented.

Whether or not CASE support with graphical description techniques rather than using dedicated IDEs like Forte or Eclipse is the right choice, is not obvious. In a model based setting, IDEs for language subsets like Spark or Ravenscar in addition to test tools may turn out to be the more practicable approach. The arguments that graphical description techniques facilitate the understanding of a system loose validity with increasing complexity of the system under development. In fact, misuse of hierarchic statecharts makes system designs foggy, as does misuse of inheritance in class diagrams.

The step from models to implementations may involve adding technical details that are not relevant in early development phases. Real time issues demonstrate, however, that low level technical details may have to be considered right from the beginning. We are convinced that in many areas, it is possible to achieve a seamless integration of abstract models and low level technical issues (for instance, this is certainly true for PLCs as considered in the case study of this paper, or for smart cards). If, in general, this turns out be an illusion, then model based development boils down to a philosophy of the activities of requirements engineering, and clearly remains most valuable in that it allows for intellectually mastering the complexity of large systems.

We are convinced of the necessity of a transparent, precise semantics. However, simplicity should be a key factor when formalizing it – otherwise, there is a formal semantics, but engineers will not have the time to deeply understand it. A clear understanding of the meaning of an artifact is the prerequisite for transformations, be they refinements (Stauner, 2001, Stauner, 2002) or refactorings. They are also necessary for code generation and validation techniques like test case generation and execution. Formal semantics for the two tools have been defined but are not part of this paper.

MaSiEd was presented, a tool for modeling and simulation of hybrid systems specifically targeting at the application field of process automation. MaSiEd integrates the ROOM virtual machine with Matlab block diagrams. The modeling concepts, an extension of ROOM, have been described and demonstrated along the lines of an example system taken from an industrial case study.

In terms of ROOM based modeling, (Pretschner et al., 2000) as well as the case study in this paper showed that the clear distinction between structure and behavior results in the need of copying the same set of states from one component to another in the same subsystem. This problem is alleviated by the use of MaSiEd's inheritance mechanism, but the general problem still persists (it does not in statecharts for there is no clear differentiation between structure and behavior as well as no concept of interfaces).

Finally, AutoFocus was presented. Due to the simplicity of its semantics, it is possible to derive test sequences for discrete or discretized systems. The idea is to use a combination of symbolic execution and state space exploration with heuristic search on the grounds of Constraint Logic Programming. The embedding of this approach into an incremental model based development process was described. This technique is a complement to the generation of HyROOMCharts from HySCs (i.e., test harnesses from scenarios) since in this latter case, complete information about signals and their temporal dependencies have to be provided. The AutoFocus based approach aims at computing this complete information.

In industrial practice, test cases are seldom developed systematically. If they are, engineers often use coarse discrete abstractions (e.g., "quickly accelerate" or "slowly accelerate") of a system in order to identify interesting scenarios. Clearly, (mis)using condition or state boxes of HySCs to this end directly lends itself to the specification of test cases with HySCs. The test case generation procedure profits from this abstractions since the model becomes less complicated.

Future work includes machine support for sound refinements and refactorings of hybrid systems. The integration of hybrid class diagrams into MaSiEd is the subject of current work. In terms of the test case generator, we currently assess its applicability in various industrial projects. The question of how to automatically extract "good" test suites is yet unsolved; we consider the analysis of error classes in a particular domain a first step in the right direction.

Acknowledgment. We would like to thank Lingxiang Xu for providing the original discrete case study. In numerous discussions, J. Philipps, B. Schätz, F. Huber, W. Schwerin, and P. Braun provided valuable insights into the nature of model based development.

Hybrid Modeling of Complex Process Control Function Blocks

Ansgar Münnemann[1], Udo Enste[2], and Ulrich Epple[1]

[1] Chair of Process Control Engineering, RWTH Aachen, Germany
[2] LEIKON GmbH, Aachen, Germany

Abstract. The extensive description of complex functionalities in function block systems, using hybrid modeling methods can be simplified by identifying an internal structure of these function blocks. For that a formal model of 'function block components' is needed. Witch such a model the internal structure of batch-oriented blocks and advanced control blocks as two typical examples for complex functionalities in process control engineering is seperated in unified and generic and typespecific describable components, whose dynamic behaviour runs from discrete to continous and frequently is hybrid.

1 Component Model for Function Blocks

Function block models are discussed in several normative standards (IEC TC65 WG6, 1999, IEC SC 65C WG7, 1999, IEC 61131-3, 1992, Fieldbus DDLS, 1996, PNO, 1999). A function block is a software unit, describing a method and its data structure. Depending on the necessity of the data exchange between several function blocks, state variables of a block are distinguished into input data, output data and hidden state variables. Besides the principle of encapsulation, a single-level class concept is characteristic for the function block technology. The algorithm and the data structure are fixed in a 'function block type'. Using these function block types, control structures can be realized by a net of self-sufficiently, asynchronously and quasi-continuously working function blocks instances. These instances process their own data sets and by referencing the method of their corresponding function block type. The function block technology provides an intuitive way to configure and modify process control architectures online. Depending on the dynamic characteristic of the function block types and the types of signals exchanged between the blocks purely discrete, quasi-continuous, as well as hybrid control nets can be realized. Looking in detail at the algorithms of complex function blocks which are used in process control applications, the function blocks themselves can be identified as hybrid systems as well. The function block methods show no internal structure (although one exists) and thus a systematic analysis is extremly difficult. The usage of a component model with formal defined language elements allows to structure the complexity of function block algorithms (see Fig. 1). Such a language with well-defined static semantics (in consideration of hybrid modeling elements) is defined by means of the graph rewriting system PROGRES, see (Schürr, 1994), in (Enste and Kneissl, 2000).

S. Engell, G. Frehse, E. Schnieder (Eds.): Modelling, Analysis and Design of Hybrid Systems, LNCIS 279, pp. 53–65, 2002.

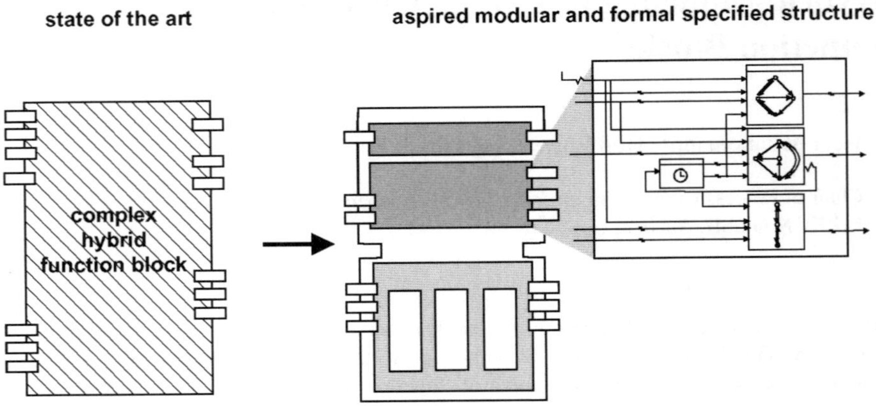

Fig. 1. Identification of internal structures of function blocks

So-called 'internal components' and the idea of templates were introduced to the function block description language (Enste and Epple, 2001, Enste, 2001) in order to describe a function block composed of smaller parts. Well-structured templates shall improve the engineering activities to design function block applications by increasing the reusability and reducing the testing effort. Using components, specific function block templates can be developed, which offer a specific structure and unified functionalities inside the function blocks of a specific class (see Fig. 2). In consequence, the single-level class concept (types and instances) can be replaced by a multi-level class concept, where function block types can be derived from templates. Because of the multiplication effect when using templates and multi-used components, it is useful to specify the dynamic of the encapsulated functionalities inside these templates by hybrid modeling methods.

Each internal component has a well-defined boundary called 'capsule' which encapsulates the component's implementation from its environment. Thus, the capsule can be considered as a black box view of the component. Components communicate by signals, which are exchanged along explicitly modelled connections between ports. Ports belong to the capsule of a component and mediate access to the component. No direct manipulation of the state of a component is possible, except by communication via connections across ports. In addition, the function block model offers a hierarchical composition of components.

The implementation of a component is described by composing and connecting capsules as black box views of subcomponents. The results of this intermediate component construction step are called templates. A template can be instantiated forming a component by substituting a component for each of its capsules. This mechanism can be explained by comparing the capsule with a socket where a component can be plugged in. A template is like a printed circuit board. Plugging components into all the sockets results in a new component, which can be used thenceforth.

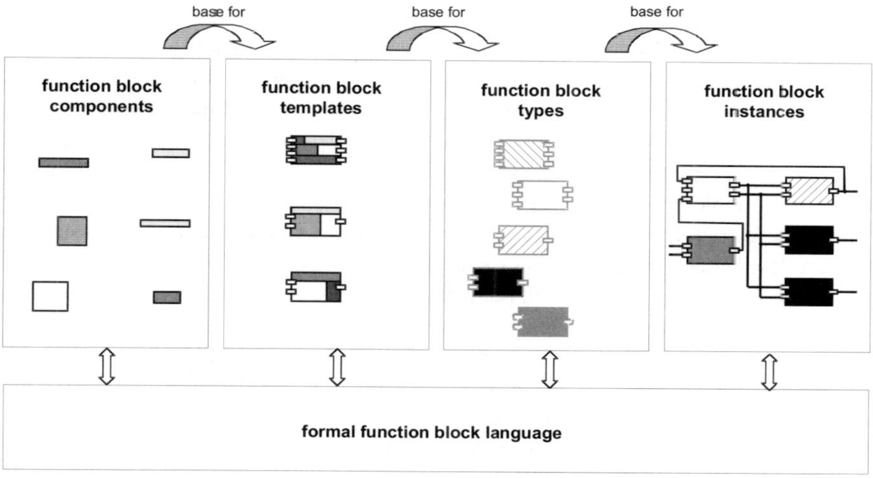

Fig. 2. Concept of developping process control applications

At all, this new description language includes many features of the normative function block languages and is similar to the object oriented modeling language for realtime systems ROOM (Selic et al., 1994).

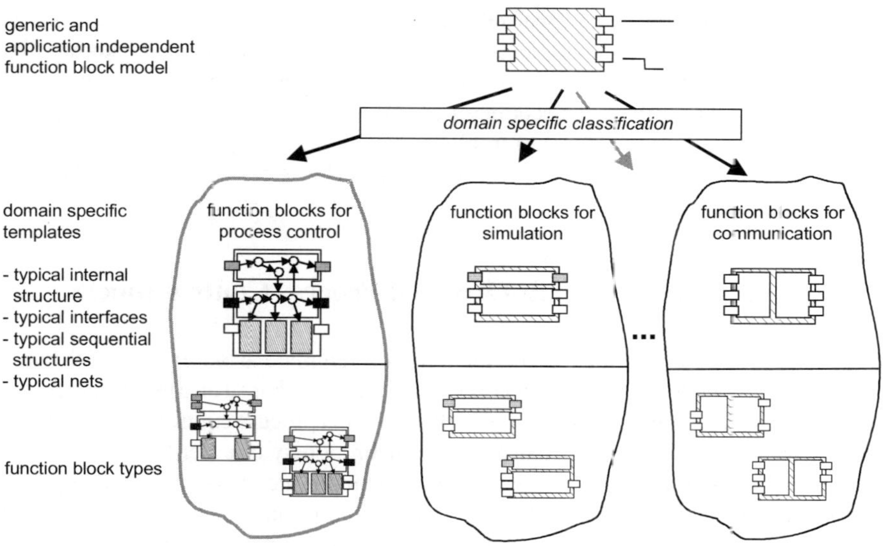

Fig. 3. Domain specific classification of function block templates

The next step is to investigate how to use templates in an adequate manner. Looking to function block applications, several classes of function blocks can be categorized (see Fig. 3). The classification results in groups of function block types which agree in similar interfaces, internal structures, related functionalities and in typical dynamic behaviours. These characteristic features can be implemented by developing a template for all function blocks of this group. Exactly these templates are reasonable objects to apply the formal dynamic modeling methods to. The effort of a formal specification is justifiable because of the generic and re-usable software level. Each template shows a modular architecture. The modules (realized by 'components') can be classified regarding their dynamic characteristics (discrete, continuous, hybrid) and their degree of unification (typespecific, generic, unified). To guarantee a practicable engineering process and an analyzable functionality it is important for the success of the dedicated dynamic validation to maximize the unified and generic components. Based on this classification suitable modeling methods to specify the defined components and templates can be applied. In the past, several approaches of hybrid modeling methods were introduced, e.g. (Nicollin et al., 1992, Tittus et al., 1994, Lynch et al., 1996, Krogh, 1993, Alur et al., 1993, Chouikha and Schnieder, 1998b, Nenninger and Krebs, 1998). From this huge amount of possible modeling methods, the condition/event systems (c/e-systems) introduced by Krogh (Krogh, 1993) and the hybrid automata introduced by Lynch (Lynch et al., 1996) were chosen to describe specific aspects of standardizable components and their interaction inside a function block. Because of its modular concept, the modeling method of c/e-systems is preferred.

On the one hand a specification of function blocks using more granular elements separating specific functionalities inside a block is possible. On the other hand, more generic function block specifications can be developed using the idea of templates. Because of the separation of encapsulated components inside a block one effect is that purely discrete components can be pulled out and analyzed using well known discrete modeling methods. The hybrid part shrinks to an analyzable subsystem inside a block. Considering these aspects in an engineering phase seems very important, instead of taking a complex function block design and trying to translate it into a formal language. Thus, a reduction of complexity can be performed while engineering a function block.

2 Hybrid Model of Batch-Oriented Process Control Blocks

In the following we will present a typical hybrid template as an example for batch oriented function blocks for process control. The design is based on a hierarchical process control model introduced in (Epple, 1994). In this process control model, several control units interact self-sufficiently and asynchronously in order to generate control values. The control architecture is organized hierarchically (see Fig. 4). Superior control units send control instructions to inferior control units. This kind of forward driven information exchange can be realized by standardized telegrams (Enste and Fedai, 1998). The handling of such control instructions, in particular the checking mechanisms to verify incoming instructions (syntactical and semantical checks) is a typical functionality which is worth to standardize by developing a generic component. This component considers also access rights of operators vs. automatic units.

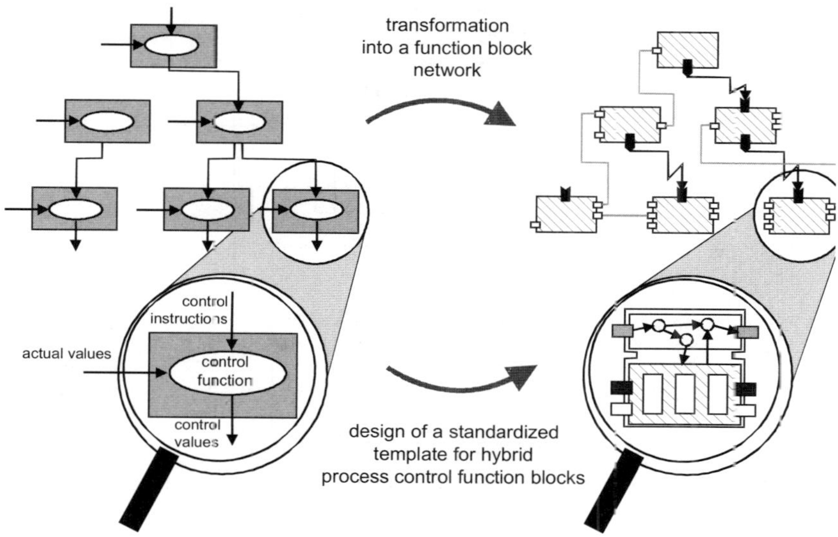

Fig. 4. Hierarchical process control model and its transformation into a function block network

2.1 The Template

The internal structure of a batch-oriented process control block can be outlined as follows, see Fig. 5 and (Enste and Epple, 1998): the transaction control unit is the interface to the tasking of the function block system. Activating this module means starting the algorithm of this function block. The transaction control represents a centralized control module inside a function block. It coordinates the control flow between all components inside the block. The checking mechanisms and verification of the acceptance of incoming control instructions are treated by a checking unit. Based on its result and on information about the operating conditions fed in by typespecific signals, a generic state machine will be activated. The link between the typespecific signals and the generic state machine is handled by a typespecific component (specified inside the template as a capsule), where the typespecific signals must be mapped to the standardized signals of the generic component with predetermined semantics. The signals may prohibit the execution of an incoming instruction. The state vector of the generic state machine causes at least the activation of one of several capsulated process control logics inside the function body.

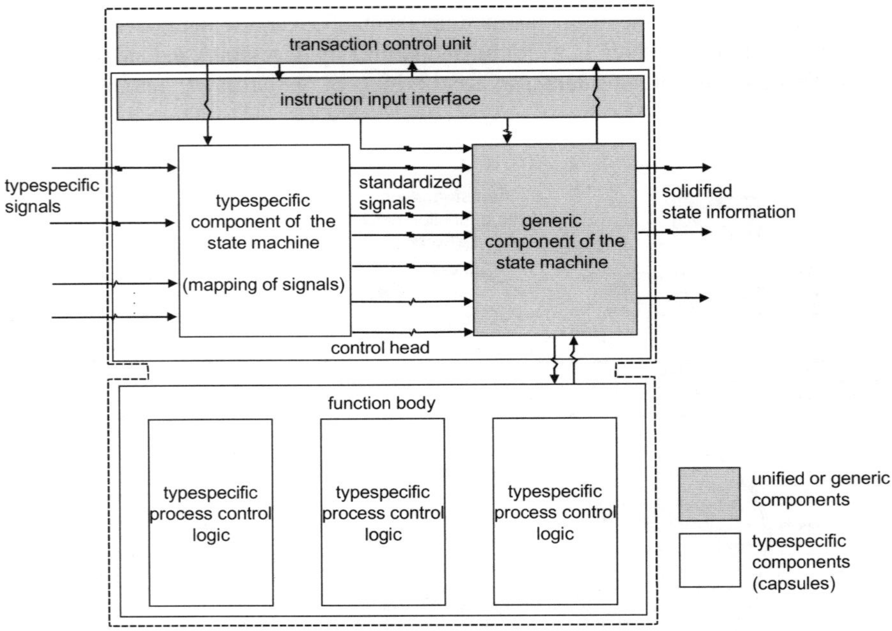

Fig. 5. Template for batch-oriented process control units

To specify the process control logic, an underlying structure is proposed. A control logic has to be implemented into one of three hybrid function charts (see Fig. 6). These hybrid function charts respectively represent the starting, the steady and the take-off phase of the controlled process unit. The hybrid feature of the function body is established by the specification of the steps inside the function charts. Each step is separated into three sections: the 'entry', 'do' and 'exit' operations, similar to (Rumbaugh, 1991). From the hybrid modeling point of view it is important to distinguish between the control actions (specified as 'entry' and 'exit' operations) and the control activities (specified as 'do' operations). Actions are switching procedures with negligible time requirements, whereas activities are continuous operations.

All unified or generic components of the template and their interaction were specified using hybrid modeling languages. Without going into detail, two components are presented as examples. First the 'instruction input interface' and second the 'generic state machine', both modeled by c/e-systems. All the generic or unified components show mainly a discrete dynamic. Except time conditions must be considered. Therefore, clocks were used, which were introduced in Kowalewski (Kowalewski, 1996). A more complex hybrid structure will occur when the typespecific components must be integrated.

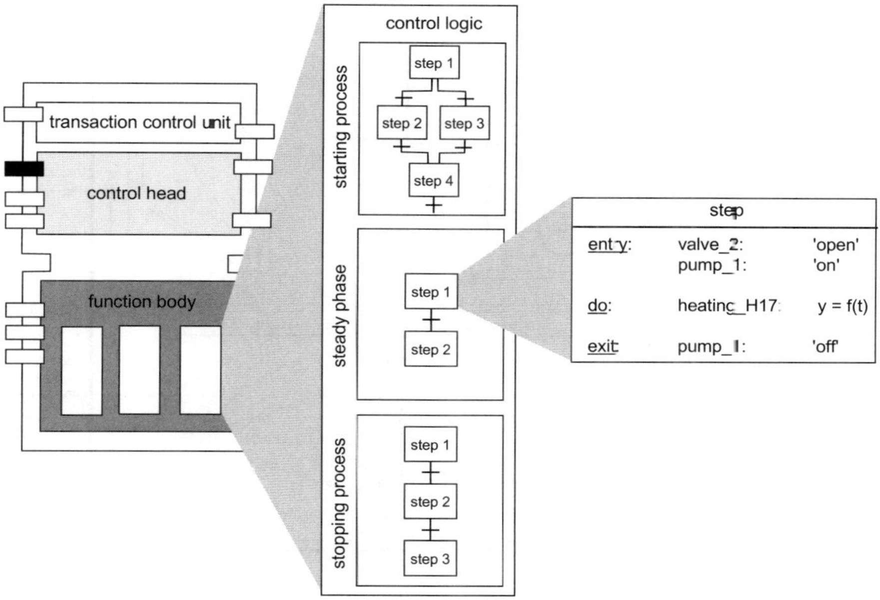

Fig. 6. Internal structure of the function body including hybrid sequential function charts

2.2 The Instruction Input Interface

The 'instruction input interface' represents a typical generic hybrid component. Regarding Fig. 7, the upper part of the component is specified in a unified manner and the lower part represents the generic functionality which can be extended by a number of states 'start x' which represent each a specific control logic inside the function body.

The unified part includes a standardized occupying mechanism and a time-based supervisory mechanism. The occupying mechanism garuantees that in one control phase just one superior unit (operator, a specific automatic unit or a local operator in the field) is allowed to send instructions to this control unit. Thus the ID of the sender of an instruction must match the actual occupying state, except the state is 'free'. Because the control units can be used in distributed control systems, a clock supervises the timespan between incoming instructions. The aim is to detect failures in communication. Therefore, each superior control unit which has occupied an inferior unit, has to send at least occupying instructions periodically to prove its existence. Finally, a semantical check of the incoming instruction has to be modelled. Therefore each instruction input interface includes a generic state diagram, which contains possible instruction types and possible sequences of instructions. At all, this component signals the actual occupier, the actual valid instruction and an event which normally initiates a switch of the controllers functionality.

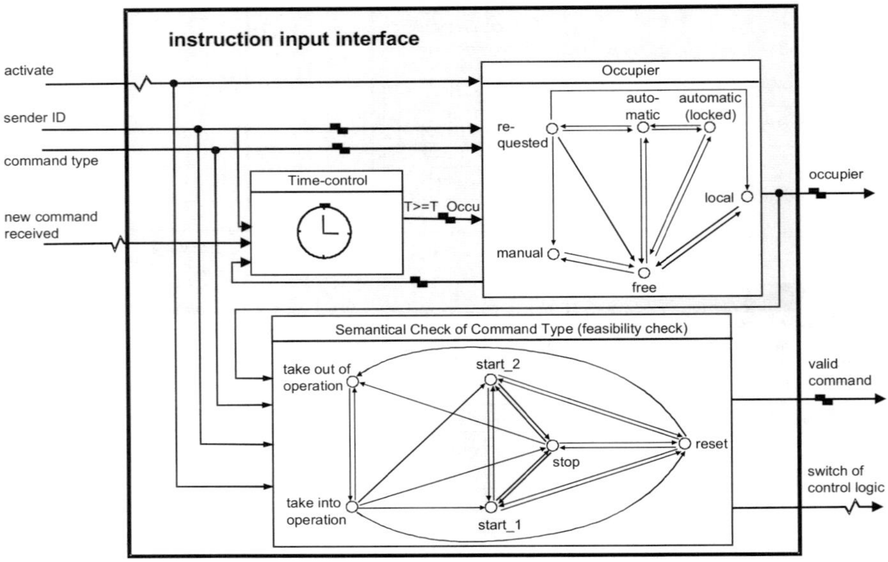

Fig. 7. The instruction input interface

2.3 The Generic State Machine for Batch-Oriented Process Control Units

Next to the instruction input interface, three more generic components generate informations for the administrative state space (Enste, 2001). These components are combined in a generic state machine in the control head (see Fig. 8). First of all you have the 'operating state logic', which describes the overall behaviour of a process control unit as one of the following operating states: out of operation, basic state, active function mode X. The operating state 'out of operation' apears, when the associated procedural unit is disused. This implies that the validity of the outputs is not guarenteed. In the 'basic state' the behaviour of the unit is supervised and as a consequence it's outputs must be valid. But in contrast to an 'active function mode', the associated procedural unit is kept in a secure state (e.g. a valve is closed) and not active controlled. The state 'active function mode X' is just a generic description of typespecific detailable function modes (e.g. a valve can 'open' or 'close' or 'position').

Each operating state - except 'out of operation' - is realized by a typespecific function body, which is internal structured in 'starting process', 'steady phase', 'stopping process' (see Fig. 6). The actual processed sequence is represented by the working state logic whose states are named similar to the sequences. The runtime of the starting process and the stopping process is supervised. If the time limit is exceeded before the startup respectively the shutdown process could be completed, the actual sequence of the associated function mode is interrupted and the working state is forced in the 'off' state. In addition the 'error state' changes to the 'bad' state and the occured error must be explicit confirmed.

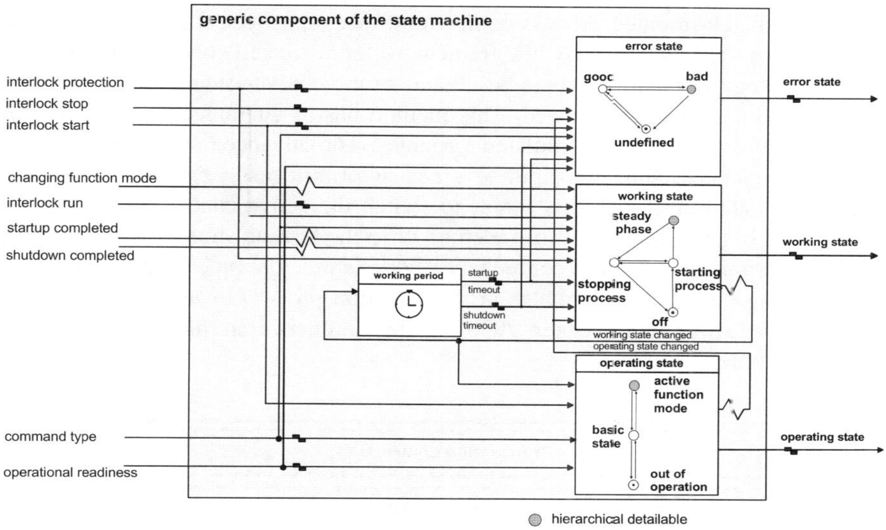

Fig. 8. Generic State machine of a batch-oriented process control unit

The behaviour of the generic state machine respectively of the process control unit is also influenced by so called interlocks. An active ‘interlock start’ forbids the starting process of a special function mode (the set of interlocks is defined for each function mode), an active ‘interlock stop’ inhibits the stopping process. The typespecific interlocks must be mapped to these generic interlocks in order to realize the required behaviour. For a more detailed description of the internal structures and sequences in a process control unit see (Enste, 2001).

3 Hybrid Model of Function Blocks for Advanced Control

The possible proceedings and problems with the integration of advanced control methods in process control systems are discussed in several papers (Kurz, 1990, Pearson, 1984, Schuler, 1992, Nöth, 1998, Münnemann and Enste, 2001). It is an wellknown fact, that the realisation of control methods as a function block includes more than the pure control algorithm. An obvious approach is to define a framework for the embedding of advanced control algorithms, which fits in the structures of a hierarchical process control model (see also Sect. 2).

3.1 The Template for an Advanced Control Block

The control head of the template shown in figure Fig. 9 is similar to that for batch-oriented process control units. You find the same instruction input interface and the same methodology with the mapping of typespecific signals to generic signals and their effect on a generic state machine. The state machine by itself differs from that of the batch-oriented process control unit (see next Sect. 3.2). Instead of the sequences in the function body, the framework for advanced controller consists of a multi strategy mechanism where beside one or more advanced control algorithms a safe backup strategy is realised. This method has to guarantee a (normally not optimal) valid output of manipulated variables. The advanced strategies are time controlled and exceeding the limit or a bad algorithm result cause a reaction of the generic state machine. This forces to switch the active control block strategy either to an other advanced control method or to the backup strategy. Certainly the template contains some more components. For example the including of validation information of measured variables in the selection process of alternative process variables, see (Enste and Uecker, 2000), or their influence on the working state of the active strategy.

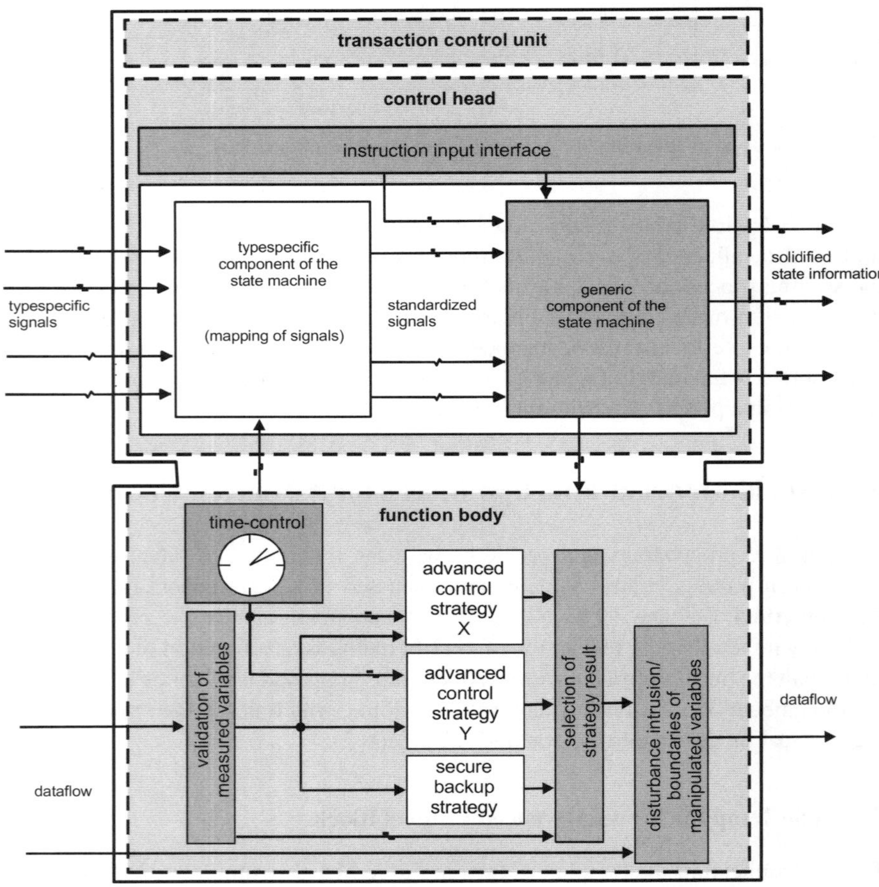

Fig. 9. The template for an advanced control unit

3.2 The Generic State Machine for Advanced Control Blocks

Similar to the state machine for batch-oriented process control blocks in the state logic for advanced controller the overall behaviour of the unit is described by an operating state. But instead of only one working state logic in the control head, an own working state is assigned to each control strategy (including the safe backup strategy) as shown in figure Fig. 10. The working state logics of an advanced control strategy and the backup strategy differ in some points. The latter has no 'initialisation' and 'synchronisation' state. Also the backup strategy may only be in the states 'standby' or 'active' after the startup process of the whole function unit. A changing from 'active' to 'standby' may only occur when one of the advanced control strategies controls active the process. This reflects the safety aspects of the backup strategy. For the most part the operating state logic reminds of the batch-oriented process control unit. Every control strategy equals a function mode, although the internal behaviour is different. While the operating state 'function mode X' marks the activated component in the function body of the unit, the operating state 'strategy X' specifies the algorithm component which controls active the process. That means the different control strategies run quasi parallel. Looking closer to the working state logic six generic states can be identified:

- 'off': the particular control strategy is not executed.
- 'initialisation': the time independent parameters are set and all other necessary initilisations are done.
- 'ready': the strategy is ready.
- 'synchronisation': the time dependent parameters are set and the strategy is adapted to the actual process situation.
- 'standby': the strategy is prepared to control the process active and tracks the process.
- 'active': the strategy controls active the process.

The condition signals which determine the working state logic behaviour are defined generically as well. The real mapping process from typespecific signals to generic signals occurs in the state logics for initialisation, synchronisation, standby mode and active mode.

4 Industrial Applications

The method of developing function blocks by reusing standardised components with a formal specified behaviour was applied to several industrial applications. As one example the automation of a membranous filtration process for the wastewater treatment is outlined. Figure 11 shows the pipes and instrumentation flow chart (P+I flow chart) of a pilot plant and the organisation of the process control. The wastewater is filtered by membrane modules, which periodically must be cleaned from sediments by compressed air. The process control structure is organised in a hierarchically way. On the base level you find standardised *single control units* for valves, pumps and

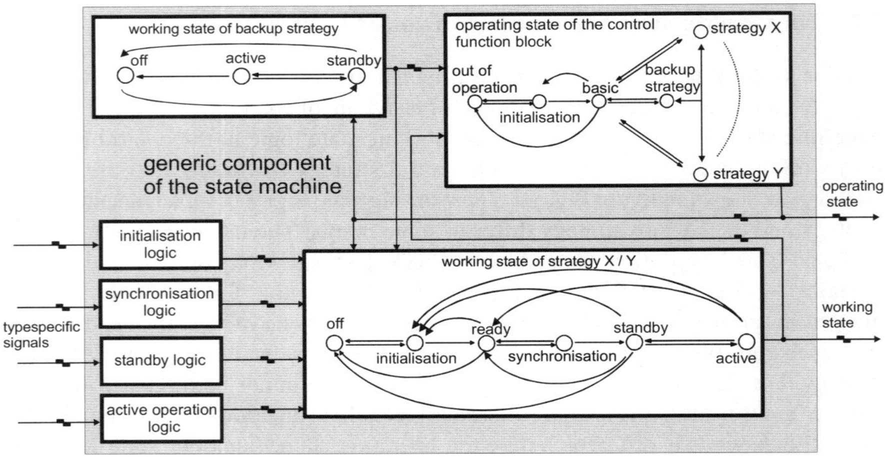

Fig. 10. State machine of an advanced controler

so on. In the second level, the functional grouped units like "Tank" and "Cleaning" are represented by *group control units*. Such a combining of functionalities can be continued in the hierachical process control model via an arbitrary number of levels. In every control unit (single or group) the internal structure is equal, defined by the template for batch-oriented process control. The function block engineering is reduced to the definition of control modes by sequential function charts. Using the instruction mechanism of the process control units, the sequence programming becomes process engineering oriented.

At top level, only one control unit is defined, which is responsible for the whole filtration process. During normal work the operator interacts via instructions with this control unit to manage the process. If necessary, trouble shooting must be done at a lower level, but remains in most cases the communication with a standardised and well known process control unit.

As expected the amount of faults and time could be reduced during the development process, and the handling for the operator was simplified.

5 Summary

The component model for function blocks includes hybrid modeling aspects to shrink the gap between the function block technology and the hybrid modeling methods. Concerning the multiplication effect using templates, it is worthwhile to specify the dynamic of the unified and generic capsulated functionalities inside these templates in a formal way. To specify the behaviour, c/e-systems were used. The specification of a modular template for function blocks in batch-oriented process control and for advanced controller allows a reuse of a lot of components. This simplifies the development of function blocks in several industrial control applications. As a next step

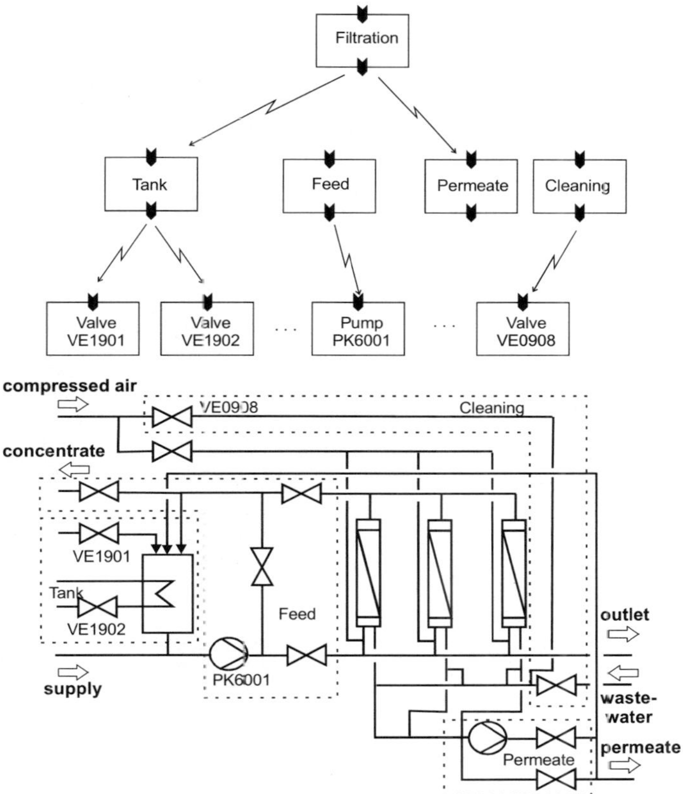

Fig. 11. Membranous filtration plant and it's process control structure

the level of detail of the components in the function body of an advanced controller should be increased and the whole component model could be transfered from a descriptive model to a dynamic realized model, with that structural changings of the implemented function modes respectively the control strategies during runtime would be possible.

Discrete Models for Hybrid Systems

Jan Lunze[1] and Jörg Raisch[2]

[1] Lehrstuhl für Automatisierungstechnik und Prozessinformatik
Ruhr-Universität Bochum
D-44780 Bochum, Germany
email: lunze@esr.ruhr-uni-bochum.de
[2] Lehrstuhl für Systemtheorie technischer Prozesse
Otto-von-Guericke Universität Magdeburg
Postfach 4120, 39016 Magdeburg, Germany
email: raisch@mpi-magdeburg.mpg.de

Abstract. This contribution provides an introduction to the topic of discrete models for hybrid systems. It motivates the use of discrete approximations of hybrid or purely continuous dynamics and comments on two complementary approaches. The article explains the core property of any suitable approximation and describes its major implications.

1 Introduction

Hybrid systems consist of two interacting subsystems – a purely continuous system with input, output and state signals evolving on dense subsets of Euclidean spaces, and a purely discrete system, with all signals "living" on discrete sets. Interaction between both components may be realized by quantization and injection. The former maps continuous-valued signals into discrete-valued signals, the latter uniquely translates discrete-valued signals into piecewise constant continuous-valued signals (Fig. 1). Hybrid systems are ubiquitous in engineering as most physical phenomena are continuous and many control devices are discrete by nature and/or implemented on digital computers.

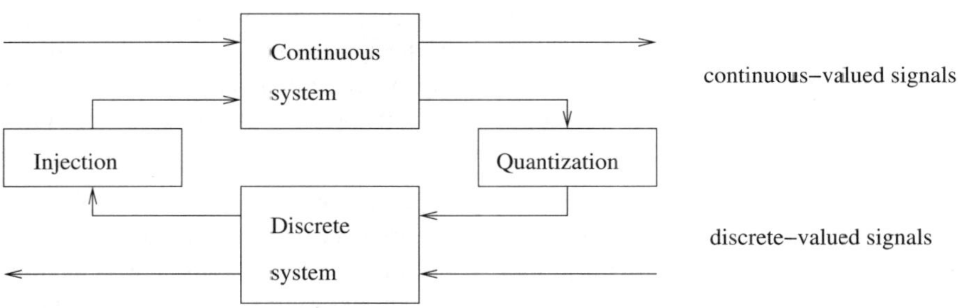

Fig. 1. Generic hybrid system model

Direct mathematical treatment of hybrid systems is notoriously difficult, the basic reason being the structure of their state sets: purely continuous systems mostly exhibit

S. Engell, G. Frehse, E. Schnieder (Eds.): Modelling, Analysis and Design of Hybrid Systems, LNCIS 279, pp. 67–80, 2002.

a nice (vector space) structure; this implies that a rich set of analysis tools can be applied to investigate continuous system dynamics. Purely discrete systems can be described by discrete, and in most cases finite, sets. Hence, the dynamical behaviour of finite discrete systems can, at least in principle, be completely investigated by finite enumeration type methods. The state set of a hybrid system is the product of the state sets of its constituent components. In general, it is therefore neither finite nor does it exhibit vector space structure, ruling out approaches that work for individual discrete or continuous subsystems. This is the main reason for the current research on hybrid systems, some of which has been reported in this volume.

A natural approach to avoid these methodological problems is to approximate a given hybrid system by a suitable discrete system. This facilitates both analysis and synthesis tasks with respect to the discrete-valued input and output signals, as a range of well-known methods from the area of discrete event systems (DES) theory can be applied to the resulting discrete problem. From Fig. 1, it is immediately clear that computing a (suitable) discrete approximation for the continuous subsystem will suffice: in combination with the given discrete subsystem, it will constitute an approximation of the overall hybrid system.

Note that for quite a few applications it also makes sense to approximate purely continuous systems by discrete models. This is especially true in the context of hierarchical control and supervision: High-level specifications are often described in much less detail than low-level tasks as, e.g., set-point control. Accordingly, high-level measurement information is mostly qualitative, or discrete. For example, from a high-level point of view, it may only be important whether the water level in a vessel is above or below certain critical thresholds whereas the exact value may be irrelevant. Similarly, control inputs on higher hierarchy levels are often discrete-valued and can only be switched between a fixed number of points. In this situation, it seems natural to approximate the given continuous model relating discrete-valued inputs and outputs by a suitable discrete approximation.

Apparently, the crucial aspect with regard to approximations is a proper definition of "suitability". We adopt a pragmatic point of view and call an approximation suitable if any solution that is found on the basis of the approximation can be guaranteed to hold for the underlying continuous or hybrid system. Obviously, this definition is independent of the particular model purpose – whether it is intended for controller synthesis, for verification, or for process diagnosis.

This contribution is organized as follows: in Sect. 2, we describe two complementary approaches to generate discrete approximations for a given continuous system. We also provide references to these approaches. In Sect. 3, we discuss the behaviour of continuous systems with discrete-valued input and output signals and some of its fundamental properties. In Sect. 4, we describe their implications for discrete models of continuous systems and recall a simple and intuitive condition for an approximation to be suitable. In Sect. 5, we discuss various ways of how to realize such approximations. In Sect. 6, we conclude by briefly hinting at possible applications and how to approach them on the basis of discrete models.

Finally, a remark concerning notation: signals are interpreted as maps; the domain of a signal is time, its codomain is referred to as the "signal space". Time can be continuous ($T = \mathbb{R}^+$) or discrete ($T_d = \{t_0, t_1, \dots\}$). Signals are represented by lower case letters, their codomains by the corresponding upper case letters. Discrete-valued signals – irrespective of their domain – are characterized by the subscript "d". For example, $y_d : T_d \rightarrow Y_d$ is a discrete-valued measurement signal which is defined on the sampling grid $T_d = \{t_0, t_1, \dots\}$ and "lives" in the (discrete) set Y_d. The codomains of all discrete-valued signals are assumed to be finite, their elements (the possible values the signal can take at each instant of time) are characterized by superscripts: the i-th element in the set Y_d, for example, is denoted by $y_d^{(i)}$.

2 Event-Driven vs. Time-Driven Sampling

We now turn to the problem of coming up with a discrete approximation for the continuous subsystem in Fig. 1. More precisely, we want to generate a discrete model that describes how the discrete-valued output signal provided by the quantization block depends on the discrete-valued input signal fed into the injection block. For simplicity, we assume that the external input signal for the continuous subsystem is completely known and can therefore be neglected (Fig. 2). The setup in Fig. 2, i.e. the series of injection block, continuous system and quantization block, will henceforth be referred to as the *quantized continuous system*.

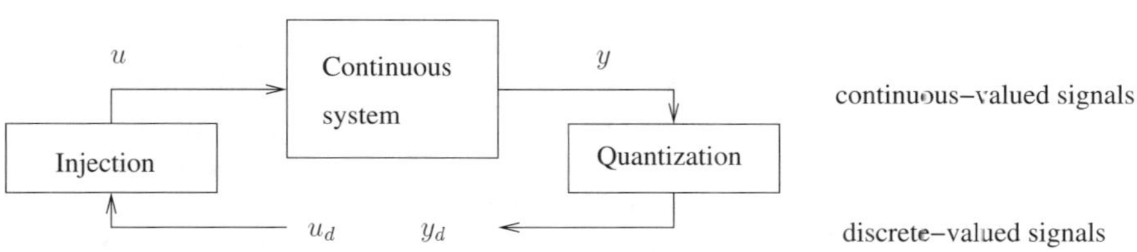

Fig. 2. Continuous system with discrete-valued input and output signal (quantized continuous system)

There is an important distinction to be made when "translating" continuous-valued into discrete-valued signals. This is related to the notion of *sampling*. Consider a signal $y : T \rightarrow Y$, i.e. a continuous-valued signal living in continuous time. A straightforward approach is to use a predefined, usually equidistant, sampling grid $T_d = \{t_0, t_1, \dots\}$ with $t_{k+1} - t_k = \Delta = \text{const.}$, $k \in \mathbb{N}_0$. As sampling instants are solely determined by the progress of time, this is referred to as *time-driven sampling*. The quantized signal is defined by $y_d : T_d \rightarrow Y_d$ by $y_d(t_k) := \text{quant}(y(t_k))$, $k \in \mathbb{N}_0$, where quant $: Y \rightarrow Y_d$ is a surjective function. An example for the output quantization function quant is shown in Fig. 3.

Another approach is based on a more sophisticated notion of sampling: as before, consider a continuous signal $y : T \rightarrow Y$ and a given surjective map quant $: Y \rightarrow Y_d$ partitioning Y into a finite number of subsets $Y^{(i)} := \{\nu \in Y \mid \text{quant}(\nu) =$

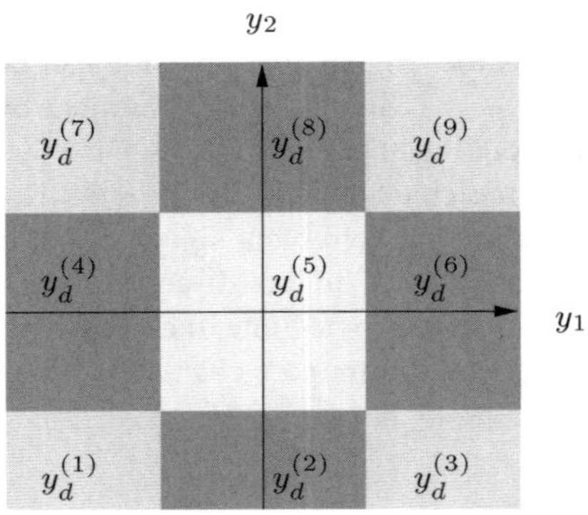

Fig. 3. Partition of output space defining output quantization (example)

$y_d^{(i)}, y_d^{(i)} \in Y_d\}$. Sampling instants are now, loosely speaking, triggered by the continuous signal y passing from a partition cell $Y^{(i)}$ into a neighbouring cell[1] $Y^{(j)}, j \neq i$. This is interpreted as a discrete event, and this type of sampling is therefore called *event-driven*. The discrete signal value may be defined as $y_d(t_k) = y_d^{(j)}$ if y "moves" into the partition cell $Y^{(j)}$ at the sampling instant t_k or, alternatively, as the event $e^{(j_i)}$ (where the subscript i indicates the partition cell about to be left). Both time-driven and event-driven sampling of a given continuous signal are illustrated in Fig. 4.

In the time-driven case, as the sampling interval Δ is known, the time instant t_k is completely determined by its index k. We can therefore replace the discrete time axis $\{t_0, t_1, \dots\}$ by the set of nonnegative integers, $\mathbb{N}_0$, without sacrificing information. This is obviously not true for the event-driven case. Here, we have to distinguish between the *logic* signal $y_d : \mathbb{N}_0 \to Y_d$, which only contains information on the order of events, and the *timed* signal $y_d : \{t_0, t_1, \dots\} \to Y_d$, which also provides information about absolute time.

In the time-driven case, all discrete-valued signals are usually defined on the same sampling grid – they are *synchronized*. In the event-driven case, this may not be true: discrete-valued input and output signals may be synchronized (for example, by restricting changes in the input signal to the sampling instants provided by the output), but can also be *asynchronous*.

The above terminology carries over from signals to models. Hence, in the event-driven case, we have to distinguish between logic and timed discrete models, and between synchronous and asynchronous models. In the time-driven case, these distinctions are not necessary. Discrete approximation of continuous or hybrid systems has been treated, for example, in (Lichtenberg et al., 1999a, Lunze, 1994, Raisch and

[1] Note that there is a subtle mathematical difficulty related to the question of which of the partition cells is open or closed. This can be avoided by *covering* instead of partitioning the continuous signal space Y, i.e. by using a "nondeterministic" map quant : $Y \to 2^{Y_d}$, where 2^{Y_d} represents the power set (the set of all subsets) of Y_d.

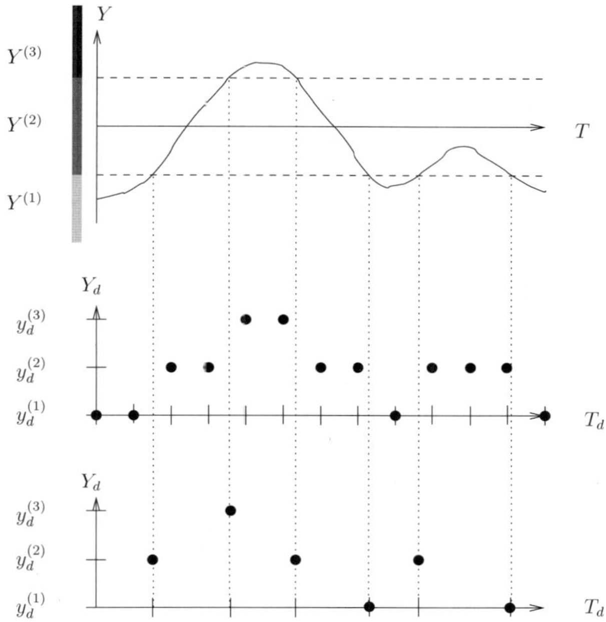

Fig. 4. Time-driven and event-driven sampling

O'Young, 1997, Raisch and O'Young, 1998). The event-driven case has been described, e.g., in (Chutinan and Krogh, 1999a, Cury et al., 1998, Franke et al., 2000, Moor and Raisch, 1999a, Stursberg et al., 1997).

3 The Behaviour of the Quantized Continuous System

As the main purpose of this contribution is to discuss the general concept of discrete approximations for continuous or hybrid systems, it will suffice to consider the synchronized case. This will also help to keep notation reasonably simple. Hence, we will assume that the discrete-valued input and output signals in Fig. 2, u_d and y_d, evolve on the same sampling grid T_d. There is no need yet to distinguish the time-driven and the event-driven case, i.e. $T_d = \{t_0, t_1, \dots\} \subset \mathbb{R}^+$ can either be an equidistant sampling grid or an irregularly spaced set of time instants defined by the occurrence of certain events.

Let the quantized continuous system in Fig. 2 be modelled by

$$\frac{\mathrm{d}x(t)}{\mathrm{d}t} = f(x(t), u(t)) \tag{1}$$

$$y(t) = g(x(t)), \tag{2}$$

$$u(t) = \mathrm{inj}(u_d(t_k)), \; t_k \leq t < t_{k+1}, \tag{3}$$

$$y_d(t_k) = \mathrm{quant}(y(t_k)), \tag{4}$$

where $t \in T = \mathbb{R}^+$, $k \in \mathbb{N}_0$; $u : T \to \mathbb{R}^m$, $x : T \to \mathbb{R}^n$, and $y : T \to \mathbb{R}^p$ are continuous-valued signals in continuous time; $u_d : T_d \to U_d$ and $y_d : T_d \to Y_d$ are discrete-valued signals in discrete time; inj : $U_d \to \mathbb{R}^m$ is an injective map and quant : $\mathbb{R}^p \to Y_d$ is a surjective function. While quant defines a partition on Y, the function $q :=$ quant $\circ\, g$ induces a (finite) partition of the continuous plant state space and maps continuous state values to discrete output values:

$$y_d(t_k) = q(x(t_k)). \tag{5}$$

Clearly, $q : \mathbb{R}^n \to Y_d$ can be interpreted as "state quantization".

Denote the set of all functions from T_d into $(U_d \times Y_d)$ by $(U_d \times Y_d)^{T_d}$. Then, the *behaviour* of the quantized continuous system, $\mathfrak{B}_c \subseteq (U_d \times Y_d)^{T_d}$, is the set of all pairs (u_d, y_d) of discrete-valued input and output signals which are compatible with the model equations (1),(3),(5):

$$\mathfrak{B} := \{(u_d, y_d) \mid \text{(1),(3),(5) hold}\}. \tag{6}$$

For a survey on "behavioural systems theory" see, e.g., (Willems, 1989, Willems, 1991). In the following, we will also need the notion of signals and behaviours restricted to the interval $[t_0, \ldots, t_k]$:

$$y_d|_{[t_0,t_k]} := (y_d(t_0), \ldots y_d(t_k)) \tag{7}$$

$$u_d|_{[t_0,t_k]} := (u_d(t_0), \ldots u_d(t_k)) \tag{8}$$

$$\mathfrak{B}|_{[t_0,t_k]} := \{(u_d|_{[t_0,t_k]}, y_d|_{[t_0,t_k]}) \mid \text{(1),(3),(5) hold}\}. \tag{9}$$

Non-determinism of quantized system behaviour: If the initial state $x(0)$ of the continuous system is known and (1) has a unique solution on $\mathbb{R}^+$ for each input signal provided by the injection block, the discrete-valued output signal y_d can be unambiguously predicted. In general, however, the first assumption is not true: in most cases, a-priori information on the initial state will be limited or completely absent. We will then not be able to uniquely predict $y_d(t_{k+1})$ from the available information, i.e. the past output $y_d|_{[t_0,t_k]}$ and the input signal u_d. This phenomenon is often referred to as *non-determinism of the quantized system behaviour*, as it explains why discrete abstractions for quantized continuous systems are, in general, nondeterministic. It is illustrated in Fig. 5. The figure depicts an example where X, the continuous state space, is $\mathbb{R}^{+2}$ and $Y_d = \{y_d^{(0)}, y_d^{(1)}, \ldots, y_d^{(9)}\}$. Hence the quantization map q partitions X into 9 quantization boxes, labelled $y_d^{(1)}$ to $y_d^{(9)}$, and the area outside these boxes, labelled $y_d^{(0)}$. Sampling is time-driven, and the input signal is fixed. Suppose that $q(x(t_0)) = y_d^{(7)}$, implying that $x(t_0)$ lies within the grey shaded box. The set of possible values for the system state at time instant t_1, according to the plant differential equation (1), is shown as a dark grey wedge in Fig. 5. It clearly intersects more than one quantization box, generating several possible values for $y_d(t_1)$. Hence, if $x(t_0)$ is only known to lie in the grey box labelled by $y_d^{(7)}$, the value of $y_d(t_1)$ cannot be uniquely determined.

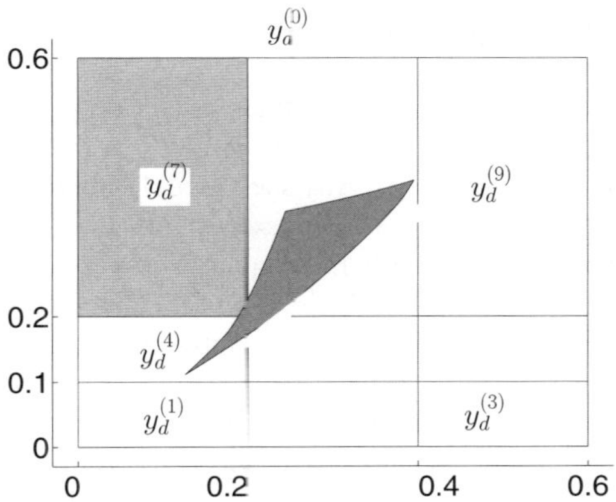

Fig. 5. Set of reachable continuous system states if $x(0) \in q^{-1}(y_d^{(7)})$.

Stochastic properties of the quantized system: Since the quantized system behaviour is non-deterministic in the above sense, it is interesting to ask for the probability that a certain output sequence $y_d|_{[t_0,t_k]}$ occurs when an input sequence $u_d\,_{[t_0,t_k]}$ has been applied. As the quantized system (1),(3),(5) is causal,

$$\text{Prob}_{u_d}(y_d|_{[t_0,t_k]}) := \text{Prob}(y_d|_{[t_0,t_k]}\,|\,u_d) \tag{10}$$
$$= \text{Prob}(y_d|_{[t_0,t_k]}\,|\,u_d|_{[t_0,t_k]}), \tag{11}$$

where $\text{Prob}(A\,|\,B)$ denotes the probability of A if B is known to have occurred. Obviously, $(u_d|_{[t_0,t_k]}, y_d|_{[t_0,t_k]}) \in \mathfrak{B}|_{[t_0,t_k]}$ if and only if the respective input string is possible and $\text{Prob}(y_d|_{[t_0,t_k]}\,|\,u_d|_{[t_0,t_k]}) > 0$. From Bayes' law, it follows immediately that

$$\text{Prob}_{u_d}(y_d|_{[t_0,t_k]}) = \text{Prob}(y_d(t_k)\,|\,y_d|_{[t_0,t_{k-1}]}, u_d)\,\text{Prob}_{u_d}(y_d|_{[t_0,t_{k-1}]}). \tag{12}$$

If the *Markov property* were to hold for the discrete-valued output y_d, i.e. if

$$\text{Prob}(y_d(t_k)\,|\,y_d|_{[t_0,t_{k-1}]}, u_d) = \text{Prob}(y_d(t_k)\,|\,y_d(t_{k-1}), u_d)$$

were true, (12) would constitute a neat recursive formula to compute the desired probabilities. Unfortunately, this is in general not the case (Lunze, 1998a). This fact is again illustrated by Fig. 5. Recall that for this example the input sequence is fixed. Assume that we only know that $y_d(t_1) = y_d^{(4)}$, i.e. at time instant t_1, the continuous state lies in quantization box 4. On the other hand, if both $y_d(t_1) = y_d^{(4)}$ and $y_d(t_0) = y_d^{(7)}$ is known, the continuous state at time instant t_1 can only lie in the intersection of the dark grey wedge and quantization box 4. Hence, the probability of moving into a specific quantization box, e.g. box 1, at time instant t_2 is clearly different in both cases.

4 Properties of Suitable Discrete Approximations

An implication of the above properties is that, in general, there exists no discrete state model with state variable $x_d(t_k) = y_d(t_k)$ or $x_d(t_k) = y_d|_{[t_{k-l},t_k]}$ that can accurately represent the quantized system behaviour $\mathfrak{B}_c$. Any such model is therefore an approximation. This is hardly surprising, and has been extensively discussed in (Lunze, 1994) for the first case (the discrete state being a quantized version of the continuous state) and (Raisch and O'Young, 1997, Raisch and O'Young, 1998) for the second case (the discrete state being a finite string of quantized continuous states).

This brings us back to the notion of suitability: when is a discrete model a suitable approximation of a given quantized continuous system? We adopt the following pragmatic point of view: suppose we can find a solution to a particular task, e.g. controller synthesis, verification, fault diagnosis, for the discrete approximation. The approximation is called suitable, if we can guarantee that the solution will also be valid for the underlying quantized continuous system. It turns out that for all these tasks, there is basically the same simple and intuitive condition for suitability. Namely, we require that the behaviour $\mathfrak{B}_d$ of the discrete approximation satisfies

$$\mathfrak{B}_c \subseteq \mathfrak{B}_d, \tag{13}$$

i.e. every pair (u_d, y_d) of discrete-valued input and output signals that is compatible with the quantized continuous system must also be contained in the behaviour of the discrete approximation. However, the latter may also contain I/O-pairs (u_d, y_d) that are not consistent with the quantized continuous system. Such I/O-pairs are called spurious solutions. Note that (13) implies $\mathfrak{B}_c|_{[t_0,t_k]} \subseteq \mathfrak{B}_d|_{[t_0,t_k]}$ for all $k \in \mathbb{N}_0$.

A discrete approximation satisfying (13) is called an *abstraction*[2]. Clearly, $\subseteq$ provides a partial order on the set of all abstractions of a given quantized continuous system. If two abstractions A_1 and A_2 with behaviours $\mathfrak{B}_{d_1}$ and $\mathfrak{B}_{d_2}$ are ordered in the sense of $\mathfrak{B}_{d_1} \subseteq \mathfrak{B}_{d_2}$, we say that A_1 is *at least as* accurate as A_2. Hence, a hierarchy of abstractions can be defined, e.g. (Raisch and O'Young, 1997, Moor and Raisch, 1999b).

To motivate condition (13), we briefly discuss the tasks of diagnosis, verification and controller synthesis within a behavioural framework: In the simplest case, on-line diagnosis is about monitoring I/O-data and deciding whether an error has occured. The quantized continuous system is assumed to model "correct functioning" of the respective plant. Diagnosis is to be performed on the basis of an abstraction: an alarm "rings" if a measured string of I/O-data is not contained in the abstraction behaviour $\mathfrak{B}_d|_{[t_0,t_k]}$. If (13) holds,

$$(u_d|_{[t_0,t_k]}, y_d|_{[t_0,t_k]}) \notin \mathfrak{B}_d|_{[t_0,t_k]} \Longrightarrow (u_d|_{[t_0,t_k]}, y_d|_{[t_0,t_k]}) \notin \mathfrak{B}_c|_{[t_0,t_k]}, \tag{14}$$

[2] In the literature on qualitative modelling and diagnosis, e.g. (Hamscher et al., 1992), such an approximation is often called *complete*. In the following, we will not adopt this terminology, as completeness in the context of behavioural systems theory has a different meaning (Willems, 1989).

i.e. any string of I/O-data not consistent with the abstraction is also incompatible with the quantized system. Hence, any alarm that is sounded on the basis of the abstraction represents a true failure – "false alarms" cannot occur.

Similarly for verification. There, the task is to guarantee that a given quantized continuous system will always behave in a desired fashion in the sense of $\mathfrak{B}_c \subseteq \mathfrak{B}_{spec}$ (i.e. only well-defined pairs of I/O-signals are allowed to occur). If (13) holds,

$$\mathfrak{B}_d \subseteq \mathfrak{B}_{spec} \Longrightarrow \mathfrak{B}_c \subseteq \mathfrak{B}_{spec}. \tag{15}$$

Hence, if the abstraction meets the specifications, the same will be true for the underlying quantized continuous system.

Let us finally turn to the task of controller synthesis. Suppose we design a causal discrete feedback controller (with input signal y_d and output signal u_d) and hook it up to the discrete abstraction. Obviously, the controller behaviour $\mathfrak{B}_{controller}$ is also a subset of $(U_d \times Y_d)^{T_d}$; the feedback loop consisting of controller and abstraction exhibits behaviour $\mathfrak{B}_d \cap \mathfrak{B}_{controller}$ – only those pairs of input/output signals "survive" that are compatible with both abstraction and controller equations. In other words: in the closed loop, each signal pair (u_d, y_d) has to be compatible with both abstraction and controller dynamics. From the abstraction condition (13), it follows immediately that

$$(\mathfrak{B}_d \cap \mathfrak{B}_{controller} \subseteq \mathfrak{B}_{spec}) \Longrightarrow (\mathfrak{B}_c \cap \mathfrak{B}_{controller} \subseteq \mathfrak{B}_{spec}) \ ; \tag{16}$$

in other words: if the controller forces the discrete approximation to obey the specifications, the feedback loop consisting of discrete controller and quantized continuous system will also meet the specifications. Note that for the problem of controller synthesis, we also have to address the issue of "blocking". This refers to the case where the controller designed for the abstraction will lead to $\mathfrak{B}_c \cap \mathfrak{B}_{controller} = \emptyset$. Hence, technically speaking, it will still enforce the specifications by disallowing all pairs of I/O-signals that are not contained in $\mathfrak{B}_{spec}$, but the only way to achieve this is by "turning off" the system. In general, blocking can be easily avoided in the context of time-driven sampling, but may pose a problem in the event-driven case.

It should be obvious that an abstraction may be useless for any of the purposes described above, if it is "too inaccurate" in the sense of containing "too many" spurious solutions. A diagnoser based on such an abstraction may not ring *any* alarm, the desired behaviour may not be verified for the abstraction, and a controller enforcing the specifications for the abstraction may not exist.

5 Discrete Models

We will now briefly describe, how discrete models satisfying the abstraction condition (13) can be set up. We start with the simplest case, nondeterministic automata.

5.1 Nondeterministic Automata

In this subsection, we concentrate on the time-driven case, i.e. T_d is a fixed equidistant sampling grid. Obviously, as the abstraction behaviour $\mathfrak{B}_d$ is a subset of $(U_d \times Y_d)^{T_d}$, abstraction and quantized continuous system share the same input and output sets, U_d and Y_d. The abstraction is realized as a finite nondeterministic automaton A_l, i.e. as a tuple $(X_d, U_d, Y_d, f_d, g_d, X_{d_0})$, where X_d is a (finite) state set, U_d and Y_d are the (finite) input and output sets, $f_d \subseteq X_d \times U_d \times X_d$ is a transition relation, $g_d : X_d \to Y_d$ an output function, and X_{d_0} the set of possible initial conditions for the discrete state variable. $(x_d^{(i)}, u_d^{(j)}, x_d^{(n)}) \in f_d$ if and only if the discrete state can go from $x_d(t_k) = x_d^{(i)}$ to $x_d(t_{k+1}) = x_d^{(n)}$ if $u_d(t_k) = u_d^{(j)}$ is applied.

In (Raisch and O'Young, 1997, Raisch and O'Young, 1998), it was suggested to define $x_d(t_k)$ as the string of the l most recent pairs of input and output values, where l is a nonnegative integer. More precisely,

$$x_d(t_k) := \begin{cases} y_d(t_0) \ \text{ if } k = 0, \\ (u_d|_{[t_0,t_{k-1}]}, y_d|_{[t_0,t_k]}), \ \text{ if } k = 1, \ldots, l, \\ (u_d|_{[t_{k-l},t_{k-1}]}, y_d|_{[t_{k-l},t_k]}), \ \text{ if } k > l. \end{cases} \tag{17}$$

Hence, $X_d \subseteq Y_d \times \ldots \times (U_d^l \times Y_d^{l+1})$, and

$$x_d^{(i)} = \left(\left(u_d^{(i_{k-\rho})}, \ldots, u_d^{(i_{k-1})}\right), \left(y_d^{(j_{k-\rho})}, \ldots, y_d^{(j_k)}\right)\right), \ 1 \le \rho \le l, \tag{18}$$

is an element in the state set X_d if and only if the input string

$$u_d|_{[t_0,t_{\rho-1}]} = (u_d^{(i_{k-\rho})}, \ldots, u_d^{(i_{k-1})}) \tag{19}$$

can cause the quantized continuous system (1),(3),(5) to respond with the output string

$$y_d|_{[t_0,t_\rho]} = (y_d^{(j_{k-\rho})}, \ldots, y_d^{(j_k)}). \tag{20}$$

To check this, we need to perform two operations: (i) compute the evolution of quantization boxes under the passage of time and (ii) intersect the result with other quantization boxes. In this way, we can determine the set of continuous states that are compatible with the quantized continuous system dynamics (i.e. (1),(3),(5)), the string of discrete inputs (19), and the string of discrete outputs (20). Denote this set by $X(x_d^{(i)})$. Clearly, $x_d^{(i)} \in X_d$ if and only if $X(x_d^{(i)}) \neq \emptyset$. For general quantized continuous systems (1),(3),(5), the necessary operations can only be approximated. Exceptions are linear systems, see (Raisch and O'Young, 1997); moreover, for the class of nonlinear monotone dynamical systems, there exist "safe" approximations for these operations (in this volume, (Moor and Raisch, 2002)). The same type of operations is needed to check whether $(x_d^{(i)}, u_d^{(j)}, x_d^{(n)}) \in f_d$ (details can be found in (Raisch and O'Young, 1997)).

It only remains to specify the output map, g_d, and the set of initial states, X_{d_0}. Both is straightforward: $g_d : X_d \to Y_d$ just picks the rightmost (most recent) output symbol from a state (18). If no a-priori information on the continuous state is assumed, the initial state set for the automaton is given by $X_{d_0} = Y_d$. This reflects that, at time instant t_0, the continuous state could be anywhere in $\mathbb{R}^n$, and any discrete measurement symbol can occur at this time instant. On the other hand, if a-priori information on the continuous state exists, this may be readily incorporated by suitably restricting the set of possible initial states of A_l.

Note that for each nonnegative integer l, the resulting automaton A_l has been shown to realize the *strongest* $l+1$*-complete approximation* $\mathfrak{B}_{d_{l+1}}$ of the quantized continuous system (Moor and Raisch, 1999b). Formally, $\mathfrak{B}_{d_{l+1}}$ is characterized by

$$\mathfrak{B}_{d_{l+1}} := \{(u,y) |\, (u,y)\big|_{[k,k+l+1]} \in \mathfrak{B}_c\big|_{[0,l+1]}\ \forall\, k \in \mathbb{N}_0\}. \tag{21}$$

From (21), it follows immediately that $\mathfrak{B}_c \subseteq \mathfrak{B}_{d_l} \subseteq \mathfrak{B}_{d_{l'}}$ for $l' < l$. Hence, by increasing l, approximation accuracy can be increased, and the number of spurious solutions decreases.

5.2 Stochastic Automata

The nondeterministic automaton described above only provides information on whether certain pairs of I/O-signals are possible or not. It does not assign any probability to I/O-pairs that are deemed possible. If this is desired, we need a *stochastic automaton*. A stochastic automaton S is a tuple $(X_d, U_d, Y_d, L_d, P_{d_0})$, where, as before, X_d is a (finite) state set and U_d and Y_d are the (finite) input and output sets. $L_d : Y_d \times X_d \times U_d \times X_d \to [0,1]$ assigns a probability to any pair of transition and output symbol. More precisely,

$$\begin{aligned} &L_d(y_d^{(m)}, x_d^{(i)}, u_d^{(j)}, x_d^{(n)}) = \\ &\qquad \text{Prob}\left(y_d(t_k) = y_d^{(m)}, x_d(t_{k+1}) = x_d^{(n)}) \,|\, x_d(t_k) = x_d^{(i)}, u_d(t_k) = u_d^{(j)}\right). \end{aligned}$$

Finally, $P_{d_0} : X_d \to [0,1]$ assigns to any element in the state set the probability of being an initial state. To ensure that the stochastic automaton is an abstraction, we require the following: $L_d(y_d^{(m)}, x_d^{(i)}, u_d^{(j)}, x_d^{(n)}) > 0$ if and only if there exists a continuous state $x(t_k) \in X(x_d^{(i)})$ such that $y_d^{(m)} = q(x(t_k))$ and $x(t_{k+1}) \in X(x_d^{(n)})$ if $u_d(t_k) = u_d^{(j)}$ is applied to the quantized continuous system.

For a given input sequence, we can now compute the probability of any discrete state and output sequence by a straightforward application of Bayes' law. Hence, we can also calculate the probability of any pair of I/O-signals.

The core problem when setting up a stochastic automaton is to determine the function L_d. Loosely speaking, for this the probability of the continuous state going from $X(x_d^{(i)})$ to $X(x_d^{(n)})$ while generating the output $y_d^{(m)}$ if the input $u_d^{(j)}$ is applied has to be calculated for every $(y_d^{(m)}, x_d^{(i)}, u_d^{(j)}, x_d^{(n)}) \in Y_d \times X_d \times U_d \times X_d$. Details are omitted here, and the interested reader is referred to (Schröder, 2002) (for the case of time-driven sampling) and (Lunze and Nixdorf, 2002) (for the case of event-driven sampling). There, an approximate procedure is discussed for the case when X_d is a quantized version of the continuous state space, i.e. for the case $l = 0$.

5.3 Semi-Markov-Processes

To provide another example for discrete abstractions, we turn to the case of event-driven sampling. Recall that the map $q : \mathbb{R}^n \to Y_d$ induces a partition of the continuous state space, and any change of the value of the discrete output signal corresponds to the fact that the continuous state variable "crosses" into a neighbouring partition cell. This is interpreted as a discrete event e, and any state partition therefore defines a set E of possible events. To keeps things reasonably simple, we restrict ourselves to the synchronous case, i.e. the discrete input signal "lives" on the sampling grid defined by the output signal. However, we want our model to provide information on the timing of events, hence we are after a synchronous timed discrete model. As a candidate, we consider a semi-Markov process SM, i.e. a tuple (E, U_d, f, e_0), where E is the set of discrete events, U_d the discrete input set, e_0 the initial event, and f a probability density with

$$f(e^{(i)}, e^{(n)}, \tau, u_d^{(j)}) = \frac{d}{dt}\text{Prob}\left(e(t_{k+1}) = e^{(n)}, t_{k+1} - t_k \leq \tau \mid e(t_k) = e^{(i)}, u_d(t_k) = u_d^{(j)}\right). \quad (22)$$

The semi-Markov process is an abstraction if the following requirement is satisfied: $f(e^{(i)}, e^{(n)}, \tau, u_d^{(j)}) > 0$ if and only if the quantized continuous system may generate the event pair $e^{(i)}, e^{(n)}$ with a temporal distance of at most τ if the input $u_d^{(j)}$ has been applied. The right-hand side of (22) can be determined by means of the quantized system model (1),(3),(5) (Lunze, 1999). Clearly, the quantized system cannot generate the same event twice in a row; hence, the given relations hold for $e^{(i)} \neq e^{(n)}$. However, the complete definition of the semi-Markov process includes the function f for $e^{(i)} = e^{(n)}$, which is given by $f(e^{(i)}, e^{(i)}, \tau, u_d^{(j)}) = -\sum_{e^{(i)} \neq e^{(n)}} f(e^{(i)}, e^{(n)}, \tau, u_d^{(j)})$.

With the semi-Markov process, the probability $\text{Prob}(e(t_k) = e^{(j)}, t_k - t_{k-1} \leq \tau)$ of the occurrence of any event $e^{(j)}$ until time τ can be determined for a given input sequence, which provides a timed description of the events that may be generated by the quantized system.

6 Conclusions

In this section, we briefly hint at possible applications and show how to solve them on the basis of discrete approximations. We will always assume that discrete approximations for continuous subsystems satisfy the abstraction condition (13). The discrete approximation of the overall hybrid system (consisting of discrete subsystems and discrete approximations of continuous subsystems) will then also satisfy (13). This, in turn, will guarantee that results obtained for the approximation will carry over to the underlying hybrid problem.

The following (closely related) tasks for hybrid systems have been approached on the basis of discrete abstractions:

Prediction: As the abstraction condition holds, a discrete approximation will generate any string of discrete events or discrete outputs that the underlying hybrid model is able to generate for a given input string. It may additionally produce spurious solutions, though. Nevertheless the abstraction will be useful for checking whether forbidden strings of events or outputs occur: if this is not the case for the abstraction, we can guarantee that it will also not occur for the underlying hybrid system. This is the motivation for the simulation method described in (Pawletta et al., 2002) (p. 107 in this volume).

Reachability analysis: Here, the task is to investigate whether a certain region in the hybrid state set will be reached for given input signals. As each element in the abstraction state set corresponds to a region in the hybrid state set (see, for example, Sect. 5.1 of this contribution), this question can be treated on an abstraction basis. If the abstraction state reaches a certain subset of X_d, we know for sure that the hybrid state signal will have "moved" into a well-defined part of its state set (Lunze and Nixdorf, 2003).

Control: If controller and hybrid plant can only interact via discrete-valued measurement and control signals, it makes perfect sense to design a discrete controller on the basis of a discrete abstraction. For this, well-known supervisory control methods, e.g. (Ramadge and Wonham, 1987, Ramadge and Wonham, 1989), can be used. In Sect. 4, we have argued that – provided the problem of blocking is properly addressed – any discrete controller enforcing a given (dynamic) specification for the abstraction will also "work properly" for the hybrid plant model. The use of discrete abstractions for the synthesis of discrete supervisory control for continuous or hybrid systems has been treated extensively during the last few years, e.g. (Alur et al., 1996, Antsaklis et al., 1999, Antsaklis et al., 1995, Antsaklis et al., 1997, Antsaklis, 2000, Antsaklis and Nerode, 1998b, Benedetto and Sangiovanni-Vincentelli, 2001, Evans and Savkin, 1999, Lynch and Krogh, 2000, Maler, 1997). Quite a bit of work in this area has been linked to the KONDISK-project, e.g. (Franke et al., 2000, Lunze, 1995, Moor et al., 2001a, Moor and Raisch, 1999b, Moor et al., 2002, Raisch et al., 2001, Raisch and O'Young, 1997, Raisch and O'Young, 1998).

Verification: The task of verification is to guarantee that a given hybrid system will always behave in a desired fashion. If the latter is exclusively defined in terms of discrete variables, the problem can again be addressed on the basis of a discrete abstraction. Namely, if the abstraction behaviour is shown to be contained in a set of acceptable signals (the specification behaviour), this will also be true for the hybrid system behavior. References on this topic can be found, e.g., in (Kowalewski, 2002) (p. 153 in this volume).

Diagnosis: For fault diagnosis of hybrid systems, a number of different models are set up, each representing either a particular failure or "proper functioning". If measured input and output data is discrete-valued, we can again work with discrete abstractions: for each (hybrid) model, a discrete abstraction is generated. If a recorded string of input and output symbols is not contained in a particular abstraction behaviour, it will also not be contained in the behaviour of the underlying hybrid system, and the corresponding fault (or proper functioning) can be ruled out. Methods for abstraction-based fault diagnosis have been reported in (Förstner, 2001, Lunze and Schröder, 1999, Philips, 2001, Schröder, 2002).

Reconfigurable control: In (Lunze and Steffen, 2002) (p. 267 in this volume), it is described how a feedback controller can be reconfigured after a fault has occurred in a hybrid system. An important step in the reconfiguration task concerns the determination of input sequences to the faulty system such that the qualitative state of the system is brought back to the nominal operation point. This step can also be carried out with a discrete model of the plant.

Part II

Simulation

An Environment for the Integrated Modelling of Systems with Complex Continuous and Discrete Dynamics

Manuel A. Pereira Remelhe[1], Sebastian Engell[1], Martin Otter[2], André Deparade[1], and Pieter J. Mosterman[2]

[1] Process Control Laboratory, Department of Chemical Engineering, University of Dortmund, 44221 Dortmund
e-mail: {M.Remelhe, S.Engell, A.Deparade}@ct.uni-dortmund.de

[2] Institute of Robotics and System Dynamics, DLR Research Center Oberpfaffenhofen, P.O.Box 1116, D-82230 Wessling
e-mail: {Martin.Otter, Pieter.J.Mosterman}@dlr.de

Abstract. The modelling and simulation of sophisticated technical systems is a demanding task. On the one hand, the physical part consists of a large number of subsystems which exhibit predominantly continuous dynamics, sometimes with (infrequent) discontinuities. On the other hand, the distributed computerised control systems constitute complex discrete-time and discrete-event systems that require completely different modelling and simulation methods. For an evaluation of the behaviour and the performance of the overall system, both types of models have to be combined and simulated efficiently. This contribution presents the requirements for a modelling environment for such systems and discusses an approach that consists of object-oriented modelling and efficient simulation of the physical part using the physical systems modelling language MODELICA, a software environment for the definition of discrete-event models using various formalisms, and the integration of both parts of the system via model translation. The coordination of both parts is performed by the MODELICA simulator. The modelling environment called DES/M (discrete-event systems for Modelica) supports the interoperation of different domain specific discrete-event formalisms. To illustrate the usage of the environment, a laboratory batch plant model is presented. A more elaborate example is described in another contribution in this volume (Mosterman et al., 2002).

1 Introduction

Sophisticated technological systems such as chemical plants, cars, and aircraft consist of a large number of physical components, numerous low-level set-point controllers and interlocks, and interacting complex supervisory controllers which may be organised in a hierarchical manner. On the supervisory control level, trajectory optimisation, fault detection, redundancy management, and sequence control e.g. for start-up and shutdown are performed and the interaction with the user is managed. The dominant part of the functions on this level consists of logic operations that are triggered by thresholds or events in the environment, including user commands. The physical part of the system and the supervisory control system put high demands on the power and the user-friendliness of the modelling techniques. In order to study the overall behaviour of such systems, a simulation model has to incorporate both parts

S. Engell, G. Frehse, E. Schnieder (Eds.): Modelling, Analysis and Design of Hybrid Systems, LNCIS 279, pp. 83–105, 2002.

and an integration is required that enables efficient and at the same time accurate simulation.

1.1 The Physical Part

The physical part of the system consists of a large number of interconnected components. The behaviour of these components is determined by the laws of physics and chemistry. The overall system may consist of subsystems from various domains: electrical circuits, pneumatic and hydraulic actuators, mechanical transmission, fuel cells, combustion chambers, tanks, gas transport systems, chemical reactors, etc. These submodels are usually developed by teams of domain experts who take a lot of technological details and domain knowledge into account. Each modelling domain has specific graphical representations and modelling traditions, but in most cases the final models are algebraic and differential equations involving continuous variables that depend on (continuous) time. The models of the physical components may contain discontinuities that strictly speaking are caused by model simplifications which are made in order to avoid models with largely different time scales. Examples are friction and contact in mechanical systems, thermodynamic phase changes, ideal switches, e.g. diodes, in electronic systems. Other discontinuities occur when physical limits are reached (overflow of a tank, rupture of a vessel) or inputs to the physical system change abruptly. At these points in time, the movement of the system trajectory in the state space may abruptly change its direction and its velocity, or very fast transients occur that can be regarded as jumps in the state space. At points of discontinuity, the number of independent state variables may change, e.g. if two rigid bodies make contact. In consequence, the physical part of the system itself may exhibit hybrid behaviour, i.e. mixed discrete/continuous dynamics.

The complexity of modelling and simulation of the physical part of the system is exacerbated if several components with hybrid behaviour interact with each other, e.g. electronic circuits with several diodes. This calls for powerful modelling and simulation techniques.

1.2 The Supervisory Controllers

Supervisory control is used for many different purposes. For instance, sequential control is needed for the execution of recipes in chemical batch plants, redundancy management is crucial for the safety of aircraft, and resource booking systems are needed for coordinating several interacting sequential controllers, e.g., to avoid collisions of robots or to prevent the mixing of batches running in parallel in chemical plants. Start up, shut down, and emergency procedures are further examples for the necessity of supervisory controllers. In decentralised or redundant automation architectures, autonomous supervisory controllers interact in order to achieve the performance goals. Other functions of supervisory control are trajectory optimisation and user interaction.

In general, a supervisory controller is a reactive discrete-event system. The states and the outputs of such a system change discontinuously according to discrete state transition sequences that are performed when a reaction to external stimuli from the user or from the physical system is required. For example, in the case of a tank that is being filled, a controller may have to close the inlet valve when the desired level is reached. The events that trigger the instantaneous reactions are determined by logical expressions containing analog and binary input variables that carry information on the current state of the physical system as well as internal variables that belong to the state of the controller and of other controllers in a distributed control system. Hence, the reactions depend on the current discrete state whereas the event times depend on the evolution of the state variables of the physical system and on clock variables in the discrete system. If the duration of a specific process, e.g. the duration of the filling of the tank, is known a priori and corresponding measurements, e.g., a sensor for the tank level, are not available, timers have to be used for triggering the transitions. Thus, time events occur that anticipate corresponding state events.

Even though supervisory controllers are mostly implemented as sampled data systems, their behaviour can adequately be described as reactive, i.e., driven by external state events. This is because the sampling intervals in the logical part are normally very short in comparison to the continuous dynamics so that at most sampling instants the controller does nothing but evaluating the triggering signals and waiting. Consequently, the sampling rate has a subordinate significance for the overall behaviour.

The difficulties for modelling and simulation arise from the fact that a reaction of a supervisory controller that appears as a monolithic state transition to the outer system may be the result of very complex inner iterations including hierarchical execution schemes as well as concurrency and synchronous and asynchronous communication.

1.3 Modelling and Simulation

The overall behaviour of a technical system is generated by the interaction of the physical components, discrete-event controllers and regulators. A precise comprehensive simulation model has to incorporate all these components and their relations if the purpose of the model is to evaluate the overall behaviour. Simulation goals may be, e.g., testing of the reaction to failures, the estimation of throughput or power consumption, a feasibility check for a specific production plan, or operator training.

Independent of the way of modelling, the physical part generally is solved by standard numerical integration methods such as Runge-Kutta methods or *backward differential formulae* (BDF) (Brenan and Campbell, 1996). This implies that the modelling process results in the generation of a consistent and uniquely solvable set of equations, either of explicit ordinary differential equations (ODE form) or of general *differential and algebraic equations* (DAE form). If hybrid phenomena have to be considered, special facilities have to be provided, because the inequalities that define the physical limits or the thresholds of a supervisory controller generate discontinuities, but the numerical integration methods usually require equations with a certain degree of continuity.

A usual approach is to ignore these inequalities during the numerical integration process and to use any efficient integration scheme, usually with a variable step size. This guarantees continuity of the equations. In order to handle the discontinuities, the values of the variables that enter into the trigger inequalities are monitored, and when a threshold is crossed, the integration is stopped and the time instant of the state event is localised up to a certain precision by backtracking. In case the event is dependent on time only, a time event, the integration simply stops directly at the predetermined time. When the integration is stopped, the discrete changes are performed and, afterwards, the integration is restarted.

The embedding of set-point controllers into the physical model is relatively straightforward since regulators are usually described by the same type of equations as the physical systems, and an overall ODE or DAE system results. Sampling effects often can be neglected because the sampling intervals are of the same order of magnitude as the integration step size. If this is not the case, time events have to be used in order to stop the integrator at every sample time. This is not convenient for multistep methods because these schemes must be restarted after every discontinuity which significantly decelerates the numerical integration (Brenan and Campbell, 1996).

In contrast to the domain of predominantly continuous dynamics where there is a standard system representation and general purpose numerical algorithms can be used, discrete-event simulation algorithms are specific to the modelling formalism used, and rather different from continuous integration methods. Popular formalisms are automata, statecharts, Petri Nets, dataflow diagrams, synchronous languages, or programming languages such as sequential function charts and function block diagrams as specified in the IEC 61131-3 standard for programmable logic controllers (IEC 1131, 1993). Each formalism has a specific syntax and semantics that closely matches users' training and expertise and that are well suited to the particular application. The transformation of formal models from one formalism into another is complicated and often leads to inefficient models, even for formalisms with equivalent expressive power (Huuck et al., 1997). Thus for a general purpose simulation environment, it is preferable, if not indispensable, to offer various modelling formalisms and even to allow the user to define new or specific formalisms with little effort. The use of domain specific formalisms results in models that are elegant, intelligible to the user, and closely correspond to the documentation formalism and/or the implementation language. This keeps the modelling effort low and makes it less error prone than a transformation into one general, tool-specific formalism.

2 Requirements for the Modelling Environment

Due to the complexity of both the physical part and the supervisory control system in large technical systems, it is evident that a powerful modelling environment and efficient simulation methods are indispensable to support the design process.

2.1 Intuitive and Effortless Modelling of Physical Systems

The physical part should be modelled as intuitively as possible. From the modeller's point of view the optimum would be to assemble the whole model using predefined building blocks that correspond to the technical components. The graphical connection of these elements would result in composition diagrams that look like familiar engineering notations, e.g. electrical circuit diagrams, flow charts, and other conventional notations.

In most cases, physical systems do not have explicit inputs and outputs; whether an external variable is input or output depends on the environment. For instance, the pressure drop in a pipe may be caused by a prescribed flow or be the cause of a certain flow rate. Thus the building blocks of larger models should have non-causal, undirected interfaces.

Due to the potential variety of components in technical systems, only a limited number of standard elements can be predefined and stored in component libraries. The remaining elements have to be defined by the modeller. For basic elements a convenient approach is to enter the underlying physical equations, possibly taken from the relevant literature, without transformation to a specific mathematical format, e.g. a system of explicit ODEs. Of course, the number of equations must match the number of unknowns. This approach is called declarative modelling, because the modeller states that these equations have to be satisfied, without determining how to perform the calculations. The model acts as a set of constraints on the coupling variables, but it is not explicitly stated how to compute outputs from inputs and initial states.

Larger elements should be defined as a composition of smaller building blocks. This leads to a hierarchical structuring of the model, which is crucial for the effective handling of large models. Since one has to deal with many different building blocks, it should be possible to establish user-defined libraries in addition to the standard libraries. Additionally, the concept of inheritance supports the modelling effort and reduces the likelihood of errors. Component model classes then can be derived from basic model classes by adding more detail. If the basic model class is modified, this modification also effects the derived classes and the models that will be instantiated from the derived classes.

2.2 Adequate Modelling of Discrete-Event Systems

The requirements for the modelling of discrete-event systems are different from those for physical systems in many respects. Discrete-event models are more diverse with respect to syntax and semantics than quantitative simulation models of physical systems. Physical systems can be treated in a uniform way using DAEs as an underlying semantic basis. The syntax of the graphical representation is also simple: the blocks have uniform ports and the building blocks are coupled by simply connecting these ports. In case of the modelling language MODELICA, the coupling semantics is that all (generalised) flow variables must add up to zero or that the (generalised) potential variables, such as voltage, pressure etc., assume the same value. In contrast, each discrete-event formalism has its specific graphical syntax that does not simply refine a common basic syntax so that specific graphical editors have to be provided for each formalism that is supported.

Furthermore, no established semantic standard form, comparable to the DAE-system, exists for discrete-event formalisms, and transformations to a basic formalism are often inconvenient, sometimes due to relatively small semantic differences. Consequently, for the simulation of DES models, specific algorithms must be used.

Regarding the complexity of real supervisory control systems which may consist of a large number of modules that are specified by different designers from different domains, it is necessary to support heterogeneous discrete-event models including hierarchical execution schemes as well as concurrency with synchronous and asynchronous communication systems, i.e., it should be possible to model different parts of a controller with different formalisms and to connect these parts in a consistent manner. If different simulators are used for different formalisms, these discrete-event simulators have to interact with each other and have to be synchronised with the numerical integration of the continuous part of the system.

2.3 Integration of Continuous and Discrete-Event Models

For a seamless integration of discrete-event formalisms and physical models, on the *syntax* level, the coupling should reflect the actual hierarchical relations. Since components of the supervisory control system often are related to particular subsystems of the continuous part, the corresponding discrete-event model should be represented by a block that can be inserted into a composition diagram of the physical model. The inputs and the outputs of the discrete blocks can be connected with ports of other building blocks, continuous or discrete.

On the *semantic* level, the coupling of a discrete-event model with the physical model is more involved. Some numerical integrators evaluate the model equations several times in order to do one step (Brenan and Campbell, 1996). This can cause unpredictable behaviours if the discrete-event system is called at intermediate points without considering the fact that the simulation of the continuous system has not yet converged. The semantics of the discrete-event formalisms must not become corrupted by the integration into the physical system. Conversely, transitions in the discrete-event part occur while time in the physical system does not progress. If iterations in the discrete-event part are performed, the intermediate states must not be transmitted to the continuous system but the simulation must be stopped until the discrete part has reached a stable state. This stable state may imply switchings not only of variables but also of the structure of the continuous part which may trigger new events in the discrete system. Even worse, the overall state of the continuous system, composed of the discrete inputs and the past state of the physical part may not be consistent such that a new initialisation has to be computed. So a nested loop of computations must be performed with frozen physical time until the overall system has reached a stable and consistent state from which the simulation can be continued.

The localisation of state events inevitably leads to increased simulation times because iteration or other additional computations are required. If the discrete part contains timers which trigger transitions, it is advantageous to propagate this information to the continuous simulator such that the simulation stops precisely at the event time and an iteration is avoided. Finally, discrete-event formalisms require an adequate visualisation of the simulation results using the graphical formalism itself typically in the form of animation. The usual plots of variables over time are not sufficient.

2.4 The State-of-the-Art

Some general-purpose commercial software tools exist for modelling and simulation of hybrid systems. Among these, the MATLAB package with SIMULINK and STATEFLOW is the most widely used tool (Matlab, 2002). In consideration of the requirements postulated above one has to realise that the state of the art is not satisfactory.

Block diagrams have a fixed causality and are not really intuitive to model large systems. The use of block diagrams results in an abstract mathematical representation of the modelled system as shown in Fig. 1. This block diagram corresponds to an electrical circuit, but it is not evident how it is related to the structure and to the parameters of the circuit. Furthermore, if e.g. a voltage source is replaced by a current generator many modifications are required in the overall model, since the cause and effect relations have to be inverted in several locations. This poses serious problems for the reuse of aggregated building blocks.

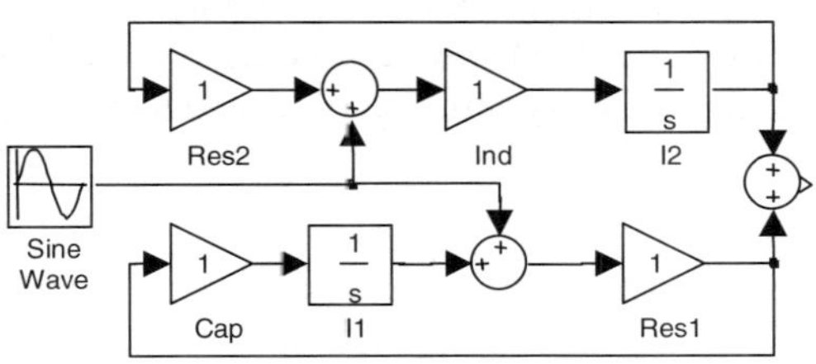

Fig. 1. A block diagram

The STATEFLOW formalism is a variant of statecharts (Harel, 1987). Figure 2 illustrates this with a STATEFLOW model of a relay mechanism. Statecharts are an intuitive and powerful formalism to model reactive behaviour and exist in many slightly different flavours. But besides statecharts, many other formalisms, such as High Level Petri Nets or GRAFCET, and programming languages, such as Sequential Function Charts, exist that have their specific strengths and can not be mapped easily onto statecharts.

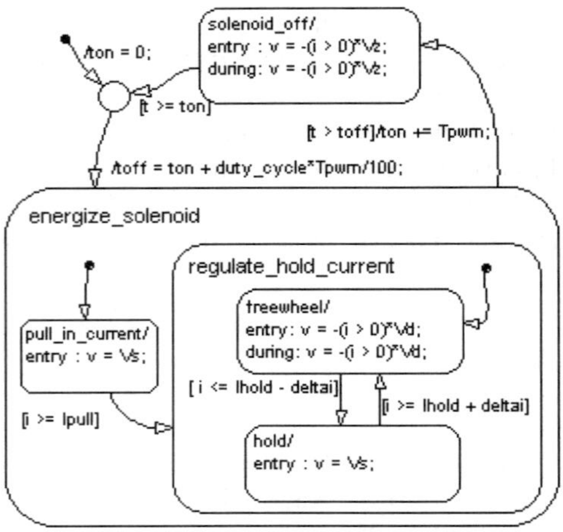

Fig. 2. A stateflow diagram (screenshot)

3 The DES/M Approach

The proposed solution for the modelling and simulation of large, complex systems with continuous and discrete-event dynamics consists of two major elements. The object-oriented equation-based modelling language MODELICA is used for the modelling of the physical part and of the regulators. A newly developed software tool for the modelling of discrete-event systems called DES/M (discrete-event systems for MODELICA) supports various formalisms (at present statecharts and SFCs) and modular, hierarchical and heterogeneous models. The discrete-event part of the model is automatically transformed into a MODELICA algorithm. Any simulator that can process MODELICA code can then be used to solve the overall system. The interaction of the continuous and the discrete part of the system is coordinated by the event handling mechanism of the MODELICA solver.

3.1 Object-Oriented Modelling Using MODELICA

In object-oriented modelling, the model elements correspond to physical components of the real system and the composition of the elements to the overall model is in accordance with the physical structure of the system. The elements have undirected interfaces and their behaviour is normally described declaratively. An element can be a composition of other elements and it can contain equations for the behavioural description. These equations need not to be solved explicitly for a particular variable. Another common feature of object-oriented modelling languages is that the equations are processed symbolically. The overall mathematical model is constituted by all the equations that describe the model elements and their connections. This usually leads to a large but sparse system of algebraic and differential equations (DAE). By means of automatic symbolic manipulations this large set of equations is transformed into a sorted DAE where as many derivatives and algebraic variables as possible are computed explicitly and redundant variables are removed. This enables efficient simulation using standard integration methods.

The most prominent object-oriented modelling languages are MODELICA (Modelica Design Group, 2000), VHDL-AMS (Heinkel, 2000) and gPROMS (gPROMS, 2002). MODELICA is best suited for our purposes because it is not tailored to a specific application domain, and it is standardised by a non-profit organisation, the MODELICA Association, and freely available. Very important are the class concepts of MODELICA that include class definition, object instantiation, partial classes, inheritance, and more, which facilitate the creation of model libraries. These features are well known from object-oriented programming languages, but they are not always supported by object-oriented modelling languages. The meaning of the term 'object-orientation' depends on the context, here the essential property is the construction of large models from building blocks which can be used freely because they are formulated in a general, context-independent fashion. For MODELICA, many free libraries exist for different domains such as electrical systems, rotational and translational mechanics, multibody systems, and others.

For the definition and simulation of MODELICA models we use the commercial software DYMOLA (Dymola, 2002). This tool provides a graphical editor for composition diagrams so that systems can be modelled visually. The graphical representation of the library components mimics conventional engineering notations. The main reason to use DYMOLA, however, is the powerful symbolic engine that transforms the set of equations into a form that can be solved efficiently. This permits the simulation of very complex physical systems including hybrid phenomena (Otter et al., 1999).

In Fig. 3 it is shown how simple it is to build a model of a hydraulic actuator using given library components. The resulting model resembles the engineering notation and can be aggregated to a new composed building block that can be incorporated into a library as well.

To illustrate how hybrid phenomena can be modelled in an equation-based declarative style, consider an ideal electrical diode (Fig. 4). Due to the idealisation a sharp discontinuity is introduced at $u = 0$. In order to achieve an equation-based description, the diode characteristic is parameterised by a parameter s so that u equals s if s is less than zero, and i equals s if it is nonnegative. This results in the following set of equations:

$$\mathit{off} = s < 0 \tag{1}$$

$$u = \text{if } \mathit{off} \text{ then } s \text{ else } 0 \tag{2}$$

$$i = \text{if } \mathit{off} \text{ then } 0 \text{ else } s. \tag{3}$$

Due to this declarative formulation, the interaction of several diodes in an electrical circuit needs not be modelled explicitly. The network behaviour is defined implicitly by the composition of the component equations and of the connection equations (Otter et al., 2000).

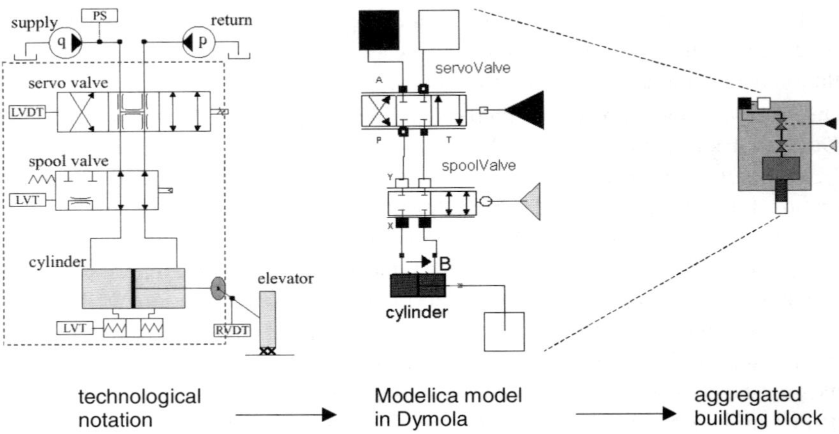

Fig. 3. Modeling a hydraulic actuator using standard components

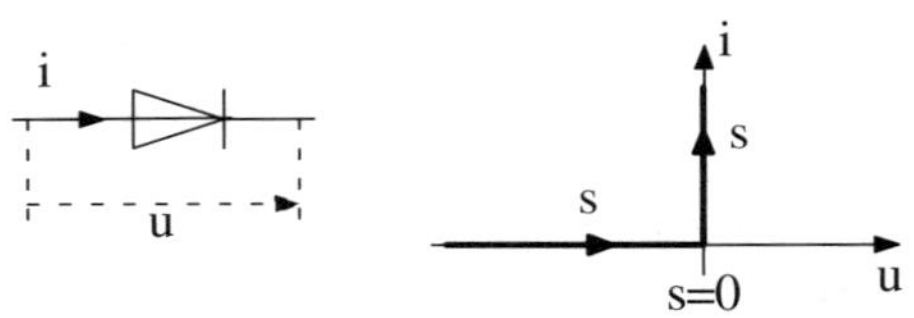

Fig. 4. Ideal diode model described as parameterized curve

Basic discrete event formalisms can also be expressed in an equation-based fashion, e.g. simple Petri Nets and automata (Mosterman et al., 1998). For instance, in a Petri Net model, the places and the transitions are represented by components that are defined in the corresponding MODELICA library. The graph structure is constituted by the connections of the ports of the components. Since each object and each connection just add equations to the overall set of equations, the behaviour of a Petri Net model is defined as the mathematical solution of the subset of equations given by the Petri Net model.

Unfortunately, this object-oriented modelling technique is not suitable for the modelling of complex discrete-event systems. The first reason is that the syntax of composition diagrams based on blocks with ports is not powerful enough for the graphical representation of complex formalisms such as statecharts. The second reason is that certain semantic elements such as local iterations can not be represented adequately by a set of equations. For instance, in certain statechart variants (Harel et al., 1987) a *step* of a statechart, i.e., its reaction to external stimuli, is defined as a sequence of *micro-steps*. Each micro-step consists of a set of concurrently taken transitions. At a micro-step, the firing transitions may generate events that trigger the transitions of the subsequent micro-step. In this manner a kind of event iteration is performed that ends when no further transitions are triggered (improper statecharts may result in infinite iterations). Micro-steps are considered just as an internal mechanism to compute the reaction of a statechart so that the micro-steps should be hidden from the environment of the statechart. Therefore, an adequate realisation would use this operational semantics to generate the behaviour of a statechart and omit an interleaved execution with the physical system. Unfortunately this is not possible with an equation-based realisation, since the equations of a statechart would have to be solved simultaneously with the equations of the physical system. Thus, each micro-step would be connected to the evaluation of the overall set of equations so that side-effects possibly can take place in the physical system.

3.2 A Compatible Modelling Environment for Discrete-Event Systems

For the reasons stated above, the DES/M modelling environment has been developed that provides dedicated editors for several discrete-event formalisms and allows to insert the discrete-event models consistently into the overall model. By this approach the restrictions on semantics, syntax and graphical appearance are circumvented, and the object-oriented modelling principles for continuous systems are not enforced in a domain where they are not appropriate. By suitable transformations, the models of the discrete-event part can be inserted into the overall MODELICA model and can be solved using standard techniques for the manipulation and the numerical solution of continuous systems.

For the definition of the discrete-event part of the models, there are two different possible options. The first is to compose the model from discrete-event building blocks, the behaviour of which is specified declaratively based on equations, similar to the procedure that is followed for the continuous part. However, these blocks would have to be quite complex because a large number of interacting variables may be required. Therefore the blocks should not simply be merged but a code optimisation step should be performed. Thus there would be two transformations before an executable model is obtained; first the transformation of the individual blocks into MODELICA code, then the construction of the overall model. The second approach is to construct the discrete-event part of the model completely on the graphical level using the chosen formalisms and the respective graphical editors, and then to perform an automatic translation into a single MODELICA-algorithm and to wrap it into a MODELICA class. We prefer the second approach. For reasons discussed above, all discrete-event subsystems that interact directly via events or messages must be represented as a monolithic block in an imperative fashion. The transformation of the complete system into an algorithm leads to a clear structure – first an overall discrete-event model is composed from subblocks that can be structured hierarchically and may even be defined using different formalisms, e.g. statecharts and SFCs, and then the transformation into an algorithm is performed following clearly specified semantics. Actually, in the end a problem specific discrete-event simulator is inserted into the MODELICA model of the physical system. This MODELICA component can be easily connected to physical components because it interacts via standard ports.

The main advantage of using a MODELICA-algorithm is that the handling of the state events is done by MODELICA automatically. The MODELICA compiler discovers all potential sources of discontinuities in the algorithm and makes sure that discontinuities are handled appropriately, i.e., when a threshold is reached and a discrete state transition or any other discontinuity occurs, the integrator will be stopped in order to perform the discrete changes. If the discrete-event model would be simulated by an external program, the conditions that trigger the state transitions in the discrete-event model still would have to be inserted into the MODELICA model in order to stop the continuous simulation when the discrete-event part causes state events. If the discrete system is specified in a different environment, this task has to be performed manually by copying the transition conditions or guards and invariants, which is tedious and error-prone. In contrast, the DES/M environment generates automatically a complete MODELICA simulation algorithm for the discrete system parts from the graphical specification.

The modelling environment supports heterogeneous and hierarchical discrete-event models by means of a special block editor. Model-reuse is enabled using an archetype concept, i.e., each block that is used in a model is an instance of an archetype that defines the ports and the general properties of the block type and one or several alternative implementations. These implementations define the behaviour of the instantiated blocks and can be specified using again block diagrams or another formalism.

In order to reduce the effort for the implementation of several editors, the DES/M environment is based on the meta-modelling tool DoME (DoME, 1999). DoME was designed as a tool for the automatic generation of complex graphical editors based on a formal syntax description and parameters that control the graphical appearance. A partially graphical language called DoME Tool Specification Language is used for specifying the graphical entities, their properties and relations, structural constraints as well as their visual appearance. More advanced features such as more complex syntactical constraints and code generation can be implemented with DoME's Lisp-like extension Alter or using Smalltalk. Besides the block diagram editor, up to the present, two further editors have been realised: a statechart (SC) editor and an editor for *sequential function charts* (SFC) (Deparade et al., 2001).

3.3 Formalism Interoperation via Special Block Diagrams

As already mentioned, a special hierarchical block diagram formalism has been implemented for supporting the interoperation of different formalisms. The main idea is rather straightforward: Certain blocks of a block diagram may contain either another block diagram or a reactive model that is specified with a state transition formalism such as statecharts or sequential function charts. Consequently, it is possible to use different formalisms within one model.

The idea to use a block diagram formalism arose from the modelling of the aircraft elevator described in detail in (Mosterman et al., 2002). The main feature of this control system is that 8 concurrent state machines, each modelled by a statechart, interact tightly in order to achieve a safe configuration of the redundant elevator actuators when failures occur. The statecharts have the same structure and their transition conditions are large logical expressions that reference the states of the other statecharts and the failure signals. The goal of the block diagram formalism was to separate the large and complex logical expressions from the statecharts, so that the statecharts become identical (and clearer) and can be instantiated from the same class. Therefore, the block diagram formalism distinguishes static blocks that are depicted with a dashed border, from dynamic blocks that have a solid border (Fig. 5).

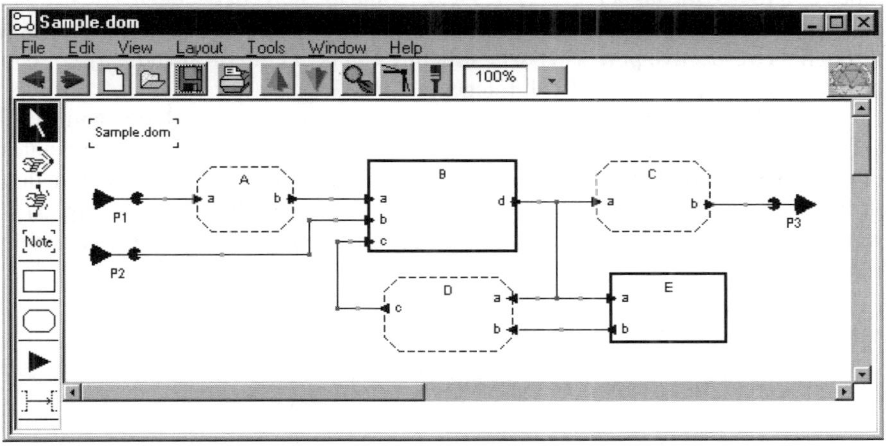

Fig. 5. A sample block graph

A static block contains an algorithm or just a set of assignments and is used to compute the current output values y_i directly from the current input values u_i of the block. Hence the behaviour of a static block can be represented by a function:

$$y_i = f_{stat}(u_i) \tag{4}$$

Such a static block is applied to, e.g., the computation of the logical expressions of the redundancy controller.

The dynamic blocks have internal state variables x_i and a quasi-synchronous semantics is applied, i.e., the blocks are evaluated synchronously, but without simultaneous data exchange:

$$x_i = f_{dyn}(x_{i-1}, u_i)\,, \tag{5}$$

$$y_i = g_{dyn}(x_{i-1})\,. \tag{6}$$

The state transition function f_{dyn} and the output function g_{dyn} impose an iterative computation scheme for the block graph such that the response of such blocks to new changes of the inputs becomes effective in the next iteration step. As long as the outputs of these blocks are changing, all blocks have to be reevaluated synchronously. This quasi-synchronous semantics is analogous to the internal computation of statechart behaviour: if a statechart contains orthogonal parts (modelled with and-states), the consequences of concurrently and independently taken transitions of a micro-step, i.e., events and the new states, only become effective in the subsequent micro-step. Thus, in the DEFORM approach, local event iterations are not only performed inside of the statecharts, where a step can be computed by a sequence of micro-steps, but also on the block diagram level.

Further elements in Fig. 5 are the outer ports that represent the interface of the block diagram to the higher level ($P1$, $P2$, $P3$) and the ports of the blocks (a, b, c, d). Each port has an associated port type that defines the structure of the data transmitted through the respective port. This data-structure can be hierarchical and may contain different basic types such as Real, Integer and Boolean.

At a higher level, the block diagram in Fig. 5 is itself a dynamic block with ports $P1$, $P2$ and $P3$. The state of this enclosing block is the Cartesian product of the states of the dynamic blocks B and E. For the computation of the state transition function of the enclosing block an iteration at the level of the inner block graph (Fig. 5) is started during which the following constraints have to be satisfied at each iteration step:

$$
\begin{aligned}
A.a_i &= P1 \\
B.b_i &= P2 \\
E.a_i &= D.a_i = C.a_i = B.d_i = g_B\left(B.x_{i-1}\right) \\
D.b_i &= E.b_i = g_E\left(E.x_{i-1}\right) \\
A.b_i &= f_A\left(A.a_i\right) \\
C.b_i &= f_C\left(C.a_i\right) \\
D.c_i &= f_D\left(D.a_i,\ D.b_i\right) \\
B.a_i &= A.b_i \\
B.c_i &= D.c_i \\
B.x_i &= f_B\left(B.x_{i-1},\ B.a_i,\ B.b_i,\ B.c_i\right) \\
E.x_i &= f_E\left(E.x_{i-1},\ E.a_i\right).
\end{aligned}
\tag{7}
$$

After this iteration has converged to a stable state, the outputs $P3$ of the enclosing block are updated and the computation of the transition function of the enclosing block is finished.

It should be noted, that for a specific block it does not make a difference whether its behaviour is specified as a block diagram or as a statechart, since both formalisms are transformed into a state transition function that hides the inner processes. Hence arbitrary other reactive formalisms and communication paradigms can be incorporated as well, as long as they can be transformed into a compatible state transition function.

3.4 The Modelling and Simulation Process

The approach described above leads to a tool architecture that consists of two main cooperating tools: DYMOLA is used for physical system modelling, whereas the DES/M environment is used for modelling discrete-event systems (Fig. 6). By means of the editors for the various discrete-event formalisms, the complete supervisory control system is described. Then it is compiled into a MODELICA class that is stored in the file system so that it can be retrieved by DYMOLA and instantiated in the model of the physical system. The MODELICA classes created in DYMOLA are stored in the file system as well. For simulating the overall model, the corresponding class has to be compiled into an executable. The transformation of the set of equations into a preferably explicit representation is performed automatically. The simulator executable generates the trajectory for a given set of parameters that can be changed without the need to recompile the model. Every time when the supervisory controller has to react, the integrator stops because a state event is generated due to inequality expressions in the MODELICA-algorithm. The execution of the algorithm at these times realises a discrete state transition and the corresponding change of the outputs. The internal processes during such a state transition do not become visible to the model of the physical system, but they are saved in a log file. This permits the visualisation of the internal processes of the discrete-event model in the DoME tool for debugging purposes.

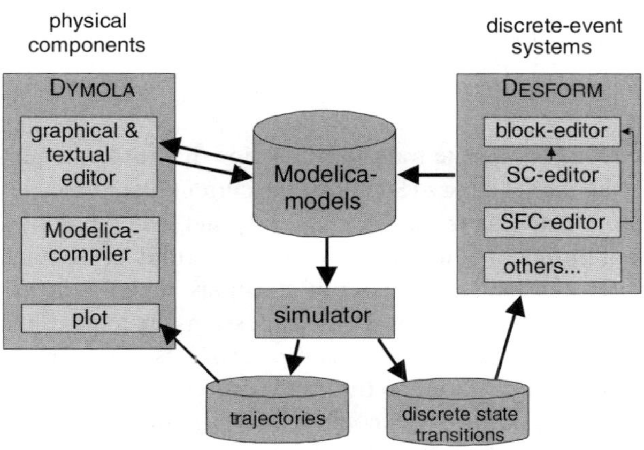

Fig. 6. The modeling and simulation process using two tools

4 Realising Discrete-Event Dynamics in MODELICA

A discrete-event model that was composed within the DES/M environment is translated into a MODELICA component that contains one algorithm for the computation of the reactions of the corresponding supervisory controller. This algorithm is a simulator for the specific discrete-event model and is possibly very complex. In the following, two simple examples are discussed in order to illustrate how the continuous integration and the discrete-event dynamics are combined using the MODELICA language. The actual code generation is intricate, in essence it is the realisation of the operational semantics of the formalisms supported by DEFORM using the MODELICA language.

4.1 Models with State Events

The synchronisation of the discrete-event dynamics and the continuous integration is straightforward (Pereira Remelhe et al., 2001). To illustrate this, consider a simple supervisory controller that fills a tank up to a certain level `h_high`, after a specific low level `h_low` was reached. For safety reasons, an additional limit sensor is installed that indicates whether the tank is full. This controller has two input variables: the current level `h` and the binary signal `limit_h_full`, as well as a binary output variable `v` for the inlet valve. The corresponding discrete-event dynamics can be described by a model with two states `S1` and `S2`, and two Transitions `T1` and `T2` (Fig. 7). An algorithm that exhibits the desired behaviour can be formulated as follows:

```
T1_fires := pre(S1) and (limit_h_full or(h>h_high));
T2_fires := pre(S2)and (h<h_low);
S1 :=(pre(S1) and not T1_fires) or T2_fires;
S2 :=(pre(S2) and not T2_fires) or T1_fires;
v := S1;
```

Such an algorithm corresponds to a state transition function that maps the previous controller state $x_{prev} = \{\texttt{pre(S1)}, \texttt{pre(S2)}\}$ and the current input variables $u = \{\texttt{h}, \texttt{limit\_h\_full}\}$ into the new state $x_{new} = \{\texttt{S1}, \texttt{S2}\}$ and the current output variables $y = \{\texttt{v}\}$. During the continuous integration this algorithm is executed simultaneously to the evaluations of the complete set of equations, but all state variables and outputs remain constant, since the inequality expressions are fixed. If the state `S1` is active, in the algorithm `pre(S1)` is true and `pre(S2)` is false. Therefore only the first logical expression can become true, and, consequently, only the inequality expression $(\texttt{h} > \texttt{h\_high})$ needs to be monitored during continuous integration. When this expression becomes true, the integrator is stopped and the whole set of equations including the algorithms is re-evaluated including the unfixed inequality expressions. Now the value of `T1_fires` becomes true, `S1` becomes false, `S2` becomes true, and `v` becomes false, i.e., the state changes from "filling" to "waiting". In a second discrete evaluation only the transition variable `T1_fires` becomes false again, since `pre(S1)` is now false. Because the discrete state variables did not change this time, the integration is started again. Now $(\texttt{h} < \texttt{h\_low})$ is monitored.

4.2 Models With Time Events

As an alternative, the limit sensor could be replaced by a time-out corresponding to the known maximum duration of the filling process. This idea is realised in the

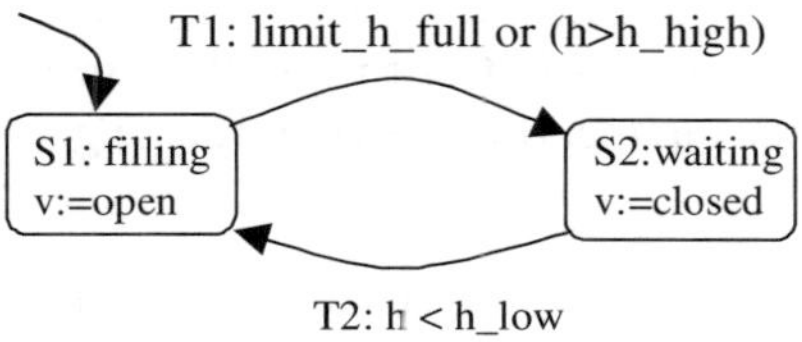

Fig. 7. Discrete-event model using only state events

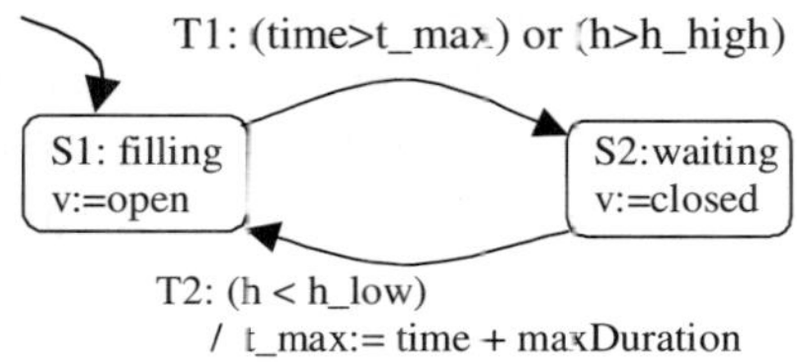

Fig. 8. Discrete-event model using state events and time events

diagram shown in Fig. 8. When the transition T2 is taken, an action is performed that assigns a new value to the variable `t_max` that stores the point in time, when the state `S1` has to be left. Additionally, the transition `T1` makes sure that the filling activity stops when this time elapses. A corresponding algorithm is as follows:

```
T1_fires := pre(S1) and ((time>pre(t_max)) or (h>h_high));
T2_fires := pre(S2) and (h<h_low);
t_max := if T2_fires then time+maxDuration else pre(t_max);
S1 :=(pre(S1) and not T1_fires) or T2_fires;
S2 :=(pre(S2) and not T2_fires) or T1_fires;
v := S1;
```

Hence, an additional state variable `t_max` is needed that, in contrast to the other state variables, is a real valued variable. If the controller is in state `S1` the inequality expression `(h>h_high)` has to be monitored in order to generate a state event, but as long as the choice of `t_max` is correct, the expression `(time>t_max)` is used to generate a time event and the simulation stops exactly at the corresponding time without the need to localise a state event.

5 An Illustrative Application Example

To illustrate how the DES/M environment can be applied, a model of a laboratory batch plant is presented that incorporates hybrid physical dynamics and a supervisory controller. The plant is a slightly simplified variant of one of the benchmark examples in this volume and was already described in (Kowalewski and Preußig, 1996). The physical part of the plant has been modelled in an object-oriented and equation-based fashion using the MODELICA language. A library has been developed that provides the classes *Valve*, *Pump*, *Condenser*, *Sensor* and 4 different types of tanks. These were graphically composed in the DYMOLA tool resulting in a process flow chart (Fig. 9) that resembles the graphics of a standard piping and instrumentation diagram.

The supervisory controller model is also included in the plant model, but the sensor objects and the actuator objects are not connected visually to the controller component inputs or outputs respectively, in order to keep the model clear. Instead, on the top level of the model, additional equations are used that relate the current values of the sensors to the input variables of the input port of the controller, e.g.:

```
controller.sensors.LIS_101 = LIS_101.value;
```

or that relate the input signals of the actuators to the outputs signals of the controller, e.g.:

```
V1.open = controller.actuators.V1;
```

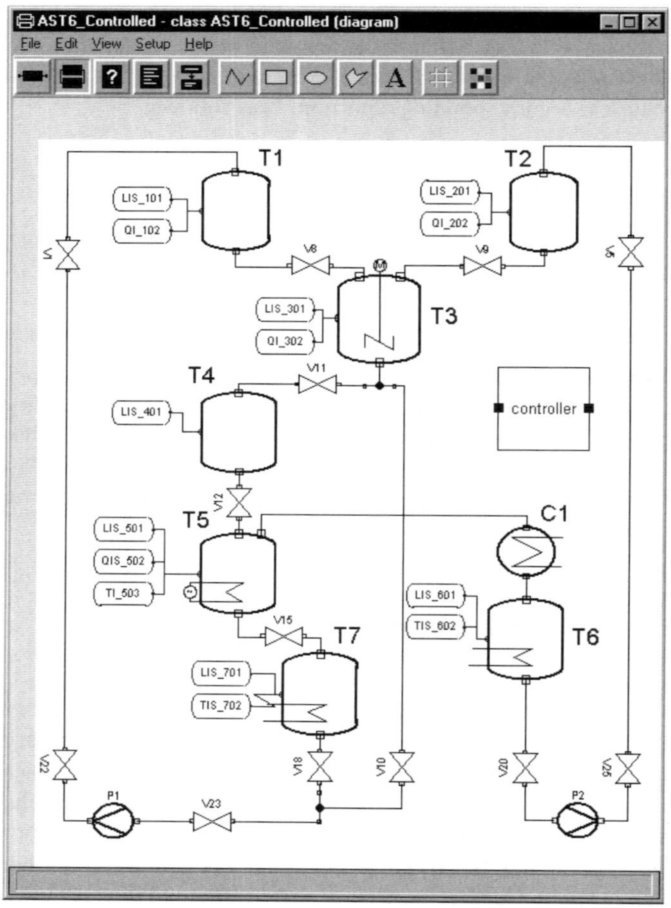

Fig. 9. The MODELICA model of the batch plant

The DES/M environment generated the MODELICA class of the supervisory controller from a graphical specification that includes sequential function charts (SFC) and the block graph formalism. Figure 10 shows the overall structure of the controller model. The objective of this controller is to run 2 recipes in parallel on the plant. As a rudimentary means of coordination, the idle tanks are determined from the sensor and actuator values using simple logical expressions such as:

```
idleTank.T7_idle := (sensors.LIS_701<=0.001) and
not(actuators.V15) and not(actuators.V18);
```

This means that a tank is idle if the tank is empty and the connected actuators are passive. Since both recipes drive the same actuators, their outputs have to be superposed. This can be realised by simple logical expressions that activate an actuator, if one recipe (or both recipes) set this actuator active, e.g.:

```
actuators.V1 := act1.V1 or act.V2.
```

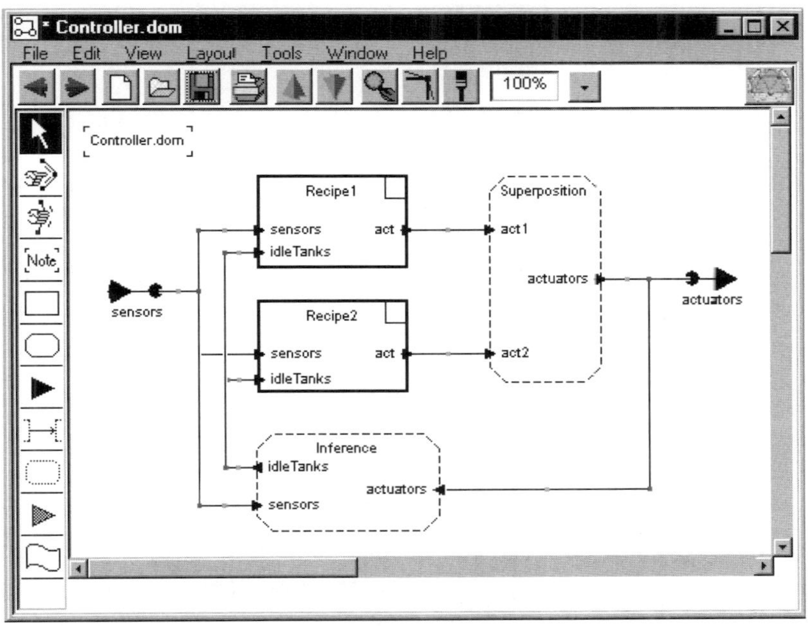

Fig. 10. The top level block graph of the controller

These logical expressions were entered as static blocks. The only dynamic discrete-event behaviour results from the recipes that are contained in the dynamic blocks. Both recipe blocks run the same recipe and only differ in their parameters such as the start time. Therefore, both recipes are described by the same SFC, see Fig. 11. When the start time of a recipe elapses, an infinite loop is started where a concentrated salt solution of tank T1 is drained into tank T3 and then mixed with pure water from tank T2 until a certain concentration `w_dilution` is obtained. This

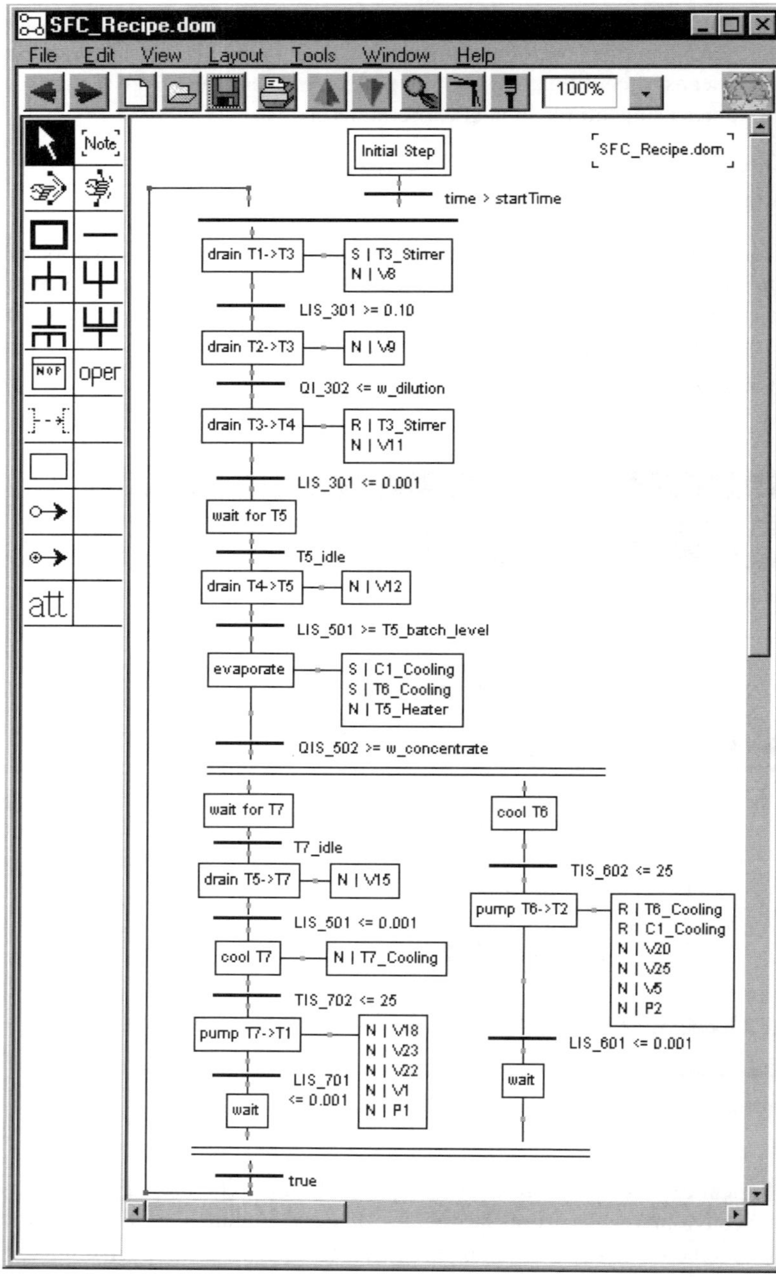

Fig. 11. The sequential function chart of the recipes

dilution is buffered in tank T4 and then concentrated in the evaporator T5 up to a certain concentration `w_concentrate`. The condenser and the cooling in tank T6 are started at the same time at which the electrical immersion heater of T5 is started. When the evaporation is stopped, the recipe splits into 2 parallel threads in which the concentrate is drained into tank T7 and cooled and then fed back to tank T1, and the condensate in tank T6 is cooled and fed back to tank T2. Then the recipe returns to the first step and repeats this procedure.

In order to omit redundancies in the model, types were declared for ports and blocks. A port type defines which variables are transmitted through a port and their numerical types (Real, Integer, or Boolean). In this example, only three port types are needed: *Sensors*, *IdleTanks*, and *Actuators*. Each port in the model must have an assigned type and connected ports must have the same type. The connections are checked automatically before the translation to MODELICA.

Since all recipes in this controller framework must have the same interface, an archetype is defined for the recipe blocks. Figure 12 shows the interface of the archetype in the centre. The interface defines the ports and their port types for all possible recipe blocks. In the upper right region of the window, both instances of this archetype are enumerated corresponding to the blocks of Fig. 10. In the lower right region, the behaviour specifications that are possible implementations of this archetype are listed. In this case only one SFC is given. But more SFCs can be specified for the usage of different recipes. In principle, also statecharts and block graphs are possible as implementations of a recipe block.

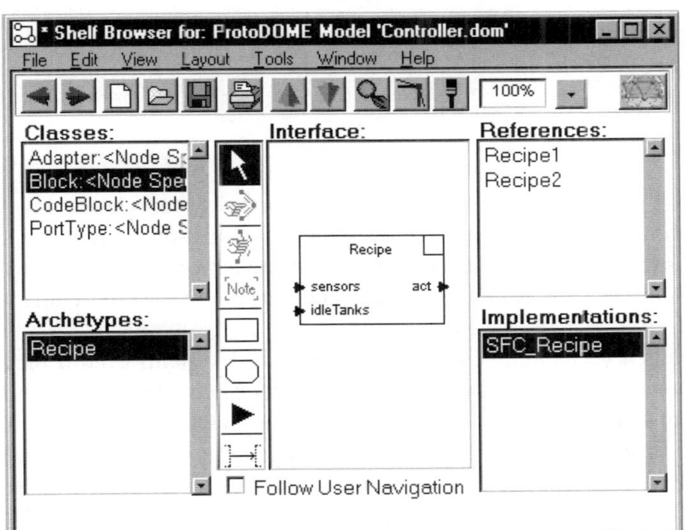

Fig. 12. The Shelf Browser in DoME with the archetype for the recipe blocks

After finishing the overall controller model, the MODELICA class can be generated automatically by selecting a corresponding item from a pull down menu. The MODELICA class contains a data structure that corresponds to the structure of the controller and an algorithm that defines the behaviour. This class can be instantiated in the plant model just like all other components. For simulation, the MODELICA model is compiled with DYMOLA resulting in a simulator executable. The model parameters can be changed conveniently for each simulation run without the need to recompile the model. This includes also the parameters of the controller such as the start times and the concentration parameters of each SFC. Thus, parameter studies can be performed efficiently. Figure 13 shows the plot of the mass hold-up of the evaporator as it results from a simulation run. From this plot one can conclude that the alternating recipes operate the evaporator at full capacity, i.e., there are no gaps where the evaporator is waiting for the next batch.

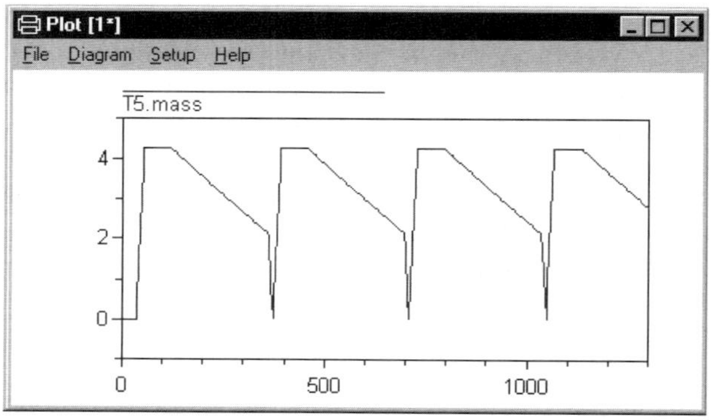

Fig. 13. A plot of the mass hold-up of the evaporator T5

6 Conclusions

The different nature of physical systems and supervisory control systems leads to two distinct sets of requirements for an integrated modelling and simulation methodology. In this work, the physical part of a system is captured by the object-oriented modelling language MODELICA is used, and for the discrete-event parts the DES/M environment has been developed.

MODELICA permits a very intuitive modelling of physical systems even if they include hybrid phenomena, and meets the demands for object-oriented modelling and effective and efficient simulation. The DES/M environment provides dedicated editors for domain specific discrete-event formalisms (and programming languages), and supports heterogeneous models including hierarchical structures and interoperation.

The connection of both parts of a model is done by translating the complete discrete-event model into a MODELICA algorithm that is wrapped in a MODELICA class. This is advantageous in several respects:

- both modelling environments are integrated seamlessly,
- the operational semantics of discrete-event formalisms can be formulated in an arbitrary way and this semantics is realised without any disturbance by the interaction of the continuous and the discrete part of the model or the solution algorithm of the continuous part (e.g. the step size of the integrator is independent of the execution of discrete transitions which occur at frozen simulation time),
- state events are accessible from the discrete-event part.

The latter point is crucial for an accurate synchronisation of continuous integration and discrete-event simulation.

At present, a first prototype of the DES/M environment is available. The current work aims at the improvement of the translation framework towards a more general approach and on the automatic support for graphical data visualisation. Then more formalisms will be implemented. Since MODELICA is based on algebraic and differential equations also other hybrid formalisms such as hybrid Petri Nets can be considered.

Acknowledgement. We thank Hilding Elmqvist and Johann Bals for the helpful discussions. The research reported here was performed in the context of the focused research program (Schwerpunktprogramm) "Continuous-Discrete Dynamic System" (KONDISK) and sponsored by the Deutsche Forschungsgemeinschaft under the grants OT174/1 and EN152/22. This support is most gratefully acknowledged.

A DEVS-Based Approach for Modeling and Simulation of Hybrid Variable Structure Systems

Thorsten Pawletta[1], Bernhard Lampe[2], Sven Pawletta[1], and Wolfgang Drewelow[2]

[1] Wismar University, Department of MVU, D-23952 Wismar, Germany
[2] University of Rostock, Institute of Automatic Control, D-18051 Rostock, Germany

Abstract. The modular-hierarchical modeling subdivides a system into dynamic describing atomic, and structure describing coupled systems. Due to the usally static structure of coupled systems, it can sometimes be hard to realize a component-oriented mapping of real system elements in model components. Hence, several approaches to model structural changes are developed, which are roughly classified in the paper.

Starting with the classical DEVS-theory and a specific extension for hybrid systems, this paper presents a modular-hierarchical modeling approach, that supports the specification of time-varying structures on the coupled system level. The possible structural variability comprises changes of the component structure and the coupling relations. The basic modeling principle is demonstrated using the example of an automated manufacturing cell. Because variable structure modeling of hybrid systems brings advanced requirements to the simulation methods, the general structure of a possible simulation approach is presented.

1 Introduction

Complex dynamic systems are often characterized by the feature that their temporal behaviour contains elements that are most appropriately modeled by structural changes. This can be applied to automation systems, which change their operational mode, the automation goal or the equipment. In biological systems a mutation of the inner functionality could be treated this way. Finally, the problem for which the model was developed can require a structure variable modeling solution. One example is the problem of adaptive modeling depth (Jähnichen and Klein-Robbenhaar, 2000) where several parts of the system need more detailed models in critical situations while for the rest of the simulation a simpler approximation is accurate enough. Moreover, such problems arise in structure optimizing experiments, where we have to find an optimal structure for the system under certain restrictions (Pawletta et al., 1997, Kaiser and Beaumariage, 1997). Typical applications in this direction are scheduling or control of manufacturing processes or plants. The given examples show that there are various kinds of variable structure systems. Hence a variety of modeling and simulation approaches exist, which are roughly classified in Sect. 2.

The main topic of the paper is the development of a modelling and simulation approach for modular-hierarchically structured systems with hybrid behaviour and complex structural changes. For discrete event systems a system theoretical formalism for modular-hierarchical modeling was developed in (Zeigler, 1976, Zeigler, 1984, Zeigler, 1990) [1], and a corresponding simulation method was derived. The

[1] Discrete event system specification theory (DEVS)

S. Engell, G. Frehse, E. Schnieder (Eds.): Modelling, Analysis and Design of Hybrid Systems, LNCIS 279, pp. 107–129, 2002.

modeling approach distinguishes between atomic and coupled systems. Atomic systems own a state space that is not further refined. Coupled systems[2] specify an invariable structure consisting of a set of subordered atomic and coupled systems and interconnections between them. Complex structural changes, such as the breakdown or exchange of a complete aggregate, cannot be directly described by changes of components and their coupling structure on the level of a coupled system. All kinds of structural changes must be modeled on the level of atomic systems. The result are often atomic systems with a very complex state space and with a restricted reusability. In many cases the transformation of complex structural changes into the atomic system level prevents a component oriented mapping of real systems into adequate model components.

Starting with the classical DEVS-theory, in (Prähofer, 1991) extensions for modeling of hybrid systems have been developed, and for discrete event systems the papers (Barros, 1996, Prähofer, 1996, Pawletta et al., 1994, Pawletta et al., 1996, Prähofer and Zeigler, 1992, Thomas, 1996) offer approaches for improved specification of complex structural changes. Using these works, the present paper develops a modeling method that supports the mapping of hybrid system dynamics and modeling of complex structural changes on the coupled system level.

After a formal representation of the modeling approach, the basic modeling principle is demonstrated using an example application. In order to make our approach comparable with other results, we choose the KONDISK benchmark of an automated manufacturing cell. This example is a hybrid, but not explicitly structure variable system. However, it will be shown that this hybrid system is mapped with benefits by the structure variable modeling approach.

For simulation purposes the model specification has to be transformed into an executable computer program. From event oriented simulation we know object oriented modular simulation approaches. During simulation they contain all information necessary for complex structure changes. Critical points of these approaches for hybrid systems are considerable limitations with respect to numerical methods or a loss of accuracy and longer computation time in comparison with monolithic computation models (Prähofer, 1996, Mosterman and Biswas, 1999, Wolf, 2001), because continuous coupled model parts can be distributed over different program modules. A simulation method based on a combined modular-monolithic computation model overcomes these problems, like presented in (Wöllhaf, 1995) in connection with the simulation of batch plant processes. Section 5 of this paper presents a general approach of a combined modular-monolithic computation model, that contains all necessary information for complex structural changes at simulation runtime and that encapsulates the continuous model description in a monolithic program object. Hence the continuous model relations can be preprocessed according to numerical and runtime specific requirements and standard numerical methods for solving ODE or DAE-systems can be applied. Of course this processing must be possibly repeated after structural changes at simulation runtime. Finally, the paper summarizes important aspects and gives a short view to some further works.

[2] Coupled systems are also called networks.

2 A Rough Classification of Modeling and Simulation Approaches for Variable Structure Systems

The term variable structure system comprises various modeling and simulation techniques. The aim of this section is a rough classification of the different approaches.

2.1 Local Structural Changes

On the one hand, the term variable structure system denotes local structural changes that have an effect on the global causality structure, for example contact problems or slip-stick-motion in mechanics, or the model of an ideal diode in circuit modeling. This form of structure variability can be treated by *non-causal modeling*, where by means of equation-based, declarative modeling the models' causality is not explicitly fixed (Elmqvist et al., 1993). The structural change is here realized by switching equations. This kind of structural variability is supported by many available simulation environments (Mosterman, 1999).

2.2 Agent Based Approaches

Another approach is based on autonomous agents building up the dynamics self-organized. During a simulation agents are created, copied and terminated. Each agent reserves his specific state space. The agents' environment is modeled in a separate state space. The communication between agents is based on dynamical relations. Agent based models are modular but mostly not hierarchical.

This modeling approach has developed from individual-oriented modeling of discrete event systems and makes use of AI-based techniques. It is used mostly on simulation of biological or sociological systems, e.g. (Uhrmacher and Arnold, 1994), and autonomous robot systems, e.g. (Lüth, 1998). Applications in the field of hybrid or coupled continuous systems are not known.

2.3 Structural Changes in Modular-Hierarchical Systems

A third variant of structural variability operates on structures that are available for modular-hierarchically modeled systems. In this formalism the complete system is decomposed into modular subsystems (Zeigler, 1990). Concerning the hierarchical formation one distinguishes between *atomic systems*, which own a state space, and *coupled systems*, which contain atomic and coupled systems and define interconnections between them. The classical modular-hierarchical modeling approach (Zeigler, 1976, Zeigler, 1984) supports only local structural changes in atomic systems. This can be modeled by a case-based separation of the state space.

Complex structural changes on the coupled system level are defined by the following operations operations (Pawletta et al., 1997, Pawletta et al., 1996, Prähofer, 1996, Barros, 1996):

- Insertion or deletion of subsystems,
- Insertion or deletion of interconnections,
- Replacement of a subsystem by another one,
- Cloning of subsystems,
- Activation or deactivation of subsystems and
- Exchange of subsystems between coupled systems.

Structural changes on coupled system level are realized relatively easily after a transformation of the dynamics into atomic subsystems by equation switching, assuming that all possible model structures are known in advance. However, the drawback of this approach is that model structuring cannot be made only by a component-oriented view of the technical units, which restricts the reuse of models considerably.

The replacement of subsystems was already discussed in (Pawletta et al., 1994) and (Thomas, 1996) from the simulation viewpoint. This consists not only of a simple replacement of subsystems with a compatible input-/output interface. When replaced subsystems belong to a common family, i.e., they have a set of common state variables, it is possible to initialize parts of the state space of the inserted system by that of the replaced system. This modeling feature allows for example a proper mapping of evolutionary and adaptive processes, as it was presented in (Pawletta et al., 1996) for an ontogenesis model. Models with adaptive modeling depth are an additional application area (Jähnichen and Klein-Robbenhaar, 2000).

The exchange of dynamic systems between coupled systems is in context of hybrid system models often useful, also for not explicitly structural dynamic systems. Among them are dynamic systems with movable components, for instance with continuous self-dynamics. This class of systems was discussed in detail in (Pawletta et al., 2001). In Sect. 4 the example of an automated manufacturing cell shows, how such kind of systems can be modelled with the benefits of structure variable approaches.

Starting with the classical DEVS-theory, two principal approaches for purely discrete-event systems have been developed to map structural variability in modular-hierarchical systems.

The first approach follows (Prähofer and Zeigler, 1992), and is based on a multi-level concept. In parallel to the behavioural model a structural model is introduced, that also can be constructed modular and hierarchical. The atomic subsystems of the modular and hierarchical behavioural model are extended by a special input and output interface. Over this interface the structural model sends requests to the behavioural model and receives answers. For that interlevel-couplings between the behavioural and the structural models have to be installed. The structural model stores the structural state of the behavioural model and specifies structural state transitions in the behavioural model. The strong seperation in a behavioural and a structural model leads to a relatively high modeling effort.

The papers (Barros, 1996, Prähofer, 1996, Pawletta et al., 1996, Pawletta et al., 1994) start with a direct extension of the coupled systems by a dynamical description. On basis of this works, the next section presents a modeling approach for hybrid modular-hierarchical systems.

2.4 Structural Changes on the Experiment Level

The extended model connecting concept (Plank, 1997) and structure investigation experiments (Pawletta et al., 1997) can also be seen as variants of structural variability.

The extended model connecting concept is a result of investigations into efficient mappings for complex equation switching, mainly between continuous models. The main idea is the fusion of the modeling and the experimental level. The dynamic behaviour of a hybrid system is described by a set of monolithic models that specify continuous dynamics. The connection between the models are realized by events that are described on the experimental level. Using these connections, models are initialized, started, stopped, reinitialized etc. The mapping of structural variability, however, is restricted on parameter and equation switching with the goal of a better model structure and reuse. Besides, further goals were of interest, such as hardware-in-the-loop simulations, simulator coupling, and parallelizing of simulations.

The approach in (Pawletta et al., 1997) for structure investigation experiments extends the classical experiment level in such a way that the model structure is seen as the topic of the experiment. In analogy to the above presented approach with a behavioural model and corresponding structural model, the experimental level takes, in a wider sense, the functionality of the structure model.

3 A Modeling Approach for Modular-Hierarchical Hybrid Systems with Complex Structural Changes

The modular-hierarchical modeling subdivides a system into dynamics describing atomic, and structure describing coupled subsystems that interchange information by coupling relations, see Fig. 1.

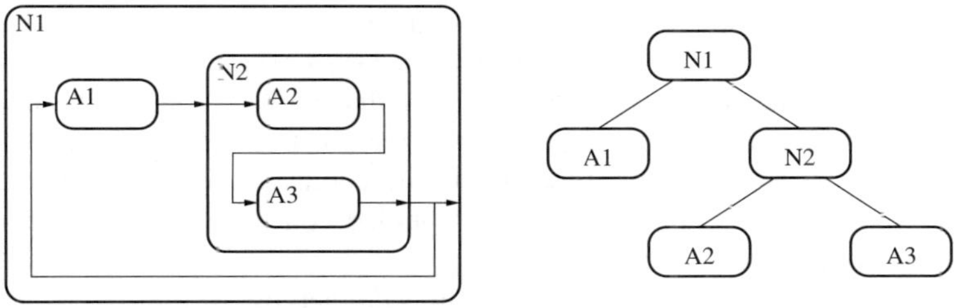

Fig. 1. Modular-hierarchical system and its composition tree

Our modeling approach for modular-hierarchical hybrid systems with structural variability builds upon the system theoretical works (Zeigler, 1976, Zeigler, 1984, Zeigler, 1990, Prähofer, 1991, Barros, 1996). The basis of these works is the classical DEVS-theory. In (Chow, 1996) a modified DEVS-approach (ParallelDEVS) was developed. From the system theoretical point of view both approaches are equivalent (Zeigler and Prähofer, 2000). The decision for the classical DEVS-basis was made with respect to its existing extension for hybrid systems (Prähofer, 1991). Presently a transfer of this approach to the ParallelDEVS-approach does not seem problematical.

3.1 Hybrid Atomic Systems

The paper (Prähofer, 1991) builds upon (Zeigler, 1976, Zeigler, 1984) and develops a system theoretical formalism for specification of hybrid atomic systems. It is shown that purely event-oriented discrete and continuous models are special realizations of the hybrid system type. Therefore, it is sufficient to consider the hybrid system type.

Following (Prähofer, 1991), a hybrid atomic system is defined by the tuple

$$A_{hybrid} = (X, Y, S, f, \lambda_c, \delta_{x\&s}, \delta_{int}, \lambda_d, ta) \quad (1)$$

where X, Y and S designate the set of inputs, outputs and states, respectively, that may be continuous or discrete. The continuous dynamics are mapped by the rate of change function f and the output function λ_c. For discrete events we distinguish internal events [3], external events [4] and state events [5]. External events and state events induce state transitions using the function $\delta_{x\&s}$. Internal events activate the discrete output function λ_d and also the state transition function δ_{int}. After each discrete state transition the internal events are recalculated by the time advance function ta.

Local structural changes of the continuous dynamics can be modeled by a separation of the dynamic description by logic state variables inside the rate of change function f, and the continuous output function λ_c (equation switching).

3.2 Coupled Invariable Structure Systems

According to the classical approaches in (Zeigler, 1976, Zeigler, 1984) coupled systems describe an aggregation of atomic and/or coupled systems, and so they permit the construction of hierarchical structures. The defined system structure is static. The dynamic behaviour of a coupled system results from the coupling relations between the state spaces of the subsystems. The coupling relations may be defined by arbitrary state-less functions. For the following considerations we restrict ourselves to equivalence relations as couplings, such that a coupled system consists of a set of subsystems and couplings.

Following (Prähofer, 1991), a coupled system is defined by the tuple [6]

$$N = (X_N, Y_N, D, \{M_d \mid d \in D\}, EIC, EOC, IC, select) \quad (2)$$

[3] time driven events

[4] events at discrete input variables

[5] threshold events of continuous quantities

[6] The notation and index N came from "network for coupled system".

and X_N, Y_N designate the set of input resp. output quantities. Furthermore, D stands for the name set of the dynamic subsystems, and M_d represents a dynamic subsystem. EIC, EOC and IC are the sets of continuous and event-oriented coupling relations, separated into external inputs, see Fig. 1: $N2 \rightarrow A2$, external outputs, see Fig. 1: $A3 \rightarrow N2, N2 \rightarrow N1$, and internal couplings, see Fig. 1: $N2 \rightarrow A1, A1 \rightarrow N2, A2 \rightarrow A3$. Finally, $select$ is a selection function to give priority to subsystems for simultaneous internal events. It is sufficient to sequentialize the internal events, because only they are able to generate output events.

3.3 Coupled Variable Structure Systems

Mapping structural changes on aggregation level is not supported by classical modular-hierarchical modeling approaches. Therefore, the model cannot be structured exclusively from the viewpoint of technical units (e.g. plant and controller). All structure changing operations have to be modeled in atomic systems. This leads to complex atomic systems and their restricted reuse.

Knowing this problem, different approaches have already been developed for purely event-oriented systems for modeling structural variability on the aggregation level. The subject of this contribution is to develop a corresponding modeling approach for hybrid systems on basis of the work by (Barros, 1996, Pawletta et al., 1994, Pawletta et al., 1996, Prähofer, 1996).

The fundamental idea consists of the direct extension of the definition of coupled systems by functions for the mapping of structural dynamics. The variety of structural changes in modular-hierarchical systems that have to be taken into account, was already discussed in Sect. 2. According to the above definition of coupled invariable structure systems, their structure consist of a set of subsystems and coupling relations. Structural changes mean modification of these sets. Obviously, the subsystems and the coupling relations should be interpreted as *structure state*. Hence, a structure variable coupled system can have different structure states $s_0, s_1, \cdots, s_n \in S_N$.

In order to store special structure dynamic informations, such as the number of realized structure changes, it is suitable to introduce an additional set of structural state variables H_N. Furthermore, also the selection function *select* can depend on the structure state. Consequently, the set of sequential structure states S_N of a structure variable coupled system has the form

$$S_N = H_N \times \{M_d \mid d \in D\} \times EOC \times EIC \times IC \times select\,. \tag{3}$$

Structure changes can be induced by external, internal or state events. Structure affecting state events can be caused by

- Output events of subsystems,
- Thresholds of
  - continuous outputs of subsystems,
  - continuous inputs of coupled systems or
  - structure related states of the set H_N.

Structural changes in coupled systems can be specified in analogy to event-oriented dynamics of atomic systems. Events that are influenced by the internal structure (time events) are planned by a time advance function ta, and its proposed structure changes are specified with a structure state transition function δ_{int}. For the generation of structure related output events of the coupled model caused by internal events, a discrete output function λ_d is introduced. Hence, for instance, it is possible to send subsystems to other coupled systems.

In order to avoid ambiguity at instantaneous internal events of subsystems and structure related internal events of the coupled system itself, the coupled system must occasionally be included into the selection function $select$ to sequentialize internal events.

In analogy to the dynamics of atomic systems, external events and state events are able to induce structure state changes, but they do not immediately generate output events. Their structure state transitions are defined by the transition function $\delta_{x\&s}$.

In contrast to the dynamics of atomic systems, it is not sufficient to sequentialize the simultaneous internal events only, if one would avoid ambiguous event handling. Figure 2 shows for instance a structure variable system. At the initial state every

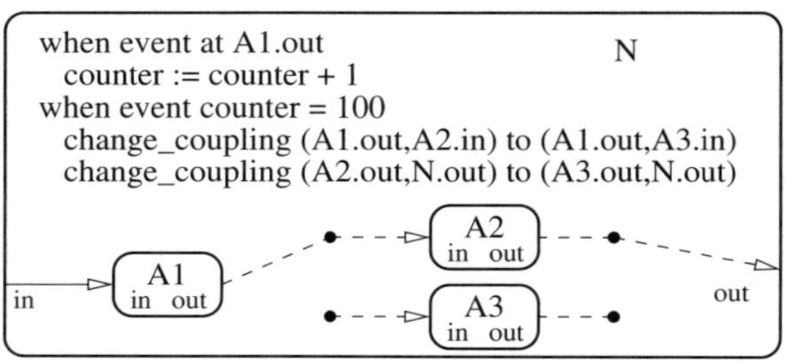

Fig. 2. Example of a coupled variable structure system

output event of the subsystem $A1$ presents an external event for the subsystem $A2$, and a structure related state event of the coupled system N. If the state variable $counter \in H_N$ reaches the value 100, the coupling between systems $A1$ and $A2$ should be substituted by a coupling between systems $A1$ and $A3$. Analogously, the external output coupling of N has to be modified. In this case it is not clear, whether a structure change, which induces an external event for the subsystem $A3$, has to be performed first, or whether the external event by the system $A2$ needs to be processed first.

To avoid the definition of a further $select$ function, it is appropriate to make changes in the set of subsystems or their coupling relations only with the internal state transition function δ_{int}. Analogously, modifcations to structure related variables in set H_N are to be made by the state transition function $\delta_{x\&s}$. The changes in structure related state variables can induce at the same time internal events, that imply a modification of the set of subsystems or the coupling relations. The sequentializing of instantaneous internal events is controlled by the above introduced *select* function.

The following dynamics specification of the coupled system N gives a unique event handling for the example from Fig. 2:

$EIC = \{(N.in, A1.in)\}$
$IC = \{(A1.out, A2.in)\}$
$EOC = \{(A2.out, N.out)\}$
$H_N = \{(counter, \sigma) \mid counter \in \mathbb{N}, \sigma \in \mathbb{R}^+\}$
$\delta_{x\&s}$: `when event at` $A_1.out$
    $counter := counter + 1$
    `if` $counter = 100$ `then` $\sigma := 0$
δ_{int}: $IC := \{(A1.out, A3.in)\}$
    $EOC := \{(A3.out, N.out)\}$
    $\sigma := \infty$
ta: `return` σ
$select$: `priority_order`$(N, A1, A2, A3)$

The considered output event of $A1$ is sent as external input event to $A2$, and induces a state event in N. Hence, both systems execute their $\delta_{x\&s}$ functions. The state transition $\sigma := 0$ in N induces by $ta := \sigma$ at the same time an internal event in N. In this case, the next event to be processed is chosen according to the specified priority in *select*, and the execution of δ_{int} changes the structure.

On basis of the definition for coupled systems without structure variability, and the above introduced extensions, we find the following formal definition for coupled variable structure systems:[7]

Definition 1. A variable structure system N_{dyn} is a 9-tuple

$$N_{dyn} = (X_N, Y_N, D, \{d_N\}, S_N, \delta_{x\&s}, \delta_{int}, \lambda_d, ta)$$

where

X_N – set of input signals with $X_N = X_c \times X_d$
Y_N – set of output signals with $Y_N = Y_c \times Y_d$
D – set of names for subsystems
$\{d_N\}$ – name of the coupled system
S_N – set of sequential structure states with
  $S_N = H_N \times \{M_d \mid d \in D\} \times EOC \times EIC \times IC \times select$
  and $select : 2^{D \cup \{d_N\}} \rightarrow D \cup \{d_N\}$
$\delta_{x\&s}$ – state transition function for external and state events with
  $\delta_{x\&s} : H_N \times X_N \underset{d \in D}{\times} Y_d \rightarrow H_N$
δ_{int} – internal state transition function with $\delta_{int} : S_N \rightarrow S_N$
λ_d – discrete output function with $\lambda_d : S_N \underset{d \in D}{\times} Y_d \rightarrow Y_N$
ta – time advance function with $ta : S_N \rightarrow \mathbb{R}_{0,\infty}$

[7] The subscript c remembers to continuous, and d to discrete.

and, moreover

Q_N – set of all structure states

$Q_N = \{(s_N, e) \mid s_N \in S_N, 0 \leq e \leq ta(s_N)\}$

with e being the elapsed time since last structure transition.

The dynamic behaviour of a coupled variable structure system can be verbally summarized as follows:

- The time advance function ta determines after each discrete state change the time advance of the next internal event, that follows from the dynamics of the subsystems and the structure dynamics of the coupled system. If up to this time instant no external event or state event has occured, the corresponding subsystem will be activated at the planned time instant, or by the output function λ_d, and after that the internal structure state transition function δ_{int} of the coupled system is executed. Simultaneous internal events are sequentialized by the $select$ function.
- If an external event occurs, it will be sent to the affected subsystems. If it influences the structure dynamics of the coupled system, its state transition function $\delta_{x\&s}$ is executed. Finally, the time advance for the next internal event is recalculated by the time advance function ta.
- If a state event occurs, the state transition function $\delta_{x\&s}$ of the coupled system is executed, after that, the time advance of the next internal event is determined by the time advance function ta.

4 A Modeling Example

A number of different well known reference examples, here partially extended by structure-variable properties, were modelled, e.g. chemical multi reactor system, ingots production line, engine control (Schildmann, 2000), to verify the modelling approach. Moreover the example of an automated manufacturing cell (advertised within the framework of the KONDISK programme) was dealt with. The system structure as well as the investigation aims of this example are relatively similar to applications in laboratory automation currently using this modelling approach To be able to distinctively compare the modelling approach with other work published within the KONDISK programme, a partial task of the reference example according to the description in (Nixdorf and Lunze, 2000b) should be considered.

4.1 Application Problem

Figure 3 shows schematically the system structure of the considered part of the manufacturing cell. The manufacturing task comprises the ejection of workpieces from the buffer and their transport using a robot to a heating facility, where they are

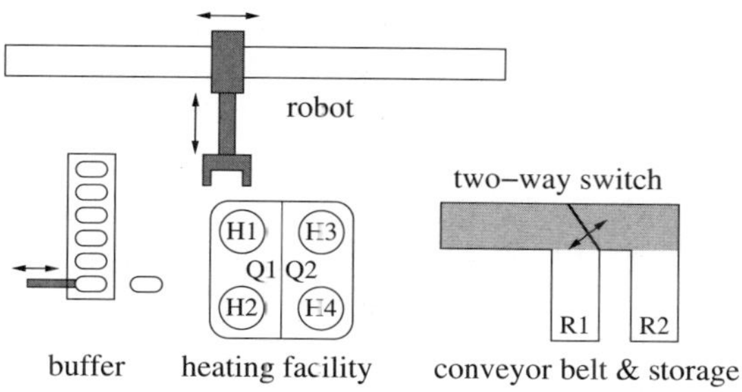

Fig. 3. System structure of a manufacturing cell

warmed up on one of the heating plates. The robot then transports the workpieces to a conveyor belt. Using a two-way switch they are sorted according to their temperature into repository $R1$ resp. $R2$.

Concerning the permanent system elements it must be noted, that the transport facility (robot) can only carry one workpiece at the time, and that the heating plates have different heating capacities ($Q1$ and $Q2$).

The workpieces, being temporary elements in the system, undergo a permanent temperature exchange with the environment. Extending the original formulation of the task, workpieces are to have different target temperatures as well as different thermodynamic behaviour [8].

The aim of the system investigation is to design an optimal process control strategy, which guarantees maximal throughput of workpieces by avoiding them being under- and overheated. Regarding the model generation it is required, that the model design is modular and component oriented. This enables successive extensions of the model to deal with more complex manufacturing tasks by incorporating further system elements.

4.2 Modeling Using Coupled Invariable Structure Systems

The conventional modular-hierarchical modeling formalisms only support a description of coupled systems with a static structure. The information exchange between different systems is limited to static data objects. Hence it is not possible in the example to model the workpieces as dynamic objects. They can only be modeled as simple data objects without any dynamics. The workpieces specific differential equations, which describe their temperature dynamics, have to be specified separated from the workpiece data objects. Each dynamic system of the manufacturing cell, which comes in contact with workpieces has to map workpiece specific differential equations.

[8] for example compound materials.

The basic modeling principle is represented as a block diagram in Fig. 4. To allow a clear view it takes into account only two system components and illustrates only the separation of the state space in the continuous rate of change function f.

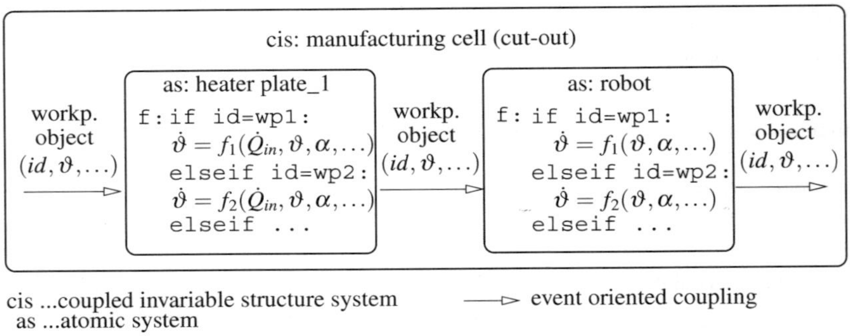

Fig. 4. Modeling approach with a coupled invariable structure system

This kind of modeling leads to a relatively complicated model. The real system components are only virtually mapped onto model components. Therefore model extensions can be expensive. The introduction of new workpiece types for example requires an update of all atomic systems to specify the new differential equations.

4.3 Modeling with Coupled Variable Structure Systems

The introduced modeling formalism of coupled variable structure systems supports the generation and termination of dynamic systems and their exchange between coupled systems. In the case of the example application, it is possible to model each workpiece as independent atomic system with its own dynamic behaviour.

The specification of a specific workpiece type as an atomic system is depicted in Fig. 5. [9] The workpiece system interface is built by a continuous input variable in_Q for the heat inflow, a continuous output variable out_ϑ for the current workpiece temperature and a discrete output variable out_ϑ_{sp} for the target temperature. Based on this specification it is possible to generate an arbitrary number of dynamical workpiece systems of this specific type.

Figure 6 shows the basic modeling idea by means of a small manufacturing cell cut-out. In this case, the permanent system components buffer, robot and heater facility are modeled as coupled variable structure systems. Atomic workpiece systems are generated in the buffer and coupled with a heat inflow $\dot{Q}_{in} = 0$. Due to events they are sent to other coupled systems, where they are temporarily integrated according to the defined coupling conditions.

The full model structure of the manufacturing cell is depicted as block diagram in Fig. 7. A more detailed representation of the heater facility is shown in Fig. 8. Both

[9] The set θ specifies the system parameters and the $init$ defines initialization operations.

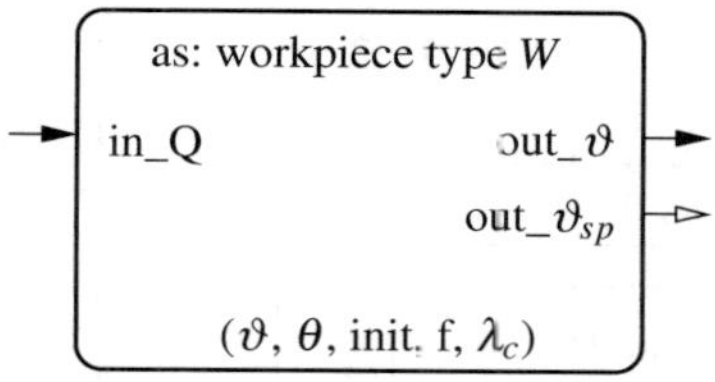

$$W = (X_c, Y_c, Y_d, S_c, \theta, init, f, \lambda_c)$$
$$X_c = \{in\_Q\},\ Y_c = \{out\_\vartheta\}$$
$$Y_d = \{out\_\vartheta_{sp}\},\ S_c = \{\vartheta\}$$
$$\theta = (\alpha, A, m, c_p, \vartheta_{sp}, \vartheta_{env})$$
$$init: \quad \vartheta := \vartheta_{env}$$
$$f: \quad \dot{\vartheta} := [in\_Q - \alpha * A * (\vartheta - \vartheta_{env})] \ / \ [m * c_p]$$
$$\lambda_c: \quad out\_\vartheta := \vartheta,\ out\_\vartheta_{sp} := \vartheta_{sp}$$

Fig. 5. Atomic system of a specific workpiece type

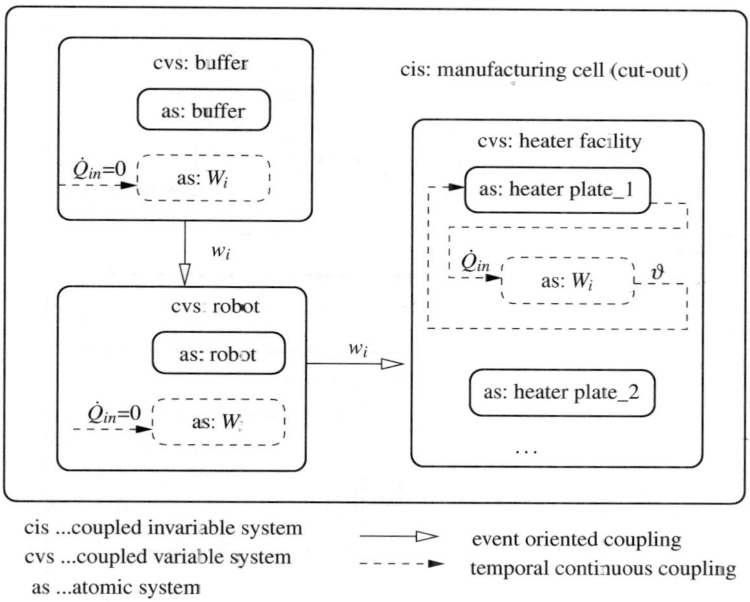

Fig. 6. Modeling approach with coupled variable structure systems

block diagrams define event numbers at the couplings. A description of the events is given in Table 1.

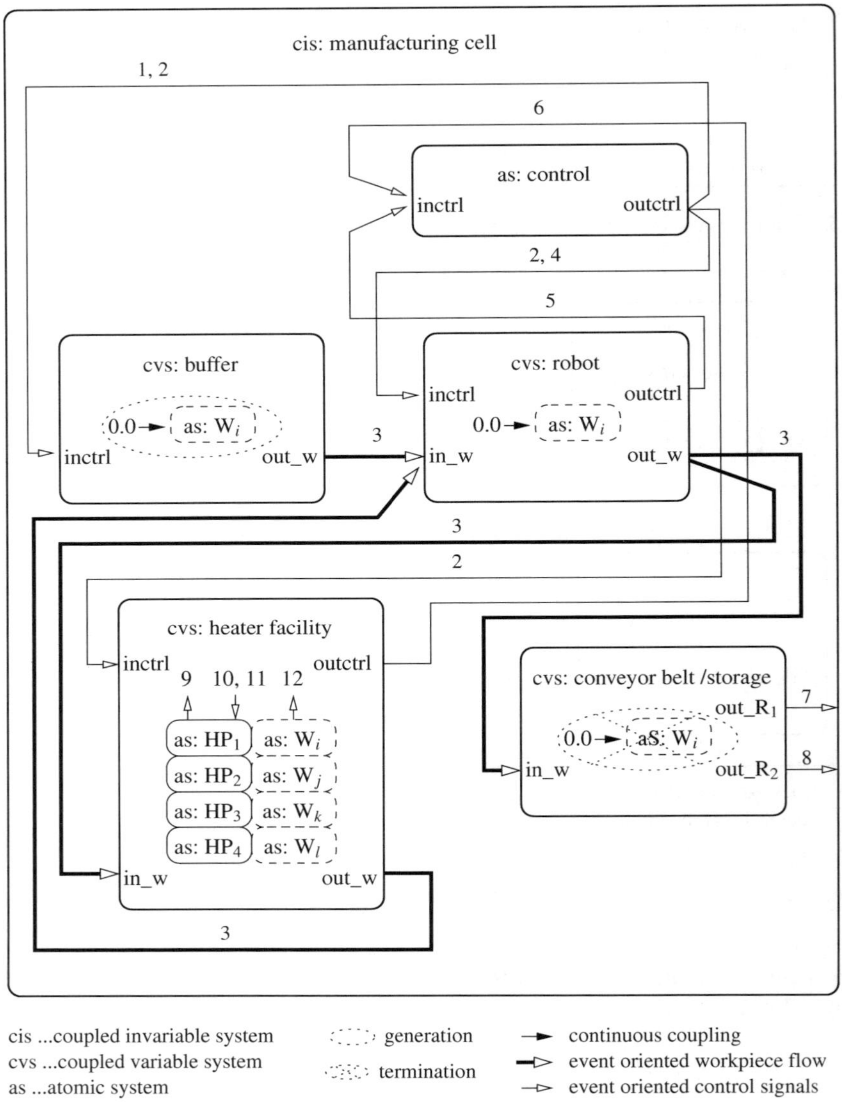

Fig. 7. Model structure of the manufacturing cell

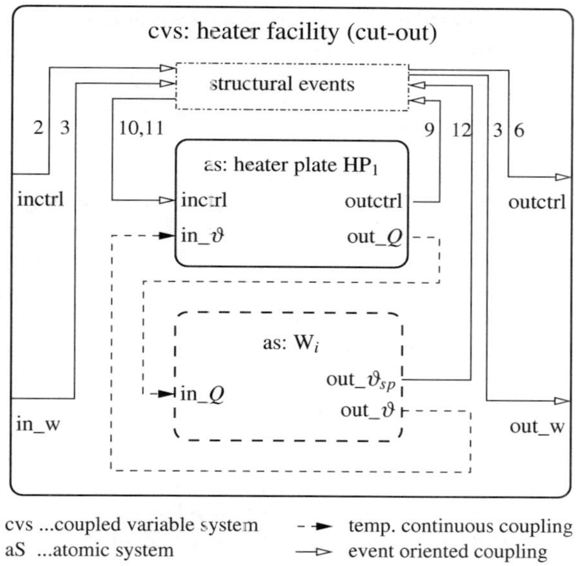

Fig. 8. Model structure of the heater facility (cut-out)

Table 1. Events of the manufacturing cell system

event no.	event type	description
1	w_buffer	generate a new workpiece
2	w_sent	signal to sent a workpiece
3	W_i	Departure/Arrival of a workpiece
4	targ_position	next target position of robot
5	targ_position_reached	robot has reached target position
6	w_finished	workpiece has reached ϑ_{sp}
7	notify_R1	notify number of workpieces in R1
8	notify_R2	notify number of workpieces in R2
	internal events of the heater facility	
9	w_finished	workpiece has reached ϑ_{sp}
10	(on, ϑ_{sp})	heater plate on and notify ϑ_{sp} to heater plate
11	off	heater plate off
12	ϑ_{sp}	notify workpiece specific ϑ_{sp} to heater facility

Based on the model structure represented in Fig. 7, the coupled variable structure system buffer can be specified in the following manner:

Dynamic specification of coupled system *buffer (B)*:

$B = (X_N, Y_N, H_N, D, IC, init, \delta_{x\&s}, \delta_{int}, \lambda_d, ta)$
$X_N = \{inctrl \mid inctrl \in \{\text{w\_buffer, w\_sent}\}\},\ Y_N = \{out\_w \mid out\_w \in Workpieces\}$
$H_N = \{(phase, \sigma, tmp_w) \mid phase \in \{\text{w\_in\_buffer,buffer\_empty}\} \wedge \sigma \in \mathbb{R}^+ \wedge$
$\quad tmp_w \in Workpieces\}$
$D = \{W_i \mid i := 1, \ldots, n \wedge W_i \in Workpieces\}$
$init : D := \emptyset,\ IC := \emptyset,\ phase := \text{buffer\_empty},\ tmp_w := \emptyset,\ \sigma := \infty$
$\delta_{x\&s}$ `when external event` $B.inctrl = \text{w\_buffer}$ `then`
$\quad phase := \text{w\_in\_buffer},\ \sigma := 0$
`when external event` $B.inctrl = \text{w\_sent}$ `then`
$\quad phase := \text{buffer\_empty},\ tmp_w := W_i,\ \sigma := 0$
$\lambda_d :$ `if` $phase = \text{buffer\_empty}$ `then` $B.out\_w := tmp_w$
δ_{int} : `if` $phase = \text{w\_in\_buffer}$ `then`
$\quad D := D \cup \{W_i\}$ #generate atomic system W_i of type $workpiece$
$\quad IC := IC \cup \{(B.0, W_i.in\_Q)\}$ #couple heat flow $\dot{Q} = 0$ to in_Q of W_i
$\quad \sigma := \infty$
`if` $phase = \text{buffer\_empty}$ `then` #delete workpiece system and couplings
$\quad IC := IC \backslash \{(B.0, tmp_w.in\_Q)\},\ D := D \backslash \{tmp_w\},\ tmp_w := \emptyset,\ \sigma := \infty$
$ta :$ return σ

The dynamic system buffer contains a discrete input port $inctrl$, a discrete output port out_w and the structural state variables $phase, \sigma, tmp_w$. According to the specification in function $init$ no workpieces are inside the buffer at start time, that means the component and coupling sets are empty.

When an external event w_buffer is received at the input port, the state transition function $\delta_{x\&s}$ will be called, which modifies the state variables $phase$ and σ. Due to the state $\sigma = 0$ the time advance function ta triggers an internal event at the same time. Thus the discrete output function λ_d and the structure state transition function δ_{int} are called, but because of the state $phase \neq buffer\_empty$ no output event is generated in λ_d. In function δ_{int} the state of $phase$ is w_in_buffer, therefore a new atomic workpiece system of type W_i is generated and coupled with a heat inflow of $\dot{Q} = 0$. Furthermore the state variable σ is set to ∞. That means, the time advance of the next internal event is set by function ta to $t = \infty$.

When an external event w_sent is received at the input port, the state transition function $\delta_{x\&s}$ will be called, which modifies the state variables $phase$ and σ. Furthermore, it assigns to state variable tmp_w a reference of an atomic workpiece system W. Due to $\sigma = 0$ the time advance function ta triggers simultaneously an internal event. Because of the state $phase = buffer\_empty$, in output function λ_d an output event W_i ($departure\ of\ a\ workpiece\ system$) is generated. This is realized by means of the reference stored in state variable tmp_w. After that the component and coupling sets as well as the state variables tmp_w and σ are updated by the structure transition function δ_{int}. Finally, the next internal event is advanced by the function ta to $t = \infty$, because of $\sigma = \infty$.

According to the application problem, the modeling approach of coupled variable structure systems allows a real component oriented mapping of natural system components to model components. Therefore, model maintanance and extension will be much easier in comparison to the classical modeling approaches. One example is the introduction of a new workpiece type, which can be included in the model by a simple specification of a new atomic system.

A full specification of the manufacturing cell example can be found in (Pawletta and Lampe, 2001).

5 A Simulation Approach

Before running on a computer, a model specification has to be transformed into an executable simulation program. For that, it is connected with a simulation runtime system, that organizes the necessary numeric methods. The transformation of the model specification and the linking of the simulation runtime system is called computing model generation. Therefore, the executable simulation program also is named computing model. Its construction can be monolithic or modular.

5.1 Monolithic Versus Modular Simulation Programs

In monolithic computing models the model specification is transferred into numerical and runtime optimal data structures. As simulation runtime systems are mostly used differential equations solver from suitable numerical libraries. A problem of this approach is the handling of complex structure changes, because different structure states of the model can only be realized in form of equation switches. But this means, that all possible structure states of a model have to be known at the time of the computing model generation.

For modular computing models the model specification corresponding to the model structure is transferred into modular or modular-hierarchical program structures. Then, the computing model normally consists of interacting objects, reflecting the components and coupling relations of the model specification. This approach suffices all requirements to execute arbitrary structure changes during simulation runtime. However, problems arise for simulation of continuous model parts, if they are distributed in different program objects. This leads either to an enormous communication and coordination overhead between the program objects, or to relevant numerical errors. Another drawback is that tools of numeric libraries are not yet directly usable, and a numerically optimized preparation of models (e.g. sorting of equations) is more difficult.

5.2 A Combined Modular and Monolithic Simulation Approach

The analysis of both general computing model approaches shows that for simulation of hybrid variable structure systems a combination of both approaches makes

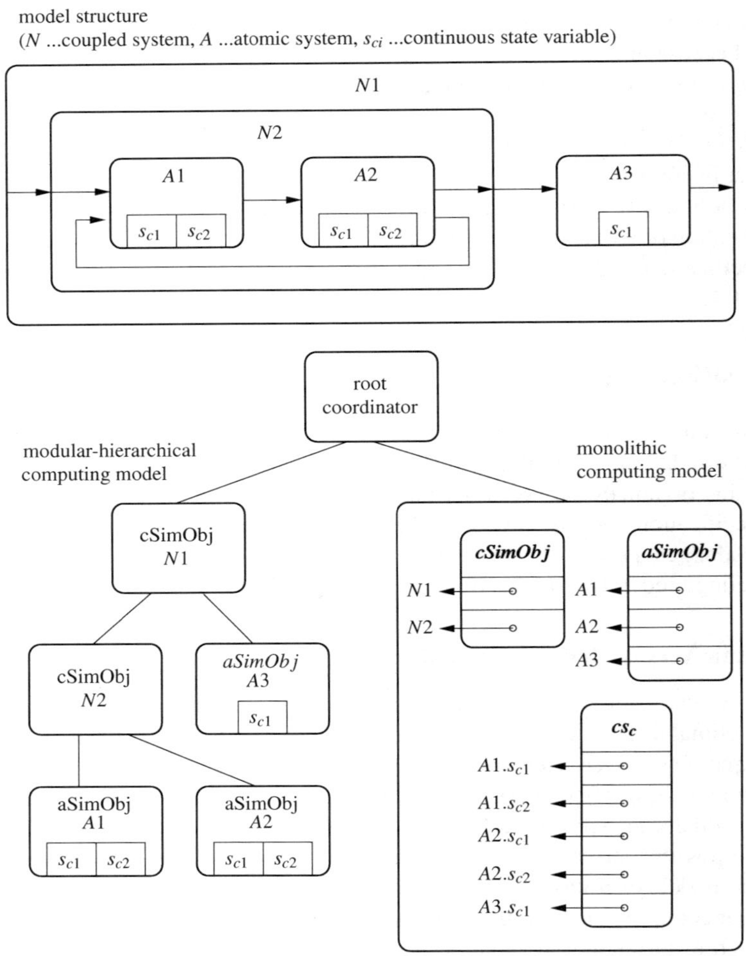

Fig. 9. Principal approach of a combined modular-monolithic computing model

sense. A practicable variant is shown in the schematic Fig. 9. The computing model consists of a modular-hierarchical part and a monolithic part. The structure of the modular-hierarchical computing model corresponds to the composition tree of the model structure – each atomic resp. coupled system component is mapped into the computing model as a corresponding atomic resp. coupled simulation object. A simulation object contains a model specification and the necessary simulation methods. The communication between the various simulation objects is done by exchange of messages along the edges of the object tree.

The monolithic computing model is built by references on the continuous model parts in the modular-hierarchical computing model. In Fig. 9 only one monolithic computing model is presented. In principal it is possible to generate any number of monolithic computing models. This is of advantage if there are no interrelations between the continuous state variables. In this case, the monolithic computing models, for instance, can operate with different integration methods. For the sake of clearness we restrict ourselves in the following considerations to one monolithic computing model.

Then, the whole simulation is controlled by a so-called root-coordinator. Event-oriented simulation steps are initiated by the root-coordinator by sending of messages to the root object of the modular-hierarchical computing model. Between the event-oriented simulation steps a continuous simulation phase takes place. This is also activated by the root-coordinator, and is made by means of the monolithic computing model. If structure changes happen in the modular-hierarchical computing model, the root-coordinator then sends a message to the monolithic computing model that arranges its reconfiguration as described in the following.

The computing model generation happens in two phases, see Fig. 10. During

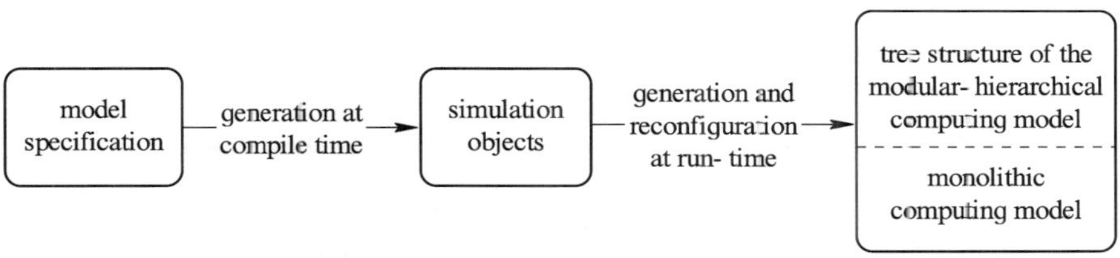

Fig. 10. Phases of computing model generation

compilation time the atomic and coupled simulation objects are generated from the model specification. The modular-hierarchical and monolithic computing model is constructed during simulation runtime. Algorithms for the generation of a modular-hierarchical computing model are not explained here in detail. We refer to the literature, e.g. (Zeigler, 1990). The initial monolithic computing model is generated after the construction of the modular-hierarchical computing model. It is realized in form of a program object, which provides a special configuration method (z-method). This method also controls the reconfiguration of the monolithic computing model after structure changes.

z-method(t) of the monolithic computing model

1. Traverse the modular-hierarchical computing model and
   (1.1) configure the global state variable vector $\boldsymbol{cS_c}$
   (1.2) configure the reference vector $\boldsymbol{aSimObj}$ and $\boldsymbol{cSimObj}$
   (1.3) generate direct references between the continuous input and output variables that have to be coupled
2. Sort the reference vector $\boldsymbol{aSimObj}$ with respect to references to atomic simulation objects of type $Moore-$ resp. $Mealy - Automata$ and sort the references to objects of type $Mealy - Automata$ according to their dependent relations

The configuration and reconfiguration of the monolithic computing model are based on multiple traversions of the modular-hierarchical computing model. The atomic and coupled simulation objects implement a z-method that is called object related by a top-down traversing. A presentation of the z-method of atomic and coupled simulation objects can be found in (Pawletta and Lampe, 2001).

As a result of the traversion the three vectors $\boldsymbol{cS_c}$, $\boldsymbol{aSimObj}$ and $\boldsymbol{cSimObj}$ are configured in the monolithic computing model, see Fig. 9. The vector $\boldsymbol{cS_c}$ stores references to all continuous state variables of the atomic simulation objects, and the vectors $\boldsymbol{aSimObj}$ as well as $\boldsymbol{cSimObj}$ store references to the atomic resp. coupled simulation objects. A second result of the traversing is building direct references between the continuous input and output variables, where the input variables obtain references to the output variables.

After traversing the algebraic relations are sorted. For this, the references in the vector $\boldsymbol{aSimObj}$ are arranged with respect to atomic simulation objects of type $Moore - Automata$ $(\lambda_c : S_c \rightarrow Y_c)$ resp. of type $Mealy - Automata$ $(\lambda_c : X_c \times S_c \rightarrow Y_c)$. In a second sorting step the references to objects of type $Mealy - Automata$ are arranged with respect to their dependent relations.

5.3 Simulation Messages of the Combined Simulation Approach

After the general function and essential aspects of computing model generation of the combined computing model approach were discussed in the previous subsection, we will now consider the messages during simulation runtime, see Fig. 11.

The i-, x- and *-messages are constructed in analogy to the event-oriented simulator concept (Zeigler, 1976). The s-message handles state events of continuous quantities. The c-message activates a continuous simulation phase in the monolithic computing model up to occurrence of the next discrete event. The z-message serves to the reconfiguration of the monolithic computing model after structure changes in the modular-hierarchical computing model.

The tasks of the several types of messages can be listed as follows: [10]

start-msg(t_{end}): The start-message activates a simulation run up to the time instant specified in the parameter t_{end}.

i-msg(t): After receiving an i-message, the simulation object is initialized to the time instant t. Moreover, coupled simulation objects transfer i-messages to their subordinated simulation objects.

x-msg(t, x): A simulation object receives a x-message to signal an external event x. For atomic simulation objects a x-message realizes an external state transition by calling the object specific function $\delta_{x\&s}$. Coupled simulation objects transfer an x-message in dependence on the actual structure state to the corresponding

[10] The last operation of each i-,*-,x- and s-message is the computation of the object specific next internal event time. Atomic simulation objects do that by calling the object specific time advance function ta. Coupled simulation objects analyse the next event time of their subordered objects and the value of their own time advance function ta.

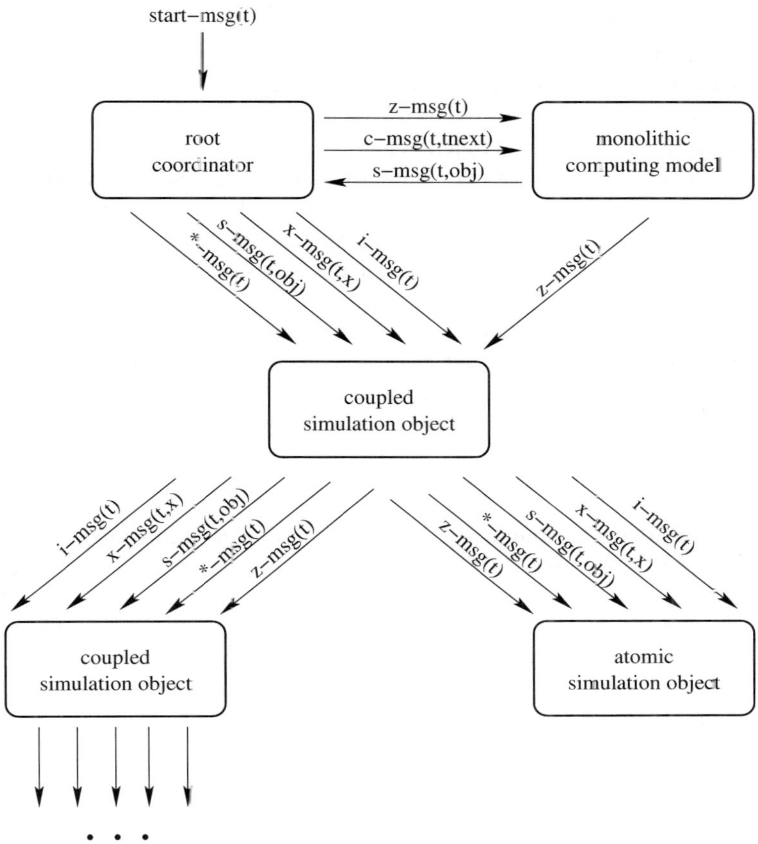

Fig. 11. Messages during simulation runtime

input ports of their subobjects, and/or execute their object specific state transition function $\delta_{x\&s}$.

***-msg(t):** A *-message induces the processing of an internal event (time event). For atomic simulation objects a *-message invokes the object specific functions λ_d and δ_{int}. Then, the generation of output events and/or a internal state transition follows. Coupled simulation objects pass a *-message to that subordinated simulation object which is planned to have the next internal event, or call their object specific structure state transition function δ_{int}. Simultaneous internal events are selected by the object specific $select$ function.

s-msg(t, obj): The s-message serves the processing of state events. These occur if during the continuous simulation phase in the monolithic computing model continuous variables reach thresholds. If a state event is detected and localized, a reference to the corresponding simulation object is stored in the parameter obj, and a s-message is sent to the root-coordinator, which actualizes the global simulation time and sends the s-message to the root object of the modular-hierarchical computing model. Coupled simulation objects transfer a s-message corresponding to the reference in parameter obj. After receiving the s-message the simulation object that is referred to by the parameter obj, changes its state by evaluating the object specific function $\delta_{x\&s}$.

c-msg(t**,** t_{next}**):** The c-message starts a continuous simulation phase. The parameter t denotes the global simulation time, and the parameter t_{next} is the time instant for the next scheduled internal event. A state event during the interval $[t, t_{next}]$ interrupts the continuous simulation phase, and a s-message is sent to the root-coordinator.

z-msg(t**):** If structure changes in the modular-hierarchical computing model are occured during a discrete simulation phase, the root-coordinator sends a z-message to the monolithic computing model, before the continuous simulation phase is restarted. This reconfigures the monolithic computing model. For the necessary structure information, the modular-hierarchical computing model is several times recursively traversed by z-messages.

With respect to a detailed algorithmic presentation of particular simulation methods we refer to (Pawletta and Lampe, 2001).

In summary, we stress that for the combined computing model approach all structure information of the model specification are available during runtime, and due to the monolithic computing model no additional numerical errors occur. Moreover, for the numerical solution of the continuous parts we can immediately use verified tools from standard libraries.

5.4 Simulator Implementation

The previously shown modeling and simulation approach was implemented as a simulator prototype using the programming system MATLAB. The prototype is designed as an object oriented framework. Basic functionality of the dynamic system types (hybrid atomic system, variable coupled system) and of the simulation specific objects (root coordinator, monolithic computing model) are implemented as specific MATLAB classes. User defined model objects have to be implemented as derived classes of their predefined system types. Hence only model specific aspects have to be implemented in the derived classes. Of course it is possible to define general model classes and to generate specific model objects by parameterization. Furthermore some additional model generation tools were developed, by which the MATLAB classes can be generated from a specific model description. The description language is based on the set theoretic notation as used in Sects. 3 and 4 .

According to the interactive nature of MATLAB, simulation experiments can be described and performed interactively from the command line or from a specific graphical user interface. For complex experiments, that exceed simple simulation runs, the simulator can be linked into complex program structures in the same way as an ordinary MATLAB function. This makes an automatic handling of complex structure investigations possible, as it was described in (Pawletta et al., 1997).

In order to verify the prototype, and for evaluation of the whole concept, a variety of example models were implemented. A number of these examples are summerized in (Schildmann, 2000).

6 Conclusions

The presented modeling approach extends the application of structure variable modeling methods, that are known from modeling of event-oriented systems, to hybrid systems. The main advantage of the approach can be seen in a substantially improved model structuring for a class of complex models, i.e., those with complex structural changes. Especially, it simplifies a successive extension of such models.

The developed simulation methods are characterized by the fact that the model structure can be arbitrarily modified during runtime, and in spite of the necessary modular construction of the executable computing model, no additional numerical errors arise for processing of continuous model parts in comparison to pure monolithic approaches.

With respect to the performance of the simulation approach we appraise that the simulation method itself, i.e. for models without structure variability, generates no remarkable runtime overhead. The structure variable modeling approach, however, leads depending on the number of realized structure changes to longer simulation runtimes, because every structure change requires a reorganization of the computing model. In return, it opens new application fields to simulation, because the variation of complete system structures can be investigated relatively easily. In connection with an optimizing algorithm, e.g. an optimal system structure can be determined analogously to the simulation aided parameter optimizing.

The new modeling and simulation possibilities are interesting in various fields, e.g. in manufacturing and laboratory automation, such as process engineering.

Object-Oriented Development of Simulation Models for Complex Hybrid Systems

André Nordwig

Technische Universität Berlin,
Institut für Softwaretechnik und Theoretische Informatik,
Sekr. FR 5-6, Franklinstraße 28/29, D-10587 Berlin, Germany
nordwig@cs.tu-berlin.de

Abstract. This paper presents an approach for the development of simulation models for hybrid systems adapting conventional software engineering techniques and methodologies. The object-oriented paradigm for simulation models is extended by well-studied concepts of object technology, namely restricted genericity and structural dynamics. Finally, a prototypical tool, which implements the development process is presented.

1 Introduction

Hybrid systems are networks of components with discrete and continuous behavior. Typical examples of hybrid systems are physical processes, along with their discrete controllers. Besides these explicit discrete components, models of the processes themselves often exhibit hybrid behavior. The reason for this phenomenon lies in the physical model abstraction often done during modeling: real continuous behaviors are partitioned into a set of stepwise continuous behaviors. Thus, the resulting trajectories lose their smoothness at the interval boundaries.

As such systems are often involved in safety-critical applications, their analysis plays an important part in current research. It is generally accepted that, in addition to *verification* and *testing*, *simulation* plays an important role in analysis of this kind of systems. Our HYFOS project focuses on the development of object-oriented simulation models for hybrid systems[1].

At this point, a short overview about the special challenges in the development of simulation models for hybrid systems is given. They lead to requirements for the description techniques as well as for the development processes.

- Detailed simulation models tend to be highly complex, multi-domain and strongly coupled. Therefore, appropriate structuring techniques must be provided by modeling and simulation environments.
- Hybrid dynamics requires support for switched differential algebraic equations (DAEs) and an appropriate event mechanism.

[1] This project was part of the main focus program "Analysis and Synthesis of Continuous-discrete Technical Systems" (KONDISK) and was sponsored by the German Research Foundation (DFG).

S. Engell, G. Frehse, E. Schnieder (Eds.): Modelling, Analysis and Design of Hybrid Systems, LNCIS 279, pp. 131–149, 2002.

- At the modeling level, events can occur at the same time instant. Therefore, chain-reactions in discrete phases are possible and must be handled meaningful.
- Due to the fact that most of the continuous phases are not analytically traceable, in general, the reachability of (discrete) states is not proof-able. Thus, simulation is necessary.
- Consistent initial values must be found after each discrete phase. Therefore, suitable numerical methods must be offered by the environment. Since not all of these initializations can be automatically done after mode switches, an infrastructure for user intervention must be present in the description techniques too.
- The precision of numerical solutions depends on the models and the overall requirements of the experiments. Therefore, solvers should be exchangeable.

Hence, the crucial questions in simulation modeling are then: How do we construct the *right model*? How do we construct it *correctly*? And what about the *languages* used for modeling?

One goal of our work is to answer these questions by developing a comprehensive methodology, which offers guidance through the hybrid system's engineering process. Further, we want to develop a set of well-defined complementary description languages that in their whole are both able to describe *all aspects* of the considered simulation models and *practically usable*. The object-oriented paradigm is a good base for the development since it guarantees smooth phase transitions between development phases. Further, it offers powerful structuring mechanisms and it is well suited for the integration of different views to a model. Therefore, the purpose of this article is twofold. On the one hand, some methodological aspects and a tool support will be discussed. On the other hand, object-oriented description languages used within the process will be extended by well-defined concepts.

This paper is organized as follows. Section 2 explains our approach to adapt conventional software engineering processes. A short description and two extensions of the object-oriented paradigm in the context of simulation of continuous and hybrid systems are given in Sect. 3. A prototypical tool is described in Sect. 4. Section 5 contains some concluding remarks.

2 Development Processes

The development of complex systems is a difficult task which requires a high degree of expert knowledge from the involved engineers. These experiences include knowledge about the domain as well as about the process itself. In this section we want to focus our attention to the second aspect. To motivate our approach we sketch the major challenges to methodologies for the development of software systems:

- Optimization of process cycles. This requires that model-errors must be detected as early as possible within the process. Therefore, suitable description techniques must be applied in each phase.

- Offering guidance to the developers. This makes it possible to transfer the above mentioned knowledge from senior developers to new engineers in a project. Again, description techniques must be applied that enable an easy communication of modeling artifacts.
- Support for distributed cooperative work. Thus, a team of developers is able to work in parallel but nevertheless cooperative on one model. Therefore, developers are assigned with dynamical roles and responsibilities within the process.
- Support for feasibility studies and cost estimation. This is a matter of resource planning.

Methodologies are not developed by invention. They are developed by abstractions from concrete completed projects. The process model discussed below is a model for several case studies we have done in the field of simulation of energy systems.

If we look at conventional software engineering processes, like the waterfall model (Royce, 1970), prototype based models (Millington and Stapleton, 1995) or the V-model (Bröhl and Dröschel, 1995), and compare them to the activities in simulation development, we can identify a lot of similarities between them:

- All of these processes are characterized by a *high complexity* of their development artifacts as well as their processes them-self.
- These complexities require a high measure of *communication* between the developers.
- Therefore, the descriptions of the artifacts must be *abstract* enough to be easily communicated respective to the development phase. Not until the late stages in the development process, they are implemented in a suitable implementation language.
- This rises questions about the inter phase breaks within the development process. The description languages used in adjacent phases should be smoothly integratable. The object-oriented paradigm is a good basis to reduce these breaks.
- The process artifacts should be highly reusable. Thus, libraries of these artifacts can be built to increase the efficiency of processes and the quality of products.
- The last fact leads to further requirements like traceability and the ability to document the process.

In the last years there was a lot of work done in the field of method engineering for software. In the area of hybrid system's simulation there exist comparatively less approaches considering development processes. Some methodological aspects we have considered in (Ernst et al., 2000). Another approach, a variant of the V-model, is given in (Merz and Litz, 2000). (Broenink et al., 1998) describe a macro-process for the development of controllers using simulation cycles for some analysis done in each phase. It this sense, simulation development can be conceived as a micro-process.

Due to the fact of the above-mentioned similarities an adaption of software-engineering processes for the area of simulation engineering seems obvious. If we apply an object-oriented software process to the simulation development we can identify five main activities: *requirements definition*, *design*, *specification*, *implementation* and *integration*. Hence, we can adapt the V-model of (Merz and Litz, 2000) by the introduction of new early phases, see Fig. 1.

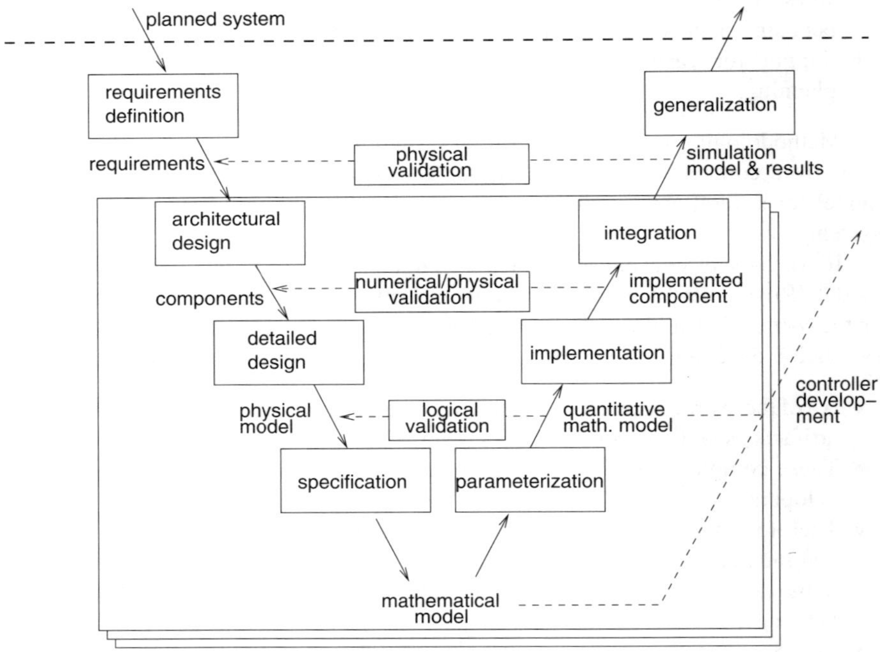

Fig. 1. V-model of the simulation development process

Starting with the collection of informal requirements to the simulation model, we can partition these results into sub-structures by hands of some rules and map them to components during design. Then, we can derive mathematical models for sub-components during specification. These components will be parameterized and implemented in a suitable implementation language. The following step by step integration of the sub-components is interleaved by validation cycles. The process cycle is enclosed by the evaluation of the results and their application to the system under consideration.

It is worth mentioning that the specifications of some of the discrete sub-components are good candidates for base models for the controller development in the above-mentioned sense of (Broenink et al., 1998). The following sections give detailed information about these phases. Special attention is given to the first phase and the validation cycles since they are significant for the model development.

2.1 Requirements Definition

At this stage, important requirements to the simulation model to be developed and to its environment are collected. Usually, these information will be described in a natural language at several levels of abstraction. One goal of this phase is to structure these requirements. Therefore, we suggest four classes of requirements:

- Process requirements
- Model requirements
- Environment requirements
- Requirements to the numerical methods

Process related requirements contain artifacts which influence the further process itself. If they are explicitly stated, a better documentation of the models becomes possible. For instance, it enables the communication of the model's intention, which makes specific design decisions clearer. In detail they comprise:

- availability of the system:
  If there exists a real system under study, then it possibly might be used to generate data for validation or identification. In the case of developing new systems, similar ones could be used for that purpose, if it is clearly stated, for which modes of operation the data are representative. The quality of these reference data should be critically estimated.
- availability of documentation:
  If experiments with real systems are not feasible, possibly documentations about their behaviors are available. These could be sources of data for validation too.
- kind of experiment:
  It should be determined, which kind of analysis should be done. Possible kinds are *ordinary simulation experiment*, *parameter identification*, *optimization* and *hardware-in-the-loop application*. The character of the analysis strongly influences the further development process as well as the reusability of library artifacts, e.g. optimization requires broader validity ranges of the artifacts.
- fixed parameter sets:
  Typically, the model to be developed is embedded within a fixed environment. This environment can be qualified and quantified by parameters. Examples are fundamental physical constants or geometric dimensions of a building during thermodynamic building simulation.

Model related requirements reflect abstract properties of the model. Here we can distinguish:

- quality of physical models:
  The overall goal of this class is to fix the degree of detail of the model and thus, of its sub-models. Therefore, the goal of the experiments must explicitly stated. Further, all relevant physical effects together with their validity ranges must be collected. If there is any doubt if an effect influences the results of the experiment, it must be considered. This can be done by an estimation at this stage.

Nevertheless, the results of this estimation should be included in the set of model artifacts for documentation purposes. Besides the effects, balances must be stated with the goal of experiments in mind. It is worth mentioning, that a major part of (often late detected) modeling failures is based on wrong assumptions about the quality of the models (Cellier et al., 1996). Hence, particular attention must be given to these requirements.
- domains:
  The previous point is complemented by the identification of all domains, e.g. mechanics or thermodynamics, which are relevant to the intended experiments and thus, for the model and the reuse of libraries. Based on this information, the boundaries of adjacent domains must be investigated.
- mode changes:
  Because of the physical model abstraction and the possible introduction of controllers done above, modes of operation can be established. The possible transitions between these modes must be investigated to match the balances given above. Here, conservation laws will be added.
- lifetime of modeling artifacts:
  If it is intended to develop library components for later reuse, special attention must be given to the interfaces, validity ranges and documentation of the components. In addition there is an obligation for extensive testing.
- observation:
  These requirements contain all information about the variables of the model that should be observed when experiments are running. Possible information are value, name, unit, accuracy of the values or time resolution.

Environmental requirements influence the choice of the target simulation environment and thus, possibly the model description techniques used within the further process.

- kinds of necessary equations:
  Most of the proposed simulation environments support at least three types of equations: algebraic, differential and discrete ones. Nevertheless, the interpretations differ in detail. Some languages, e.g. Smile, only allow explicit equations for all types. Other approaches support highly implicit definitions. Both variations have their benefits and drawbacks that are reflected within the semantics of the structuring concepts. Thus, they influence the refactoring of the model during detailed design.
- structural dynamics:
  If it is intended that the model structure changes at simulation runtime, an infrastructure for that special dynamics must be present. If dynamic object creation is not required, structural dynamics can be modeled as discussed in Sect. 3.2. During implementation such dynamics could be partly automatically transformed into equations. If object dynamics is not avoidable, then runtime support is necessary.
- interfaces for data processing:
  This class contains requirements for the input and output of data, of the model

state or parts of the model. Simulation environments vary in the support of that kind of interfacing.
- distribution:
For some time consuming experiments distributed computing is required to reduce the costs by orders of magnitude.
- simulator coupling:
Especially for heterogenous experiments, there exists a lot of sub-models represented by different simulation languages. In that cases often a coupling of a set of simulators is prefered instead of model transformation or redevelopment. Thus, special properties of the interfaces of these simulators are required.

Requirements for the numerical methods are:

- quality of mathematical solutions:
Efficiency and quality of solutions are conflicting requirements. A tradeoff must be found considering time consumption, numerical stability and accuracy of solutions. If a HIL-simulation is intended, realtime requirements must be satisfied.
- time domains:
The time constants of the sub-components influences the choice of the numerical methods and solver strategies at later stages. For example, if the time constants of several sub-components differ by orders of magnitude (stiffness), special solver strategies must be used during simulation.

2.2 Architectural Design

At this stage, the model to be developed is broken down into sub-models by a top-down decomposition. This partitioning is dominated *by structural properties* of the real system. Therefore, a set of structure diagrams will be developed supported by a graphical tool.

A further important aspect of this decomposition is the parallel development of a suitable infrastructure for the later successive integration of the components. Therefore, visibility and thus connectability aspects must be considered.

Here, standardized *design notations* help to develop complex models gradually and clearly. Thus, they support developers, especially during the early stages, and potentially improve communication among them. Therefore, we provide an object-oriented notation UML^h (Friesen et al., 1998b), an extension of a subset of the *Unified Modeling Language* (UML) (Rational, 1999). On the one hand, these diagrams visualize the inter-class relationships such as inheritance or composition. On the other hand, they describe the interfaces of the classes partitioning the state variables in attributes and ports. Thus, the intra-class structure can be specified. The distinction between discrete, continuous and hybrid classes, which is proposed by the UML^h, enables a smooth transition to the next development phase. If it is impossible to classify some of the objects and thus their classes at this early stage, these decisions can be deferred to later phases by using abstract classes.

2.3 Detailed Design

Starting with the leaf classes in the developed composition hierarchy, the physical models for the identified components are developed. Therefore, the model specific aspects found during requirement compilation, must be taken into consideration. Thus, the identified effects and balances will be assigned to the classes found during the previous phase. Thereby, often the model structures will be extended and changed by a comprehensive refactoring. In contrast to the above-mentioned structuring, this refactoring is driven *by physical effects*. Here, exhaustive developed artifacts of existing class libraries will be first plugged in the design. In this phase, non-observance of validity ranges and of intentions of the reused sub-models are major sources of failures. The artifacts developed in this phase are of a mixed formal and informal nature. Physical principles can be given by mathematical (in)equations or characteristic curves. In addition, mode transitions found during requirements definition and event types must be considered. Thus, the mapping to the classes is likewise informal. This weak coupling will be strengthened in the next phase.

2.4 Specification

Now, is easy to move to an object-oriented specification, since identified structures can be directly adopted. The effects and balances of the subsystems can be mapped to precise mathematical behaviors. In addition, weak relations like empirical dependencies or characteristic curves will be formalized. Thus, the kind of the classes will be fixed as mentioned above. At this stage, a description in a concrete simulation language would be glutted with implementation details; specifications are superior owing to their abstract nature. Here we provide the later in Sect. 3.1 discussed wrapper to the specification language ZimOO (Friesen, 1995, Friesen, 1997), an object-oriented extension of Z, which is based on a continuous environment-based logic (CEL) (Friesen, 1998) and permits:

- the definition of pre- and post-conditions of reactive operations
- the definition and partitioning of continuous state spaces by the means of Boolean predicates and algebraic-differential equations
- actuator and sensor coupling of components
- extensibility of the formalism

2.5 Parameterization

In this phase, two kinds of parameterization must be done. On the one hand, generic classes possibly found in the model specification, must be actualized. Due to the genericity mechanism described later, both the source and target specifications can be described by (wrapped) ZimOO. These transformations could be done automatically. On the other hand, until yet, there is nothing said about validity ranges of the components. They must be fixed to restrict the possible set of behaviors.

Further definitions of constants and initializations of state attributes and hence start values of differential attributes can be established. Due to the compositional style of the specifications, only class level attributes could be described within *axiomatic descriptions* and the *INIT-schema* of the ZimOO classes. Thus, parameterization of sub-components is a matter of integration.

2.6 Logical Validation

Now, it is possible to start a first validation cycle for the developed sub-components. The goal of these concurrent cycles is to assure that all abstractions made during specification are valid respective to the physical sub-models.

2.7 Implementation

At this stage, the classes will be implemented in a concrete object-based or object-oriented simulation language. Therefore, the abilities of the target language must be taken into consideration. These comprise the offered structuring concepts, support for non-causal equation based modeling, types of equations and connectors and further, support for structural dynamics. Thus, a specified behavior is implemented by possibly several equations. The introduction of these new equations requires further initialization and parameterization and validation. Likewise, new implementation classes will be developed extending the library of already defined classes.

Usually, it is impossible to transform a specification in an implementation owing to the highly implicit nature of specifications. Nevertheless, this procedure can be partly automated. We provide the full-featured simulation environment Smile (Kloas et al., 1995) at this stage.

2.8 Numerical/Physical Validation

Then, the implemented classes should be inspected during a numerical validation. Here, systematic errors, e.g., improper initializations of (new) equations, range mismatches or wrong functional abstractions, will be eliminated. In particular the choice of the numerical solver should be validated. Further, it is necessary to validate the behavior of the implementation against the intended one of the component. Here comparisons with analytically derived solutions help to estimate the quality of the model.

2.9 Integration

When all of the leaf classes found in the composition hierarchy are refined, a successive bottom-up integration can be performed. This will be done in a cyclic manner. In one integration step, the classes of sub-components are parameterized within the composition classes. Then, all connections must be made respectively to the above-mentioned integration-strategy. Thus, a new specification – implementation cycle interleaved by the validations is entered for the composition class. The mentioned validations can be supported by the simulation environment, if simple experiments and possibly model stubs are developed for the sub-model under consideration. Some of these simulation results might not be meaningful in its raw representation. These must be refined by appropriate post-processing and visualization.

2.10 Physical Validation

The last integration step leads to an overall simulation model. At this stage, the results of the experiment must be validated against the requirements made at the beginning of the process. It is worth mentioning, that besides data, the knowledge won during development is one of the most important results. Therefore, it should be documented together with the associated modeling artifacts.

2.11 Generalization

If it is intended to collect parts of the model in a library, often some generalizations must be done to increase the reusability of the artifacts. This is a matter of refactoring.

By the identification of the outlined phases, transitions and model artifacts, we meet the first three goals for methodologies mentioned above. This is due to the fact, that causal connections are established between abstract modeling decisions and the associated validations. Thus, an early error detection and correction becomes possible. Furthermore, the decomposition of the model by the abstraction level as well as by physical properties leads to work packages that are easy to distribute among a developer team. Finally, the process is easy to communicate. As mentioned above, a lot of case studies were done along this process regarding the energy domain. Some examples are photo-voltaic and hydraulic systems, building simulation and air-conditioning technology.

3 Object-Oriented Structuring Concepts

The object-oriented approach would appear to be a very good basis for dealing with the mentioned complexity. That is the main reason why recently proposed simulation languages for continuous and hybrid systems, like Omola (Andersson, 1994), Dymola (Elmqvist et al., 1993), Smile (Ernst et al., 1997) or Modelica (Modelica Design Group, 2000), all offer object-oriented structuring concepts. Another advantage of the object-oriented paradigm is the adequacy of modeling physical properties and components as objects, which leads to model components that are more reusable.

All of the above mentioned object-oriented languages use two crucial concepts in their semantics, the *objects* themselves and *connectors*.

An *object* is an encapsulation of a vector of *state attributes*. Objects have a unique identity in the semantic space, e.g. the runtime structures in memory, and are described by *class* definitions. Thus, classes are abstractions of structurally identical objects. The model is fully described by classes. Hence, the modeler uses only *intentional artifacts* during development. The classes are uniquely identified by their names.

Within the class, invariant properties of these attributes could also be expressed as *DAEs* or *discrete update functions*. Furthermore, *initial values* of the attributes, *subcomponents* and some of their properties could be stated in the associated classes. *Subcomponents* are used to express a pairwise directed composition relation between the associated classes. To navigate to subcomponents, the common infix dot notation is used. To ensure encapsulation, several of the languages forbid deep navigation. In these cases, some kind of variable lifting must be applied.

Connectors relate several attributes of objects to certain properties. There are typically two kinds of properties, equality and zero-sum (Modelica Design Group, 2000).

If the model is instantiated, e.g. at runtime, beginning at a defined root class, each class that is referenced by a composition class is instantiated and linked to this composition. The resulting object structure spans a tree, as shown in Fig. 2.

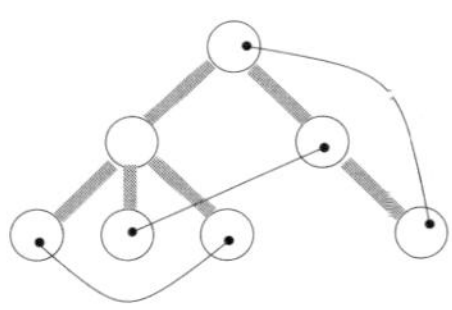

Fig. 2. Object structure with connected attributes

Thus, the object structure does not change during computation – the model description is purely declarative. Hence, all model dynamics have to be expressed within the equations of the classes. While this interpretation has benefits due to some error classes, e.g. void references, it prohibits crucial flexibility of the object-oriented paradigm. In the following, we want to extend this interpretation by two well-defined concepts, namely genericity and structural dynamics.

3.1 Integration of Restricted Genericity

Besides composition and generalization/specialization, some object-oriented modeling and implementation languages offer a further structuring mechanism, namely genericity. In the following, an extension of the specification language ZimOO is proposed, which allows generic class definitions.

The introduction of genericity was motivated by comparison of simulation runs of coupled building and plant models. Therefore, it is necessary to exchange the type of special components in a complex model structure in a uniform and easy manner.

Technically, the extension will be made by wrapping the ZimOO specification. Hence, the ZimOO core is not modified. This results in an encapsulation of ZimOO within the development process (see Fig. 3).

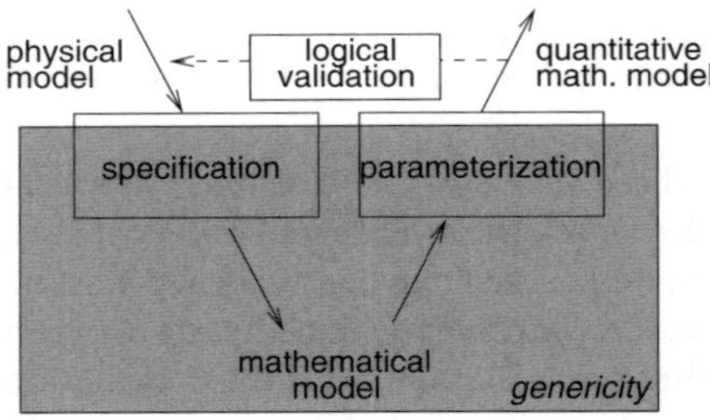

Fig. 3. Integration of genericity into the development process

The above-mentioned simulation languages offer different kinds of parameterization. The easiest one is the *parameterization by constants* of base types. This kind of parameterization is directly used during reuse of components. The flexibility will be increased by separate experiment descriptions, which allows overwriting such model specific parameterizations.

Further, some object-oriented simulation languages offer *parameterization by objects*, which are instances of complex data types. Therefore, different conformity criteria will be established. For instance, the type system of Modelica is based on the approach of (Abadi and Cardelli, 1996) and is restricted to a weak kind of similarity of signatures between formal and actual parameters.

It is generally accepted that strict type systems are a major contribution to the development of complex systems, since they allow to eliminate several classes of modeling errors regarding to the logical validity of the models. Furthermore, type errors could be automatically detected. Therefore, we prefer a strict type system (Meyer, 1997) to introduce genericity as an intentional structuring concept. The idea is based on the restricted genericity offered in Eiffel (Meyer, 1992).

Figure 4 t.r. shows the concept of (non-restricted) genericity. The goal is to ease the reuse of types and classes which model often used data structures. On the one hand, we are interested in fine-grained modeling artifacts. On the other hand, we do not want to lose conformity and encapsulation of the artifacts. The central idea is the separation of the definition of a data type from to behavior of its components. This is implemented by *type parameterization* of the data type with the type of its components. In the example, $X[I]$ defines a generic data type. If it is reused in another declaration, the formal parameter I is actualized with an concrete one, here B or D. It is worth mentioning, that this is a purely static concept. The generic specifications can be automatically expanded to plain ZimOO specifications.

This kind of genericity is very useful for the definition of abstract data types. But, in the context of continuous and hybrid systems it would not bring any benefit. Nevertheless, a variation of the concept, namely restricted genericity, does (see Fig. 4 b.r.).

The generic data type requires from its parameters some properties, which are expressed by an interface (A, in the example). All valid actual parameters must satisfy this interface, which is implemented by a direct or indirect specialization relation.

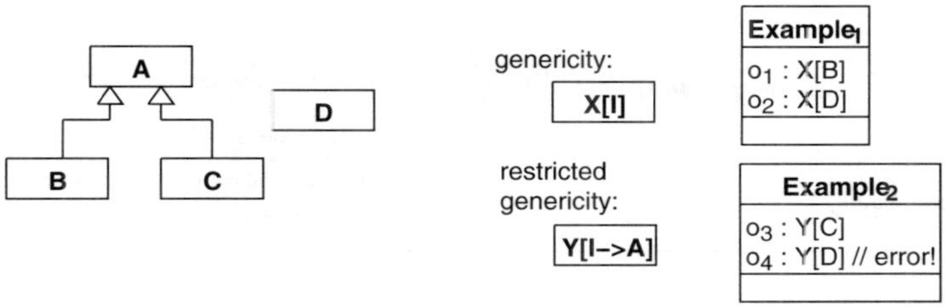

Fig. 4. Genericity

Thus, it is possible to access attributes of the formal parameters. The concept is unfolded in the same way as done with non-restricted genericity.

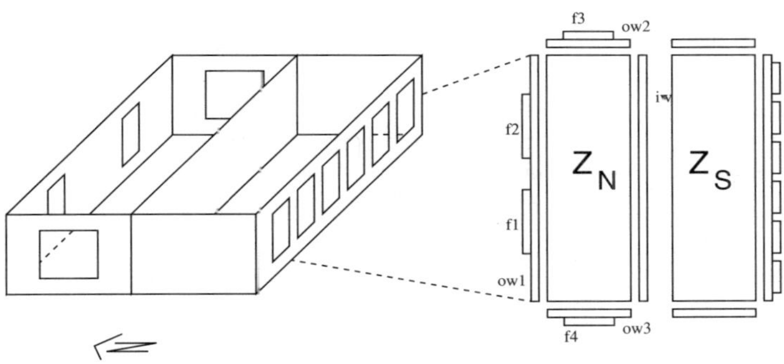

Fig. 5. Building model – extensional view

An easy example should illustrate this concept. The goal is to perform several experiments with a building, varying its wall and window types (Nytsch-Geusen, 2001). A concrete question could be the temperature distribution in an unheated building on sunny days. An important restriction is to use only the same kind of outside wall and window within one building instance. In this example we use a building with one floor containing two zones with walls and windows (see Fig. 5).

At the south side there is a glass facade. The development of the class model is based on an existing class library. Thus, we are able to select special wall and window types. A conventional class model is shown in Fig. 6.

A typical specification would define these components and would establish connections between the components.

Due to the lack of space, we will restrict our attention to a few of these classes. Figures 7 and 8 show only a part of the zimoo specification: The class $Zone\_North$ is a specialization of the class $Zone$ (which is not further specified here). $Zone\_North$ has two vector valued components ow and $window$ that model the outer walls and windows of the zone. The Attribute iw models the inner wall which connects both zones. The (static) connections between associated attributes of the related compo-

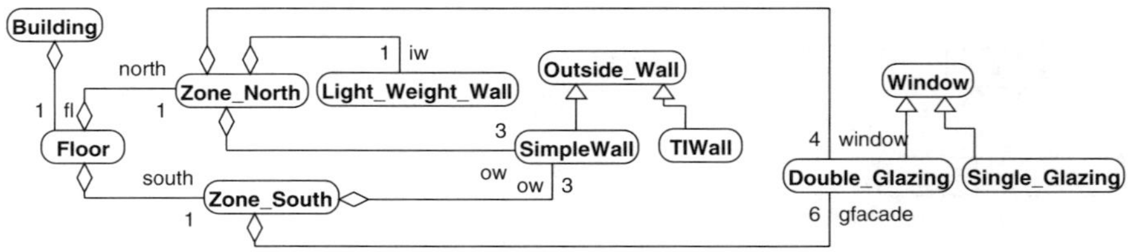

Fig. 6. Building model – intentional view

Building
fl : *Floor*
...

Floor
north : *Zone_North*
south : *Zone_South*
...

Zone_South
Zone
$ow : 1..3 \rightarrow SimpleWall$
$gfacade : 1..6 \rightarrow Double\_Glazing$
...

Zone_Nord
Zone
$ow : 1..3 \rightarrow SimpleWall$
$window : 1..4 \rightarrow Double\_Glazing$
$iw : Light\_Weigth\_Wall$

$ow(1).fdim(1) = window(1).dim \wedge ow(1).fdim(2) = window(2).dim$
$ow(2).fdim(1) = window(3).dim \wedge ow(3).fdim(1) = window(4).dim$
$iw.eport(1) = vec(1)$
$\forall\, i \in 1..3 \bullet ow(i).eport(1) = vec(i+1)$
$\forall\, i \in 1..4 \bullet window(i).eport(1) = vec(i+4)$
...

...

Fig. 7. Building model – conventional specification

nents are established in the state schema of the class. In the example, dimensions and energy ports are connected beside the specification of other properties.

Figure 8 illustrates a part of the wall class library. The abstract interface for all outer walls is specified by the class $Outside\_Wall$. As an example, we focus on the marginal and surrounding temperatures. Two specializations $SimpleWall$ and $TIWall$ are concrete wall types, that could be used in our building model. The characteristics of these variants are specified by the equations stated within their state schemas. As an example, a simple heat conduction model for a wall $SimpleWall$ and a wall with thermal isolation $TIWall$ is specified (Nytsch-Geusen, 2001).

We are interested in the type declaration of the attributes of the building classes. If another model configuration with different wall and window types should be studied, these declarations must be identified and changed in a uniform manner. If the model is considerably complex, this would be error prone. Restricted genericity offers an elegant solution to this problem (see Fig. 9). The relevant sub models will be replaced by generic class definitions. $Zone\_North$ for example, is parameterized by two generic parameters W and F which must conform to their established inter-

Outside_Wall

$T_i, T_o \mathbb{R}$ [inner and outer temperatures]
$T_{o1}, T_{o2} : \mathbb{R}$ [marginal temperatures]
$A : \mathbb{R}$ [surface]
...

SimpleWall
Outside_Wall

$T_w : \mathbb{R}$ [wall temperature]
$k_1, k_2 : \mathbb{R}$ [heat conduction coeff.]
$m : \mathbb{R}$ [mass]
$c : \mathbb{R}$ [specific heat capacity]
...

$$m \cdot c \cdot \dot{T}_w = A \cdot (k_1 \cdot T_o + k_2 \cdot T_i - (k_1 + k_2) \cdot T_w)$$
$$T_w = T_{o1} \wedge T_w = T_{o2}$$
...

TIWall
Outside_Wall

$T_{ti}, T_w : \mathbb{R}$ [isolation and wall temp.]
$k_1, k_2, k_3 : \mathbb{R}$ [heat conduction coeff.]
$m_{ti}, m_w : \mathbb{R}$ [masses]
$c_{ti}, c_w : \mathbb{R}$ [specific heat capacities]
...

$$m_{ti} \cdot c_{ti} \cdot \dot{T}_{ti} = A \cdot (k_1 \cdot T_o + k_2 \cdot T_w - (k_1 + k_2) \cdot T_{ti})$$
$$m_w \cdot c_w \cdot \dot{T}_w = A \cdot (k_2 \cdot T_{ti} + k_3 \cdot T_i - (k_2 + k_3) \cdot T_w)$$
$$T_{ti} = T_{o1} \wedge T_w = T_{o2}$$
...

Fig. 8. Building model – part of the wall class hierarchy

faces $Outside\_Wall$ and $Window$. These generic parameters are used to declare the attributes of the classes. As mentioned above, we can use all features of their interfaces. The actual parameters of the generic model are specified in the root class $Building$. Thus, they are propagated trough the composition hierarchy leading to a different model.

Zone_South[$W \to$ *Outside_Wall*, $F \to$ *Window*]
$ow : 1..3 \to W$
$gfacade : 1..6 \to F$
...

Zone_North[$W \to$ *Outside_Wall*, $F \to$ *Window*]
$ow : 1..3 \to W$
$window : 1..4 \to F$
iw : *Light_Weigth_Wall*
...

Floor[$W \to$ *Outside_Wall*, $F \to$ *Window*]
north : *Zone_North*[W, F]
south : *Zone_South*[W, F]
...

Building
fl : *Floor*[*SimpleWall*, *Double_Glazing*]
...

Fig. 9. Building model – generic specification

Until yet, we have focused our attention to static parameterization mechanisms. The rest of this section outlines a kind of *dynamic parameterization*.

3.2 Integration of Structural Dynamics

A generally accepted technique in modern software engineering is the combination of different models (or views) to describe software systems. The benefit of this approach is that it focuses on related aspects like structure or behavior during modeling. For this principle, called *separation of concerns*, different specialized techniques have been developed.

In the context of hybrid systems, we can distinguish four aspects: *structure*, *continuous behavior, discrete behavior* and *structural dynamics*. Static properties of components, like its internal structure, can be described as mentioned above. For the last to aspects, we adapt *object-oriented statecharts* (OOSC) (Harel and Gery, 1996) because of the following reasons:

- It is easy to integrate OOSCs into the metamodel of class diagrams (Geisler et al., 1998), as shown in (Klar and Mann, 1998).
- OOSCs are well-suited to describe complex discrete dynamics.
- Hybrid dynamics can be considered as special discrete ones: When events occur, the actual vector of DAEs will be modified. Thus, a suitable separation between switches and behavioral descriptions must be established. Using OOSCs to describe structural dynamics associated to classes, we are able to perform this step.

The idea is based on the encapsulation of exchangeable behaviors within objects (Fig. 10). These objects are described by behavioral classes containing the associated equations. Thus, (configurable) objects can be decorated with behavioral objects by receiving events. The concrete reaction for incoming events is specified within the objects statemachine modeled by a statechart. Therefore, an adaption of the Decorator pattern (Gamma et al., 1995) can be applied (Fig. 11).

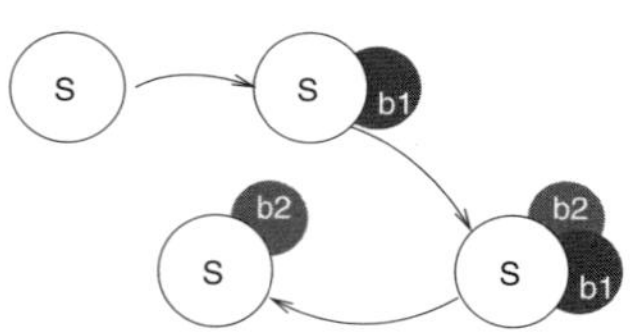

Fig. 10. Exchange of behaviors

The modification of objects are mapped to dynamic connections of attributes with the same name. Therefore, an abstract interface must be declared which ensures the compatibility of base and behavioral object. The integration of the driving statemachines and the class declaration is made by an easy reconfiguration language. This infrastructure is implemented in a framework class $Exchanger$ which is the base of all reactive classes. The operations, given in this class, can be associated with transitions of the statemachine. Furthermore, events can be propagated into sub-components. Thus, complex model reconfigurations can be modeled.

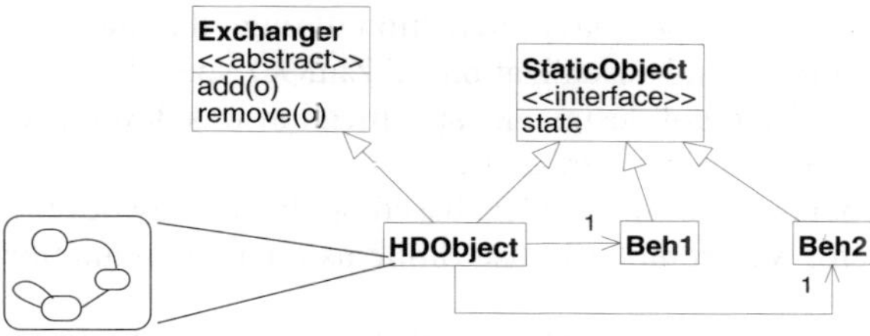

Fig. 11. Exchange of behaviors (Decorator)

The proposed approach has several benefits compared to conventional techniques. At the modeling level, it is an abstract concept which allows partly automatic implementation of models in several object-oriented simulation languages. We reach a better locality of definitions of behaviors, which enforces easier reuse and exchangeability. Further, we are able to decompose the model by the physical structure of the real system *and* by physical effects at the same time. Using conventional techniques, often we have to prefer one criteria neglecting the other one.

There are also some drawbacks. One the one hand, we end up in a lot of fine-grained behavioral classes, which requires a more elaborated management of class libraries. On the other hand, we have to deal with new (runtime) error types which requires informative debugging and logging support during the interpretation of models. Nevertheless, it seems useful for building up libraries of reusable (behavioral) components. A detailed description of this approach, its formal semantics, especially related to the used dynamic connectors, and a discussion about typical application scenarios is given in (Nordwig, 2002).

The rest of the paper outlines a modeling tool which was built within the above-mentioned project.

4 Tool Support

To evaluate the methodology given in Sect. 2, a graphical tool `zooed` (Nordwig, 2000) was developed, which implements the described process. The third release of the prototype offers full support for all phases from the architectural design to the integration and post-processing. Thus, it can be seen as a development environment for object-oriented simulation models. An important overall requirement to the tool was its future extensibility. This goal was reached by an open model-view-controller (MVC) architecture, which extensively incorporates design patterns (Gamma et al., 1995). In the following we give a short overview about the yet realized features `zooed`.

- Features related to architectural design, detailed design and specification are:
  - implementation of class diagrams of the UML^h, which serves as a type system
  - support for modules to break complex systems into manageable pieces; Hence, the set of all relevant modules constitutes the overall model.

  - hierarchical class browser to support easy navigation through the libraries
  - support for the development and visualization of ZimOO specifications; Therefore, a smooth integration of design and specification was developed.
- Implementation and integration level properties are:
  - support for the implementation of the models based on the already developed specifications; Here, we exemplarily use Smile as a target simulation language.
  - incremental compilation of submodels; This results in a considerable speed-up during validation cycles.
  - definition and start of experiments
  - visualization/animation[2] of results
- Features related to interfacing and usability are:
  - import and export of models using a tool-independent representation; Here XML (W3C, 1998) was used as description language.
  - export of ZimOO-specifications
  - import and export of Smile classes; Hence, a reverse-engineering of existing models becomes feasible.
  - persistence of models
  - undo/redo mechanism with infinite depth
  - support for greek letters and indices often used in this domain
  - implementation of a user-friendly object-oriented interaction interface (Collins, 1995); Figure 12 shows a screen-shot of the tool at work. There, a fully specified and implemented part of a steam boiler (class HeatSource) is shown at three different levels of abstraction.

As a target language we choose Java with the Java Development Kit (JDK 1.1) (Kramer, 1997). Thus, the tool is platform-independent. It is available via our homepage (Nordwig, 2000).

5 Related Work and Conclusions

As outlined in Sect. 2, there exist a lot of domain specific methodologies for the development of discrete models. In contrast, generic approaches, related to simulation models of hybrid systems, are rare. Besides the approaches presented in this book, our goal was to contribute some ideas about such a generic methodology.

Considering simulation tools for hybrid models, in addition to the approaches presented in the other articles of this book, `HYBRSIM` (Mosterman, 2000a) is an interesting approach for modeling structural dynamics. This tool is based on an extension of bond graphs (Paynter, 1961). Thus, a visual object based modeling is supported. Nevertheless, object-oriented abstraction techniques settled at the intentional level, like inheritance or genericity, are not supported. We believe, that these complementary concepts can be easily integrated in a sound way.

In this paper, we have outlined an approach for the development of simulation models for hybrid systems adapting conventional software engineering techniques

[2] The animation is based on the possibilities of `python` and its extensions.

and methodologies. Two extensions of the conventional object-oriented paradigm in simulation engineering were discussed. A graphical tool that implements the proposed methodology was outlined.

Further work involves the formal integration of the mentioned views as well as the maintenance and evolution of the tool. Respective to the formalization, most elements of the metamodel considering the type system, connectors, action language and the mathematical model we have already formally specified. Future work is the formal semantics of the adaption of object-oriented statecharts for hybrid systems and its integration into the whole metamodel. There is still a lively discussion about semantic issues of OOSCs in literature (John, 2001).

At the tool level, we are working on the project integration considering version and change control. Further, we work on the integration of other views which allow specifications of the network structure and of reactive dynamics. In parallel, a model representation, an appropriate incremental compiler and a runtime environment which directly support structural dynamics, as discussed above, will be developed. Then, the abstract models can be simulated directly.

Acknowledgment. I would like to thank Christoph Nytsch-Geusen for the helpful discussions providing insight into the domain of energy systems.

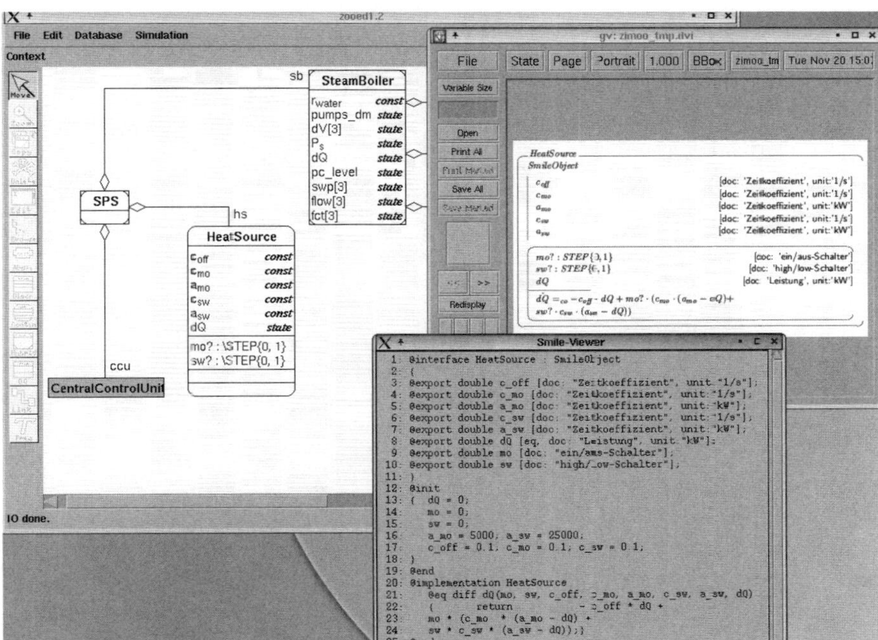

Fig. 12. Screen-shot of the tool

Part III

Analysis and Verification

Introduction to the Analysis and Verification of Hybrid Systems

Stefan Kowalewski

Robert Bosch GmbH, Corporate Research and Development, Frankfurt am Main, Germany,
stefan.kowalewski@de.bosch.com *

Abstract. This contribution provides an introduction to the formal analysis of hybrid systems. It highlights different directions from which hybrid models and their analysis have been approached in computer science and control theory. Fundamental problems arising from the combination of discrete and continuous dynamics are discussed and related articles in this volume are put in relation to the different basic approaches.

1 Introduction

The combination of interacting discrete and continuous dynamics in one model brings up the need for appropriate analysis methods. While such methods for purely discrete or for purely continuous systems have been existing for decades, the formal analysis and verification of hybrid systems is a relatively recent research problem and standard procedures are not yet available. It was therefore one of the main tasks in the focussed research program *Analysis and Synthesis of Technical Systems with Continuous-Discrete Dynamics (KONDISK)* of the German Research Council to develop new analysis and verification approaches or extend the existing methods such that they can be applied to hybrid systems. The four papers in this chapter by Nenninger et al., Wolter et al., Simon et al. and Huuck et al. as well as further contributions to this volume (e.g., Lunze, Müller et al., Buss et al., and Decknatel et al.) present different solutions to this problem. They illustrate the broad spectrum of problem classes, models and analysis procedures in KONDISK.

The purpose of this introductory article is to put the contributions of this chapter into the perspective of a broader view on the formal analysis of hybrid systems. We highlight the different historic starting points and directions and discuss the main problems which have to be dealt with. Based on this, the contributions are put in relation to the different approaches.

The paper is organized as follows. In the following section *Modelling of Hybrid Systems* we revisit the problem of modelling hybrid systems because of its effects on the analysis. The nature of hybrid systems as a model property is discussed and three different starting points and directions are identified to create hybrid modelling frameworks from discrete and continuous ones. In the section *Analysis of Hybrid*

* The insights and opinions expressed in this paper were developed while the author was with the Process Control Laboratory in the Department of Chemical Engineering at the University of Dortmund, Germany.

S. Engell, G. Frehse, E. Schnieder (Eds.): Modelling, Analysis and Design of Hybrid Systems, LNCIS 279, pp. 153–171, 2002.

Systems we discuss different levels of rigor of the analysis methods and computational issues. An example is presented to illustrate the theoretical computational limitations and the need for abstraction which arises from these results. A summary of the current state of the art of hybrid systems analysis concludes the paper.

2 Modelling of Hybrid Systems

2.1 The Nature of Hybrid Systems

Before we take a closer look at the analysis and verification methods, we discuss different approaches to modelling hybrid systems and their relation to analysis problems and procedures. A hybrid system is usually defined as a system which combines continuous and discrete dynamics. This definition is superficial. To be more precise, the term "hybrid systems" refers to models, not systems as such. A system is not hybrid by nature, but it becomes hybrid by modelling it this way. Whether it makes sense to build a hybrid model depends not only on the system, but also on the application and the purpose of the model. The latter most often concerns the analysis that shall be performed. So, there is a strong relationship between modelling and analysis of hybrid systems.

Hybrid systems arise and must be analyzed whenever both abstraction levels – continuous and discrete – have to be considered to solve a particular problem. However, this does not exclude that during the analysis the problem is mapped to a single abstraction level and solved there: As we will see later, in many cases it is either not possible or not appropriate to perform the complete analysis with a hybrid model. In this case it is often helpful to abstract from the concrete problem, e.g. by discretizing the continuous part and solving the problem using discrete analysis techniques. Thus, hybrid systems analysis and verification is not only concerned with hybrid models, but also with the problem of how to map hybrid problems into spaces where they can be solved better. This will be discussed in more detail in Sec. 3.3.

2.2 Different Approaches to Hybrid Modelling Frameworks

In principle, there are three different ways to create a modelling framework for hybrid systems:

1. The first option is to take existing discrete formalisms, e.g., finite automata, Petri Nets or logics, and extend them by continuous variables which evolve according to differential equations associated with discrete states. Discrete transitions can then switch between continuous modes, and the continuous variables can be reset when a transition takes place. Resulting frameworks of this kind are *hybrid automata* (Alur and Dill, 1990) or *hybrid Petri Nets* (David and Alla, 1992).

2. The opposite direction is to extend models for continuous systems by discrete mechanisms like switching or resetting time dependent continuous variables. The resulting frameworks consist of differential equations, algebraic equations and/or inequalities with both continuous and binary variables. The latter are used to activate and deactivate terms by multiplication, e.g. to switch the right hand side of the state equation. This class of systems is therefore often referred to as *switched continuous systems* (Liberzon and Morse, 1999).
3. The third approach is not to extend a formalism but to employ an existing discrete model and an existing continuous model as they are and couple them by appropriate interfaces. Prominent representatives of this concept are commercial simulation tools like *Simulink/Stateflow* or *Matrix-X/Betterstate*.

In the KONDISK program, all three approaches were employed. Hybrid extensions of discrete formalisms are used, among others, by Decknatel et al., Simon et al. and Huuck et al. in this volume. In the first two cases, the model is based on Petri Nets, in the third case the authors chose hybrid automata. It is interesting to note that in Decknatel et al., the hybrid dynamics was realized solely by applying the modelling mechanisms already offered by the Petri Net tool *DesignCPN*. The underlying model did not have to be extended. A good example for the second approach is the contribution by Buss et al. in this volume. The third approach can be found in Müller et al. and Nenninger et al. in which a continuous state space model with piecewise linear (or affine) dynamics is connected to a Petri Net. The coupling of discrete and continuous formalisms has also been pursued in KONDISK by approaches to the simulation of hybrid systems (see Pawletta et al.and Remehle et al.).

The three fundamental modelling approaches listed above are equivalent in the sense that the models are equally expressive as long as comparable assumptions about number spaces and permitted mathematical operations are made. The choice of one of the approaches is therefore usually determined by the scientific discipline: With only few exceptions, the first approach has been followed by computer scientists whereas control theorists have preferred the second and third approach. This is not surprising since the original domains of interest in these two fields were on the opposite ends of the hybrid dynamics spectrum – purely continuous dynamic systems in control theory, discrete state systems in computer science. (And in the third approach, although appearing to be a union of equally privileged formalisms, the continuous dynamics often was first and is still dominant at closer look. See, for instance, the triggering of discrete transitions in *Stateflow* by the integration steps in *Simulink*.)

The awareness of the opposite starting points and perspectives on hybrid modelling in computer science and control is important for the perception of research results from the other field. For example, in computer science it was natural to choose timed models like *timed automata* (Alur and Dill, 1990) as the first class of hybrid systems to investigate. Time can reasonably be modelled as a continuous variable, and, from a computer science point of view, extending discrete models by continuous time is the simplest way to obtain hybrid models. In the beginning of the KONDISK program, I often experienced the misconception by control theorists that this would not be genuine hybrid systems research. Looking from the continuous systems' end of the hybrid dynamics spectrum, they felt that "real" hybrid systems require more complex differential equations than $\dot{x} = 1$. In the meantime, this prejudice vanished and I believe that it was the fruitful exchange in the KONDISK program which helped to achieve this understanding. It is now commonly agreed on that timed systems are an interesting class of hybrid systems, both with respect to theoretical limitations of computer analysis (see Sec. 3.2) and practical usefulness as an abstraction of more complex hybrid dynamics (see Sec. 3.3). In the modeling part of this volume, this is demonstrated by the papers by Huuck et al. and by Simon et al. in which timed automata or timestamp Petri Nets, respectively, are used to analyze hybrid systems.

A different issue in modelling hybrid systems is how to deal with uncertainty. Often, for various reasons, there is not sufficient information available to determine the exact next state of a transition or the exact time of switching. One way to deal with this issue is to use transition relations instead of transition functions, like in nondeterministic automata. Another possibility is to assign probabilities to competing transitions. Stochastic automata are a model of this kind (see Lunze in this volume). It is also possible to specify probability distributions for the switching times of the discrete transitions, like in stochastic Petri Nets. In this volume, Wolter et al. present an extension to stochastic Petri Nets, so-called *fluid* stochastic Petri Nets, in which a subset of the places carries a time-dependent continuous value instead of discrete tokens.

3 Analysis of Hybrid Systems

3.1 Simulation and Verification

Model-based analysis of hybrid systems can be performed with different levels of rigor. For control engineers it is customary to build a *simulation model* and use it to simulate scenarios of interest. This means that the input values and open decisions in the model are fixed before the model is executed. Then properties of the system are inferred from the resulting output or state trajectories. The shortcoming of this approach is that all the properties analyzed by this procedure are only proved for the considered scenarios. It cannot be excluded that there are other inputs for which the result would be different. However, it has to be noted that in many applications this problem does not occur because the relevant scenarios are easy to identify.

The activity of proving system properties for every possible choice of free inputs and decisions is called *formal verification* (or, in the following, simply *verification*) (Clarke and Kurshan, 1996). The term originates from computer science where two different directions of formal verification are distinguished, *algorithmic* and *deductive* verification. Algorithmic verification, often called *model-checking*, means that a computer algorithm is used, which receives a model of a system and a specification of its required behavior as input and then checks whether the requirements hold for all possible behaviors of the system model. This is done basically using search techniques, very often by computing the reachable states of the system. Each verification algorithm is applicable to a particular class of systems (e.g., finite transition systems).

In principle, algorithmic verification is only possible for such classes for which it is guaranteed that the search procedure terminates. In the case of hybrid systems, this is only true for very restricted classes (see Sec. 3.2). It is therefore often necessary to find a finite abstraction of a hybrid system before algorithmic verification can be applied. On the other hand, the advantage of algorithmic verification is that the user only has to provide the system model and the requirements and can leave the rest to the algorithm. No further expertise and knowledge of the analysis technique is needed. However, a major shortcoming in comparison to simulation is the computational complexity resulting from the exhaustive search in discrete spaces.

The most important algorithmic verification procedure for hybrid systems is *reachability analysis*. It answers the question whether for a given hybrid system a certain hybrid state (discrete state and a region in the continuous space) is reachable from the initial hybrid state. This problem is so important because many problems can be reduced to a reachability problem.

When applying deductive verification, also called *theorem proving*, the question whether a system has certain desired properties is answered by creating a proof. For this purpose, the user not only has to specify the system behavior and the requirements in an appropriate logic, but also has to find a suitable sequence of arguments. Although this is often supported by a set of proof-rules, which can be applied in a schematic way, in the end, the success of the verification depends much more on the intuition, creativity and experience of the user than it does for algorithmic verification. However, one main advantage of deductive verification is that the application domain is not restricted to systems with finite search spaces or finite approximations. If a suitable theory is available, infinite systems (i.e., systems with infinitely many states) can be handled also. This is important in the case of hybrid systems, which are infinite by definition due to the continuous part of their state.

In this volume, several papers deal with algorithmic verification. Müller et al. and Nenninger et al. present procedures for reachability analysis. In both cases, the analysis is used to design controllers. In the contribution by Simon et al. reachability is solved by analyzing whether those transitions which switch to the relevant states (or markings, as this procedure is defined for Petri Nets) can not be blocked by the timing conditions. Again, this is used for design, in this case to determine valid timer parameters of a controller. The paper by Huuck et al. presents an approach in which algorithmic verification and deductive verification is combined. The purpose of the combination is to overcome the complexity problem of algorithmic verification by using deduction to structure the problem into smaller, more feasible subproblems (Kowalewski et al., 2001a). Finally, the contribution of Decknatel et al. is a representative of the simulation approach to the analysis of hybrid systems. The authors simulate different scenarios in a railway system to determine the performance of the supervision system.

3.2 Computational Issues

It was mentioned before that exact algorithmic reachability analysis is only feasible for very restricted classes of hybrid systems. For the other classes, it is impossible to formulate an algorithm which computes the exact reachable state space for any system from this class in finite time. In other words, the reachability problem is *undecidable*. The major part of hybrid systems research in computer science has been concerned with identifying decidable and undecidable classes of hybrid systems (with respect to reachability, which is sometimes not mentioned explicitly). The control community, in contrast, has much less interest in this issue. As a matter of fact, two papers in this volume, by Müller et al. and by Nenninger et al., present reachability algorithms for a class of hybrid systems for which reachability is actually undecidable. To resolve this apparent contradiction, this section will provide a short introduction to the kind of problems which are looked at in computer science research on decidability of hybrid systems. It may help to understand not only the fundamental issues but also the practical implication that undecidability should not necessarily prevent engineers from developing reachability algorithms for the respective class of systems.

Hybrid Automata. The most popular hybrid systems model in computer science is the *hybrid automaton* (HA) (Alur et al., 1995), see p. 230 for a formal definition. Roughly speaking, the HA model complements (discrete) finite automata by time-dependent continuous variables. While the system is in a certain discrete state, these variables evolve according to differential equations, called *flows*, which are assigned to each discrete state. Conditions can be formulated which have to be true while the system remains in a discrete state. They are called *invariants*. When an invariant evaluates to false, the discrete state must be left or must not be entered, respectively. The continuous variables can be *reset* by the discrete transitions, and finally, so-called *guards* represent conditions for taking a transition between discrete states.

Figure 1 shows an example of a HA which is borrowed from (Henzinger et al., 1998b). It belongs to the simplest and historically first class of HA, i.e., *timed automata* (TA) (Alur and Dill, 1990). In a TA, the continuous variables are called *clocks* and their value is always increasing with a rate of one [1]. Resets can only be assignments of zero, and the invariants and guards are independent inequalities or equalities with rational constants for each clock. In the case of Fig. 1, we have two discrete states, s_1 and s_2, and three clocks, a, b, and c. There are guards at each of the three transitions, but only with transition t_1 a reset (of the clocks b and c) will be performed. The arrow symbol is used to separate the guard from the reset. The invariant in s_1 is $a \leq W \wedge b \leq W \wedge c \leq W$, and in s_2 it is true, meaning that there is no condition restricting the entrance and the visiting time in s_2.

To understand the behavior of TA (or, in computer science terms, the *operational semantics* of this model), it is helpful to know that TA were originally introduced

[1] Like in Fig. 1, the corresponding flows $\dot{x} = 1$ are usually omitted in the graphical representation of a TA.

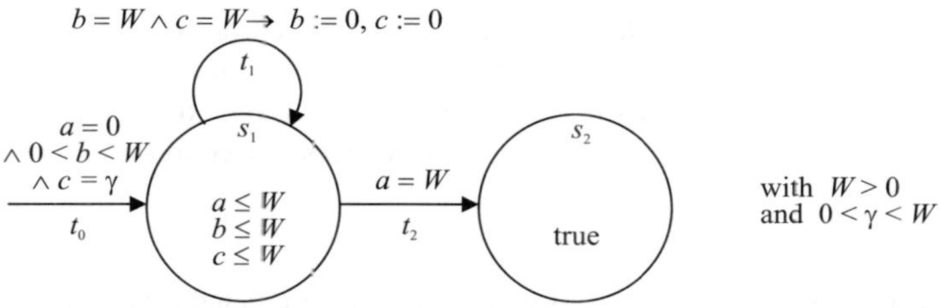

Fig. 1. Example for a timed automaton

by Alur and Dill in (Alur and Dill, 1990) as a generalization of *ω-automata*. ω-automata were developed to model non-terminating systems like, for instance, data base servers. They are automata over infinite words, which means their language consists of infinite sequences of symbols. Their acceptance criteria are based on accepting states which have to be visited infinitely often while reading an accepted word. So, acceptance can only be decided by looking at the infinite behavior.

The same is true for TA, only that the notion of infinity is not applied to sequences but to time. The behavior of a TA is given by a set of *runs* which are infinitely lasting trajectories in which discrete transitions (possibly with resets) and time intervals with continuous growth of the clock values alternate. Figure 2 shows a valid run for the TA from Fig. 1. It starts with transition t_0 at which the clocks b and c have the same value: $b = c = \gamma$ (note that this is not implied by the guard which would permit any value between 0 and W for b). When b and c increase to W, the invariant of s_1 will no longer permit to stay in this discrete state. Transition t_1 is possible because the guard is true, whereas t_2 cannot be taken. Performing t_1 triggers the reset of b and c, and s_1 can be entered again. When $a = W$ becomes true, transition t_2 must be taken. After that, the system remains in s_2 and the clocks can increase to infinity. Therefore, the run in Fig. 2 is infinitely lasting and, thus, is part of the behavior of the TA from Fig. 1.

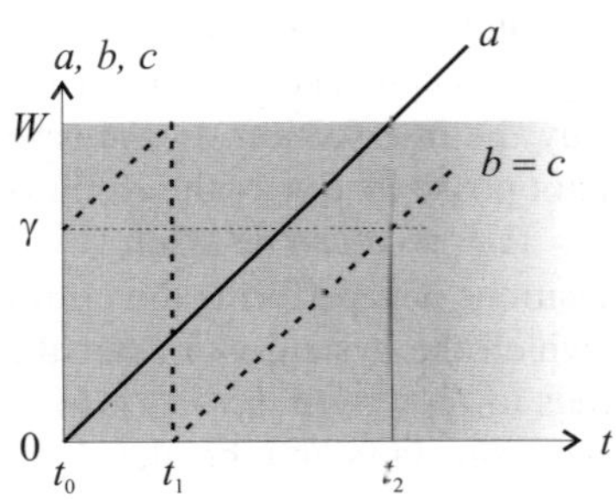

Fig. 2. A valid run for the example from Fig. 1

What would happen, if a different value for $b(t_0)$ is chosen, e.g., $b = \gamma/2$? In this case clock c would reach W first, the invariant would not permit further residence in

s_1, but none of the two transitions could be taken because the guards are false (note that a and c are still less than W). Obviously, the behavior up to this point can not be extended to an infinitely lasting run – in computer science jargon, "time stops" or "a *time deadlock* occurs". In fact, $b(t_0) = c(t_0) = \gamma$ is a necessary condition for the existence of a run in the example[2].

Reachability Analysis of Hybrid Automata. The (forward) reachability analysis of hybrid automata has to determine all possible runs and check whether the hybrid target state is visited during at least one of these runs. Hybrid states are pairs (s_i, R_j) consisting of a discrete state s_i and a set R_j of values of the continuous variables, often called a *region*. The analysis algorithm roughly works as follows.

1. The starting point is the initial discrete state s_0 and the initial region R_0 which is determined by intersecting the regions corresponding to the guard of the entering transition and the invariant of the initial discrete state. In our example, this would be $(s_0, R_0) = (s_1, \{(a,b,c) \in Q | a = 0 \wedge 0 < b < W \wedge c = \gamma\})$.
2. The next step is to let the region grow according to the flow assigned to the discrete state but neglecting the invariant (or in other words, assuming that the system could remain in the discrete state forever). In a TA, for instance, this means that all continuous variables increase by a rate of one. In the example, the resulting region R_1 is unbounded: $R_1 = \{(a,b,c) \in Q | c = a + \gamma \wedge a \leq b \leq a + W\})$.
3. Obviously, not all of R_1 is actually reachable because the invariant would force the system to leave s_0 as soon as it is violated. To determine the values actually possible during this visit of s_0, R_1 is intersected with the invariant (in the example: $R_2 = \{(a,b,c) \in Q | (a,b,c) \in R_1 \wedge a \leq W \wedge b \leq W \wedge c \leq W\})$.
4. At this stage, the algorithm would take the list of previously visited regions and check whether R_2 (or a subset of it) had been computed for s_0 before. If this is case, it would abort the current search branch.
5. If not, the hybrid state (s_0, R_2) is added to the reachability set and the algorithm will continue this search branch. Now, all possible transitions from s_0 have to be determined. In our example, there are only t_1 and t_2. To find out which of them are viable, we have to check whether the guards can become true for the values of a, b and c in R_2. This is computed by the intersection. In the case of t_2 the result is empty, thus, the transition is not possible. For t_1 the result is $R_3 = \{(a,b,c) \in Q | a = W - \gamma \wedge b = W \wedge c = W\}$, so it can be taken.
6. The algorithm chooses one of the possible transitions and performs the corresponding reset. The result is the region with which the system can enter the new discrete state. In the example, the reset leads to $R_4 = \{(a,b,c) \in Q | a = W - \gamma \wedge b = 0 \wedge c = 0\}$. Now, the second iteration starts and the algorithm goes back to step 2.
7. The algorithm terminates when no more new transitions can be traversed.

[2] For this reason, the TA from Fig. 1 can be found in proofs to model a function which checks the equivalence of two values (Henzinger et al., 1998b). Note that $b = c = \gamma$ holds also at t_2.

There are tools available in which this algorithm or variants of it are implemented. The most prominent are Kronos (Yovine, 1997), Uppaal (Larsen et al., 1997), and Hytech (Henzinger et al., 1997). The classes of systems which can be handled by these tools differ, but they have in common that the regions R_i can be represented by polyhedra.

For TA, this algorithm will always terminate. This means that the reachability of TA is decidable (Alur and Dill, 1990). However, even small generalizations can lead to undecidability. This shall be illustrated by the example in Fig. 3 which is inspired by similar examples in (Henzinger et al., 1998b). The only extension to the model is that we allow one clock to be stopped and to be started again with the value at stopping time. In other words, the flow can be either $\dot{x} = 1$ or $\dot{x} = 0$. Such a clock is called a *stopwatch* (Henzinger et al., 1998b), the resulting model is a *stopwatch automaton*. In Fig. 3, d is a stopwatch, while a, b, and c are clocks as in the example before[3].

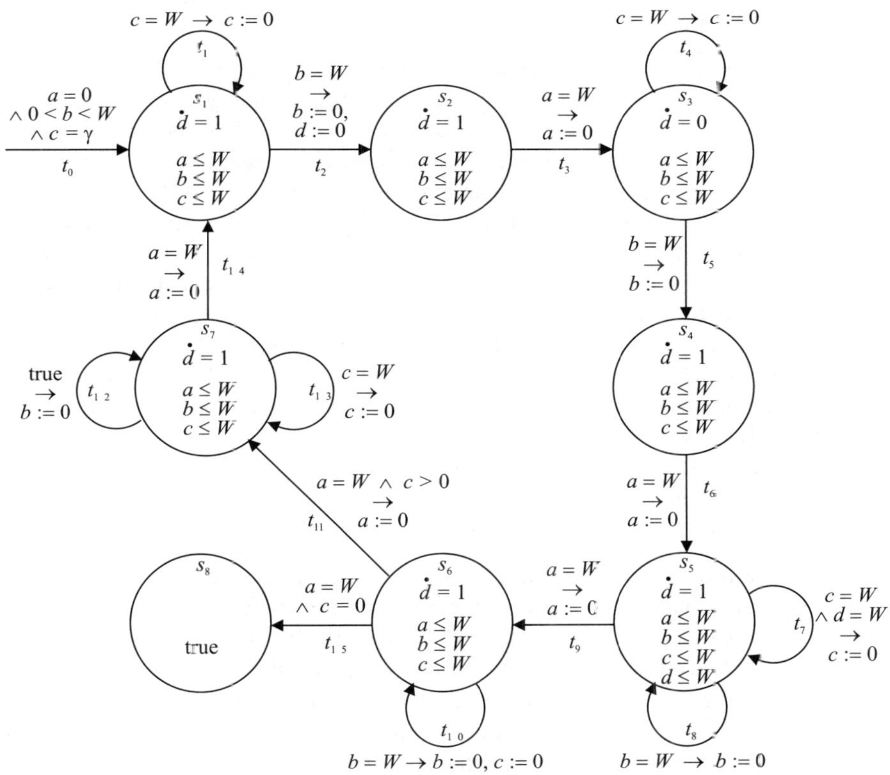

Fig. 3. Example for a stopwatch automaton with non-terminating reachability analysis

[3] As in Fig. 1, the flows of the clocks are omitted and only the flows of the stopwatch are presented.

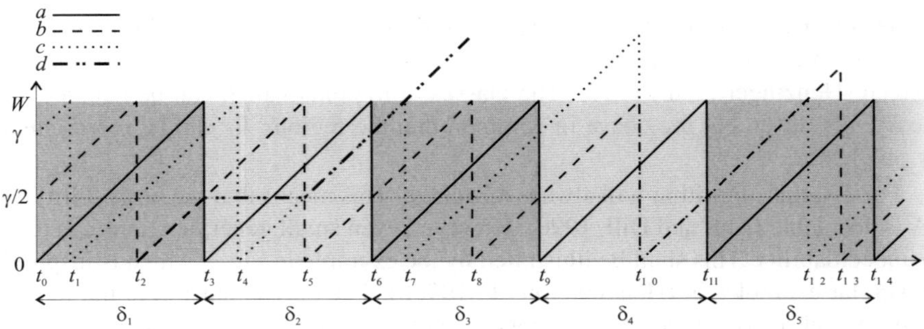

Fig. 4. A fragment of a run for the stopwatch automata from Fig. 3

Figure 4 will help to understand the behavior of the example. It shows a fragment of a run with the choice of $b(t_0) = \gamma/2$ [4]. Clock a can be regarded as providing a clock tick with a constant frequency $1/W$. In the fragment of Fig. 4 it defines five time intervals δ_1 to δ_5.

- The purpose of δ_1 is to synchronize clock b and stopwatch d, which is done at t_2. Note that apart from this change, all clocks have equal values at t_0 and at t_3.
- When δ_2 starts by entering s_3, d is stopped. With t_5, it is started again and, because we chose $c(t_0) = 2{\cdot}b(t_0)$, it is now synchronized with c. At the end of δ_2 $d = \gamma$ holds and the clocks again have the same values as at t_0.
- In δ_3 it is checked whether $c(t_6) = d(t_6)$. This is done by applying the construction of Fig. 1. Transition t_7 can only be taken if $c(t_6) = d(t_6)$, otherwise there would be a time deadlock. So, at the end of δ_3, the run can only proceed if $c(t_6) = d(t_6)$. Since this requires $c(t_3) = 2{\cdot}b(t_3)$ in δ_2, the whole part of the TA from t_0 to t_9 can be regarded as a test of $b(t_0) = c(t_0)/2 = \gamma/2$.
- In δ_4, c is synchronized with b. The result is $c(t_{11}) = b(t_9)$. Now, if we consider the value of c at t_0 and at t_{11}, it becomes apparent that our stopwatch automata did nothing else but the assignment $c := c/2$.
- The purpose of δ_5 is to choose an arbitrary new value for b before s_1 is entered again and the next cycle begins. Of course, the time deadlock in s_5 can only be avoided, if the choice is such that $b(t_{14}) = b(t_{11})/2$.

When the run is continued, this cycle will repeat and every time t_{11} is taken c will increase to half of its value at the beginning of the cycle. This leads to an infinite sequence $(\gamma, \gamma/2, \gamma/4, \ldots)$ which asymptotically approaches zero but will never reach it. If we now consider the problem whether state s_8 is reachable in the automaton of Fig. 3, it is easy to see that the algorithm described above will not terminate: Every time, the possible transitions from s_6 are checked, c will have a new, smaller but positive value and t_{11} has to be taken again.

In practice, that is, when applying tools like Kronos, Uppaal, and Hytech, the algorithm will be aborted after some time. The reason for this is that the tools were

[4] Note that the t_i symbolize transitions and not points in time. However, for the fragment of Fig. 4 there is no difference because each transition is taken only once.

implemented for exact reachability analysis and therefore the rational values for storing the regions (e.g. the corner points) are represented by two integers for the nominator and denominator, respectively. As a consequence, infinite sequences, like the one for clock c at transition t_{11} in the example above, would lead to memory overflows for the integers.

Of course, finding one example for which a particular algorithm will not terminate does not prove undecidability of the general problem because the example may be analyzable by other approaches. Actually, for this particular example it can be shown by deduction that s_8 is unreachable. A proof that reachability of stopwatch automata is undecidable can be found in (Henzinger et al., 1998b) together with several other decidability results. The proofs are conducted by reducing the reachability problem to a problem which is known to be undecidable. Some of the constructions used in the examples of Figs. 1 and 3 are taken from these proofs.

Discussion. The undecidability of stopwatch automata does not mean that any class of hybrid systems that is more general than TA is undecidable. There have been other classes defined in which the flows are, for example, differential inclusions or even linear differential equations with particular properties, and decidability is achieved by certain restrictions on the guards or resets (Henzinger et al., 1998b, Lafferiere et al., 1999). But the fact remains that the decidability boundary is far beyond the classes usually considered for control systems. So, what are the practical implications of this theoretical result? The answer has four aspects.

- The first aspect still follows from theory: Even for undecidable classes of hybrid systems, the reachability algorithm may well terminate for the particular problem under consideration. It is also possible that backward reachability will terminate while forward analysis does not, or vice versa.
- The second aspect is that it is often possible to find abstractions, which still are sufficient models for the analysis problem, but fall into decidable classes (see Sec. 3.3).
- The third aspect is more pragmatic: It is usually a much bigger problem to cope with the exponential complexity of the algorithm with respect to the number of continuous variables (which adds to the discrete state space explosion problem), so that even for decidable problems it may be impossible to wait for the analysis result or memory overflows occur.
- Finally, from an engineering point of view, absolute exactness of the analysis is not appropriate because the models and the requirements already are of limited accuracy. If, for example, s_8 would represent a dangerous state in a technical process and reachability analysis should check whether it will be avoided, $\dot{c} = 1$ would only be an approximation of the real clock speed. It would therefore be sufficient for the analysis, if the sequence would be stopped when the value for c is rounded down to zero. Thus, in applications, it is reasonable to use approximate analysis with numerical rounding as long as the error is bounded (as it has been practice in control engineering for a long time).

For these reasons it can be justified to ignore the undecidability issue when developing analysis procedures for hybrid systems.

3.3 Abstraction

The computational problems of reachability analysis (undecidability, complexity) are the motivation for a very active area in hybrid systems research which is concerned with abstraction techniques. The basic idea is the following: Instead of trying to analyze the original system under investigation, the analysis problem is mapped into a class of problems which is easier to solve. The mapping consists of two steps. First, a substitute model is created by omitting details from the original model (for example by replacing the exact continuous state by a discretized one). This model will include the behavior of the original system but, in general, allow additional behavior because of the less concrete specification. In the second step the original property to check is generalized so that it can be reformulated for the substitute model. This kind of mapping (and its result) is called an *abstraction*.

The first advantage of abstractions is that the resulting model is rougher and, thus, often simpler and analyzable with less computational effort. Of course, this gain of efficiency must be paid for by a loss of accuracy. The consequence for dynamic models is that the degree of uncertainty will increase. For instance, state space discretization of a deterministic continuous system in general will lead to a nondeterministic discrete system. This means for the analysis that the results can be inconclusive. If, for example, reachability analysis of an abstracted system shows that an abstracted state region is reachable, this could be because the original target region is reachable in the concrete system. But it may as well be that it is just one of the additional trajectories in the abstracted model which reaches the target region, or that the reachable part of the target region was just added by the abstraction. However, if the abstract region is *not reachable*, we can be sure that the corresponding concrete region is not reachable in the original system, neither. This is the second advantage of abstractions (which general estimations do not have): Problems can be posed such that one of the possible analysis results is conclusive and provides a guaranteed solution to the original problem.

Because of the potential inconclusiveness and the loss of accuracy, abstractions are only useful if two conditions are fulfilled: First, the properties of interest must be formulated such that conclusive results are possible, e.g., reachability of forbidden states for a conservative safety analysis. Second, the level of uncertainty of the dynamics must not be so low that only inconclusive results exist (e.g., the whole state space becomes reachable).

This problem and further issues concerning abstractions of hybrid systems are discussed, for example, by Lunze and Raisch in this volume for the case of discrete systems as substitute models. In (Alur et al., 2000b), Alur et al. provides a survey on fundamental theoretical results from computer science on this topic. Particular methods to generate a discrete abstraction for a given hybrid or continuous system are presented, for instance, in (Chutinan and Krogh, 1999a, Dang and Maler, 1998, Greenstreet and Mitchell, 1999, Raisch and O'Young, 1998). Further approaches can be found in the special issue (Engell, 2000). The idea of representing continuous dynamic systems by discrete abstractions, however, is older than the recent research on hybrid systems. It was already the basis of the work on qualitative simulation, see for example (Kuipers, 1986).

Abstraction by Hybrid Automata. The abstraction of switched continuous systems with linear or even nonlinear differential equations by discrete automata is a relatively rough approximation. All the quantitative information about the dynamics is lost and replaced by a qualitative description. One way to save quantitative information is to capture the arising uncertainty in stochastic models (see Wolter et al. and Lunze et al. in this volume). Another possibility is to use hybrid automata as abstractions (Henzinger et al., 1998a, Stursberg, 2000a, Stursberg and Kowalewski, 1999, Stursberg and Kowalewski, 2000, Stursberg et al., 2000). The remainder of this section is devoted to the latter approach.

As in most work on abstraction of switched continuous systems, the basis of this approach is an orthogonal partitioning of the continuous state space. This means that each dimension is divided into bounded or unbounded intervals. The result are hyper–rectangles as partition cells and hyper–planes as boundary manifolds. The abstraction by hybrid automata then consists of three steps: First, the discrete state space is defined based on the partitions. Second, the continuous dynamics in each partition cell is abstracted so that it complies to the desired class of hybrid automata. In the third step, the discrete transitions are determined by analyzing which partition cells are connected by trajectories in the abstracted dynamics.

For the first step, it is straightforward to map each partition cell into a discrete state. If we choose TA as the target model, however, this construction has the disadvantage that the state can move from a cell to a neighbored one in zero time. Thus, zero will be the lower time limit for each transition and, consequently, each trajectory in the abstracted state space could be traversed with zero time consumption. To avoid this undesired idealization, the discrete states can be defined to lie on the boundary hyper–planes between cells (Stursberg and Kowalewski, 2000, Stursberg et al., 2000, Stursberg, 2000a).

The abstraction of the continuous dynamics depends on the chosen class of hybrid automata. For TA, we have to determine the upper and lower limit of the time that the continuous state can reside in a partition cell or, in the case of mapping discrete states to boundary hyper–planes, of the time that is needed to move from one boundary hyper-plane to the next.

The orthogonal partitioning, however, suggests a further class of hybrid automata for abstraction, namely *Rectangular Automata* (RA) (Stursberg and Kowalewski, 1999). Roughly speaking, in a RA the invariants, guards, reset sets and flows are *rectangular predicates*, which means they are specified by intervals (possibly degraded to points) for each continuous variable or its time derivative, respectively. For a formal definition see (Henzinger et al., 1998b). In the work described here, the guards only check the equivalence of one variable to one of its bounding values, and the resets are always assignments of a bounding value to a variable.

The abstraction of the switched continuous system with orthogonal partitioning to a RA is straightforward: The invariants, guards, and resets are given by the partition boundaries. The only problem left to solve is to find upper and lower limits for the flow intervals. For non-trivial systems, the corresponding optimization problem cannot be solved analytically. In these cases the flows can be approximated conservatively by interval arithmetics (Stursberg and Kowalewski, 1999, Stursberg, 2000a).

Approximate Analysis of Rectangular Automata. At this stage the question arises how the resulting RA can be analyzed algorithmically. In Sec. 3.2 it was demonstrated that reachability of stopwatch automata is undecidable, and RA obviously are more general than stopwatch automata. Moreover, the numerical problems due to the integer arithmetics of the mentioned tools for HA apply to RA, too. In this section, an algorithm is sketched which overcomes these problems by conservatively approximating the reachable regions in a RA (Preußig, 2000). The first version of the algorithm was introduced in (Preußig et al., 1998), a more complex version with smaller over–approximations was presented in (Preußig and Wong-Toi, 2000). In the following, only the first version is sketched in order to provide an impression of the basic idea.

The algorithm is based on the concept of faces. A *face* is a rectangular predicate with one dimension fixed to a certain value. The rationale for introducing faces is to use rectangular faces to represent non-rectangular sets. A *face-region* $\mathcal{F}$ is a set $\{F_1, \ldots, F_q\}$ where each F_i is a face. The semantics of $\mathcal{F}$ is the convex hull over its q faces, i.e. $\langle\mathcal{F}\rangle = convexhull\{\langle F_1\rangle, \ldots, \langle F_q\rangle\}$. This is shown for an example in Fig. 5 where a face-region $\mathcal{F}_1$ is represented by the two faces F_1 and F_2. In practice, the faces of a face-region over n variables are derived from $2n$ constraints of the form $x_j = l_1$ or $x_j = l_2$. In the example, the face F_1 corresponds to $x_1 = 1$ and the face F_2 to $x_2 = 7$, with the empty faces for $x_1 = 7$ and $x_2 = 1$ being omitted.

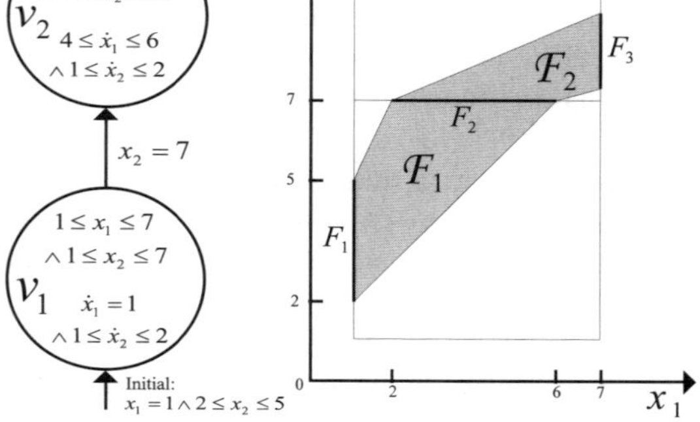

Fig. 5. Reachability analysis of RA using *faces*

A reachable face-region within the invariant can be represented by faces that lie on the invariant's bounds. Let $\mathcal{F}_1$ be a reachable face-region in a discrete state v_1. Now we want to compute the new face-region $\mathcal{F}_2$ in another discrete state v_2 that is adjacent to v_1 in terms of the invariant conditions. Then we can first check if any face of $\mathcal{F}_1$ is within the invariant condition of v_2. In our example this holds for F_2. So, this face can be used to determine a reachable region $\mathcal{F}_2$ in discrete state v_2. This is done by determining for each boundary l of an invariant of v_2 a face as the part of invariant l that can be reached starting from F_2 according to the possible flow in v_2. Here, only for the boundary $x_1 = 7$ a face can be found, namely F_3.

We use the computation of F_2 from F_1 in the example in Fig. 5 to show how an outgoing face can be computed from an ingoing face. First we determine a time interval in which any point within F_1 will be moved to F_2 according to the flow in dimension x_2. The distance between F_1 and F_2 in dimension x_2 ranges between 2 (=7-5) and 5 (=7-2). With a flow $1 \leq \dot{x}_2 \leq 2$ in v_1 this distance can/must be cleared within a time interval $T = [1; 5]$. Since the flow in each dimension is independent from the other dimensions, we can now use this time interval to compute how any point in F_1 will be shifted in the other dimensions while moving towards F_2. In our example, the only other dimension is x_1 for which we have a fixed flow $\dot{x}_1 = 1$. So in the time interval $T = [1; 5]$ a point starting from $x_1 = 1$ can flow to values ranging from 2 to 6. This yields F_2 with $2 \leq x_1 \leq 6 \wedge x_2 = 7$.

The complete reachability analysis is performed by considering all outgoing faces of an initial discrete state as ingoing faces to adjacent discrete states to which control switches exist. For these incoming faces then the outgoing faces within the invariants of the adjacent discrete states are computed. In the next step these newly computed faces are considered as ingoing to all adjacent discrete states again and so an iteration evolves. This iteration terminates when all reachable faces of a given automaton are found. The termination is guaranteed, since RA are always defined over a finite discrete state space and our analysis is approximate. Due to rounding in the approximative analysis there is only a finite number of points considered in the continuous state space. Thus, there is also only a finite number of faces that the algorithm can find within this state space.

Example. Using the algorithm described in the previous subsection, the abstraction of switched continuous systems by RA can be practically applied to the analysis of switched continuous systems. We illustrate the results of this kind of analysis by a small example taken from (Preußig et al., 1999). The example is a two-tank system in which the first tank is filled by a fixed input flow F_{in} and is emptied into $Tank$ 2 through a connecting pipe (see Fig. 6). The outflow of $Tank$ 2, which is located on a lower level than $Tank$ 1 (height difference: H), is denoted by F_{out}. The flow in the connecting pipe depends on F_{in}, the liquid levels h_1 and h_2 in both tanks, and the setting of the valve controlling the flow F_{12}. The latter can be in two positions, *half–open* and *open*.

The dynamical behavior of this switched continuous system can be described as follows. The state vector is (h_1, h_2), and the variable $valve$ denotes the input of the

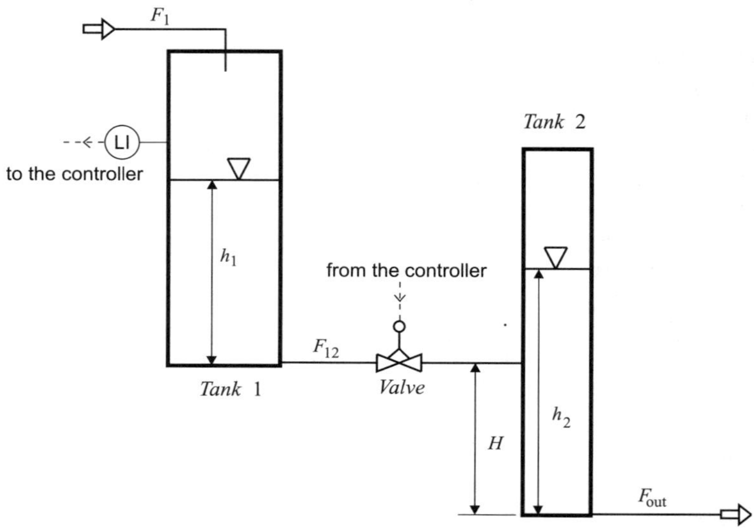

Fig. 6. Scheme of the two-tank system

system. Changes of the gradient field defined by 1 occur when either $valve$ is switched to another discrete value, or when h_2 exceeds H. The normalized parameters are: $A_1 = 1.14 \cdot 10^{-2}$, $A_2 = 1.98 \cdot 10^{-3}$, $H = 0.4$, $F_{in} = 1.11 \cdot 10^{-4}$, $K_1^1 = 1.2 \cdot 10^{-4}$, $K_2^1 = 3.4 \cdot 10^{-4}$, and $K_2 = 1.5 \cdot 10^{-4}$.

$$\begin{aligned} \dot{h}_1 &= (F_{in} - F_{12})/A_1, \\ \dot{h}_2 &= (F_{12} - F_{out})/A_2, \\ h_2 < H \ &: \ F_{12} = K_1 \cdot \sqrt{h_1}, \\ h_2 \geq H \ &: \ F_{12} = K_1 \cdot \sqrt{h_1 - h_2 + H} \quad \text{if} \quad h_1 \geq h_2 - H, \\ & \quad\ F_{12} = 0 \quad \text{else}, \\ F_{out} &= K_2 \cdot \sqrt{h_2}, \\ valve &= \begin{cases} '\text{half}-\text{open}' : K_1 = K_1^1 \\ '\text{open}' \qquad\quad : K_1 = K_1^2 \end{cases} \end{aligned} \tag{1}$$

The analysis is concerned with the following scenario. We assume that the initial liquid heights are $h_1 = [0.2, 0.3]$ and $h_2 = [0.2, 0.3]$ and that $valve = '\text{half}-\text{open}'$ applies. Since F_{12} is smaller than F_{in} at this setting, h_1 will rise. To prevent an overflow of $Tank$ 1 the controller switches the value of $valve$ to 'open' as soon as it receives the information that h_1 has reached the value $h_{1,S} = 0.8$. As a consequence, h_1 will start to decrease immediately and h_2 will increase. The analysis shall check whether opening the valve can lead to a situation in which the limit $h_2 > 0.9$ is exceeded.

The abstraction is performed according to the procedure described above. The range of h_1 and h_2 is divided into 10 intervals each of equal length which leads to 100 discrete states in the RA. The result of the reachability analysis using the presented algorithm is shown in Fig. 7. The grey-shaded area marks the region which is determined as reachable from the dark-shaded initial region. To provide a better understanding of the analysis result, the continuous trajectories starting at the corners of the initial region are drawn additionally. The plot reveals that the *critical* region with $h_2 > 0.9$ is found to be reachable, i.e. the switching value $h_{1,S}$ was not chosen correctly to avoid an overflow of $Tank$ 2. [5]

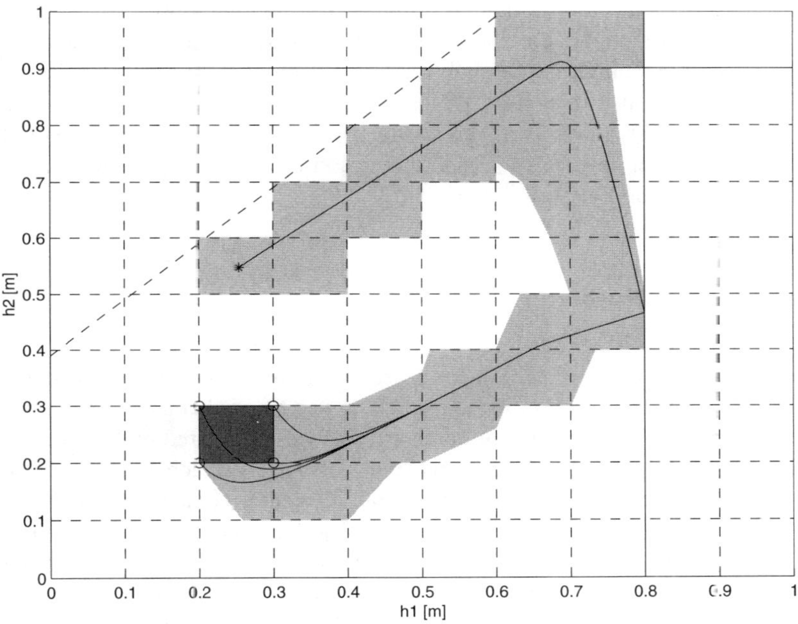

Fig. 7. Analysis results

The example demonstrates the degree of over–estimation which is the price for the simpler analysis model. It also hints at a problem arising from the orthogonal partitioning: In the first part of the trajectory, the abstraction results in reachability of complete partition cells. This is because the gradient field approaches the trajectories to the equilibrium point from both sides and change the sign of direction in both dimensions. This affect can lead to bad over–approximations which make the results inconclusive. For a more thorough analysis of this example, the reader is referred to (Preußig, 2000).

[5] Obviously, in this example the reachable region can easily be determined by simulation. However, for more complex systems exhaustive simulation will become impossible.

Experiences. Empirical data about the efficiency of this and other approaches to the verification of hybrid systems can be found in (Preußig, 2000, Stursberg, 2000a, Treseler, 2001). Experiences are reported for several examples with different complexity of the discrete part and the continuous part as well as for different abstraction and analysis methods. The abstraction of switched continuous systems into TA or RA by interval arithmetics was performed for three–dimensional systems and discretizations with roughly 800 cells or discrete states, resp. (Stursberg, 2000a). Computing time on a PC (Pentium II, 266 MHz) was in the order of 20 minutes for RA and 40 minutes for TA. In (Preußig, 2000) the reachability analysis of RA was applied to a more complex version of the two–tank example in the previous subsection. The system was three–dimensional (a continuous valve was added) and the state space was partitioned into 600 rectangles. The analysis took 11 minutes. In (Treseler, 2001), the problem, which is also treated by Wolter et al. in this volume, is solved using a TA model. The largest model was three–dimensional (three workpieces) with a discretization into seven intervals for each dimension (i.e., the temperatures of the workpieces). Computing time was 3 hours.

3.4 Alternative Approaches

An interesting alternative approach in the research on analysis of hybrid systems is the application of optimization techniques. The use of mathematical programming for the analysis of switched continuous models was suggested by Dimitriadis et al. (Dimitriadis et al., 1996a, Dimitriadis et al., 1997). The reachability problem is reformulated as an optimization problem in the discrete time domain which can be solved by mixed integer programming. Basically, the optimization determines the worst possible behavior, meaning that the system is most often in an undesired region of the continuous state space. The approach is general in the sense that it can be applied to hybrid systems as well as to purely discrete or purely continuous systems. Its strength lies in the ability to take advantage of well tested and efficient optimization procedures. A limitation is given by the fact that the size of the mixed integer program grows with the product of the number of discrete time steps and the number of equations and logical expressions describing the plant and the controller, respectively. A similar approach has been followed by Bemporad and Morari (Bemporad and Morari, 1999b). Here, an iterative scheme is used to perform conventional reachability analysis. This scheme avoids setting up a huge one-step optimization problem which is most likely not tractable. It can therefore be applied to larger problems than the approach of (Dimitriadis et al., 1996a, Dimitriadis et al., 1997). The verification method is part of a comprehensive modelling and analysis approach to hybrid systems, including a scheme for model-predictive control (Bemporad and Morari, 1999a). Further representatives of the mathematical programming approach to verification are Park and Barton who solve purely discrete model checking problems by integer programming (Park and Barton, 1997). In this volume, Stursberg et al. employ optimization techniques to design control policies for hybrid systems.

4 Conclusions

The paper presented an overview of different approaches to the modelling and analysis of hybrid systems. We discussed the theoretical problem of undecidability and its practical implications. Approaches to overcome this and other challenges like applicability to large systems were sketched.

The current status of hybrid systems analysis can be characterized as follows. The theoretical foundations are largely established, the main obstacles on the way to practical application are identified, and first progress in this direction is made. The major challenge is still the computational complexity of the analysis procedures. The contributions to this volume provide good examples of promising approaches to move the research in hybrid systems analysis nearer to practical application.

For more information about the analysis of hybrid systems the reader is referred to the numerous proceedings volumes and to special issues of various control journals which appeared in the recent years. The main conference series are *Hybrid Systems* (Grossman et al., 1993, Antsaklis et al., 1995, Alur et al., 1996, Antsaklis et al., 1997, Antsaklis et al., 1999), *Hybrid Systems: Computation and Control* (Maler, 1997, Henzinger and Sastry, 1998, Vaandrager and van Schuppen, 1999, Lynch and Krogh, 2000, Benedetto and Sangiovanni-Vincentelli, 2001, Tomlin and Greenstreet, 2002), or *Automation of Mixed Processes* (in future: *Analysis and Design of Hybrid Systems*) (Zaytoon, 1998, Engell et al., 2000). Examples for special issues on Hybrid Systems are (Antsaklis and Nerode, 1998a, Schumacher et al., 1999, Antsaklis, 2000, Maler, 2001), a survey on the control of hybrid systems can be found in (Lemmon et al., 1999). A monograph is also available (van der Schaft and Schumacher, 2000).

Acknowledgments. The results and opinions presented in this paper were developed while I was a member of the Process Control Laboratory in the Chemical Engineering Department at the University of Dortmund. They are the result of many discussions with colleagues and partners in several research projects. I am in particular grateful to Nanette Bauer, Paul Chung, Sebastian Engell, Holger Graf, Hans-Michael Hanisch, Oded Maler, Bruce Krogh, Yassine Lakhnech, Angelika Mader, Peter Niebert, Jörg Preußig, Olaf Stursberg, and Heinz Treseler. Apart from the KONDISK program, the following research projects contributed to the presented results and experiences: the ESPRIT LTR project *Verification of Hybrid Systems* (VHS) funded by the European Commission, see (Maler, 2001), the temporary graduate school ("Graduiertenkolleg") *Modelling and Model-Based Design of Complex Technical Systems* funded by the German Research Council (DFG), and the exchange programs *British-German Academic Research Collaboration* (ARC) with the British Council and *Project-related Exchange of Personnel* with the NSF both funded by the German Academic Exchange Service (DAAD).

Reachability Analysis and Control of a Special Class of Hybrid Systems

Gero Nenninger[1], Goran Frehse[2], and Volker Krebs[3]

[1] Robert Bosch GmbH, CS/AS-ESW, P.O. Box 30 02 40, D-70442 Stuttgart
[2] Universität Dortmund, Lehrstuhl für Anlagensteuerungstechnik, Emil-Figge-Straße 70, D-44221 Dortmund
[3] Universität Karlsruhe, Institut für Regelungs- und Steuerungssysteme, Kaiserstraße 12, D-76131 Karlsruhe

Abstract. The main task in the control of dynamical systems with mixed discrete-continuous behavior is to guide its hybrid state from an actual operating point to the desired target state. One precondition for the design of an appropriate controller is a reachability analysis to determine all states which are both, reachable from the initial state and controllable to the target state.

The analysis as well as the controller synthesis closely depend on the model representing the knowledge about the process with respect to the desired objective. If the control task requires the precise discrete-continuous state, a mixed discrete-continuous modeling of the hybrid dynamics is necessary.

This article presents a hybrid control concept based on the Net State Model formalism and outlines the fundamental design steps. The mathematical method is explained for hybrid systems with piecewise affine continuous dynamics. A two-tank-system serves as an illustrative example.

1 Introduction

For all technical systems, a save, economic, and environment-friendly operation is desired, often without man acting continuously for the control of the process but rather supervising the system. For that purpose, the process and its components have to run in particular operating points or ranges, or certain sequences of events have to be ensured. New reference inputs and the influence of disturbances require a process control unit to maintain or adapt the system's state and so to exclude an undesirable or even dangerous behaviour for man and environment.

Dependent on the control objective, the type and accuracy of the model to describe the real world process under investigation is to be selected. If the system's behavior at the adequate level of abstraction is characterized by time-driven variables changing their dynamics or values depending on certain events, generally both, discrete and continuous variables are necessary to cover completely the system's state representation at a certain instant in time. Such systems are called mixed discrete-continuous or hybrid systems if different continuous dynamics are possible within the same area of the state space depending on the history of the system evolution.

If the system is to be guided from one area of the state space to another without considering the precise value of the state, a purely discrete control strategy based

S. Engell, G. Frehse, E. Schnieder (Eds.): Modelling, Analysis and Design of Hybrid Systems, LNCIS 279, pp. 173–192, 2002.

on a discrete abstraction of the mixed discrete-continuous behavior is sufficient to match the control objective (Raisch and O'Young, 1998). On the other hand, for technical systems with only a few discrete phenomena the focus is on the continuous time behavior and thus on modeling paradigms and control methods, respectively, which use special elements in the corresponding differential equations to represent the discrete behavior (Branicky, 1995). However, if at least in parts of a complex system the discrete and continuous behavior is to be treated on an equal basis, the ideas and methodologies of the approaches mentioned above must be integrated in a complementary way. In that case, the hybrid control task to be dealt with is to be solved by an adequate combination of discrete and continuous inputs. Therefore, hybrid controller design is based on a hybrid process model which either combines the dynamic parts in a common formalism like Hybrid Petri Nets (Chouikha et al., 2000) or represents the discrete and continuous behavior with its own part and links them adequately.

The last idea is used in this article. Based on the analysis of an uncontrolled process, measures to ensure a certain behavior under unknown but bounded external influences are derived and a hybrid controller is designed to implement these measures. After a brief review of the Net State Model (Nenninger et al., 1999) in Sect. 2, the essential steps for hybrid controller design with respect to the precise hybrid state's transfer are outlined. Key elements are a forward-backward reachability analysis along with a state space control method. Due to the complexity of the presented concepts, we focus on hybrid models with piecewise affine continuous dynamics and we make use of its characteristics like their left eigenvectors. In addition to the two-tank-system in Sect. 6, throughout this article, simple academic examples are used to illustrate the concepts described in detail especially in (Nenninger, 2001).

2 The Net State Model

The Net State Model (NSM) consists of an interpreted Petri Net representing the discrete event subsystem and an extended state space model describing the continuous subsystem (Fig. 1). Both system parts are coupled by means of particular interfaces. Moreover, discrete as well as continuous inputs and outputs may occur. The general form of the NSM is given in (Nenninger, 2001). In the following, we consider a simplified version, omitting jumps of the continuous state variables and assuming the output variables to be identical to the state variables. The continuous dynamics be piecewise affine due to admissible simplifications during modeling or due to a suitable approximation of the nonlinear continuous dynamics (Nenninger, 2001). Under these assumptions, the NSM is defined by the 6-tuple

$$\Sigma_{NSM} = (T, \mathcal{X}_D, \mathcal{U}_D, \underline{f}_{Dx}, \underline{f}_{\beta}, \mathcal{X}_C, \mathcal{U}_C, \underline{f}_{Cx}, \underline{f}_{CD}, \underline{f}_{DC}, \mathbf{x}_0).$$

$T = \mathbb{R}^+$ is the time scale. The continuous state $\underline{x}_C(t) \in \mathcal{X}_C \subseteq \mathbb{R}^{nC}$ changes according to the state space model

$$\begin{aligned} \dot{\underline{x}}_C(t) &= \underline{f}_{Cx}(\underline{x}_C(t), \underline{v}_C(t), \underline{u}_C(t)) \\ &= \underline{A}(\underline{v}_C) \cdot \underline{x}_C(t) + \underline{a}_0(\underline{v}_C) + \underline{B}(\underline{v}_C) \cdot \underline{u}_C\,, \\ \underline{x}_C(0) &= \underline{x}_{C0} \end{aligned} \tag{1}$$

with the parameter vector $\underline{v}_C(t) = \underline{f}_{DC}(\underline{x}_D(t)) \in \mathbb{R}^{m_C}$ and influenced from outside by continuous inputs $\underline{u}_C(t) \in \mathcal{U}_C \subseteq \mathbb{R}^{p_C}$.

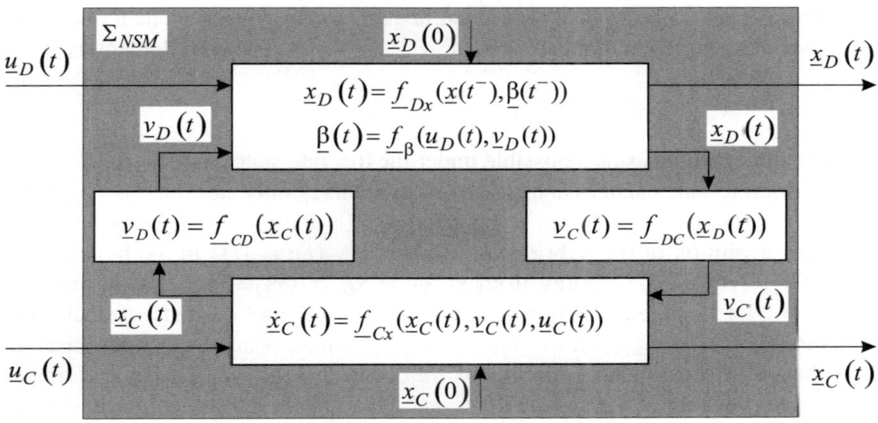

Fig. 1. Net State Model

The discrete state $\underline{x}_D(t) \in \mathcal{X}_D \subseteq \mathbb{N}^{n_D}$ is equivalent to the marking of the interpreted Petri Net with n_D places and m transitions and depends on the pre-state $\underline{x}_D(t^-)$:

$$\underline{x}_D(t) = \underline{f}_{Dx}(\underline{x}_D(t^-), \underline{\beta}(t^-)) = \underline{x}_D(t^-) + \underline{N} \cdot \underline{s}(\underline{x}_D(t^-), \underline{\beta}(t^-)), \tag{2}$$

where $x_{Di}(t)$ specifies the number of tokens in place i at time t, $\underline{N} = [\underline{\tau}_1, \ldots, \underline{\tau}_m]$ is the incidence matrix of the Petri Net with the column vectors $\underline{\tau}_i$ corresponding to the marking modification if the associated transition i fires; $\underline{s}$ is the firing vector. A transition i is enabled, iff

$$\underline{0} \leq \underline{x}_D + \underline{\tau}_i \leq \underline{\kappa}$$

holds line by line. The vector $\underline{\kappa}$ contains the maximum capacity κ_ℓ of each place ℓ. An enabled transition i fires immediately iff the corresponding boolean condition

$$\beta_i(t) = f_{\beta i}(\underline{u}_D(t), \underline{v}_D(t)) \in \{0, 1\}^m$$

is true. At first, the boolean condition depends on the discrete input vector $\underline{u}_D \in \mathcal{U}_D \subseteq \mathbb{Z}_{p_D}$. A change of the discrete state $\underline{x}_D(t)$ corresponds to a controlled event if it is directly initiated by a change of at least one component in $\underline{u}_D(t)$. An internal event occurs, if a change of the discrete state is initiated by the trigger vector $\underline{v}_D(t) =$

$\underline{f}_{CD}(\underline{x}_C(t)) \in \{0,1\}^{m_D}$, with $v_{Di}(t) = 1$, iff $\underline{x}_C(t) \in \Omega_i$. If the continuous state's reaching of a manifold $\Omega_i \subseteq \mathcal{X}_C$, a so-called trigger set, results in an internal event, Ω_i is referred to as an active trigger set under the current discrete state $\underline{x}_D$, marked as $\Omega_i^*(\underline{x}_D)$. We suppose the trigger sets boundaries denoted as $\partial\Omega_i$ to be hyperplanes

$$\partial\Omega_i = \{\underline{x}_C \in \mathcal{X}_C \,|\, \underline{\omega}_i^T\, \underline{x}_C = \omega_{0i}\}\,.$$

In order to receive a clearly defined model, a boundary common to several neighboring trigger sets Ω_i is to be assigned exclusively to one of them. The union of all trigger sets $\Omega_i^*(\underline{x}_D)$ is marked as $\Omega^*(\underline{x}_D) = \cup_i \Omega_i^*(\underline{x}_D)$. The complementary set

$$\Gamma(\underline{x}_D) = \{\underline{x}_C | \underline{x}_C \notin \Omega^*(\underline{x}_D)\}$$

contains all continuous states possible under the discrete state $\underline{x}_D$.

The discrete state changes at discrete time instants t_k only, i.e., $\underline{x}_D(t) = \underline{x}_D(t_k)$, $\underline{v}_C(t) = \underline{v}_C(t_k)$, $\forall t \in [t_k, t_{k+1})$, and thus, $\underline{x}_D(t)$, and $\underline{v}_C(t)$ are piecewise constant. The evolution of the hybrid state $\mathbf{x}(t) = (\underline{x}_D(t); \underline{x}_C(t))$ in the hybrid state space $\mathcal{X}_H = \mathcal{X}_D \times \mathcal{X}_C$ starting from $\mathbf{x}(t_0) = \mathbf{x}_0 = (\underline{x}_D(0); \underline{x}_C(0))$ in the time interval $[t_0, t_e]$ is called a hybrid trajectory $\mathbf{x}_{[t_0,t_e]}(t)$ and consists of a discrete trajectory $\underline{x}_{D[0,\ell]} = \langle \underline{x}_D(0), \underline{x}_D(1), \ldots, \underline{x}_D(\ell)\rangle$ (thus, inside $[t_0, t_e]$ the discrete state changes ℓ times) and a continuous trajectory $\underline{x}_{C[t_0,t_e]}(t)$. The hybrid trajectory evolves because of the system's internal dynamics or a suitable hybrid control $\mathbf{u}(t) = (\underline{u}_D(t); \underline{u}_C(t)) \in \mathcal{U}_H = \mathcal{U}_D \times \mathcal{U}_C$, which comprises a piecewise constant discrete control $\underline{u}_D(t)$ and a continuous control $\underline{u}_C(t)$.

3 Hybrid Control and the Importance of Reachability Analysis

The Net State Model described above is now used to design a controller which guarantees a desired behaviour of the plant under unknown but bounded external influences. In this connection, there are two main control objectives: on the one hand, in case of temporary disturbances it is to assure that the plant state will return to the previous operating point. This corresponds to the stability of an equilibrium point, a fundamental property dealt with for hybrid systems e.g. in (Li et al., 2000, Nenninger, 2001). On the other hand, the process state is to be guided from one operating point to another by means of the available inputs. Interpreting the plant state after a disturbance as initial state and the operating point to be stabilized as target state, the first objective is equivalent to the second one.

Thus, the basic control task is to lead the plant state $\mathbf{x}_0$ out of an initial set $X_0 = \{\mathbf{x} : \underline{x}_D = \underline{x}_{D0}, \underline{x}_C \in X_{C0}\}$ to a state $\mathbf{x}_e$ out of the target set $X_e = \{\mathbf{x} : \underline{x}_D = \underline{x}_{De}, \underline{x}_C \in X_{Ce}\}$. To realize more complex process runs, as in batch or cyclic processes, this basic task must be carried out repeatedly.

If the control objective requires the consideration of the precise discrete-continuous state as assumed in this contribution, the control task usually cannot be solved using either discrete or continuous input values, but a combination of both types of inputs. Therefore, a hybrid control loop as depicted in Fig. 2 consists of two NSM modules, one for the hybrid plant and the other for an appropriate hybrid controller. Discrete and continuous variables are coupled with each other exclusively to distinguish transparently between events within the process and within the controller and, moreover, all discrete information about the plant state needed by the controller is obtained from the discrete state of the plant. For that reason we assume complete state observability.

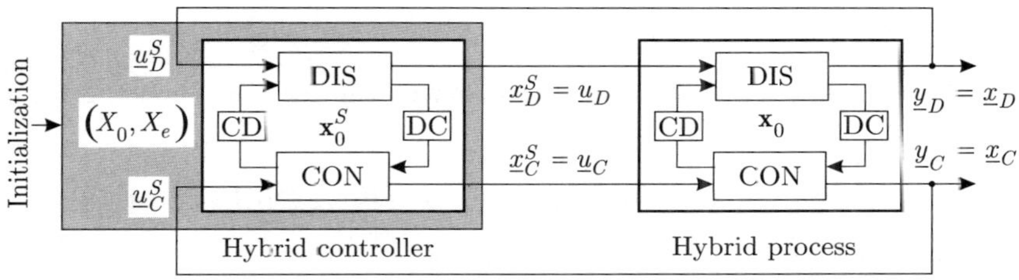

Fig. 2. Structure of a hybrid control system

Because of the closed-loop structure, the hybrid controller reduces the influence of disturbances during the state transfer, if they do not drive the system out of the desired target equilibrium point's hybrid domain of attraction. Methods for the estimation of hybrid domains of attraction can be found in (Nenninger, 2001).

Both, for controller synthesis and verification a reachability analysis is fundamental to determine which hybrid states are reachable within a finite time interval starting from certain initial states and applying the available inputs. In the end, if the set of all reachable states of the controlled system is a subset of the designer's requirements which can consist of constraints, desired or forbidden states, the controller is said to be valid.

For hybrid systems as a special class of nonlinear systems, usually there is no universally valid criterion to examine the reachability of all pairs of initial and target states at the same time. Therefore, reachability is to be carried out for special pairs of states $(\mathbf{x}_0, \mathbf{x}_e)$ or state sets (X_0, X_e) (Nenninger et al., 2001).

Definition 1 (Reachability for hybrid systems).
A hybrid system is called reachable according to a set $X_0 = \{\mathbf{x} : \underline{x}_D = \underline{x}_{D0}, \underline{x}_C \in X_{C0}\}$ of hybrid initial states and a set $X_e = \{\mathbf{x} : \underline{x}_D = \underline{x}_{De}, \underline{x}_C \in X_{Ce}\}$ of hybrid target states, iff its state can be transferred out of X_0 by suitable inputs to the set X_e within a finite time interval $[0, t_e]$.

A reachability analysis starts at the initial state $\mathbf{x}_0$ or state set X_0 and tries to find out all reachable states by means of a recursive procedure (Nenninger et al., 1999). All continuous states reachable only by continuous inputs under the discrete state $\underline{x}_{D\ell}$ without triggering a discrete state transition form the local reachability set

$$R_C^{[j]}(k; \underline{x}_{D\ell}, X_C^{[j]}(k)) ; \tag{3}$$

k denotes the step within the recursive procedure. The release set $R_C^{\partial\Omega_i}(\cdot;\underline{x}_{Dj},X_C)$ is the subset of the trigger set's Ω_i boundary $\partial\Omega_i$ active under $\underline{x}_{D\ell}$ reachable from $X_C(k)$ using the available continuous inputs. By calculation of the local reachability sets of succeeding discrete states, the hybrid reachability set $R(X_0)$ is iteratively constructed. If $X_e \subseteq R(X_0)$ applies, all states within X_e are reachable starting from X_0 with suitable chosen inputs. The representation of the reachability set as a hybrid reachability graph depicts one or more discrete trajectories

$$\underline{x}_{D[0,\ell]} = \langle \underline{x}_D(0), \underline{x}_D(1), \ldots, \underline{x}_D(\ell)\rangle \tag{4}$$

with $\underline{x}_D(0) = \underline{x}_{D0}$ and $\underline{x}_D(\ell) = \underline{x}_{De}$ combined with the corresponding local reachability sets

$$R_{C[0,\ell]} = \langle R_C(0;\underline{x}_D(0),\cdot), R_C(1;\underline{x}_D(1)\cdot), \ldots, R_C(\ell;\underline{x}_D(\ell),\cdot)\rangle$$

from the initial set to the target set (Nenninger, 2001).

However, this forward reachability analysis yields a necessary condition for the desired state transfer only. For nonlinear systems, not all states reachable from an initial state must be controllable to a certain target state. Therefore, all hybrid states along the discrete trajectories (4) with both characteristics according to X_0 and X_e have to be determined. This is done by a reachability analysis starting from X_0 up to X_e in combination with a reachability analysis starting from X_e backwards to X_0 along the discrete trajectories (4) passed through in reverse order. Within the Net State Model, for time invariant continuous dynamics the sign of the gradient in (1) reverses, and the state transitions within the interpreted Petri Net take place in reverse order. Thus, the incidence matrix $\underline{N}$ changes to $-\underline{N}$.

The local reachability sets $R_C(k;\underline{x}_{Dj},\cdot)$ of the forward reachability analysis along $\underline{x}_{D[0,\ell]}$ determine together with the corresponding sets $\bar{R}_C(\ell-k;\underline{x}_{Dj},\cdot)$ of the backward analysis – which yield in summary the backward reachability set $\bar{R}(X_e)$ – the intersection

$$\mathcal{R}_C(k) = R_C(k;\underline{x}_{Dj},\cdot) \cap \bar{R}_C(\ell-k;\underline{x}_{Dj},\cdot)\,. \tag{5}$$

At the k th step of the recursion, the discrete state $\underline{x}_{Di}$ and the corresponding set $\mathcal{R}_C(k)$ according to (5) form the pair

$$\mathcal{R}(k) = \big(\underline{x}_D(k) = \underline{x}_{Dj}; \mathcal{R}_C(k)\big)\ .$$

It is a part of the so called *transfer set* $\mathcal{R}(X_0, X_e)$, which comprises the ordered sequences $\mathcal{R}_{[0,\ell]} = \langle \mathcal{R}(0), \mathcal{R}(1), \ldots, \mathcal{R}(\ell)\rangle$ of local transfer sets calculated along the discrete trajectories (4) with $X_0 \subseteq \mathcal{R}(0)$ and $X_e \cap \mathcal{R}(\ell) \neq \emptyset$. A sufficient condition for the state transfer from X_0 to X_e is $X_e \cap R(X_0) \neq \emptyset$ and at least one trajectory $\underline{x}_{D[0,\ell]}$ according to (4) fulfills $X_0 \subseteq \bar{R}(X_e)$.

In principle, the order of forward and backward reachability analysis can be exchanged, but the backward reachability analysis for its own generally does not yield the minimal local reachability sets inside of which the hybrid trajectory is to be kept by the controller during the transfer from X_0 to X_e.

Because of the considerable effort entailed with a bidirectional reachability analysis for a pair of states, the reachability analysis always is done for sets X_0 and X_e of initial and target states. Obviously, the calculation is in general large-scaled because of the multitude of possible internal and external events and the shape of the individual local reachability sets. Local reachability sets $\mathcal{R}_C(k)$ are bounded by hyperplanes in special cases only. Even that does not guarantee that the reachability of the system relating to the state sets X_0 and X_e is decidable. If the reachability set $R(X_0)$ and so the reachability graph cannot be determined completely, one cannot definitely state whether any state set X_e is reachable from X_0 or not.

Decidability can be determined for special classes of hybrid systems by means of bisimulations. Their existence is connected with the interaction of the vector field's flow (1) within every discrete state and the possible initial state sets and active trigger sets Ω_i^* in each case. The state sets and flows of vector fields involved in must meet the o-minimal theories in mathematical logic (Lafferriere et al., 2000). However, note that for hybrid controller design the indicated procedure for reachability analysis is useful despite of possibly missing decidability: for the solution of technical problems not the entire reachability set is necessary, but any possible traces from an initial set of operating points to a desired state set are to be found. Certainly, the occurring state sets must be constructed analytically or at least numerically by approximation. In general, this is possible for special classes of hybrid systems only.

After calculating the transfer set $\mathcal{R}(X_0, X_e)$ by means of the bidirectional reachability analysis a hybrid controller can be designed which transfers all states out of X_0 to X_e according to a given specification. The state transfer task is divided in several parts. On a global level, the controller has to force a trajectory $\underline{x}_{D[0,\ell]} = \langle \underline{x}_D(0), \dots, \underline{x}_D(\ell) \rangle$ out of the transfer set. This discrete trajectory evolves on the one hand due to internal events because of the continuous trajectory's entering active trigger sets and on the other hand due to the release of controlled events caused by a change of discrete inputs $\underline{u}_D$. As depicted in Fig. 2, the discrete inputs of the hybrid process are the discrete states $\underline{x}_D^S$ of the hybrid controller. Consequently, $\underline{u}_D$ changes because of discrete state transitions within the controller dependent on the discrete part of the process via discrete state feedback or on the continuous part $\underline{x}_C^S$ of the hybrid controller. Therefore, controlled events within the process – corresponding with a change of the controller's discrete state – are connected with internal events of the process or they can be fixed at certain moments by means of integral continuous dynamics of the hybrid controller (Nenninger, 2001).

If there are several discrete trajectories from the initial to the target set as a result of the reachability analysis, one of them is to be chosen, using either heuristic knowledge about the process and its control strategies or by evaluation of individual discrete states and state transitions. Though, the development of such measures requires the design of local continuous controllers under each discrete state (Nenninger, 2001). That kind of a discrete trajectory's selection may not be confused with the determination of an optimal hybrid trajectory by minimization of a global mixed discrete-continuous cost index: for hybrid optimal control, all control inputs have to be chosen appropriately not only relating to isolated parts of continuous trajectories but to the entire hybrid trajectory (Schnabel, 2001).

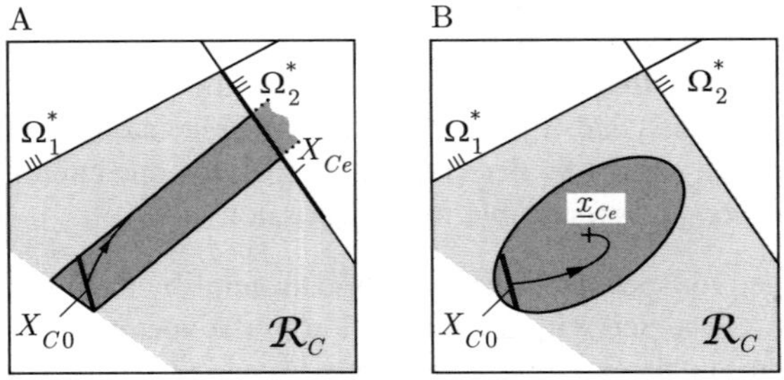

Fig. 3. Two basic situations for local control: state transfer to release an internal event (**A**) and stabilizing an equilibrium point under the discrete target state (**B**)

Besides this global problem the local task is to guide the continuous state under each discrete state by appropriate continuous inputs within the local transfer set $\mathcal{R}_C$ from the local initial set X_{C0} to the local target set X_{Ce} which is the initial set for the discrete post-state, cf. Fig. 3. These sets are the link between successive discrete states. The local control in each state is independent from the one under its discrete predecessor and successor. If the local target state is part of an active trigger set's boundary $\Omega_i^*(\underline{x}_D)$, an internal event is released with entering this set and therefore a state transition to the successive discrete state of the selected discrete trajectory takes place. To ensure that the continuous trajectory enters the active trigger set, the local control under $\underline{x}_D$ may be stable iff the closed-loop local equilibrium point is within the active trigger set $\Omega_i^*(\underline{x}_D)$. If there are several equilibrium points possible within $\Omega_i^*(\underline{x}_D)$, one of them is to be chosen as a degree of freedom during design. If not stabilizing an equilibrium point but the continuous state's transfer is on the focus of attention, local control in individual discrete states can yield an unstable closed-loop behavior as long as the continuous state does not leave the local transfer set $\mathcal{R}_C(k)$. Generally, within the discrete target state, the continuous target set X_{Ce} is an equilibrium point stabilized by the local controller.

In this way hybrid control forces the process to generate trajectories within the transfer set $\mathcal{R}(X_0, X_e)$. All in all, the hybrid controller consists of individual local controllers combined in a Net-State-Model: its interpreted Petri Net has a chain-like structure. The only token selects the relevant local controller via the DC-interface according to the place it is in.

4 Reachability Analysis in Case of Affine Continuous Dynamics

To calculate numerically the reachability and release sets in an efficient way and to process them after intersection, they must be completely determined by a finite number of edges or hyperplanes. For example, this condition holds for a continuous dynamics under each discrete state consisting of integrators only (Alur et al., 1993), but generally not for a nonlinear continuous dynamics. Nevertheless, for hybrid systems with linear or affine time-invariant dynamics (1), the reachability analysis can be applied for practical problems under certain conditions.

If we suppose an unbounded state space $\mathcal{X}_C = \mathbb{R}^{n_C}$ and unbounded input variables, there are well-known global criteria for the reachability of states (Föllinger, 1994). For linear dynamics, according to the Hautus criterion, natural boundaries in the shape of hyperplanes through the equilibrium point of the autonomous model and a left eigenvector $\underline{w}_i$ of the matrix $\underline{A}$ in (1) as normal vector can never be crossed if the control intervention runs parallel (Föllinger, 1994). A more detailed discussion of the importance of left eigenvectors can be found in (Nenninger, 2001).

However, even for dynamical systems with full state controllability, the reachability of a target state does not state anything about the course of the trajectory between the initial and the target state. With hybrid systems, the active trigger sets Ω_i^* under a certain discrete state $\underline{x}_{Dj}$ – the boundaries of which are supposed to be hyperplanes – limit the local reachability set $R_C(\cdot;\underline{x}_{Dj},\cdot)$. The existence of further bounding surfaces is evident, in particular if the number p_C of continuous inputs is smaller than the number n_C of state variables. Then the gradient in any state cannot be chosen independently. Because of the special structure of (1), the state vector $\underline{x}_C$ is located in a subspace characterized by the *internal dynamics part* $\hat{\underline{x}}_C = \underline{A}\cdot\underline{x}_C + \underline{a}_0$ and the *controlled dynamics part* $\hat{\underline{u}}_C = \underline{B}\cdot\underline{u}_C$. For any unlimited inputs $\underline{u}_C(t) \in [-\infty, +\infty]^{p_C}$, the entire subspace $U_B(\underline{x}_{C0}) = \{\underline{x}_{C0} + \hat{\underline{u}}_C\} \subseteq \mathbb{R}^{p_C}$ round about a fixed state $\underline{x}_{C0}$ is asymptotically reachable: it consists of the *control hyperplane*

$$H_S(\underline{x}_{C0}) = \left\{ \underline{x}_C \in \mathcal{X}_C \mid \underline{\nu}_B^T\,\underline{x}_C = \nu_0 = \underline{\nu}_B^T\underline{x}_{C0} \right\} \tag{6}$$

through $\underline{x}_{C0}$. The normal vector $\underline{\nu}$ of this hyperplane fulfills $\underline{\nu}_B^T\,\underline{B} = \underline{0}^T$ and we expect it to be adjusted so that $\underline{\nu}_B^T\,\dot{\underline{x}}_{C0} > 0$ holds. The internal dynamics part $\hat{\underline{x}}_C$ extends $U_B(\underline{x}_{C0})$, if the normal component of $\hat{\underline{x}}_C$ to $\hat{\underline{u}}_C$ is unequal to zero for all inputs $\underline{u}_C$. In that case, $[\underline{A}\,,\,\underline{a}_0]$ has a greater rank than $\underline{B}$. If an affine dynamics (1) meets the rank condition

$$rank\,(\underline{B}) = rank\,([\underline{A}\,,\,\underline{a}_0]) - 1\,, \tag{7}$$

the *dividing hyperplane* $H_T(\underline{x}_{C0})$ with

$$|\,\underline{A}\cdot\underline{x}_C(t) + \underline{a}_0\,,\,\underline{B}\,| = 0\,, \tag{8}$$

cuts the state space into two disjoint regions inside of which the normal component of $\dot{\underline{x}}_C$ to H_T points to the opposite direction and does not vanish. We call piecewise affine hybrid systems *linear divided systems* or briefly LD-systems iff for all their discrete states $\underline{x}_D \in \mathcal{X}_D$ the corresponding continuous dynamics meets (7). Because of the linear dependence of $\hat{x}_C = \underline{A}\,\underline{x}_C + \underline{a}_0$ and $\underline{B}$ for LD-systems, the gradient $\dot{\underline{x}}_C$ on H_T can vanish for an appropriate choice of $\underline{u}_C$; therefore, H_T is the set of all possible equilibrium points of (1).

If the boundaries of the trigger sets are hyperplanes, the set $\Gamma(\underline{x}_D)$ is convex. Assuming a hybrid LD-system with unbounded continuous inputs, the local reachability set is limited by hyperplanes and can be determined by means of the previously described restricted directions of motions. Even if the continuous dynamics (1) is completely controllable or reachable under a discrete state $\underline{x}_D$, in most cases not all continuous states $\Gamma(\underline{x}_D)$ are really reachable from an initial state $\underline{x}_{C0}$ or a state set X_{C0}, but the local reachability set according to (3) is a real subset of $\Gamma(\underline{x}_D)$ in most cases. Only if $rank(\underline{B}) = n_C$ holds, $R_C(\cdot; \underline{x}_D, \cdot)$ is equal to $\Gamma(\underline{x}_D)$ for unbounded inputs, and all active trigger sets can be reached.

LD-systems cannot be directly influenced by the control inputs in exactly one direction which is illustrated in Fig. 4 by means of an example in $\mathbb{R}^2$ with a controllable dynamics and one continuous input. A detailed description of the method for higher dimensions, too, is given in (Nenninger, 2001).

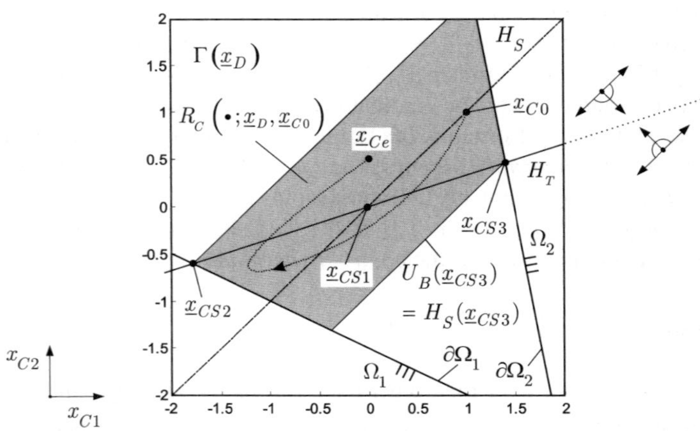

Fig. 4. Example: local reachability set for unbounded inputs

You can see the control hyperplane H_S passing through the initial state $\underline{x}_{C0}$. To reach the target state $\underline{x}_{Ce}$, the trajectory must cross at first the dividing hyperplane H_T (in $\mathbb{R}^2$ a straight line): the possible gradient directions depicted at the right side show that $\underline{x}_{C0}$ cannot be guided directly to $\underline{x}_{Ce}$ because the gradient of the corresponding vector field (1) points above H_T exclusively downwards and therefore, possible gradient directions or trajectories starting in $\underline{x}_{C0}$ are bounded by the control hyperplane $H_S(\underline{x}_{C0})$. Below of H_T, trajectories move to the top left. Crossing the dividing hyperplane is only feasible outside the active trigger sets Ω_1^* and Ω_2^*. Thus, particular parts of the state space are not reachable: for example, the subspace $U_B(\underline{x}_{CS3})$ determined by the intersection $\underline{x}_{CS3}$ of $\partial\Omega_2$ with H_T – if $\partial\Omega_2$ does not belong to the active trigger set Ω_2^* – is asymptotically reachable with unbounded inputs. $U_B(\underline{x}_{CS3})$ is extended by the internal dynamics of the system towards H_T. By analogy, the same holds for the intersection $\underline{x}_{CS2}$ of H_T with the boundary $\partial\Omega_1$ of the second active trigger set Ω_1^*. In addition to the boundaries $\partial\Omega_1$ and $\partial\Omega_2$ of the two active trigger sets, the control hyperplanes through the intersection $\underline{x}_{CS2}$ and $\underline{x}_{CS3}$ bound the local reachability set $R_C(\cdot;\underline{x}_{Di},\underline{x}_{C0})$ outlined in grey in Fig. 4. As a result, no trajectory can reach the area on the top left and at the bottom right starting at $\underline{x}_{C0}$ although these areas belong to $\Gamma(\underline{x}_D)$, because the direction of the control input and the bounded gradient due to the internal dynamics does not approve it.

Frequently, not only an initial state $\underline{x}_{C0}$ but a set X_{C0} of initial states is to consider while calculating a local reachability set. Assuming X_{C0} to be a convex set and to be determined by a finite number of edges, the entire local reachability set is derived by the combination of the reachability sets for each edge of X_{C0}.

After calculating a local reachability set under a discrete state, with help of the active trigger sets, boundaries of the release sets have to be determined and as a result one obtains the new initial sets under the possible discrete post-states. For external events, the new initial set for the successive state is the entire reachability set before the event. An internal event occurs while entering the release set $R_C^{\partial\Omega_i}(\cdot;\underline{x}_{Dj},\cdot)$ which is calculated by that subset of $\partial\Omega_i$ immediately bordering on the local reachability set $R_C(\cdot;\underline{x}_{Dj},\cdot)$ with

$$\forall\,\underline{x}_C \in \partial\Omega_i : \quad \underline{\omega}_i^T\,\dot{\underline{x}}_C < 0 ; \tag{9}$$

$\dot{\underline{x}}_C$ depends on the continuous dynamics valid under $\underline{x}_{Dj}$, even if $\partial\Omega_i$ belongs to the active trigger set Ω_i. If the boundary is not a part of Ω_i, (9) yields a small over estimation of the following local reachability set, because its boundary does not belong to the real reachability set. However, this has no practical meaning, because there is already an over estimation due to the assumption that there are unbounded inputs. Obviously, a state $\underline{x}_{Ce}$ inside a local reachability set $R_C(\cdot;\underline{x}_D,\cdot)$ of a LD-system determined by this method is only a neccessary but not sufficient condition for its reachability from X_{C0} or $\underline{x}_{C0}$, respectively, under a certain discrete state $\underline{x}_D$.

A sufficient condition for reachability requires the consideration of the real limited continuous inputs. Generally, the boundaries of the local reachability sets are no longer hyperplanes but parts of trajectories, and the shape of the exact sets makes it a large-scale problem, especially to determine the new initial set under the discrete post-state after a controlled event. Therefore, in practical use, a precise analysis considering bounded inputs is carried out in essential discrete states only, for example for systems with two continuous state variables by a graphical method (Nenninger, 2001).

In summary, a numerically effective, recursive procedure, determining individual local reachability sets, computing possible internal or external events and the initial continuous state sets under the new discrete states, yields for LD-systems at the end the transfer set, inside which a hybrid controller has to keep the system's state while transferring it from the initial set to the target $\underline{x}_{Ce}$.

5 Local Control Strategies Based on Left Eigenvectors

According to Sect. 3, the hybrid control loop is closed by the state variable's feedback (see Fig. 2). In this article, a linear feedback

$$\underline{y}_C^S = \underline{u}_C = -\underline{R}(\underline{v}_C^S) \cdot \underline{x}_C + \underline{r}_0(\underline{v}_C^S) \quad \text{with} \quad \underline{v}_C^S = \underline{f}_{DC}^S(\underline{x}_D^S) \tag{10}$$

of the continuous states dependent on the particular discrete state $\underline{x}_D^S$ of the controller and therefore of the process state $\underline{x}_D$ is considered.

With the constant part $\underline{r}_0$ the equilibrium point of the controlled system is to be determined within the set of all possible equilibria. The feedback matrix $\underline{R}$ responsible for the dynamics of the closed loop contains $p_C \cdot n_C$ free parameters. The required design of the local control strategy under each discrete state comprises the change of the local gradient field (1) to guarantee a desired course of all possible trajectories starting in the set of local initial states to the local target set within the local transfer set. With the feedback (10), the hybrid closed-loop system gets the continuous dynamics

$$\begin{aligned} \dot{\underline{x}}_C &= (\underline{A}(\underline{v}_C) - \underline{B}(\underline{v}_C)\underline{R}(\underline{v}_C)) \cdot \underline{x}_C + (\underline{B}(\underline{v}_C)\underline{r}_0(\underline{v}_C) + \underline{a}_0(\underline{v}_C)) \\ &= \underline{A}_R(\underline{v}_C) \cdot \underline{x}_C + \underline{a}_{0R}(\underline{v}_C) \ , \end{aligned} \tag{11}$$

which is piecewise affine like the open loop.[1] Therefore, the left eigenvector assignment design described in the following can be used. After the design of the local control policy, also e.g. with pole placement, the same methods – namely reachability analysis – as for the open loop are applied to check whether all continuous trajectories remain inside the local transfer set.

The left eigenvectors of $\underline{A}$ within (1) are characteristic for the corresponding continuous dynamics: they influence the course of the continuous trajectories by natural boundaries the trajectories cannot cross, as indicated in the previous section. The objective of left eigenvector assignment is to move these boundaries under each individual discrete state $\underline{x}_D$ by a specific change of the left eigenvectors in such a way that the closed-loop behavior meets the requirements. Thereby it is ensured that no trajectory starting in the local initial set X_{C0} leaves the transfer set $\mathcal{R}_C(k)$ and releases undesirable internal events, but

- triggers the expected internal events and, therefore, ensures the transition to the respective successive discrete state by entering the particular local target set $X_{Ce} \subset \Omega_i^*(\underline{x}_D)$, or guarantees alternatively that
- all trajectories within $\mathcal{R}_C(k) \subseteq \Gamma(\underline{x}_D)$ converge to an equilibrium point, if $\underline{x}_D$ is the discrete target state of the entire discrete state transfer.

For the last case, the local control strategy must asymptotically stabilize the desired operating point and has to ensure boundaries of the closed-loop continuous dynamics comprising the initial set and the equilibrium point. These boundaries have the shape

[1] For simplicity, we do not distinguish here in notation between the parameter vector $\underline{v}_C$ of the process and $\underline{v}_C^S$ of the controller.

of polyhedra or ellipsoids (Jirstrand, 1998) and must be located completely inside of $\mathcal{R}_C(k) \subseteq \Gamma(\underline{x}_D)$ (Nenninger et al., 2000, Nenninger, 2001).

The available degrees of freedom in the choice of the closed-loop continuous equilibrium point and the feedback matrix $\underline{R}$ must be used at first to fulfill the above-mentioned demands; beyond it, an optimization of the local behavior can be done.

Starting with the definition $\underline{w}_i^T(\lambda_i \underline{I} - \underline{A}) = \underline{0}^T,\ i = 1, \dots, n_C$, of the left eigenvectors, the n_C equations for the closed loop yield

$$\underline{\Lambda}_R \underline{W}_R = \underline{W}_R \underline{A}_R = \underline{W}_R(\underline{A} - \underline{B}\,\underline{R}) \tag{12}$$

with the matrix

$$\underline{\Lambda}_R = \begin{bmatrix} \lambda_{R1} & \dots & 0 \\ \vdots & \ddots & \vdots \\ 0 & \dots & \lambda_{Rn_C} \end{bmatrix}$$

of the eigenvalues λ_{Ri} and the matrix

$$\underline{W}_R = \begin{bmatrix} \underline{w}_{R1}^T \\ \vdots \\ \underline{w}_{Rn_C}^T \end{bmatrix}$$

of the left eigenvectors $\underline{w}_{Ri}$ of the controlled system. For $p_C < n_C$, (12) cannot be solved exactly for the feedback matrix $\underline{R}$ because $\underline{W}_R \underline{B}$ has the dimension $n_C \times p_C$; therefore only p_C of the n_C left eigenvectors $\underline{w}_{Ri}$ can be chosen independently by means of $\underline{R}$. Assigning p_C eigenvalues and left eigenvectors,

$$\underline{\Lambda}_{R,p_C} = \begin{bmatrix} \lambda_{R1} & \dots & 0 \\ \vdots & \ddots & \vdots \\ 0 & \dots & \lambda_{Rp_C} \end{bmatrix} \quad \text{and} \quad \underline{W}_{R,p_C} = \begin{bmatrix} \underline{w}_{R1}^T \\ \vdots \\ \underline{w}_{Rp_C}^T \end{bmatrix}, \tag{13}$$

and replacing $\underline{\Lambda}_R$ in (12) by $\underline{\Lambda}_{R,p_C}$ and $\underline{W}_R$ by $\underline{W}_{R,p_C}$, and solving for $\underline{R}$ ends in

$$\underline{R} = \left(\underline{W}_{R,p_C} \underline{B}\right)^{-1} \left(\underline{W}_{R,p_C} \underline{A} - \underline{\Lambda}_{R,p_C} \underline{W}_{R,p_C}\right), \tag{14}$$

if the condition

$$\underline{w}_{Ri}^T \cdot \underline{B} \neq \underline{0}^T, \quad i = 1, \dots, p_C$$

holds. The desired p_C left eigenvectors of the closed loop may not be perpendicular to the possible control inputs. Otherwise, there are boundaries which cannot be crossed and the controlled system is not controllable and reachable, respectively. Equation (14) reminds of the controller formula of Modal Control (Föllinger, 1994), but instead of $\underline{W}_{R,p_C}$ the matrix $\underline{W}$ of the left eigenvectors is found, because this method is based on the modal structure of the uncontrolled process.

The feedback of $\underline{x}_C$ over $\underline{R}$ according to (14) results in the p_C desired eigenvalues in (13) for the closed loop, which is to be checked by

$$|\lambda_R \underline{I} - \underline{A} + \underline{B}(\underline{W}_{R,p_C}\underline{B})^{-1}\left(\underline{W}_{R,p_C}\underline{A} - \underline{\Lambda}_{R,p_C}\underline{W}_{R,p_C}\right)| = 0\,.$$

However, for $p_C < n_C$ there does not exist inevitably suitable local boundaries for the closed-loop continuous trajectories because the left eigenvectors are not totally independent as mentioned above. This means for practical use an iterative dialog with the computer to determine appropriate left eigenvectors for the closed loop under the individual discrete states along the discrete trajectory so that all hybrid initial states get to the hybrid target set within the desired transfer set.

In the following, the left eigenvector assignment is illustrated within one discrete state with the affine continuous dynamics

$$\dot{\underline{x}}_C = \begin{bmatrix} -0.5 & 0 \\ 0.2 & -0.05 \end{bmatrix} \underline{x}_C + \begin{bmatrix} 1 \\ 0 \end{bmatrix} u_C + \begin{bmatrix} 0.4 \\ -0.15 \end{bmatrix}$$

and the equilibrium point $\underline{x}_{CR} = [0.8 \;\; 0.2]^T$ for $u_C = 0$. All states out of the initial set $X_{C0} = \{\underline{x}_C \,|\, x_{C1} = 1\,, 0.4 \leq x_{C2} \leq 0.6\}$ should be guided to the target set $X_{Ce} = \{\underline{x}_C \,|\, 1.3 \leq x_{C1} \leq 2\,, x_{C2} = 1\}$ to release there a discrete state transition. The eigenvalues and left eigenvectors of the uncontrolled system are $\lambda_1 = -0.5\,, \underline{w}_1^T = [1\,, 0]^T$ and $\lambda_2 = -0.05\,, \underline{w}_1^T = [0.4061\,, 0.9138]^T$. This yields boundaries for trajectories of the uncontrolled system starting in X_{C0}, depicted grey in Fig. 5 on the left side.

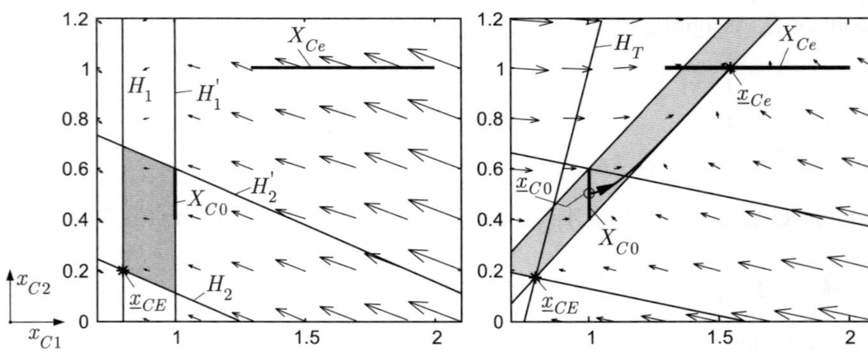

Fig. 5. Gradient field, natural boundaries as well as initial and target set for the left eigenvector assignment example, open-loop (left) an closed-loop (right) system

The dynamics of the closed-loop system with a state feedback (10) can be adapted like in Fig. 3 so that a polyhedron contains completely the initial set X_{C0} and at least partially the target state X_{Ce} and all trajectories starting in X_{C0} do not leave the polyhedron until releasing an internal event while entering the local target set. On account of the initial and target set's position, the normalized closed-loop eigenvector is set to $\underline{w}_{R1}^T = [-0.7399\,,\ 0.6727]$ with $\lambda_{R1} = -1$. According to (14), the corresponding local control is $\underline{R} = \underline{r}^T = \left[0.3182\ -0.8636\right]$, and both the left eigenvectors change their values for the closed loop – one of them as demanded, the other unintentionally but unavoidable ($\underline{w}_{R2}^T = [-0.2606\,,\ -0.9785]$). As a result there is an unstable closed-loop with $\lambda_{R1} = -1$ and $\lambda_{R2} = 0.1318$. All states out of X_{C0} reach autonomously the target set, if the polyhedron is suitable adapted by means of the constant part r_0 in (10) which moves the equilibrium point of the closed loop on the hyperplane H_T. Figure 5 depicts on the right side the result for $r_0 = 0.1$ and shows the closed-loop trajectory starting from $\underline{x}_{C0}^T = [1\,,\ 0.5]^T$.

In the context of the precise hybrid state's transfer, another approach for the local closed-loop design is to calculate e.g. a linear state feedback (10) by a standard optimization method so that the continuous trajectory moves within a certain sector from the initial to the target state (Nenninger, 2001). This second local approach can be combined with the left eigenvector assignment for the design of an hybrid controller.

6 Application Example

The essential methods for reachability analysis and controller synthesis presented in this contribution are demonstrated by means of the two-tank-system depicted in Fig. 6.

6.1 A Two-Tank-System

The process consists of two tanks with the bottom area A connected with each other by a pipe with A_0 in height h the cross-section of which is very small relatively to h. Another pipe with A_1 between the two tanks on the same latitude as the tank's ground can be opened or closed by means of the valve V the discrete input $u_D \in \{0, 1\}$ is assigned to. A change from $u_D = 0$ to $u_D = 1$ is the controlled event for opening the valve, the opposite change closes the valve. Tank 2 has a permanently open outlet with the cross-sectional area A_2. A pump P feeds liquid into tank 1, and the flow can be continuously controlled by $u_C \in [0, 1]$ up to the maximum Q_{Pmax}. If the level in tank 1 reaches h_{max} the discrete sensor LS11 switches P off to prevent an overflow of T1.

This process has a typical hybrid behavior. If the valve V is closed and the liquid levels in each tank are below h, the dynamics of each tank are independent. By contrast, if one of the levels is greater than the connecting pipe above, there is a one-sided coupling of the two states. Exceed both levels the height h even a two-sided coupling exists. Therefore, the two-dimensional state space is divided in 4 areas with two discrete states each representing the continuous dynamics for the closed and the open valve V, respectively.

Assuming an affine behavior of the liquid levels in each tank and of the flows through the connecting pipes and the outlet, the prescribed reachability analysis and

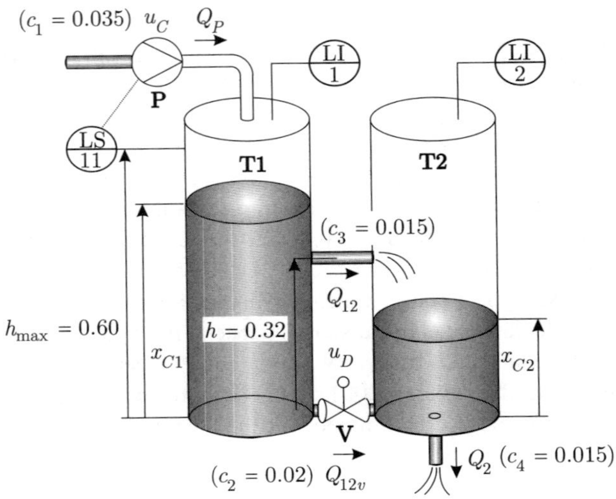

Fig. 6. Two-tank-system

controller design methods can be immediately applied. The corresponding Net State Model has the continuous part[2] according to (1) with

$$\underline{A}(\underline{v}_C) = \begin{bmatrix} -c_2 v_{C1} - c_3(v_{C3} + v_{C5}) & c_2 v_{C1} + c_3(v_{C4} + v_{C5}) \\ c_2 v_{C1} + c_3(v_{C3} + v_{C5}) & -c_2 v_{C1} - c_3(v_{C4} + v_{C5}) - c_4 \end{bmatrix}$$

as well as

$$\underline{a}_0(\underline{v}_C) = c_3\, h \begin{bmatrix} v_{C3} - v_{C4} \\ v_{C4} - v_{C3} \end{bmatrix} \quad \text{and} \quad \underline{b}(\underline{v}_C) = \begin{bmatrix} c_1 v_{C2} \\ 0 \end{bmatrix} .$$

The values of each constant c_i can be read from Fig. 6. The components v_{Ci} of the parameter vector $\underline{v}_C = [v_{C1}, v_{C2}, v_{C3}, v_{C4}, v_{C5},]$ depend on the Net State Model's discrete state, the marking of the interpreted Petri Net. It consists of 4 simple nets, one for the discrete state of the valve V, two for the qualitative levels in both tanks below or above h, and the last for the overflow protection (Nenninger, 2001). Table 1 reflects a part of the mapping within the DC-interface between $\underline{x}_D$ and the parameter vector $\underline{v}_C$.

6.2 Reachability Analysis

As an example, the following initial situation is considered: the levels in $T1$ and $T2$ do not exceed 0.1 and the valve V is closed. The corresponding initial set is

$$\begin{aligned} X_0 = (\underline{x}_D^{[1]}(0) &= \underline{x}_{D1} = [1\,0\,1\,0\,1\,0\,1\,0\,1\,0]^T; \\ X_C^{[1]}(0) &= \{\underline{x}_C \,:\, 0 \le x_{Ci} \le 0.1\,, i = 1, 2\}) \end{aligned} \tag{15}$$

[2] We use no units here.

Table 1. Part of the mapping from $\underline{x}_{Di}$ to $\underline{v}_{Cj}$ within the Net State Model's DC-interface

Discrete state	Parameter vector	Valve V	x_{C1} , x_{C2}
$\underline{x}_{D1}$	$[0\,1\,0\,0\,0]^T$	closed	x_{C1} , $x_{C2} < h$
$\underline{x}_{D2}$	$[0\,1\,1\,0\,0]^T$	closed	$x_{C1} \geq h$, $x_{C2} < h$
$\underline{x}_{D5}$	$[1\,1\,0\,0\,0]^T$	open	x_{C1} , $x_{C2} < h$
$\underline{x}_{D6}$	$[1\,1\,1\,0\,0]^T$	open	$x_{C1} \geq h$, $x_{C2} < h$
$\underline{x}_{D7}$	$[1\,1\,0\,0\,1]^T$	open	x_{C1} , $x_{C2} \geq h$

and the first steps of the reachability analysis are depicted in Fig. 7. The black arrows symbolize controlled events, the grey ones internal events. White arrows indicate further discrete state transfers; for clarity, the overflow activation is not taken into account. For the individual discrete states $\underline{x}_{Di}$ the dividing hyperplane H_T according to (8) is drawn as long as it is visible within the corresponding set $\Gamma(\underline{x}_{Di})$. In each case, on the right below H_T the preferential course of the continuous trajectory points upwards, above H_T downwards. Each control hyperplane (6) is adjusted in parallel to the x_{C1} axis and so yields boundaries of the grey marked local reachability sets, calculated as described in Sect. 4.

If the pump feeds liquid into $T1$, starting from $X_C^{[1]}(0)$ this results in the local reachability set $R_C^{[1]}(0)$ (A). The level in tank 1 increases independently from $T2$ which decreases slowly by the outlet (this is not visible in Fig. 7 because of the reachability set's over estimation!). Next, either the controlled event e_g^{1-5} is triggered by the discrete input u_D and so the valve is opened or the level in $T1$ reaches the height h and releases the internal event e_i^{1-2}. For the first case, there is already for $x_{C1} < h$ an input flow to tank 2 which yields under $\underline{x}_{D5}$ with an unbounded continuous input the local reachability set $R_C^{[1]}(1)$ depicted grey (B). Entering the release set $R_C^{\partial\Omega_1[1]}(1)$ results in a discrete state transition to $\underline{x}_{D6}$ (D), and there is $R_C^{[1]}(2)$ the next local reachability set starting from $X_C^{[1]}(2) = R_C^{\partial\Omega_1[1]}(1)$.

Alternatively, this discrete state is taken with the above mentioned second case, too. If e_i^{1-2} releases a discrete state transition under $\underline{x}_{D1}$, it is evident from the next local reachability set $R_C^{[2]}(1)$, that for a closed valve tank 2 can be filled up to 0.28 (C) only, because the overflow protection prevents tank 1 to reach a level over 0.60 and the outlet of $T2$ is permanently open. To fill tank 2 starting from this discrete state beyond the upper connecting pipe, the valve V is to be opened which is equivalent to the controlled event e_g^{2-6} and the discrete state transition to $\underline{x}_{D6}$ (D). Thus, the local reachability set $R_C^{[2]}(2)$ does not differ from $R_C^{[1]}(2)$. For example, starting from this set and avoiding the activation of the overflow protection makes an internal event e_i^{6-7} possible by entering the release set

$$R_C^{\partial\Omega_3} = \{\underline{x}_C \,|\, 0.46 \leq x_{C1} < 0.6 \wedge x_{C2} = 0.32\} \tag{16}$$

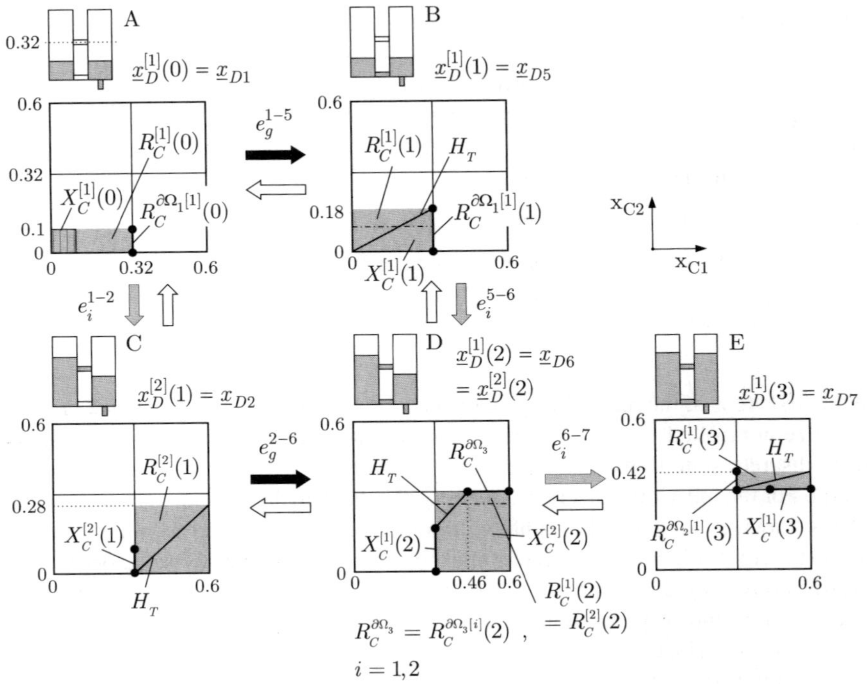

Fig. 7. Part of the two-tank-system's reachability analysis

which results in a state transition to $\underline{x}_{D7}$. Only in (16) and not within the entire range of the areas bordering each other in the continuous state space, the condition (9) holds and the continuous trajectory can enter the trigger set Ω_3 which is apparent considering the dividing hyperplane H_T.

Under the discrete state $\underline{x}_{D7}$, tank 2 can be filled at most to 0.42 determined by the reachability set $R_C^{[1]}(3)$. From there, for example, a controlled event e_g^{7-3} corresponding with the closing of V results in a discrete state transition to $\underline{x}_{D3}$ (E).

6.3 Design of a NSM Controller

In our example the hybrid controller must transfer all hybrid states out of X_0 according to (15) to

$$X_e = \mathbf{x}_e = (\underline{x}_{D7} = [0\,1\,0\,1\,0\,1\,1\,0\,1\,0]^T; \underline{x}_{Ce} = [0.57\ 0.40]^T)\,, \tag{17}$$

an equilibrium point within the reachability set $R(X_0)$ on account of the reachability analysis described in the previous section. As you can see in Fig. 7, the state transfer can be made along two discrete trajectories

$$\underline{x}_{D[0,3]}^{[1]} = \langle \underline{x}_{D1}, \underline{x}_{D5}, \underline{x}_{D6}, \underline{x}_{D7} \rangle \quad \text{and} \quad \underline{x}_{D[0,3]}^{[2]} = \langle \underline{x}_{D1}, \underline{x}_{D2}, \underline{x}_{D6}, \underline{x}_{D7} \rangle\,. \tag{18}$$

To calculate the transfer set $\mathcal{R}(X_0, X_e)$ along each discrete path a backward reachability analysis follows, which yields in this case no restriction of the forward local reachability sets $R_C(\cdot)$ and the release sets. Thus, for each step k holds $\mathcal{R}_C(k) = R_C(k)$ and the transfer set comprises the two ordered sequences

$$\langle \mathcal{R}^{[1]}(0), \mathcal{R}^{[1]}(1), \mathcal{R}^{[1]}(2), \mathcal{R}^{[1]}(3) \rangle \quad \text{and} \quad \langle \mathcal{R}^{[2]}(0), \mathcal{R}^{[2]}(1), \mathcal{R}^{[2]}(2), \mathcal{R}^{[2]}(3) \rangle$$

with

$$\mathcal{R}^{[1]}(0) = (\underline{x}_{D1}, \mathcal{R}_C^{[1]}(0)) = \mathcal{R}^{[2]}(0)\,,\ \mathcal{R}^{[1]}(1) = (\underline{x}_{D5}, \mathcal{R}_C^{[1]}(1))\,,$$

$$\mathcal{R}^{[1]}(2) = (\underline{x}_{D6}, \mathcal{R}_C^{[1]}(2)) = \mathcal{R}^{[2]}(2)\,,\ \mathcal{R}^{[1]}(3) = (\underline{x}_{D7}, \mathcal{R}_C^{[1]}(3)) = \mathcal{R}^{[2]}(3)$$

and

$$\mathcal{R}^{[2]}(1) = (\underline{x}_{D2}, \mathcal{R}_C^{[2]}(1))\,.$$

After that, local control strategies under each discrete state within (18) are designed to guide the continuous trajectories within each discrete state from the local initial set to the respective continuous target set, which is always a release set apart from the continuous equilibrium point $\underline{x}_{Ce}$ out of (17). Under $\underline{x}_{D1}$ and $\underline{x}_{D6}$, the left eigenvector assignment described in Sect. 5 is used, for the stable continuous dynamics under $\underline{x}_{D2}$ and $\underline{x}_{D5}$ an open-loop control with r_0 according to (10) is chosen and under $\underline{x}_{D7}$ the standard pole assignment method is applied.

Now the time instant the discrete input changes its value – releasing a controlled event – is to be determined: for the first trajectory out of (18), e_g^{1-5} results in a transition from $\underline{x}_{D1}$ to $\underline{x}_{D5}$, for the second trajectory, e_g^{2-6} causes a transition from $\underline{x}_{D2}$ to $\underline{x}_{D6}$. The objective is to minimize the time interval between the internal event when entering the discrete state before the controlled event and the internal event when leaving the discrete state after it. The suitable moment for the controlled event to occur is calculated by solving a standard static optimization problem (Nenninger, 2001). In this example, for the first trajectory the event e_g^{1-5} must occur immediately before e_i^{5-6}, and for the second trajectory e_g^{2-6} must follow e_i^{1-2} immediately. Obviously, there is no more difference in the effective continuous dynamics between the first and the second trajectory, because the discrete state $\underline{x}_{D5}$ and $\underline{x}_{D2}$, respectively, are taken only for an infinitely short time. So tank 1 is to be filled with a closed valve as fast as possible until the height h, afterwards, V must be opened immediately to guide the level in tank 2 as quick as possible to h, too. This discrete sequence is reproduced in the discrete part of the hybrid controller and selects via the controller's DC-Interface the appropriate continuous strategy.

Starting from the initial state $\mathbf{x}_0 = (\underline{x}_{D1}; [0.10\ 0.05]^T)$, the resulting hybrid closed-loop for the two-tank-system yields the time responses of the continuous variables depicted in Fig. 8. The event times are labeled t^{1-6} and t^{6-7} corresponding to the discrete states before and after the event occurs; they are perceptible by the discontinuities in the time response of the manipulated input u_C, too.

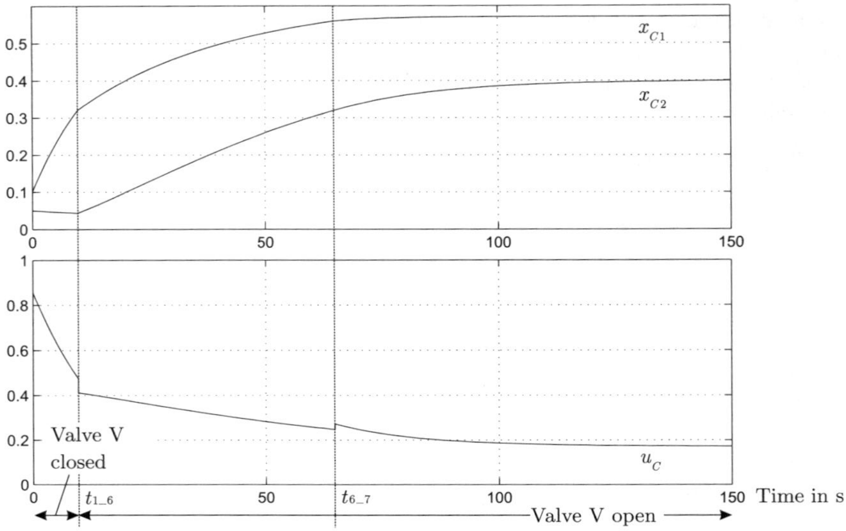

Fig. 8. State transfer by hybrid control for the two-tank-system

7 Conclusion

If the automation objective for a hybrid system focuses on individual operating points with a discrete and a continuous part, a hybrid controller must specifically choose the discrete as well as the continuous inputs of the process to maintain the actual hybrid state or to guide it along a certain trajectory to the desired target operating point. For these tasks, it is important to consider strictly the given specification and the system's dynamics.

Starting with the Net State Model formalism, this contribution proposes a systematic approach for modeling, reachability analysis and controller design. An essential part is a both forward and backward reachability analysis between the initial and the target state, to determine the so-called transfer set, inside which the hybrid trajectory under the hybrid control's influence must remain during the entire state transfer. The subsequent design of a hybrid controller comprises on the one hand the selection of a suitable discrete trajectory out of the transfer set, which fixes already the discrete inputs of the process. On the other hand, to each state of the chosen discrete trajectory, a local control strategy is assigned, which all in all forms the continuous part of the hybrid controller. For hybrid systems with piecewise affine continuous dynamics, appropriate analysis and design methods are illustrated. They make use of special characteristics of this class of systems, like the left eigenvectors of its continuous dynamics. The practical use is within the automation of particular parts of extensive processes, the behavior of which shows hybrid phenomena because of different component's interaction. Therefore, the proposed approach combines classical methods from control engineering and computer science and contributes to help man in controlling complex processes.

Performance Models for a Hybrid Reactor System

Katinka Wolter[1], Andrea Zisowsky[2], and Günter Hommel[1]

[1] Technische Universität Berlin, Sekr. EN 10, Einsteinufer 17, 10587 Berlin, Germany. Email: {katinka,hommel}@cs.tu-berlin.de
[2] Fakultät für Mathematik und Informatik, Universität des Saarlandes, Postfach 15 11 50, D-66115 Saarbrücken, Germany. Email: zisowsky@num.uni-sb.de

Abstract. In this paper we will present an improved numerical algorithm for the analysis of fluid stochastic Petri Nets with two fluid places. This algorithm uses an alternating direction implicit (ADI) discretization, whereas the previously used algorithm was a θ-discretization scheme. Especially derived discrete reflecting boundary conditions ensure that both schemes are conservative in the probability mass. In the second part of this paper we will study a model of a hybrid reactor system. This system has been studied before with very simple models that are extended here. We will use this model not only to present performance and availability measures of the model, but also to compare the two solution algorithms using part of the full model. The new algorithm is almost equal to the θ-scheme in memory usage, but roughly twice as fast.

1 Introduction

Hybrid systems have gained much attention over the last years. Formal methods for their verification, performance and reliability analysis have been investigated intensively. For the latter a number of extensions to stochastic Petri Nets have been proposed (Horton et al., 1998, Gribaudo et al., 1999, Wolter and Zisowsky, 2001).

Stochastic models often suffer from not having closed-form solutions, so numerical methods or simulation techniques have to be applied (Nicol and Miner, 1995, Ciardo et al., 1999, Bobbio et al., 1999). We use a hybrid Petri Net formalism, which we call *fluid stochastic Petri Nets* (FSPNs). We solve these models with discretization methods. In this paper we will present a new numerical solution technique for solving FSPN models with two fluid places. The new algorithm uses an alternating direction implicit (ADI) discretization scheme, which will be described and compared with the formerly used θ-scheme.

In the second part of this paper we use an earlier developed tool for the analysis of a real-world hybrid application. We study the behaviour of a hybrid automated surveyance system that exists at the University of Hamburg-Harburg (Nixdorf and Lunze, 2000a). In this system work pieces are heated up on a heater and cooled down in a water tank. A robot is used for transportation of the metal pieces. One question that arises in designing these systems is whether there is sufficient cold water so that no dangerous heat-up of the pieces can happen because there is not sufficient cold water available. Also important is whether there are enough transportation facilities to remove and replace the hot metal sticks as quickly as necessary. Other aspects rather

S. Engell, G. Frehse, E. Schnieder (Eds.): Modelling, Analysis and Design of Hybrid Systems, LNCIS 279, pp. 193–210, 2002.

concern the performance of the system: is the transportation utilized as wanted or should the system be dimensioned differently.

At first the simple model from (Wolter, 2001) will be presented, in which only one metal piece is used. The model has two continuous variables: one for the temperature of the metal piece and the other one for the temperature of the water in the tank. Two continuous variables can be conveniently handled by our solution algorithm. Increasing the number of metal pieces immediately increases the model complexity. An arbitrary number of continuous variables to represent the temperatures of an arbitrary number of work pieces can be handled by no existing software tool. In consequence the models have to be formulated differently.

We will show a more general formulation of the model, which will allow us to flexibly choose the number of work pieces used. The simple and the general model should match in the water temperature if the complex model uses only one work piece. All known formalisms for modelling hybrid systems have to go over to a discrete system or a discrete description of the hybrid system. They do so either on the level of the model, on the level of its mathematical description, or upon solving the equations describing the behaviour of the model. We decided to discretize as late as possible, that means to use the latter approach. This has several advantages. The user of our formalism can operate in a purely hybrid world and does not see the discretization that has to take place. Using an intuitive hybrid modelling paradigm is considered much more convenient by many engineers. The construction is much faster and the model more concise.

The paper is organized as follows: in the next section the modelling formalism we use is defined. In section 3 the two numerical solution algorithms are described. Section 4 introduces the system we study, Section 5 presents the models we have developed and Section 6 concludes the paper.

2 The FSPN Formalism

For the definition of FSPNs the common notations (Marsan et al., 1995, Horton et al., 1998) are taken over and extended where it is needed.

An FSPN is formally defined as a 8-tuple FSPN = $(\mathcal{P}, \mathcal{T}, \mathcal{A}, m_0, g, \lambda, r, w)$ where the set of places ($\mathcal{P} = \mathcal{P}_c \cup \mathcal{P}_d$) is divided into the *fluid (continuous)* and the *discrete* places. Discrete places (the elements of $\mathcal{P}_d$) are drawn as single lined circles and hold an integer number of *tokens*, whereas the continuous, or fluid, places (the elements of $\mathcal{P}_d$) are drawn as two concentric circles and they hold a real-valued amount of *fluid*. The set of transitions $\mathcal{T} = \mathcal{T}_E \cup \mathcal{T}_I$ is composed of the exponentially distributed, and the immediate transitions, respectively. Exponentially distributed transitions are drawn as empty rectangles and the immediate transitions as thin bars.

The set of arcs $\mathcal{A} = \mathcal{A}_d \cup \mathcal{A}_c$ is divided into two subsets, the discrete arcs (the elements of $\mathcal{A}_d$), and the continuous, or fluid, arcs ($\mathcal{A}_c$). The discrete arcs are drawn as single-lined arrows, whereas the fluid double-lined arcs are drawn like pipes.

The *marking* $\mathbf{m} = (s, \mathbf{z}) \in \mathcal{M}$ consists of a discrete part $s = (\#p_i, i \in \mathcal{P}_d)$, where $\#p_i$ denotes the number of discrete tokens in the i-th discrete place p_i, and a continuous part, a vector representing the fluid level in each fluid place, $\mathbf{z} = (z_k, k \in \mathcal{P}_c) \in \mathcal{Z}$. The initial marking is $\mathbf{m}_0 = (s_0, \mathbf{z}_0)$. The total number of discrete states is S, the set of all discrete states is denoted $\mathcal{S}$.

$\mathrm{r}_{t,p}, \mathrm{r}_{p,t} : \mathcal{A}_c \times \mathcal{M} \to \mathrm{I\!R}^2$ is the flow rate function along the fluid arc connecting the timed transition t and the fluid place p or vice versa. The flow rate $r = (\mu, \sigma^2)$ is a normally distributed random variable specified by expectation μ and variance σ^2. If it is specified by only one parameter, that one is the expectation and the variance is equal to zero.

Let Z_k be the fluid level in place $k \in \mathcal{P}_c$. The change rate of the fluid level is again a normally distributed random variable with

$$\frac{dZ_k}{dt} = \sum_{t \text{ enabled in } \mathbf{m}} r_{t,p}(\mathbf{m}) - \sum_{t \text{ enabled in } \mathbf{m}} r_{p,t}(\mathbf{m}).$$

Reflecting boundaries assure, that the fluid level in each place stays within its range $[z^{\min}, z^{\max}]$. For the sake of clearness and simplicity we write the first argument (the arc) as a subscript, defined by place and transition (source and destination).

$g_t : \mathcal{M} \to \mathrm{I\!B}$ is the guard of transition t that can be a function of the discrete and the continuous state. $\lambda_t : \mathcal{M} \to \mathrm{I\!R} \cup \{\infty\}$ is a function of both the continuous and the discrete marking. Immediate transitions have firing rate $\lambda_t = \infty$.

The weight function $w_t : \mathcal{S} \to \mathrm{I\!R}$ is defined for immediate transitions. The firing probability of each of the enabled immediate transitions in a vanishing state is $w_t(s) / \sum_{t_i \text{enabled in } \mathbf{m}} w_{t_i}(s)$.

The enabling rules for the transitions are the same as in discrete Petri Net models. The fluid arcs act only on the fluid places and do not influence the enabling conditions of the transitions.

The reachability graph, for our system represented in Fig. 6, gives a graphical representation of the stochastic process underlying the model. Augmenting it with continuous variables, one for each fluid place, gives a description of the full stochastic process which is formally defined as

$$\{(S(t), Z_1(t), Z_2(t)), t \in \mathrm{I\!R}_0^+\}, \tag{1}$$

if the model has two fluid places. $S(t)$ is a random variable for the discrete marking s at time t, and $Z_k(t)$ is a random variable, representing the fluid level in fluid place k at time t. Let $\pi_s(t, \mathbf{z}) = \partial / \partial \mathbf{z}\ P(S(t) = s, \mathbf{Z}(t) \le \mathbf{z})$ denote the transient probability of being in discrete state s with fluid levels in an infinitesimal environment around z_k at time t, for all fluid places $k \in \mathcal{P}_c$.

The stochastic process (1) can be mapped on a system of partial differential equations with appropriate initial and boundary conditions.

$$\frac{\partial}{\partial t}\boldsymbol{\pi} = \frac{1}{2}\frac{\partial^2}{\partial z_1^2}(\boldsymbol{\Sigma}_1^2\boldsymbol{\pi}) + \frac{1}{2}\frac{\partial^2}{\partial z_1^2}(\boldsymbol{\Sigma}_2^2\boldsymbol{\pi}) - \frac{\partial}{\partial z_1}(\boldsymbol{M}_1\boldsymbol{\pi}) - \frac{\partial}{\partial z_2}(\boldsymbol{M}_2\boldsymbol{\pi}) + \mathbf{Q}^T\boldsymbol{\pi}, \tag{2}$$

where the probability mass density π and the system parameter matrices $M_k = \text{diag}(\mu_{k,1}, \ldots, \mu_{k,S})$ and $\Sigma_k^2 = \text{diag}(\sigma_{k,1}, \ldots \sigma_{k,S})$ depend on the two fluid (spatial) variables z_1 and z_2. Also π depends on the time variable t. M_k and Σ_k^2 are the first two moments of the normally distributed change rate function $\frac{dZ_k}{dt}$, cf. (Wolter, 1999).

Equation (2) is a system of S differential equations, one for each discrete state. These equations are coupled by the term $\mathbf{Q}^T\pi$, that describes the transition from one state to another. The *generator matrix* $\mathbf{Q}$ contains the transition rates.

Two different discretization methods for this system of equations will be discussed in the next section.

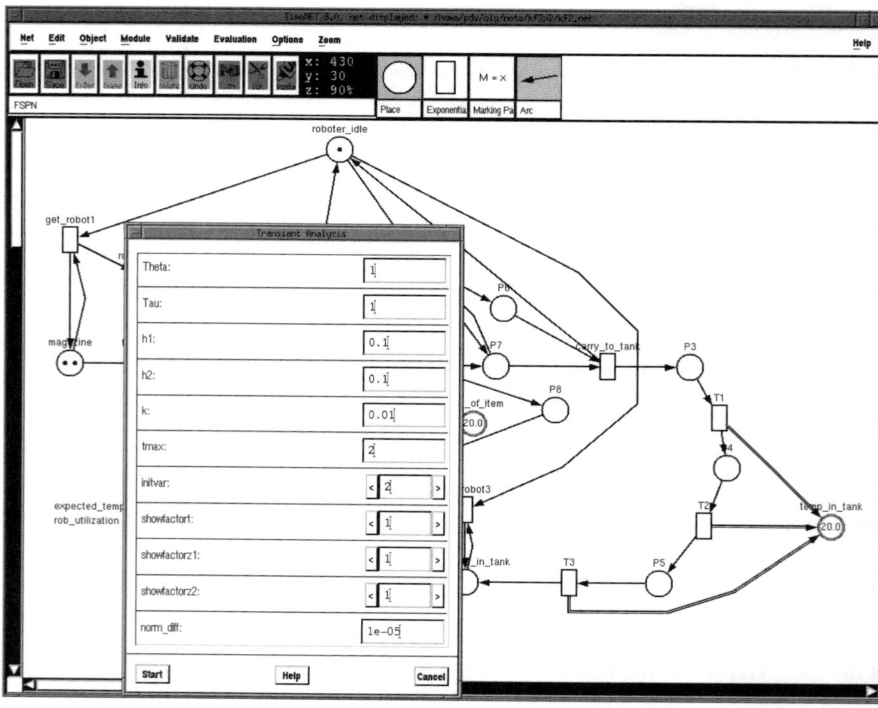

Fig. 1. Screen shot of the software tool

A screen shot of the graphical editor of the tool TimeNET (Zimmermann et al., 2000) as it has been used for the analysis of the model in this paper can be seen in Fig. 1. In the graphical editor window the general model is displayed. All the textual description in Fig. 9 is hidden in attributes of the objects. The small window shows the parameters of the transient analysis.

3 The Numerical Solution Algorithms

In this section we will describe two algorithms to solve the system of partial differential equations (2). First we introduce a discretization using a *θ-scheme* with *upwind strategy* together with boundary conditions (BCs) that results in a difference scheme which conserves the probability mass.

The discretization poses the problem to solve a system of linear difference equations given by a 5-banded matrix. In fact, no standard difference scheme for the chosen model yields a more simple matrix structure. But still the numerical solution is rather expensive: we use an iterative solution method, that works on sparse matrices, and thus has a computational effort of linear order for each iteration step (for full matrices it is of quadratic order). This means that it can be quite expensive for high accuracy; nevertheless it is less costly than direct methods.

To reduce expenses we consider a second solution algorithm: the *alternating direction implicit scheme* (ADI). This method is based on the discretization described above and splits the scheme into two successive steps with triangular matrices. These systems can be solved by a triangular Gaussian elimination, the so called *Thomas algorithm*, cf. (Thomas, 1995), with linear effort.

3.1 The θ-Scheme

The discretization of the system of differential equations is carried out on an equidistant grid with step-size $\triangle t$ in time and step-sizes $\triangle z_1, \triangle z_2$ in the two space directions z_1 and z_2. At the grid points the function $\pi_s(t, z_1, z_2)$ is approximated by the discrete function $u_{s,j}^n$ with

$$\begin{aligned} u_{s,j}^n \approx \pi_s(nk, z_1^{\min} + j_1 h_1, z_2^{\min} + j_2 h_2) \qquad &\text{for } n = 0, \dots, T \quad \text{and} \\ j_1 = 0, \dots, N_1, \quad j_2 = 0, \dots, N_2 \quad \text{with} \quad &j = j_1 N_2 + j_2. \end{aligned} \tag{3}$$

The system of equations has been discretized using a θ-scheme. The *implicitness parameter* θ weighs two neighbouring time levels. It can take on all values between zero and one. Depending on θ the discretization is explicit ($\theta = 0$), implicit ($\theta = 1$) or the well-known Crank-Nicolson scheme ($\theta = 1/2$). By $\boldsymbol{u}_j^{n+\theta}$ we will denote the weighted sum

$$\boldsymbol{u}_j^{n+\theta} := \theta \boldsymbol{u}_j^{n+1} + (1-\theta)\boldsymbol{u}_j^n . \tag{4}$$

For the first order spatial derivatives an *upwind scheme* is used, cf. (Strikwerda, 1989). 'Upwind' means that forward ($D_{z_k}^+$ in space direction z_k) and backward difference quotients ($D_{z_k}^-$) are used, weighed with the *upwind parameter* $\varrho_{k,s}$ and $(1 - \varrho_{k,s})$, respectively. The parameter $\varrho_{k,s}$ depends on the sign of the convection parameter $\mu_{k,s}$. This is motivated by the fact that if e.g. $\mu_{k,s}(z_1, z_2) > 0$, the mass moves to the right and the backward difference is more appropriate in describing this motion and thus yields a stable scheme. Hence, a reasonable choice of $\varrho_{k,s}$ is the following:

$$\begin{aligned} \mu_{k,s}(z_1, z_2) > 0: \quad & 0 \le \varrho_{k,s} < \tfrac{1}{2} \quad \text{backward difference weighs more,} \\ \mu_{k,s}(z_1, z_2) \le 0: \quad & \tfrac{1}{2} \le \varrho_{k,s} \le 1 \quad \text{forward difference weighs more,} \end{aligned} \tag{5}$$

where s is the index of the state number. For the discretization of the second order spatial derivatives the second order difference quotient $(D_{z_k}^+ D_{z_k}^-)$ and for the time derivative the forward difference quotient is used. The coupling term $\boldsymbol{Q}^T \boldsymbol{\pi}$ is discretized explicitly, using the two previous time levels, not the new time level. Therefore the set up system of difference equations can be decoupled. A separate system of linear difference equations is solved for each discrete state, and the coupling term appears in each one as an inhomogeneity.

Thus the discrete form of (2) reads

$$\begin{aligned}\frac{\boldsymbol{u}_j^{n+1} - \boldsymbol{u}_j^n}{\Delta t} &= \frac{1}{2} \frac{1}{\Delta z_1{}^2} D_{z_1}^+ D_{z_1}^- (\boldsymbol{\Sigma}_{1,j}^2 \boldsymbol{u}_j^{n+\theta}) + \frac{1}{2} \frac{1}{\Delta z_2{}^2} D_{z_2}^+ D_{z_2}^- (\boldsymbol{\Sigma}_{2,j}^2 \boldsymbol{u}_j^{n+\theta}) \\ &- \frac{1}{\Delta z_1} \left(\boldsymbol{R}_1 D_{z_1}^+ + (\boldsymbol{I} - \boldsymbol{R}_1) D_{z_1}^- \right) (\boldsymbol{M}_{1,j} \boldsymbol{u}_j^{n+\theta}) \\ &- \frac{1}{\Delta z_2} \left(\boldsymbol{R}_2 D_{z_2}^+ + (\boldsymbol{I} - \boldsymbol{R}_2) D_{z_2}^- \right) (\boldsymbol{M}_{2,j} \boldsymbol{u}_j^{n+\theta}) \\ &+ \boldsymbol{Q}^T \left((1-\theta) \boldsymbol{u}_j^n - \theta \boldsymbol{u}_j^{n-1} \right) .\end{aligned}$$

When a reflecting boundary condition is implemented, usually the analytic BC is discretized in an intuitive way. But the discrete scheme will not necessarily adopt conservation properties of the analytic equation. Therefore we use a *discrete reflecting boundary condition*, which is called discrete, because it is completely derived on a discrete level. With these BCs the overall difference scheme conserves the probability mass and makes the commonly applied normalization step obsolete. An additional normalization has three disadvantages: first of course, it increases the computational effort, second the discrete model is no good approximation of the analytic model, since it has different properties and third it needlessly produces additional computational errors. To derive the discrete reflecting BC we sum up the difference equation at each interior grid point. This sum can be replaced using a discrete form of the mass conservation law, which involves all grid points and thus yields a condition for the grid points at the boundary. For more details on the θ-scheme and the reflecting boundary conditions we refer to (Zisowsky, 1998).

3.2 The ADI Method

The two-dimensional scheme is unconditionally stable (for the Crank-Nicolson scheme or any scheme with $\theta \geq 0.5$) but the discretization of both spatial derivatives yields a 5-band matrix, that is much more difficult to invert at each time step than the tridiagonal matrices encountered in one-dimensional problems. The ADI method is a strategy to reduce the two-dimensional problem to two successive one-dimensional problems.

For the discretization we again use the θ-scheme with upwind strategy as described above. We will show how to add a 'zero' to the discrete equation, that allows the equation to be factored. By introducing an artificial intermediate time step, the scheme can be split into two steps. At each of these steps a tridiagonal system of equations has to be solved.

Let $\boldsymbol{A}_1, \boldsymbol{A}_2$ denote the operators describing the discretization of the first and second order spatial derivatives in z_1 and z_2 direction respectively, i.e.

$$\boldsymbol{A}_{k,s}\boldsymbol{u}_s = -\frac{1}{2} D^+_{z_k} D^-_{z_k} \left(\boldsymbol{\Sigma}^2_{k,s}\boldsymbol{u}_s\right) + \left(\boldsymbol{R}_{k,s}\, D^+_{z_k} + (\boldsymbol{I} - \boldsymbol{R}_{k,s})\, D^-_{z_k}\right)(\boldsymbol{M}_{k,s}\boldsymbol{u}_s) \ , \tag{6}$$

for $k = 1, 2$. The vector $\boldsymbol{u}_s$ denotes the approximation of the probability density function π_s at the grid points. The dimension of $\boldsymbol{u}_s$ is the number of spatial grid points. $\boldsymbol{R}_{k,s}$ contains the upwind parameter $\varrho_{k,s}$ at each grid point. For simplicity of notation, we will omit the index s for the operators $\boldsymbol{A}_{k,s}$.

With the abbreviation $\boldsymbol{u}_s \underset{n-1}{\overset{n}{\textcircled{\scriptsize T}}} \boldsymbol{Q}$ for the discretization of the coupling term, the discrete form of (2) reads

$$\frac{\boldsymbol{u}_s^{n+1} - \boldsymbol{u}_s^n}{\triangle t} = -\theta\,(\boldsymbol{A}_1 + \boldsymbol{A}_2)\,\boldsymbol{u}_s^{n+1} - (1-\theta)\,(\boldsymbol{A}_1 + \boldsymbol{A}_2)\,\boldsymbol{u}_s^n + \boldsymbol{u}_s \underset{n-1}{\overset{n}{\textcircled{\scriptsize T}}} \boldsymbol{Q} \tag{7}$$

with an error in time of $O(\triangle t)$ and $O(\triangle t^2)$ for the Crank-Nicolson scheme; or sorted for $\boldsymbol{u}_s^{n+1}$ and $\boldsymbol{u}_s^n$

$$\begin{aligned}(\boldsymbol{I} + \theta\triangle t\boldsymbol{A}_1 &+ \theta\triangle t\boldsymbol{A}_2)\boldsymbol{u}_s^{n+1} \\ &= [\boldsymbol{I} - (1-\theta)\,\triangle t\boldsymbol{A}_1 - (1-\theta)\,\triangle t\boldsymbol{A}_2]\,u_s^n + \triangle t\,\boldsymbol{u}_s \underset{n-1}{\overset{n}{\textcircled{\scriptsize T}}} \boldsymbol{Q}\,.\end{aligned} \tag{8}$$

The 'trick' now is to add the quadratic term $\triangle t^2\theta^2\boldsymbol{A}_1\boldsymbol{A}_2\boldsymbol{u}_s^{n+1}$ on both sides of the equation and to add the terms $\pm(1-\theta)^2\triangle t^2\boldsymbol{A}_1\boldsymbol{A}_2\boldsymbol{u}_s^n$ on the r.h.s.

$$\begin{aligned}(\boldsymbol{I} + \theta\triangle t\boldsymbol{A}_1 &+ \theta\triangle t\boldsymbol{A}_2 - \theta^2\triangle t^2\boldsymbol{A}_1\boldsymbol{A}_2)\boldsymbol{u}_s^{n+1} \\ &= \left[\boldsymbol{I} - (1-\theta)\triangle t\boldsymbol{A}_1 - (1-\theta)\triangle t\boldsymbol{A}_2 + (1-\theta)^2\triangle t^2\boldsymbol{A}_1\boldsymbol{A}_2\right]\boldsymbol{u}_s^n \\ &\quad + \triangle t\,\boldsymbol{u}_s \underset{n-1}{\overset{n}{\textcircled{\scriptsize T}}} \boldsymbol{Q} + \theta^2\triangle t^2\boldsymbol{A}_1\boldsymbol{A}_2 u_s^{n+1} - (1-\theta)^2\triangle t^2\boldsymbol{A}_1\boldsymbol{A}_2\boldsymbol{u}_s^n\,.\end{aligned} \tag{9}$$

Now the equation can be factored as

$$\begin{aligned}(\boldsymbol{I} &+ \theta\triangle t\boldsymbol{A}_1)(\boldsymbol{I} + \theta\triangle t\boldsymbol{A}_2)\boldsymbol{u}_s^{n+1} \\ &= (\boldsymbol{I} - (1-\theta)\triangle t\boldsymbol{A}_1)(\boldsymbol{I} - (1-\theta)\triangle t\boldsymbol{A}_2)\boldsymbol{u}_s^n + \triangle t\,\boldsymbol{u}_s \underset{n-1}{\overset{n}{\textcircled{\scriptsize T}}} \boldsymbol{Q}\,.\end{aligned} \tag{10}$$

The term $\theta^2\triangle t^2\boldsymbol{A}_1\boldsymbol{A}_2\boldsymbol{u}_s^{n+1} - (1-\theta)^2\triangle t^2\boldsymbol{A}_1\boldsymbol{A}_2\boldsymbol{u}_s^n$ is of order $O(\triangle t^2)$ and thus of less (or equal) order than the ordinary discretization error.

Finally we split (10) by introducing an artificial intermediate value $\tilde{\boldsymbol{u}}_s$:

$$(\boldsymbol{I} + \theta\triangle t\boldsymbol{A}_1)\tilde{\boldsymbol{u}}_s = (\boldsymbol{I} - (1-\theta)\triangle t\boldsymbol{A}_2)\boldsymbol{u}_s^n + \theta\triangle t\,\boldsymbol{u}_s \underset{n-1}{\overset{n}{\textcircled{\scriptsize T}}} \boldsymbol{Q} \tag{11a}$$

$$(\boldsymbol{I} + \theta\triangle t\boldsymbol{A}_2)\boldsymbol{u}_s^{n+1} = (\boldsymbol{I} - (1-\theta)\triangle t\boldsymbol{A}_1)\tilde{\boldsymbol{u}}_s + (1-\theta)\triangle t\,\boldsymbol{u}_s \underset{n-1}{\overset{n}{\textcircled{\scriptsize T}}} \boldsymbol{Q}\,. \tag{11b}$$

A homogeneous two step scheme of this kind is called *Peaceman-Rachford scheme* and has the same stability behaviour as the underlying spatial discretization (Strikwerda, 1989).

To verify this splitting, multiply (11a) with $(\boldsymbol{I} - (1-\theta)\triangle t\boldsymbol{A}_1)$ and (11b) with $(\boldsymbol{I} + \theta\triangle t\boldsymbol{A}_1)$ and use (a) to replace $\tilde{\boldsymbol{u}}_s$ in (b). This yields (10).

In each step of the ADI method just one z-direction is implicit, the other direction is explicit. This is alternated in the next step. For the efficient solution the exact structure of the matrices on the l.h.s. of (11) is important: (11a) is already tridiagonal, (11b) is 3-banded. To use the fast tridiagonal solver also for step (b), we reorder our data $\tilde{\boldsymbol{u}}_s$ after step (a). Reordering of data is equivalent to changing the order of numeration from row wise to column wise as illustrated in Fig. 2. In the row wise

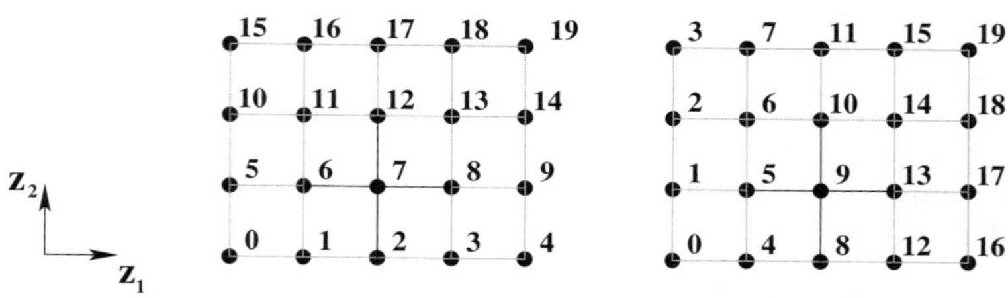

Fig. 2. Numeration of grid points for step (11a) row wise (left) and for step (11b) column wise (right)

numeration points 2,7 and 12 are involved in the spatial differences in z_2-direction in point 7. This yields a diagonal element and elements in the fifth off-diagonals. In the column wise numeration points 8,9 and 10 are involved in the spatial differences in z_2-direction in point 9. This results in entries in the diagonal and the first off-diagonals.

We used the overset tilde for the intermediate step to emphasize the fact, that it is no solution of our problem, but merely an artificial quantity.

Usually the development of boundary conditions for an ADI scheme is troublesome. For the intermediate quantity boundary values have to be prescribed, generally by extrapolation. Our aim was again to derive BCs that yield a probability mass conserving discrete scheme. With the same technique as in the 'ordinary' θ-scheme it is possible to develop discrete reflecting boundary conditions for the ADI method. These BCs conserve the probability mass in each of the two steps (11) and therewith also the mass after a complete time step.

The structural flow of the two algorithms is shown in Fig. 3. While the gain in memory for the ADI method is comparatively small, it is much faster than the θ-scheme. The computational complexity of both algorithms is linear in the number of discrete states S. But whereas the computational effort of the ADI method is independent of the model (for a fixed S), the iterative solution process of the θ-scheme terminates, when a required precision ε_{req} is reached. This depends strongly on the properties (e.g. the spectral norm) of the iteration matrix and thus on the

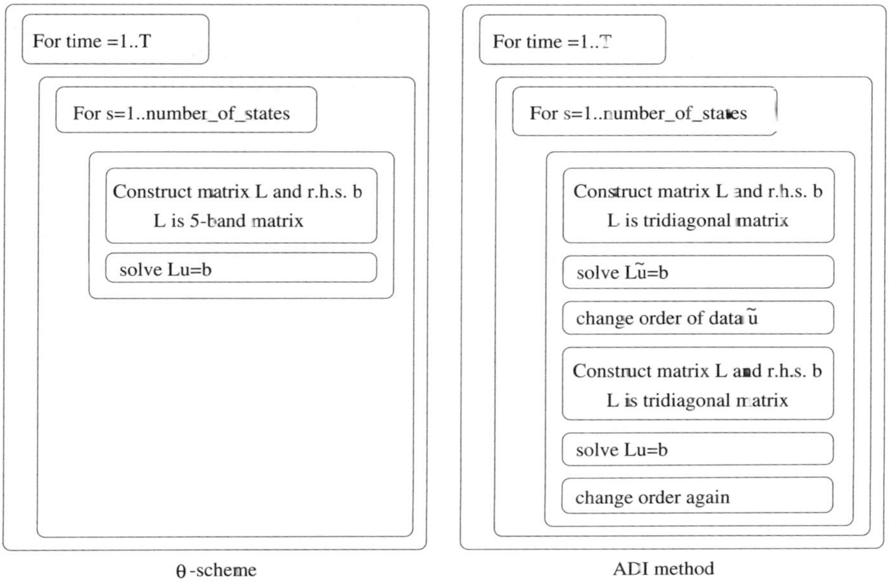

Fig. 3. Comparison of the structural progress of the θ-scheme and the ADI method

Table 1. Comparison of cpu-times of the two implemented algorithms for a test model

solution algorithm	cpu-time
ADI method	8 min 50
θ-scheme, $\varepsilon_{req}= 10^{-6}$	15 min 46
θ-scheme, $\varepsilon_{req}= 10^{-9}$	18 min 51
θ-scheme, $\varepsilon_{req}= 10^{-13}$	21 min 42

model. We compared the algorithms on a Sun Ultra 5 with 333 MHz for a small test model with only three discrete states, 65×65 grid points and 200 time steps. In Table 1 we present the cpu-time of the θ-scheme and of the ADI method. The choice of $\varepsilon_{req} = 10^{-13}$ yields the same accuracy in the probability mass as the direct solution with the ADI method, but the ADI method is more than twice as fast. Even for a reduced accuracy of $\varepsilon_{req} = 10^{-6}$ the θ-scheme is comparatively slow.

In the following section we will regard a system that is modelled and solved with the ADI method.

4 The System

The system studied in this paper is a hybrid automated surveyance system in which metal pieces are heated and later cooled down again in a water tank. The metal pieces are the discrete items in the system, whereas the temperatures of both the water in the tank and the metal pieces are continuous variables. The control of the system switches between discrete states, depending on the discrete state and the values of the continuous variables.

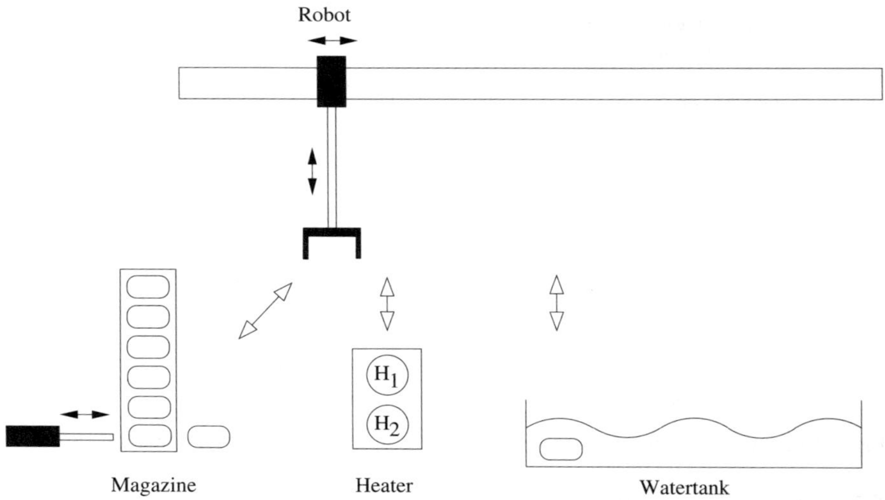

Fig. 4. Sketch of the hybrid manufacturing system

The elements of the considered system are a magazine, holding small metal pieces, a heater on which those pieces are warmed up, a water tank in which the pieces cool down again - while increasing the temperature of the water - and a robot that is used for transportation.

The system performs the following control sequence: the robot carries a piece to the heater, where this item stays until some temperature is reached, then the robot picks the work piece up from the heater and carries it to the water tank in which the metal piece is placed. When the metal piece has cooled down to some predefined temperature it is removed from the water tank and placed back on the heater. Figure 4 shows a sketch of the hybrid system.

Moving the robot to the position where it is needed (either the magazine, the heater or the tank) takes on the average 10 seconds, as does transportation of a work piece from one location to the next. The control sequence creates an action loop in the system, which will in a real nuclear power plant be ended after a known number of iterations. In our model system there is no limit to the number of iterations.

All parts have an initial temperature of 20° Celsius. Throughout this paper temperature is measured in degree Celsius. Transportation of items between locations and movements of the robot in general are assumed to be exponentially distributed with mean value of 10 seconds - that corresponds to a rate $\lambda = 0.1$, the time unit is one second, the metal pieces are assumed to be round disks with 25 mm height and a diameter of 40 mm. The water tank holds $0.5\ \mathrm{dm}^3 = 0.5$ liters. In most cases the assumption of exponentially distributed times is a simplification. In availability or reliability models where we seek very precise results (e.g. 'five nines') a wrong distribution will severely change results. Here, however, we want to point out that there is variation and slight randomness in all execution times in technical systems, so the possibly slight uncorrectness of the exponential distribution will not significantly bias the results.

5 FSPN Models of the Hybrid System

In this section we will first present a simple model, that is specially designed to represent the situation with only one work piece in the system. The main metrics of interest are utilization of the robot – that is the probability that the robot is busy – and the temperature of the water in the tank.

A general model in which an arbitrary number of work pieces can be used will be presented in the subsequent subsection. The general model should give the same results as the simple one with respect to the two regarded measures if only one work piece is used.

5.1 A Simple Model of the Hybrid System

Figure 5 shows an FSPN model of the hybrid system. The transitions *get_roboti* represent the time it takes to position the robot where it is needed. We study here a model with only one piece in the magazine. Initially this work piece is in the magazine and the robot is idle. After the robot has reached the magazine it picks up the metal piece and carries it to the heater. Here the work piece is put down and the robot is idle again.

While the piece lies on the heater transition *heat* is enabled and *temp_of_item*, the temperature of the metal piece increases as specified by rate *r1*. For this rate, as for both the other rates, only the expectation is given and the variance is assumed to be zero. Assuming a constant increase or decrease of the temperature might not be realistic, but to specify the variance of the rates is a problem we have not yet addressed. The expectation of the rates can be computed from the system description (Nixdorf and Lunze, 2000a), but nothing is known about the nature of the variances. Transition *T8* has a fluid-dependent firing rate *q1* and is activated only if the temperature of the item on the heater exceeds $70°$. We model here only one heater position although the original system has two, for one item can only lie in one position. Once the heating process is finished the robot is called to the heater again. It picks up the item and carries it to the water tank, in which the item is placed. The temperature of the metal piece decreases (as represented by *r2* since *T2* is enabled) and the temperature of the water increases (through *r3*). Transition *T2* again has a fluid-depended firing rate and is enabled only if the item has cooled down to at most $30°$. It will then still cool down a bit while the robot moves to the tank. The item is picked up and carried back to the heater to be warmed up once more.

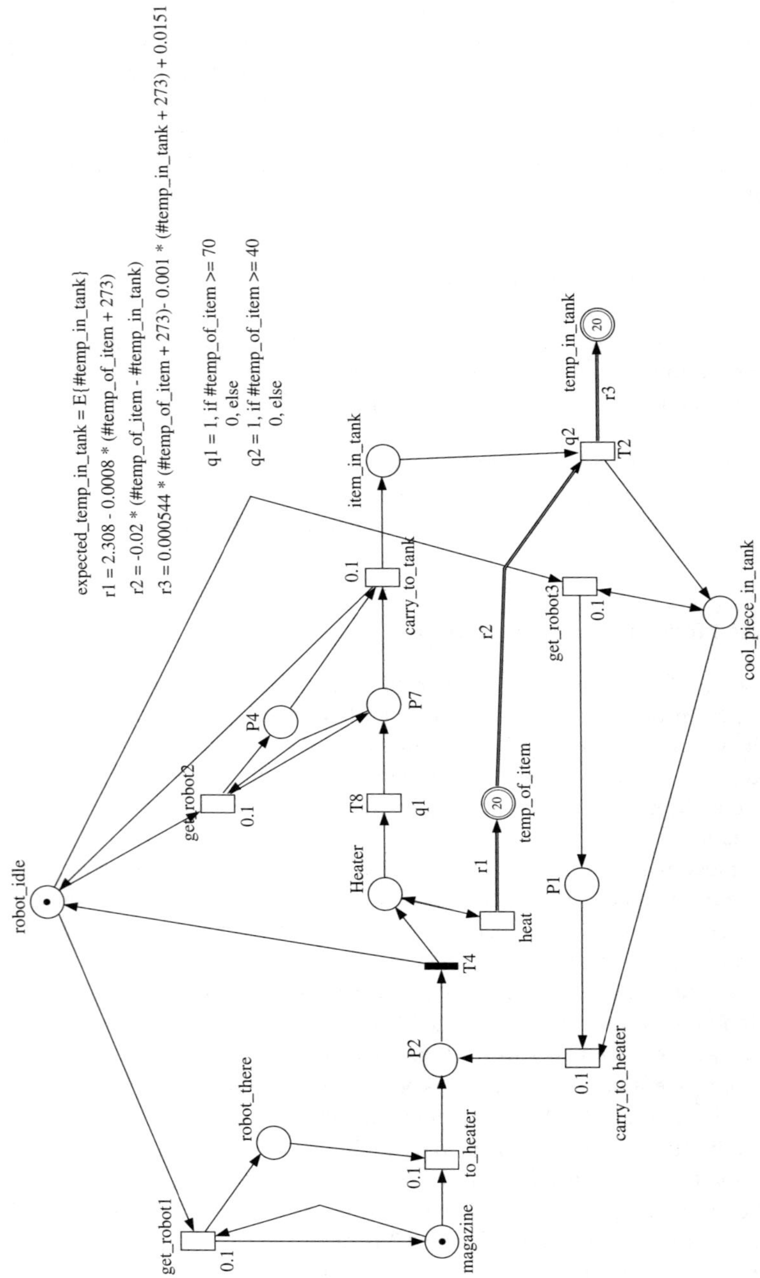

Fig. 5. FSPN model of the hybrid manufacturing system

The rates *r1, r2* and *r3* are computed following the formula for temperature changes:

$$\frac{d\vartheta}{dt} = \frac{\dot{Q}_{in} - \dot{Q}_{out}}{mc_p} \tag{12}$$

where $\dot{Q}$ is the thermal input or output stream, m is the mass of the material whose temperature change shall be computed and c_p is the specific thermal capacity of the material. The change of temperature occurs where two substances of different temperature meet. It is proportional to the size of the contact area A and a coefficient of warmth transmission α. $\dot{Q}$ is defined as

$$\dot{Q} = \alpha A(\vartheta_1 - \vartheta_2). \tag{13}$$

The needed parameters for heating up the iron metal piece on the heater ($r1$), are $c_p = 0.46$, $A = 56.55cm^2$ and $\alpha = 13.85W/Km^2$. The heater has a heating power of 200 Watts (W), which is the warmth input to the metal piece. In the technical parameters temperature is always measured in Kelvin (K), whereas in all figures we use degree Celsius[1].

Analogously, $r2$ is the temperature change of the work piece in contact with the water and $r3$ is the temperature change of the water, when being in touch with a metal piece. All rates are computed following the same principles.

Figure 6 shows the reachability graph of this simple model. The states are labelled with the discrete marking of the net and in addition they are enumerated, for simplicity in referencing.

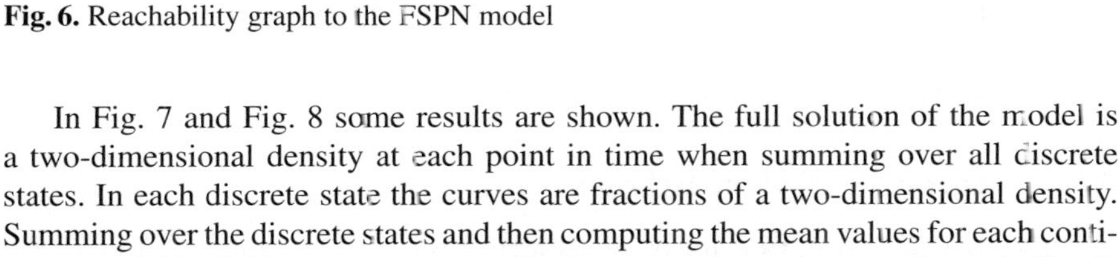

Fig. 6. Reachability graph to the FSPN model

In Fig. 7 and Fig. 8 some results are shown. The full solution of the model is a two-dimensional density at each point in time when summing over all discrete states. In each discrete state the curves are fractions of a two-dimensional density. Summing over the discrete states and then computing the mean values for each continuous variable yields a transient curve for the mean temperatures as shown in Fig. 7. Those means converge to the same and fairly low value. This shows that the amount of water is sufficient for cooling down the metal piece.

Figure 8 shows the complement of the robot utilization (that is the probability that the robot is idle, Prob($\#rob\_idle = 1$)). The robot initially is used, but on the

[1] $[^\circ Celsius] + 273 = [Kelvin]$

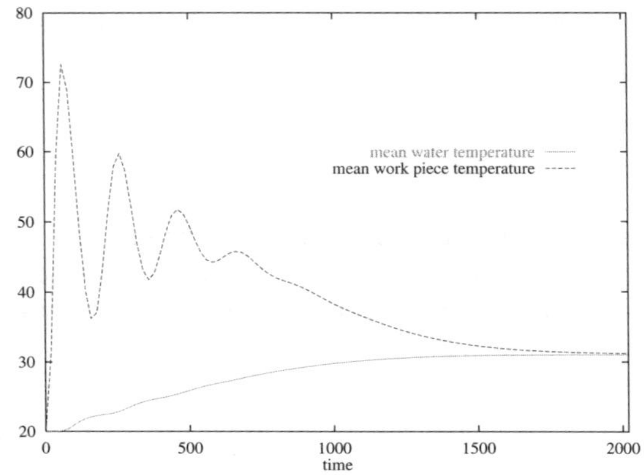

Fig. 7. Mean item temperature and mean water temperature

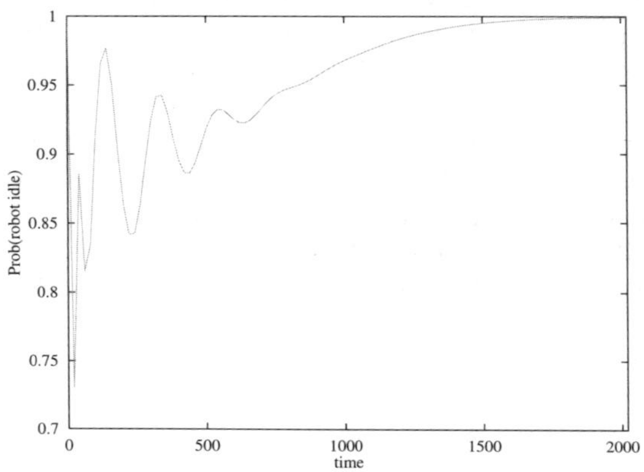

Fig. 8. Probability that the robot is idle

long-term average this usage is almost negligible. Most of the time the robot is idle. Consequently, the robot would be well capable of serving a system with more metal pieces. How many of them the system configuration can deal with well will be one of the questions posed to a more complex model.

5.2 A General Model of the Hybrid System

In this section the model from the previous section is in one way simplified so we can make it more general in a different way. We want to allow for an arbitrary number of work pieces to be present in the system. This means that we can no longer trace the temperature of each work piece, since no more than two continuous dimensions can be used in a model. Therefore we have to partially discretize the warming process of the water. We do so by splitting the heating period into three phases, one where the piece is still very hot, a second where it is medium warm, and a third where it has almost cooled down completely. The rates at which the water is heated up (r_3, r_4 and r_5) are computed by assuming a fixed temperature of the work piece in each of the three phases and evaluating rate r_3 in the simple model for each of them. These temperatures are 70° in the first phase, 45° in the second phase and 22° in the third phase. In doing so, two additional discrete states are included in the model.

The discrete part of the model remains more or less the same as in the simple model, but the continuous part has to be changed significantly. Fluid place *temp_of_item* now denotes the temperature of the item that is present on the heater. As the item leaves the heater, the temperature has to be reset, so that the fluid place will represent the temperature of the next item to be put on the heater. The reset of the heater position temperature is modelled with the fluid rate r_2. An additional discrete state for the reset is inserted. The transition at the target of the fluid arc labelled with r_2, transition q_2, is enabled only when the item has left the heater and the heater position temperature is still beyond 30°. Before the temperature of the item on the heater is reset no new work piece can be positioned on the heater.

Figure 9 shows the generalized FSPN model where K is the number of work pieces in the magazine. We first chose $K = 1$ and then this model should be equivalent to the simple model shown in Fig. 5. Equivalent here means in first place, that the temperature of the water tank increases in a similar way in both models. We would also wish the models to be equivalent in the sense that the robot utilization should be similar. With $K = 1$ the model has 15 discrete states. The reachability graph is omitted here, since it is constructed following the same principles as the reachability graph for the simple model.

Figure 11 on the right hand side shows the mean water temperature for the simple model and the generalized model with only one work piece. These means do not increase in exactly the same way, but their behaviour is fairly similar and the stationary values matched sufficiently well. With respect to the robot utilization, however, the models are not yet equivalent as can be seen in Fig. 10. It shows the probability that the robot is idle in the different models. Robot utilization in the generalized model is much higher than in the simple model. This is most likely due to the fact that in the generalized model the robot has to wait holding a work piece in front of the heater until the heater position temperature is reset. We are still working on matching the robot utilization as well. A future refinement of the model should either increase the speed of reset of the heater, or model in such a way, that the robot is only called when the heater is immediately available.

Increasing K rapidly increases the complexity of the stochastic process. For $K = 2$ the model has already 74 discrete states and for $K = 3$ the model has 216 discrete states. Since in each discrete state the full continuous state space must be stored, complexity soon reaches a limit where none of our machines will be able to store the matrices anymore and runtime increases tremendously. In this paper we only solve for $K = 1, 2$. It can be seen already, that increasing the number of work pieces in the model with severely increase the usage of the robot. The water, however, is warmed equally slowly with two work pieces as it was earlier with only one. So the cooling capacity of the water seems to be sufficient for some more work pieces.

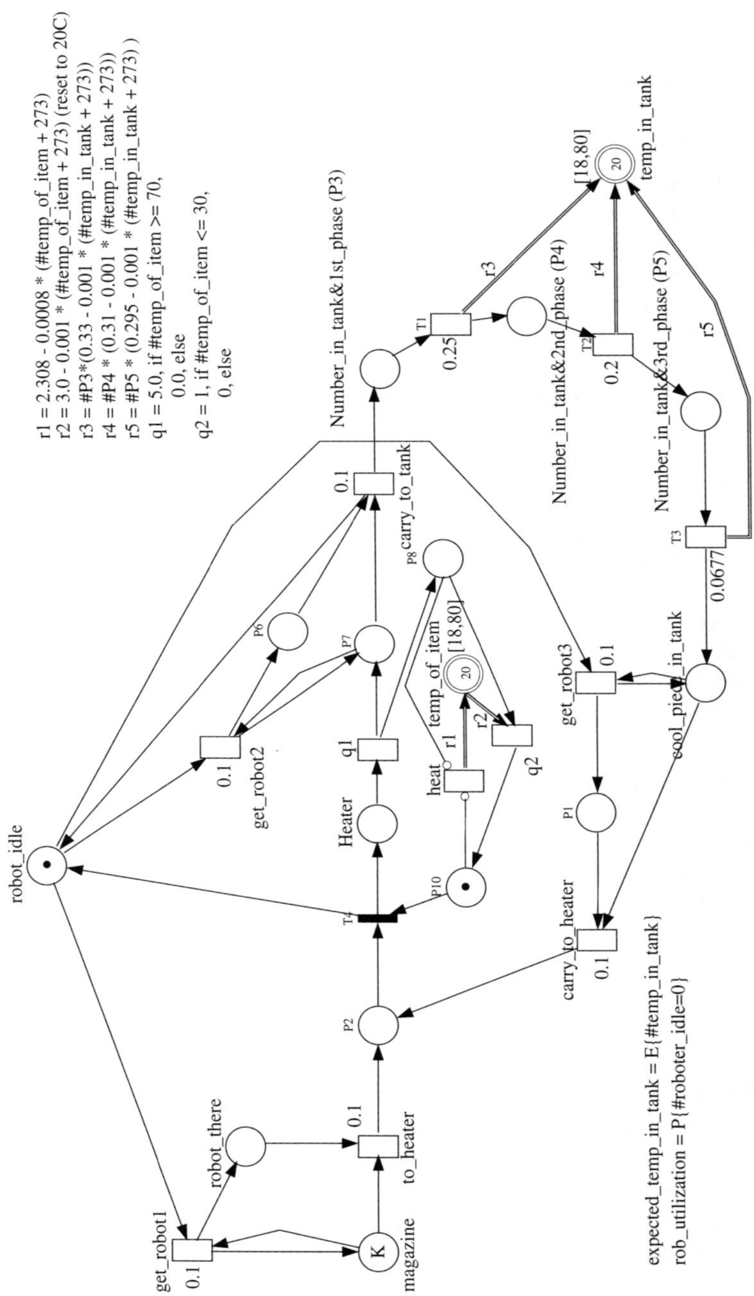

Fig. 9. General FSPN model of the hybrid manufacturing system

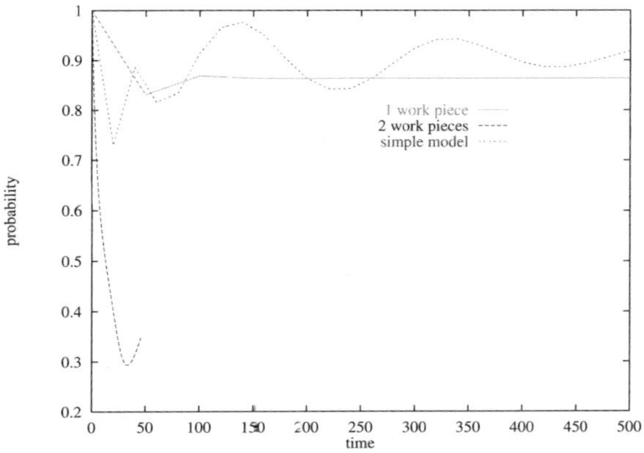

Fig. 10. Comparison of the probability that the robot is idle

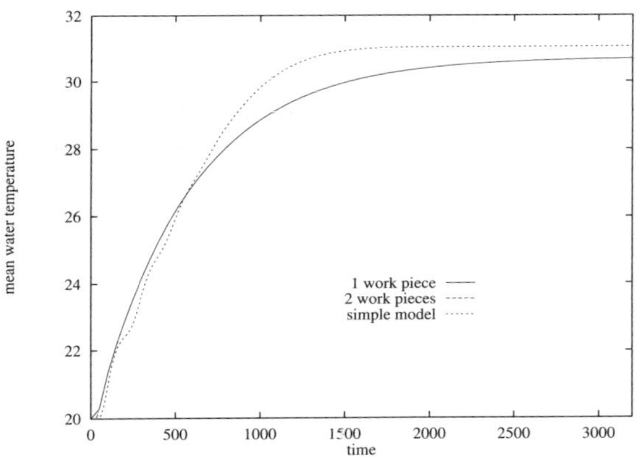

Fig. 11. Comparison of the mean water temperatures

6 Conclusions

Fluid stochastic Petri Nets are designed to intuitively represent hybrid systems. The formalism hides from the user the fact that a discrete solution is computed by the algorithm. Formulating the model on the discrete level would be highly unattractive. Not alone drawing a model that has some hundreds of places and transitions instead of one is not pleasant, but also the solution of this model would not be any faster than our algorithm, only more inaccurate. Analysis of hybrid systems using discrete models is not feasible, using FSPNs it becomes very simple.

In this paper we have introduced a new numerical solution algorithm for solving the system of equations underlying fluid stochastic Petri Nets with two continuous dimensions. The algorithm implements an alternating direction implicit (ADI) discretization scheme. Formerly we used a θ-discretization. Due to especially constructed discrete reflecting boundary conditions, both schemes conserve the probability mass.

It could be shown that the ADI method is about twice as fast as the θ-scheme, while not increasing the amount of memory needed.

We have used the new solution method to solve different models of a well-known manufacturing system, where work pieces are carried by a robot first to a heater and then to a water tank. On the heater the work piece is heated up and it is put in the water tank in order to cool down, while the water is warmed. These models allow us to obtain metrics such as the utilization of the robot and the mean temperature of the water tank. Appropriate dimensioning of the system as well as checking of security requirements can be carried out on the basis of the presented models.

Like many performability models and most hybrid models, our models suffer from numerical complexity, while the models as such are comparatively small. This problem can be addressed in different ways: by choosing even simpler models and by improving solution methods. The implementation of the ADI scheme is a step in the latter direction. If high accuracy is not a requirement one could improve the solution process still significantly by exchanging the data type *double* with the less accurate *float* in the implementation.

Acknowledgements. The first and third author's work is supported by the German Research Council (DFG) under Grant No. Ho 1257/11-3, and the second author's work is supported by DFG under Grant No. AR 277/3-1.

Using Parameterized Timestamp Petri Nets in Automatic Control

Carlo Simon[1], Kurt Lautenbach[2], Hans-Michael Hanisch[3], and Jan Thieme[3]

[1] Universität Koblenz-Landau, Institut für Management, D-56075 Koblenz
[2] Universität Koblenz-Landau, Institut für Softwaretechnik, D-56075 Koblenz
[3] Martin-Luther Universität Halle-Wittenberg, Fachbereich Ingenieurwissenschaften, D-06099 Halle (Saale)

Abstract. In the following sections, we describe a method to deal with time critical problems in the field of automatic control of manufacturing systems. The behavior of a technical system is represented by Timestamp Petri Nets. With the aid of a symbolic analysis of such nets we generate a linear optimization problem from a system of inequalities. Their solutions imply time parameters for software controllers which avoid the occurrence of dangerous situations. We show the applicability of the method with a small example where unknown time parameters of a controller have to be determined in order to avoid forbidden states of the controlled system.

1 Introduction

As a base for synthesizing a minimal restrictive controller, we need a model of the *uncontrolled plant*. We model this by means of *Net Condition Event Systems* (NCES) (Sreenivas and Krogh, 1991b, Rausch and Hanisch, 1995), which are a kind of CES (Sreenivas and Krogh, 1991a) that use *Petri Nets* to describe the inside of the modules. By assigning a timing concept to these NCES, one can speak of *Timed NCES models* (TNCES). If a TNCES model is safe one can transform it into a timed Petri Net (Thieme and Lüder, 1999, Simon and Thieme, 1999) which can be analyzed.

For Petri Nets, there exist several timing concepts. We decided to use *Timestamp Nets*, where intervals are assigned to the incoming edges of transitions (König and Quäck, 1988). Such an interval assigned to some edge (p, t) describes the permeability of this edge relative to the timestamp of a token on p. For each token its timestamp denotes the time the token was put on its place (Ghezzi et al., 1991). In Timestamp Nets, transitions which synchronize several token flows (i.e. such with more than one incoming edge) may be supplied with tokens sufficiently without being able to fire if their incoming edges are not permeable simultaneously. In this case we say the transition got *timewise stuck*.

In the following we examine conditions for transitions of Timestamp Nets getting timewise stuck by reducing these investigations to solving systems of inequalities. The method presented here can also be used to determine *parameters* for a Timestamp Net in order to prevent transitions from getting timewise stuck. The applicability of the method is shown by a small technical plant described in (Kowalewski and Preußig, 1996).

S. Engell, G. Frehse, E. Schnieder (Eds.): Modelling, Analysis and Design of Hybrid Systems, LNCIS 279, pp. 211–223, 2002.

2 Example

The technical plant we chose as an example is part of a laboratory batch plant. The flowchart of the entire batch plant is shown in Fig. 1 and was already described in (Kowalewski and Preußig, 1996). Here, we consider a modified problem which is restricted to the part of the plant shown in Fig. 2.

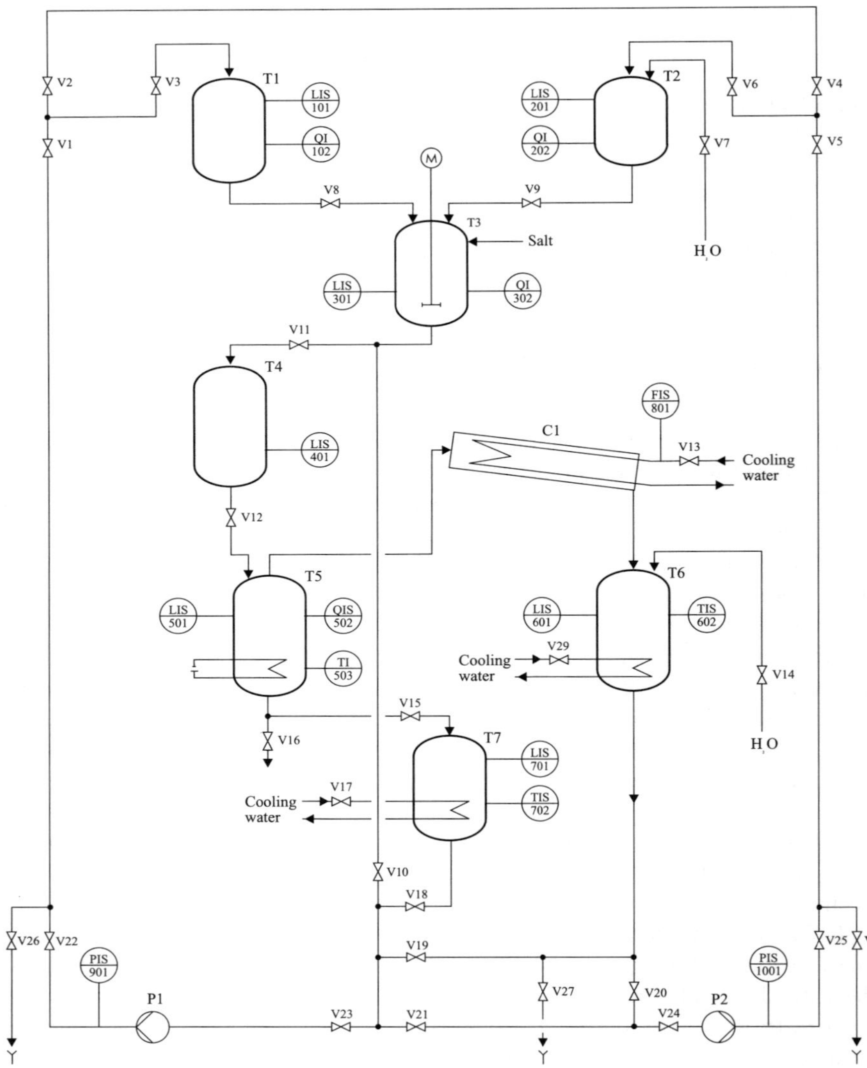

Fig. 1. Flowchart of the entire process

In tank $T5$, a solution is evaporated in order to increase its concentration. The steam is captured by condenser $C1$ which is supplied with cooling water. When the desired concentration is achieved and tank $T7$ is empty, the concentrated solution is drained off into $T7$ where a postprocessing step takes place. Finally, the medium is pumped out of $T7$. For the different processes their duration is approximately

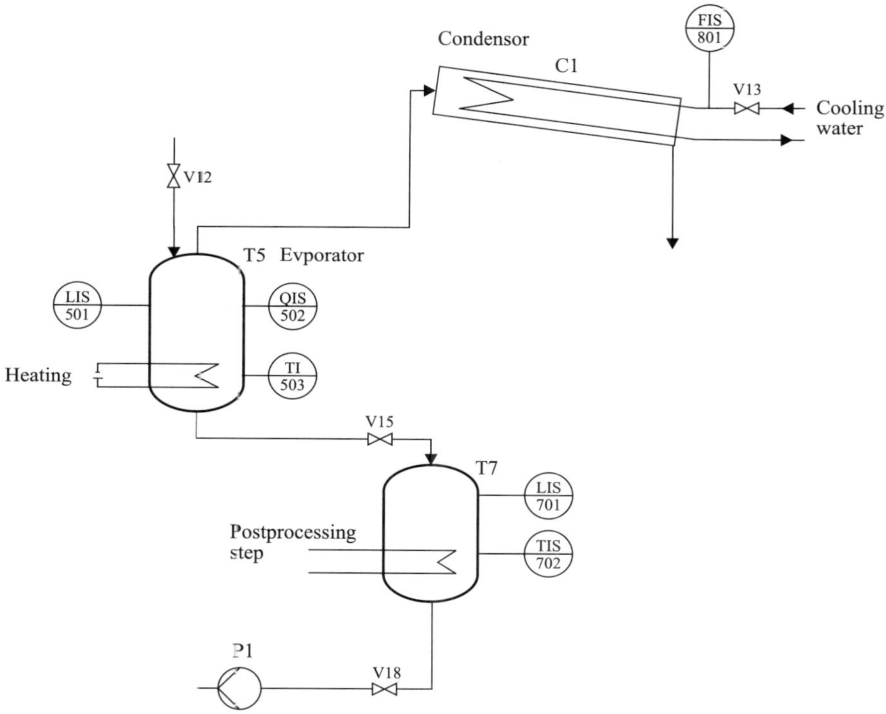

Fig. 2. Flowchart of the considered sub-process

known (Kowalewski and Preußig, 1996). To be more realistic, we permit here that they may vary (Hanisch et al., 1998b, Hanisch et al., 1998a). Table 1 gives an overview over our estimations. If in the case of a condenser's cooling system breakdown the evaporation process is continued in $T5$, it comes to a dangerously high pressure in $C1$ after a certain time which might result in the destruction of the condenser. Therefore, the evaporation process must be aborted in case of such a breakdown. However, there is a problem. If the medium in tank $T5$ is not heated up and cannot be drained off into $T7$, because the previous batch is still processed in $T7$, it will gelatinize after a certain time. This in turn would cause the destruction of the evaporator $T5$.

Consequently, we are always in a dangerous situation if the breakdown (cooling water loss) occurs and $T7$ is still occupied with processing the previous batch. Now

Table 1. Process steps and their duration

Process step	Time (min)
Filling $T5$	$3-4$
Evaporation process in $T5$ until the target concentration is achieved	$58-61$
Draining the medium off $T5$ into $T7$	$2-3$
Postprocessing step in $T7$	$40-44$
Draining medium off $T7$	$10-11$
Time between switching off the heating and gelatinizing of the medium	4
Time between loss of cooling and destruction of the condenser	8

we are interested in solutions for two time parameters in order to develop a controller that allows us to react on a condenser's cooling system breakdown.

1. The first time parameter T_V is the time between processing two batches in $T5$. It must be large enough to ensure that enough time remains after a shutdown of the heating, allowing the medium in $T5$ to be drained off into $T7$ before it gelatinizes.
2. The second time parameter T_W states the amount of time the heating can be continued without destroying the condenser.

3 Timestamp Nets

In this section, we introduce a timing concept for Petri Nets which helps us solving such problems as described in the previous section. We assume that the reader is familiar with the basic concept of Petri Nets (Baumgarten, 1990, Hanisch, 1992).

Definition 1. Given $a \in \mathbf{R}_0^+$ and $b \in \mathbf{R}_0^{+,\infty}$. $[a;b]$ is called *time interval*, and if $a \leq b$ holds $[a;b]$ is called *non-negative time interval*. $\mathbf{I}_\mathbf{R}^+ := \{[a;b] \mid (a;b) \in \mathbf{R}_0^+ \times \mathbf{R}_0^{+,\infty} \wedge a \leq b\}$ is the *set of all non-negative time intervals*.[1]

Definition 2. $\mathcal{N} = (P,T,F,I)$ is called *Timestamp Net*, iff (P,T,F) is a Petri Net, and $I : (P \times T) \cap F \rightarrow \mathbf{I}_\mathbf{R}^+$ assigns a non-negative time interval to each incoming edge of a transition describing the edge's permeability. Given $I(p,t) = [r;l]$, $I_r(p,t) = r$ denotes the *begin* and $I_l(p,t) = l$ the *end* of the permeability of edge $(p,t) \in (P \times T) \cap F$ relative to the moment place p was marked. The *marking* m of a Timestamp Net $\mathcal{N} = (P,T,F,I)$ is a function $m : P \rightarrow (\langle \mathbf{R}_0^+ \rangle \cup \{0\})$. $m(p) = 0$ means that $p \in P$ is not marked. If a place $p \in P$ is marked by $m(p) = \langle ts \rangle$ then *ts* is called the *token's timestamp* indicating the moment the token was put on p.

[1] With: $\mathbf{R}_0^{+,\infty} := \mathbf{R}_0^+ \cup \{\infty\}$, $\forall r \in \mathbf{R}_0^+ : r < \infty$ and $\forall r \in \mathbf{R}_0^{+,\infty} : r + \infty = \infty$. In the following, we will also consider the set $\langle \mathbf{R}_0^+ \rangle := \{\langle ts \rangle \mid ts \in \mathbf{R}_0^+\}$

Definition 3. Let $\mathcal{N} = (P, T, F, I)$ be a Timestamp Net and m a marking of $\mathcal{N}$.

$$max(m) := \begin{cases} max\{ts_p \mid p \in P \ \wedge m(p) = \langle ts_p \rangle\} & \text{if } \exists p : m(p) \neq 0 \\ 0 & \text{otherwise} \end{cases}$$

is called the *highest timestamp* or as well the *timestamp of the youngest token* of m in case that a token exists under m. Now, let $t \in T$ be a transition of $\mathcal{N}$.

$$eft(t, m) := max\{ts_p + I_r(p, t) \mid p \in \bullet t \wedge m(p) = \langle ts_p \rangle\}$$

is the *earliest* and

$$lft(t, m) := min\{ts_p + I_l(p, t) \mid p \in \bullet t \wedge m(p) = \langle ts_p \rangle\}$$

the *latest firing time* of t under m.

The enabling of a transition does not only depend on whether its preplaces are marked, but also on the timestamps of the tokens. Therefore, we have a multi-stage enabling term.

Definition 4. Let $\mathcal{N} = (P, T, F, I)$ be a Timestamp Net and m a marking of $\mathcal{N}$.

- $t \in T$ is called *marking enabled* under m, iff $\forall p \in \bullet t : m(p) \neq 0 \ \wedge \ \forall p \in t\bullet - \bullet t : m(p) = 0$.
- t is *possibly enabled* under m, iff t is marking enabled under m, $eft(t, m) \leq lft(t, m)$, and $lft(t, m) \geq max(m)$ hold.
- t *can get enabled* under m, iff t is possibly enabled under m and no other transition t' is possibly enabled with $lft(t', m) < eft(t, m)$.
- t is *enabled* under m, iff t is possibly enabled under m, and $eft(t, m) \leq max(m)$.

Now, we can define fire rule and firing sequences for Timestamp Nets.

Definition 5. Let $\mathcal{N} = (P, T, F, I)$ be a Timestamp Net and m a marking of $\mathcal{N}$, $t \in T$ might can get enabled and/or is enabled under m, and $\tau \in \mathbf{R}_0^+$. t *can fire at* τ, iff $eft(t, m) \leq \tau \leq lft(t, m)$ and $\tau \geq max(m)$. (t, τ) is called a *step*, iff t fires at τ. An enabled transition must fire at the latest at $lft(t, m)$ or is no longer enabled due to the firing of other transitions. The *follower marking* m' of place $p \in P$ which is reached after step (t, τ) ($m\,[t, \tau\rangle\, m'$) is defined by

$$m'(p) = \begin{cases} 0 & \text{if } p \in \bullet t - t\bullet \\ \langle \tau \rangle & \text{if } p \in t\bullet \\ m(p) & \text{if } p \in P - (\bullet t \cup t\bullet) \end{cases}$$

Let m_0 be the initial marking of $\mathcal{N}$, and $w \in (T \times \mathbf{R}_0^+)^*$ with $w = (t_1, \tau_1) \ldots (t_n, \tau_n)$ a *firing sequence*. A marking m'' of $\mathcal{N}$ is *follower marking* of m_0 under w, written as $m_0[w\rangle m''$, iff

- $w = \Box \wedge m_0 = m''$, or
- there exists a marking m' of $\mathcal{N}$, with $m_0[(t_1, \tau_1) \ldots (t_{n-1}, \tau_{n-1})\rangle m' \wedge m'[t_n, \tau_n\rangle m''$.

$\square$ designates the empty firing sequence.

$$[m_0\rangle_{\mathcal{N}} := \{m \mid \exists w \in (T \times \mathbf{R}_0^+)^* : m_0[w\rangle m\}$$

is called the *reachability set* of $\mathcal{N}$ under m_0.

Although marking enabled, a transition t with at least two preplaces cannot fire under a marking m, if its incoming edges are not permeable simultaneously. In this case we say t is timewise stuck and *eft*$(t, m) >$ *lft*(t, m) holds, consequently:

Definition 6. Let $\mathcal{N} = (P, T, F, I)$ be a Timestamp Net, m_0 its initial marking, and m a marking of $\mathcal{N}$ with $m \in [m_0\rangle_{\mathcal{N}}$. A transition $t \in T$ is *timewise stuck* under m, iff it is marking enabled but *eft*$(t, m) >$ *lft*(t, m) holds. t is *possibly getting timewise stuck*, iff a marking $m \in [m_0\rangle_{\mathcal{N}}$ exists and t is timewise stuck under m.

If a transition gets timewise stuck, it forces tokens to remain on places forever. In a dynamic system this is an indicator for a faulty situation and as a consequence there is a need to detect such situations and prevent them from occurring. Since in Timestamp Nets the underlying time concept is continuous, there usually exist overcountably many markings in the reachability set of such a net. As a consequence, finding all situations under which transitions might get timewise stuck cannot be computed by considering the entire reachability set.

In (Hanisch et al., 1998c) we proposed a method for estimating whether certain transitions might get timewise stuck or not. This method which has been implemented in the Petri Net tool POSEIDON (Simon et al., 1997) works as follows: Instead of having a marking where timestamps are assigned to tokens we use symbolic tokens. The domain of such a symbol describes the range of possible timestamp values that could be achieved due to a certain fire sequence. Such ranges can be defined with the aid of time intervals and their calculation bases on the non-negative time intervals assigned to the net's edges. If firing a transition produces two or more tokens simultaneously, under a symbolic marking the equality of their timestamp values is taken into account by using the same symbol for each of these tokens.

Now, determining whether a certain transition t might get timewise stuck or not is done by calculating all symbolic markings under which t is enabled, i.e. we have to consider the reachability set of the symbolic marking instead of the reachability set of the timestamp marking. For these symbolic markings we derive systems of inequalities such that if these have a solution t cannot get timewise stuck.

This approach has the advantage that it can be even extended by some few modifications: If we use parameters to describe the time intervals at the edges of our net, and if we use these parameters for calculating our symbolic marking, then a solution for these parameters guarantees that those transitions we concentrate our investigations on cannot get timewise stuck. With other words, our approach allows to find values for our parameters such that we can avoid faulty situation.

In the rest of this section, we define Parameterized Timestamp Petri Nets which have to be used for our approach. In the following section, we define symbolic markings and conditions under which we can exclude that certain transitions can get timewise stuck under a corresponding timestamp marking.

Definition 7. Let $a \in \mathbf{R}_0^+ \cup V_P$, and $b \in \mathbf{R}_0^{+,\infty} \cup V_P$ with V_P is a set of variables. $[a;b]$ is called *parameterized time interval* and $\mathbf{I}_{V_P} := \{[a;b] \mid a \in \mathbf{R}_0^+ \cup V_P, b \in \mathbf{R}_0^{+,\infty} \cup V_P\}$ is the *set of all parameterized time intervals*.

Definition 8. A *Parameterized Timestamp Net* $\mathcal{N} = (P, T, F, V_P, I)$ is a tuple, where (P, T, F) is a net, V_P a set of variables, and $I : (P \times T) \cap F \rightarrow \mathbf{I}_{V_P}$ assigns to each incoming edge of a transition a parameterized time interval. For $f \in (P \times T) \cap F$ and $I(f) = [r;l]$, both, $I_r(f)$, and $I_l(f)$ are defined comparable to definition 2. The *initial marking* $\hat{m}$ of a Parameterized Timestamp Net $\mathcal{N} = (P, T, F, V_P, I)$ is a function $\hat{m} : P \rightarrow (\langle \mathbf{R}_0^+ \rangle \cup \{0\})$ with $\hat{m}(p) = 0$ means that $p \in P$ is not marked.

We are not interested in the behavior of Parameterized Timestamp Nets but in such solutions for the variables of V_P for which certain transitions in the corresponding Timestamp Net will never get timewise stuck. As a corresponding Timestamp Net we define a Timestamp Net which results from a Parameterized Timestamp Net by replacing each occurrence of a parameter by its solution.

4 Symbolic Markings

Under a symbolic marking the places are marked with symbols (i.e. variables) instead of tokens with timestamps. The domain of these symbols is restricted by a system of inequalities. Solutions for these variables imply possible moments at which tokens can be put onto their places under a timestamp marking. If the same symbol is used to mark several places, these places were marked simultaneously by a branching transition. In preparation of definitions 10 and 14, we need rules for calculating with intervals

Definition 9. Let $[a;b], [c;d] \in \mathbf{I}_{\mathbf{R}}^+$ and therefore $a \in \mathbf{R}_0^+$.

$$\begin{aligned} [a;b] \oplus [c;d] &:= [a+c; b+d] \\ a \oplus [c;d] &:= [a;a] \oplus [c;d] \\ [a;b] \sqcap [c;d] &:= [\mathit{max}\{a,c\}; \mathit{min}\{b,d\}] \end{aligned}$$

Definition 10. Let V be a set of variables. A *term* is defined inductively as follows:

- Each value $r \in \mathbf{R}_0^{+,\infty}$ and each variable $v \in V$ are terms.
- If t_1, t_2 are terms then $t_1 + t_2$, $\mathit{max}\{t_1, t_2\}$, and $\mathit{min}\{t_1, t_2\}$ are terms, too.

An *interval term* is defined inductively as follows:

- $[r;l]$ is an interval term if r and l are terms.
- If t_1, t_2 are interval terms then $t_1 \oplus t_2$, and $t_1 \sqcap t_2$ are interval terms, too.

$\mathbf{I}_V$ is the set of all interval terms which can be build using the variables in V.

Definition 11. Let V be a set of variables. A function $c : V \to \mathbf{I}_V$, assigning intervals to the variables of V, is called a *constraint system* on V if a partition $V_1, \ldots, V_n$ of V exists, with

- $\underset{1 \leq i \leq n}{\forall} : V_i \neq \emptyset, \bigcup_{1 \leq i \leq n} V_i = V$ and $\forall i, j \in \{1, \ldots, n\}, i \neq j : V_i \cap V_j = \emptyset$
- and for $v \in V_i, 1 \leq i \leq n$: $c(v) \mapsto it,$ with $it \in I_{\bigcup_{j<i} V_j}$

Although the conditions of a constraint system c are build upon one another, Def. 11 excludes that these dependencies are cyclically defined.

Definition 12. For a set of variables V a function $\mathcal{A} : V \to \mathbf{R}_0^+$ is called an *assignment*. Let $i \in \mathbf{I}_V$ be an interval term over V and $V' \subseteq V$. An *evaluation* $i_{V'|\mathcal{A}(V')}$ of i is that interval term which results from i by substituting each occurrence of a variable $v \in V'$ by its assignment $\mathcal{A}(v)$. Finally, let $c : V \to \mathbf{I}_V$ be a constraint system. An assignment $\mathcal{A} : V \to \mathbf{R}_0^+$ is called *feasible* under c, iff for $v \in V$ $c_r(v)_{V|\mathcal{A}(V)} \leq \mathcal{A}(v) \leq c_l(v)_{V|\mathcal{A}(V)}$.

We are now using the former definitions for defining symbolic markings.

Definition 13. Let $\mathcal{N} = (P, T, F, V_P, I)$ be a Parameterized Timestamp Net. A *symbolic marking* $\tilde{m} = (V, sm, c)$ of $\mathcal{N}$ is a tuple consisting of

- a set of variables V,
- a function $sm : P \to V \cup \{0\}$ where $sm(p) = 0$ means that place $p \in P$ is not marked, and
- a constraint system $c : V \to \mathbf{I}_V$ over V.

Therefore, under a symbolic marking places are marked with variables of V that are limited due to a constraint system c.

Definition 14. Let $\mathcal{N} = (P, T, F, V_P, I)$ be a Timestamp Net and $\tilde{m} = (V, sm, c)$ a symbolic marking. $t \in T$ is called *enabled* under $\tilde{m}$, iff $\forall p \in \bullet t : sm(p) \neq 0$ and $\forall p \in t\bullet - \bullet t : sm(p) = 0$. The *symbolic follower marking* $\tilde{m}' = (V', sm', c')$, which is achieved if an enabled transition t fires under the symbolic marking $\tilde{m}$ (written $\tilde{m}[t\rangle\tilde{m}'$), is defined as follows:

- $V' = V \cup \{\nu\}$ for a $\nu \notin V$.
- $sm' : P \to V' \cup \{0\}$ with

$$sm'(p) \mapsto \begin{cases} 0 & \text{if } p \in \bullet t - t\bullet \\ \nu & \text{if } p \in t\bullet \\ sm(p) & \text{if } p \in P - (\bullet t \cup t\bullet) \end{cases}$$

- $c' : V' \to \mathbf{I}_{V'}$ with

$$c'(v) \mapsto \begin{cases} \bigsqcap_{p \in \bullet t} (sm(p) \oplus I(p, t)) & \text{if } v = \nu \\ c(v) & \text{if } v \in V \end{cases}$$

So, by firing a transition t under a given symbolic marking, all time intervals at incoming edges of t are included to specify the variety of possible values for the new symbolic tokens.

Definition 15. Let $\mathcal{N} = (P, T, F, V_P, I)$ be a Parameterized Timestamp Net and $\tilde{m}_0$ its symbolic initial marking. Furthermore, let $\omega \in T^*$ with $\omega = t_1 \dots t_n$ be a *symbolic firing sequence*. A symbolic marking $\tilde{m}''$ of $\mathcal{N}$ is a *symbolic follower marking* of $\tilde{m}_0$ under ω, written as $\tilde{m}_0[\omega\rangle\tilde{m}''$, iff

- $\omega = \Box \wedge \tilde{m}_0 = \tilde{m}''$ holds, or
- a symbolic marking $\tilde{m}'$ of $\mathcal{N}$ exists, with $\tilde{m}_0[t_1 \dots t_{n-1}\rangle\tilde{m}' \wedge \tilde{m}'[t_n\rangle\tilde{m}''$.

$\Box$ designates the empty firing sequence. $[\tilde{m}_0\rangle_{\mathcal{N}} := \{\tilde{m} \mid \exists\omega \in T^* : \tilde{m}_0[\omega\rangle\tilde{m}\}$ is called the *symbolic reachability set* of $\mathcal{N}$ under $\tilde{m}_0$.

5 Solution for the Example

We use symbolic markings to find values for the parameters of a Parameterized Timestamp Net such that certain transitions in its corresponding Timestamp Net will never get timewise stuck. To achieve this, we integrate these net parameters into the symbolic marking in accordance to the following definition:

Definition 16. Let $\mathcal{N} = (P, T, F, V_P, I)$ be a Parameterized Timestamp Net with initial marking $\hat{m}$. A symbolic marking $\tilde{m}_0 = (V_0, sm_0, c_0)$ is called *fitting symbolic initial marking* for $\hat{m}$ iff

- $V_0 = V_P \cup \{v_p \mid p \in P, \hat{m}(p) \neq 0 \text{ and } v_p \text{ is variable}\}$
- $\forall p \in P : sm_0(p) = \begin{cases} 0 & \text{, if } \hat{m}(p) = 0 \\ v_p & \text{, if } \hat{m}(p) \neq 0 \end{cases}$
- $c_0 : V_0 \to \mathbf{I}_{V_0}$, with

$$\forall v \in V_0 : c(v) \mapsto \begin{cases} [ts; ts] & \text{, if } sm_0(p) = v \wedge \hat{m}(p) = \langle ts \rangle \\ [0; \infty] & \text{, if } v \in V_P \end{cases}$$

A solution for the parameter values of a Parameterized Timestamp Net guarantees that in the Timestamp Net that results from this solution certain transitions never get timewise stuck, i.e. the moments at which tokens reach the preplaces of these transitions do not vary such that one of these incoming edges looses its permeability before another gets permeable. Therefore, we compute the moments places can get marked relative to one another by two functions lowest (*lpv*) and highest possible value (*hpv*) which are defined as follows.

Definition 17. Let V and V_P be two sets of variables with $V \cap V_P = \emptyset$, and $c : V \cup V_P \to \mathbf{I}_{V \cup V_P}$ a constraint system. For a variable $v \in V \cup V_P$ its *lowest possible value* $lpv(v)$ is defined by:

- If $c(v) = [a; b]$, $a \in \mathbf{R}_0^+ \cup V_P, b \in \mathbf{R}_0^{+,\infty} \cup V_P$, then $lpv(v) := a$.

- Otherwise, if $c(v) = \bigsqcap_{1 \leq i \leq n} (v_i \oplus [a_i; b_i])$, $a_i \in \mathbf{R}_0^+ \cup V_P$, and $b_i \in \mathbf{R}_0^{+,\infty} \cup V_P$ for $1 \leq i \leq n$, then $lpv(v) := max\{lpv(v_i) + a_i \mid 1 \leq i \leq n\}$.

For two variables $v_1, v_2 \in V \cup V_P$ the *highest possible value* $hpv(v_1, v_2)$ which can be assigned to v_1 relative to v_2 under all assignments feasible with respect to c is defined by:

- If under all feasible assignments the value of v_1 depends on the value of v_2, then $hpv(v_1, v_2) := lpv(v_2)$.
- Otherwise, if $c(v_1) = [a; b]$, $a \in \mathbf{R}_0^+ \cup V_P$, and $b \in \mathbf{R}_0^{+,\infty} \cup V_P$, then $hpv(v_1, v_2) := b$.
- Otherwise, if $c(v_1) = \bigsqcap_{1 \leq i \leq n} (v_{1_i} \oplus [a_{1_i}; b_{1_i}])$, $a_{1_i} \in \mathbf{R}_0^+ \cup V_P$, and $b_{1_i} \in \mathbf{R}_0^{+,\infty} \cup V_P$ for $1 \leq i \leq n$, then $hpv(v_1, v_2) := min\{hpv(v_{1_i}, v_2) + b_{1_i} \mid 1 \leq i \leq n\}$.

Theorem 1, proved in (Hanisch et al., 1998c), has been implemented in POSEIDON as the core of an algorithm to determine sufficient time parameters by solving a linear optimization problem. It is based on the definition of *lpv* and *hpv*.

Theorem 1. *Let $\mathcal{N} = (P, T, F, V_P, I)$ be a Parameterized Timestamp Net with initial marking $\hat{m}$, $\tilde{m}_0$ the symbolic initial marking fitting to $\hat{m}$ and $\mathcal{N}$, $t \in T$ a transition of $\mathcal{N}$, and $\mathcal{A} : V_P \rightarrow \mathbf{R}_0^+$ an assignment for the variables of V_P such that under all symbolic markings $\tilde{m} = (V, sm, c) \in [\tilde{m}_0\rangle_{\mathcal{N}}$ under which t is enabled for all $p_1, p_2 \in \bullet t$ the following (in-)equalities hold*

$$(hpv(sm(p_1), sm(p_2)) + I_r(p_1, t))_{V_P|\mathcal{A}(V_P)} \leq (lpv(sm(p_2)) + I_l(p_2, t))_{V_P|\mathcal{A}(V_P)}$$
$$or$$
$$I_l(p_2, t) = \infty$$

Then in the Timestamp Net $\mathcal{N}' = (P, T, F, I_{V_P|\mathcal{A}(V_P)})$ transition t is never getting timewise stuck under $\hat{m}$.

6 Results for the Example

The Parameterized Timestamp Net in Fig. 3 is a model of the plants dynamic. With the aid of time intervals we take account of the former specified duration of/between each single process step. Table 2 explains the names of the places in the net. The meaning of the transitions' names is explained in Table 3.

The problem to be solved is to find such time parameters for T_V and T_W which guarantee that forbidden states cannot be reached. As dangerous situations we identified a possible destruction of the condenser or a gelatinizing medium in Tank $T7$. However, these situations can be excluded if we prevent transitions $t9$, $t10$, $t12$, and $t13$ from getting timewise stuck.

With the aid of the Petri Net tool POSEIDON we computed the symbolic reachability set. From this we derived a linear optimization problem that we solved using MATHEMATICA to determine the time parameters. Here, it was problematic that

Table 2. Places of the Net in Figure 3

Component	Place	Meaning
Tank $T5$	$T5E(1)$	empty but unavailable for a new charge
	$T5E(2)$	empty and available for a new charge
	$T5F$	is getting filled
	$T5I$	heating off, waiting for $T7$
	$T5P$	heating on
	$T5R$	target concentration is achieved and heating is off, waiting for $T7$
Tank $T7$	$T7E$	is empty
	$T7notE$	is not empty
	$T7P$	postprocessing step is active
	$T7D$	is pumped empty
Tanks $T5$ and $T7$	$T5D/T7F$	$T5$ is emptied and $T7$ filled
Condenser $C1$	CR	cooling breakdown recognized by the controller
	$C1B$	cooling water loss and the heating in $T5$ is on
	$C1B_{OK}$	cooling water loss but the heating in $T5$ is off
	$C1C$	working
	$C1N$	not working

Table 3. Transitions of the Net from figure 3

Transition	Function
$t1$	Delay filling $T5$
$t2$	Start filling $T5$
$t3$	Start evaporation process in $T5$
$t4$	After termination of the postprocessing step, pump $T7$ empty
$t5$	Cooling water loss
$t6$	Usual termination of the evaporation process in $T5$
$t7$	$T7$ is pumped empty
$t8$	Usual termination of the evaporation process in $T5$ although cooling water loss (switching off the heating is not necessary)
$t9$	Premature termination of the evaporation process in $T5$ because of a cooling water loss. Medium can be drained off into $T7$ since $T7$ is empty
$t10$	Premature termination the evaporation process in $T5$ because of a cooling water loss. Waiting, that medium can be drained off into $T7$
$t11$	Repair of the cooling water supply
$t12$	Start emptying $T5$ / filling $T7$. Target concentration not achieved
$t13$	Start emptying $T5$ / filling $T7$. Target concentration achieved
$t14$	Drain off medium from $T5$ into $T7$

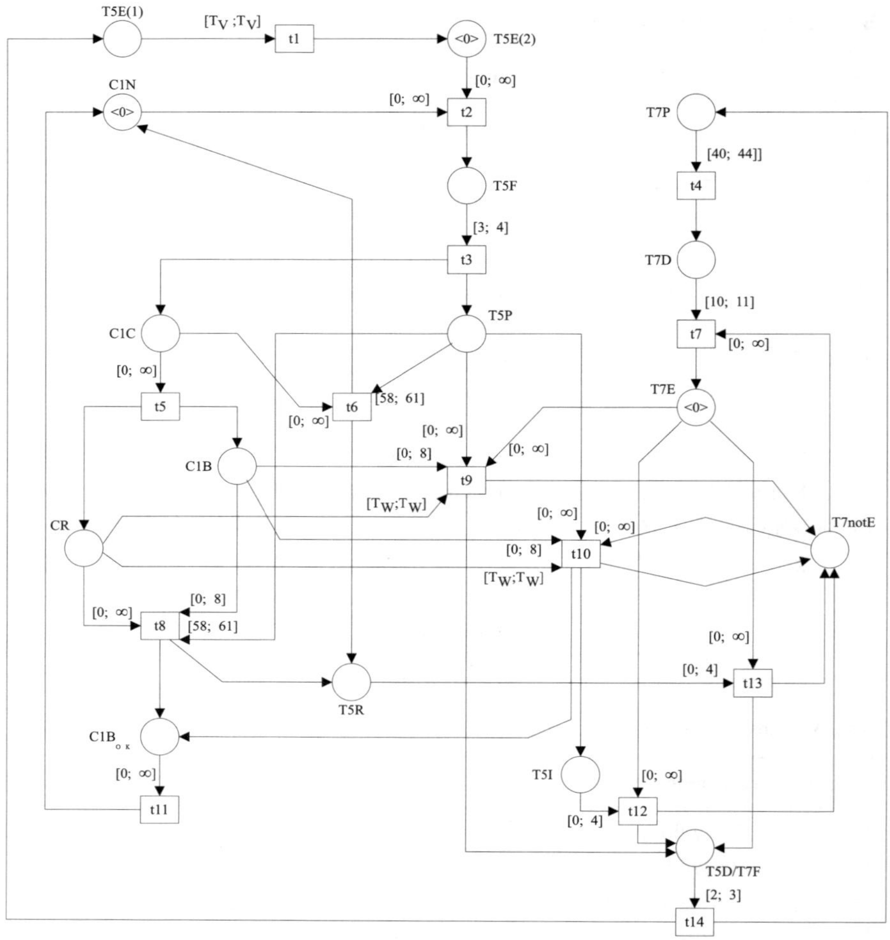

Fig. 3. Parameterized Timestamp Net to the plant of Fig. 2

the symbolic reachability set of the Parameterized Timestamp Net shown in Fig. 3 is infinite. Fortunately, we found a problem dependent criterion which allowed to cut the calculation of the symbolic reachability set without falsifying the results. This is possible whenever we reach a marking where the same places are marked compared to a former marking and where only possible time values for the symbols are deferred on the time scale with respect to the former marking. The computed results are:

- $T_V + T_W \geq 48$ must be feasible since otherwise one of the transitions $t12$ and $t13$ might get timewise stuck.
- $T_W \leq 8$ must be feasible since otherwise the transitions $t9$ and $t10$ might get timewise stuck.

7 Conclusion

Our contribution shows how to find time parameters for technical problems with the aid of (Parameterized) Timestamp Nets. For this purpose, forbidden situations are specified with the aid of transitions that get timewise stuck. For finding appropriate time parameters, we derive a linear optimization problem from the Timestamp Petri Net model of the technical system's dynamic which can be solved with standard mathematical tools. In the model, solutions for the parameters imply that certain transitions never get timewise stuck.

It is obvious that the generation of systems of inequalities terminates if the Petri Net is acyclic. In cyclic nets, cases can exist where it cannot be decided when the generation of inequality systems can be terminated without loss of information of the complete system behavior. In technical applications, however, one usually can find criteria to stop the enumeration which are based on heuristics and knowledge of the technological background of the system which is modelled. Formal criteria for termination of state space computation are developed in (Thieme, 2002).

However, this approach has some limits. Usually, the duration of activities is not static or lies within an interval but depends on other technical parameters. In the Timestamp Nets we defined in Sect. 3 time is the only parameter which can be represented. Extended Timestamp Nets (Lautenbach and Simon, 1999) avoid this problem. In accordance to Predicate-Transition Nets (PrT-Nets) (Genrich and Lautenbach, 1981) they allow to code any kind of information in addition to time information. The manipulation of this information is specified with the aid of annotations at the incoming and outgoing edges of transitions. Now, time parameters are not only statically defined but may also be defined by functions which depend on these parameters. The use of Extended Timestamp Nets allows to model the entire exemplary batch plant shown in Fig. 1 (Lautenbach and Simon, 2001).

Beside a model of the uncontrolled plant, a formal description of the technological processes is required. Here, a *Logic of Actions* introduced by (Genrich, 1978, Lautenbach and Simon, 2000) has been identified in the past as especially useful, because the formulas of this logic can be transformed into Petri Nets. After this transformation, the Petri Net model of the uncontrolled plant and the Petri Net model of the technological processes can be automatically integrated into a model of the controlled plant. These models can be used for investigation like those we introduced in the previous sections. A full description of this processes can be found in (Simon, 2001b, Simon, 2001a).

Compositional Verification of Continuous-Discrete Systems

Ralf Huuck[1], Ben Lukoschus[1], Goran Frehse[2], and Sebastian Engell[2]

[1] University of Kiel, Institute of Computer Science and Applied Mathematics, Chair of Software Technology, D-24098 Kiel, Germany, {rhu,bls}@informatik.uni-kiel.de
[2] University of Dortmund, Process Control Lab (CT-AST), D-44221 Dortmund, Germany, {g.frehse,s.engell}@ct.uni-dortmund.de

Abstract. Hybrid systems are well-suited as a design and modeling framework to describe the interaction of discrete controllers with a continuous environment. However, the systems described are often complex and so are the resulting models. Therefore, a formal framework and a formal verification to prove the correctness of system properties is highly desirable. Since complexity is inherent, standard formal verification techniques like *model checking* soon reach their limits. In this work we present several options how to tackle the complexity arising in the formal verification of hybrid systems. In particular we combine the model checking approach with *abstraction* and *decomposition* techniques such as the *assumption/commitment* method as well as deductive methods.

1 Introduction

The description of real-world physical systems has always been an issue. Such a system model not only enhances the understanding of the underlying physics but it makes it possible to actually predict the system's behavior. Since nowadays nearly every production, power generation, and logistics process is highly automated, such a prediction is extremely valuable in order to simulate the system's behavior in different environments or even to prove that certain properties are satisfied.

However, every system model is by nature an abstraction of the real world. Finding the right abstraction and, thus, developing the right system model is not an easy task. In general the abstraction and, therefore, the model is chosen according to the design level one is interested in. At times the detailed continuous physical behavior is the focus of study and other times discrete behaviors such as communication and synchronization is of main interest.

Continuous models are used, e.g., to describe movements of mechanical systems, linear circuits or chemical reactions while discrete models are sufficient to describe the collision of two objects in a mechanical system, the switching in circuits or the use of pumps and valves in chemical plants. Continuous models are generally given in the form of differential equations, possibly supplemented by a set of algebraic constraints. In contrast, discrete models are more diverse but often can be captured by some form of a state representation. Physical processes are controlled by software on digital computers. Such embedded control systems combine continuous physical behavior with discrete control algorithms and are called *hybrid systems*.

S. Engell, G. Frehse, E. Schnieder (Eds.): Modelling, Analysis and Design of Hybrid Systems, LNCIS 279, pp. 225–244, 2002.

To model and to verify hybrid systems which describe the interaction of control software with a physical plant is a very desirable goal. However, it is not easy to achieve since it requires a unified framework covering the continuous and discrete world and a verification framework which is able to cope with such systems. This is even more difficult when thinking of exhaustive verification that is as much automated as possible. This means, from a formal system description and a set of requirements the verification task can be left to a computer. Despite theoretical limitations of this approach any verification technique has to cope with the immense complexity that lies in the nature of such a system.

Then, why to use formal description methods and formal verification at all? In the current design system processes some methods to enhance the quality of the software and the overall system have already made their way into industrial standard practice. These comprise techniques such as listing software requirements and striving for a clean documentation for the design process as well as code review, system simulation and testing to check for the correctness of the implemented system.

However, techniques in common practice have various drawbacks.

- They often lack a formal basis. For instance, the specification is defined in natural language, which easily leads to misunderstanding and misinterpretation.
- The requirements are not complete, i.e., there are cases which are not taken into account. Hence, parts of the system remain unspecified and, thus, are allowed to behave differently than the designer had in mind. Especially without a formal model it is more likely to forget cases, or cases which are defined in a contradictory way remain undetected.
- The verification of the implementation might be approached in an unorganized way. For instance, testing is done with some arbitrary inputs when it is more efficient to choose data according to the boundary conditions of the model or the implementation, simulation focuses on "non-important" variables and code review neglects inter-procedural dependencies etc.
- Most informal techniques like testing or simulation are not exhaustive, i.e., the they do not cover all program executions and thus give way to subtle but often fatal flaws.

Formal methods promise to remedy the above mentioned weaknesses. However, formal methods are not fool-proof by themselves. Sometimes they require exhaustive knowledge in mathematics, logics and the understanding of the system. Moreover, they do not a priori prevent forgetting about requirements or even ensure to appropriately map the real world to the model. Especially in software design one often has a clear idea of what to achieve, but much less of how to specify this formally or even to define what is considered to be legal and what to be harmful. Formal methods provide tools to further investigate into the design and verification process and allow to enhance the quality of the system significantly, but they do not buy any guarantee that what you prove is what you have in mind.

This work concentrates on applying formal methods to hybrid systems while at the same time tackling the inherent complexity issues. The approach presented here does not take the whole system at once into account, but divides the system and, hence, the verification task, into several components and several layers of abstractions. The division and verification of single components is known as *compositional verification.* However, hybrid systems are often too complex and too interwoven to find single components that can be verified independently of the remaining systems. It is much more natural to make some *assumptions* about the behavior of the remaining system under which the selected component can be verified. The verification of the component results into some *commitment* this unit fulfills under the given assumptions. Applying this method to every component still leaves the task to combine the assumptions and commitments in a meaningful and non-contradictive way. In this work we present a framework which allows to reason in this so-called *assumption/commitment style* and supports a formal and automated verification as far as possible.

We start with a brief overview on software verification and compositional techniques in Sect. 1.1. In Sect. 2 an introduction into a standard formal verification technique called *model checking* (Queille and Sifakis, 1982, Clarke and Emerson, 1982) is given. We explain the basic terms as well as logics and techniques used. The subsequent Sect. 3 deals with complexity issues of formal verification and hybrid systems.

1.1 Historical Notes

This section gives a brief summary of the development of formal software verification approaches in general and some important compositional methods in particular. More extensive surveys can be found in (de Roever, 1998, de Roever et al., 2001).

Formal Verification of Software. From the beginnings of the computer age, verification of software has always been an issue for programmers and system developers. Pioneers like John von Neumann and Alan Turing already thought about the correctness of programs for the first computers (Goldstein and von Neumann, 1947, Turing, 1949).

In 1967 Robert W. Floyd presented the *inductive assertion method* (Floyd, 1967), a formal strategy to prove the correctness of sequential programs written as labeled transition systems. A similar approach was presented by Peter Naur (Naur, 1966) one year earlier. C.A.R. Hoare axiomatized this method into a compositional style for sequential programs (Hoare, 1969). Programs are annotated with assertions, and their correctness is proven locally. Then the local assertions can be combined in a compositional fashion to obtain a global specification.

This Hoare-style proof system was extended to concurrent shared-variable programs in 1976 by Susan S. Owicki and David Gries (Owicki and Gries, 1976). Their method however involves a so-called *interference-freedom test,* which operates on every combination of local control locations and therefore is non-compositional.

Proof systems for programs with distributed synchronous communication were independently developed by Krzysztof R. Apt, Nissim Francez and Willem-Paul de Roever (Apt et al., 1980) and by Gary M. Levin and David Gries (Levin and Gries,

1981). Here a so-called *cooperation test* is done for every combination of input and output actions, which is also non-compositional.

In 1977 Amir Pnueli developed the temporal logic approach for the verification of concurrent programs (Pnueli, 1977), for which he received the Turing Award in 1996. This method is also non-compositional.

Compositional Approaches. In 1969 Edsger W. Dijkstra was the first to publish within the computer science community the opinion that compositional reasoning is needed for the formal verification of large programs (Dijkstra, 1969a).

Cliff B. Jones developed a compositional verification approach for concurrent shared-variable programs (Jones, 1981, Jones, 1983). His so-called *rely-guarantee* formalism, which specifies a system by its desired properties (guarantee) provided that its environment behaves in a certain way (rely).

A similar compositional approach for distributed synchronous communication, called the *assumption-commitment* method, was presented by Jayadev Misra and K. Mani Chandy in 1981 (Misra and Chandy, 1981).

Within the field of process algebra – the main languages used include CCS (Milner, 1980, Milner, 1989), CSP (Brookes et al., 1984, Hoare, 1985), and ACP (Bergstra and Klop, 1984) – one has always been striving for compositional reasoning, e.g., by defining behavioral preorders which are preserved by the composition operators.

2 Model Checking

Common to every verification task is to prove that a system, a program or simply an abstract model of a problem satisfies certain requirements. Formally, this is denoted by

$$M \models \varphi,$$

where M is a model of the system, φ is the requirement and $\models$ denotes the satisfaction relation. Model checking (Queille and Sifakis, 1982, Clarke and Emerson, 1982) is an algorithmic way to decide whether M satisfies φ. Although any verification approach is based on this, the actual logic or – more general – the formalism to denote these three items varies a lot. In the following we present some formal models for each of them. We mainly focus on the ones we will use throughout this work.

2.1 System Model

In this article we concentrate on verification of *reactive systems*. These are systems which communicate with their environment and may often – like operating systems – not terminate. Hence, a model which captures their infinite behavior in a concise way is desirable. Simply specifying their input/output behavior is not sufficient, it is rather interesting to know the *states* of a system, too.

Therefore, we start by describing the behavior of a system with some state-based formalism. Such formalisms include Petri Nets (Reisig, 1985), CSP (Brookes et al., 1984, Hoare, 1985), CCS (Milner, 1980, Milner, 1989), different forms of automata, LOTOS (Bolognesi and Brinksma, 1987), SDL (SDL92, 1992), etc. In these formalisms the behavior of the system is described in terms of local state changes or events. The global behavior of the system is given as the state-space generated from the system description.

In this work we use different kinds of automata for the system description, namely discrete, timed and hybrid automata.

Discrete Automaton. A discrete automaton $A = (Q, q_0, \delta, F)$ over an alphabet Σ (events, actions) is a structure where

- Q is a finite set of control locations,
- q_0 is an initial location,
- $\delta : Q \times \Sigma \longrightarrow Q$ is a transition function, and
- F is an acceptance condition.

A sequence of actions in Σ which is produced by taking a path through the automaton, starting with the initial location and satisfying the acceptance condition, is called a *word.* The set of all words, i.e., the set of all possible sequences, for an automaton A is called the *language* of A denoted by $L(A)$. The acceptance condition can vary from a single location which indicates the end of the sequence once it is reached to a set of locations which have to be reached infinitely often. The acceptance condition mainly determines the different kinds of discrete automata which can be found in the literature.

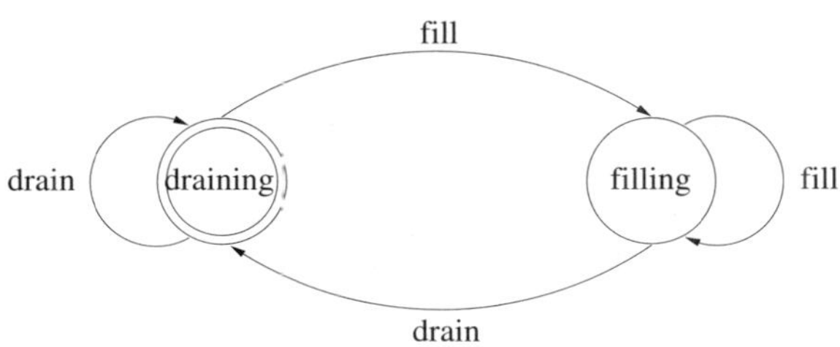

Fig. 1. Discrete automaton for a tank model

An example for a discrete automaton is depicted in Fig. 1. This automaton gives a rough model for a tank with a constant outflow, so there are only two control locations *draining* and *filling*. The double framed circle around *draining* indicates the initial location. Depending on the actions *drain* and *fill* transitions depicted by arrows are taken. We do not give an explicit acceptance condition here and note that the model is very simplified, i.e., it is not captured that the tank might run empty or might overflow.

Timed Automaton. In contrast to discrete automata, the setting of timed automata is in a dense real-time world. To express quantitative time, *clocks* are introduced which are real-valued variables evolving over time. Moreover, they can be checked against thresholds, and they can be reset when a transition is taken.

Formally, a timed automaton over an alphabet Σ is a quadruple $T = (Q, q_0, C, E)$ where

- Q is a finite set of locations,
- q_0 is the initial location,
- C is a finite set of clocks, and
- E is a set of edges of the form (q, γ, a, ρ, q'), where $q, q' \in Q$ are the source and target locations, γ is a *transition condition*, i.e., a Boolean formula over clock variables and thresholds, $a \in \Sigma$ is an action and ρ is the set of clocks that are reset when taking this transition.

The language of a timed automata is given by the set of all execution sequences over time. Traditionally, only infinite sequences are considered.

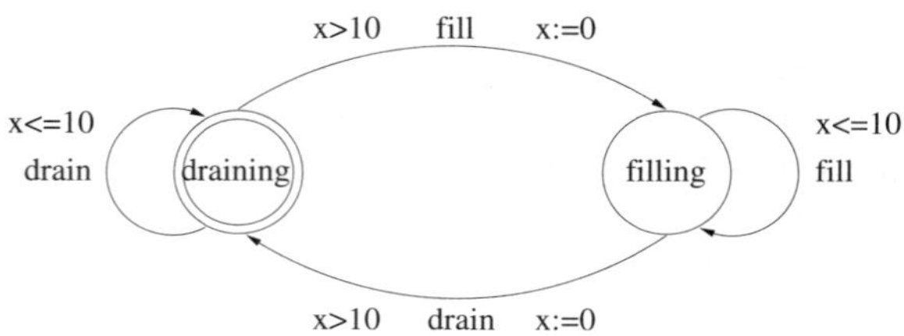

Fig. 2. Timed automaton for a tank model

A timed automaton example is depicted in Fig. 2. In contrast to Fig. 1 there is a clock x which constraints the moments transitions are taken. This clock serves as a timer for draining and filling periods. Starting from the initial location *draining* the location is changed only if x is greater than 10. If so the x is reset and control will reside in location *filling* until the clock value exceeds 10 again. In the meanwhile self-loops are possible.

Hybrid Automaton. Discrete automata do not incorporate quantitative time, but timed automata do so by the use of clocks. However, they do not model arbitrary continuous functions. This feature is covered by so-called hybrid automata. These allow do model and reason about a set of continuous variables evolving over time.

Formally, a hybrid automaton $H = (Q, q_0, \mathit{Var}, E, \mathit{Act}, \mathit{Inv})$ over an alphabet Σ consists of

- a finite set of locations Q with some initial location q_0,
- a finite set of real-valued variables Var.

- a finite set E of discrete transitions. Each transition $e = (q, \rho, a, q')$ between two locations $q, q' \in Q$ labeled by some action $a \in \Sigma$ depends on a transition condition ρ which reasons about the variables in Var,
- a labeling function Act that assigns a set of activities to each location $q \in Q$. The activities describe how the variables in Var evolve continuously as long as control resides in q.

The semantics of an hybrid automaton is defined by all trajectories of the continuous variables as well as the actions over time.

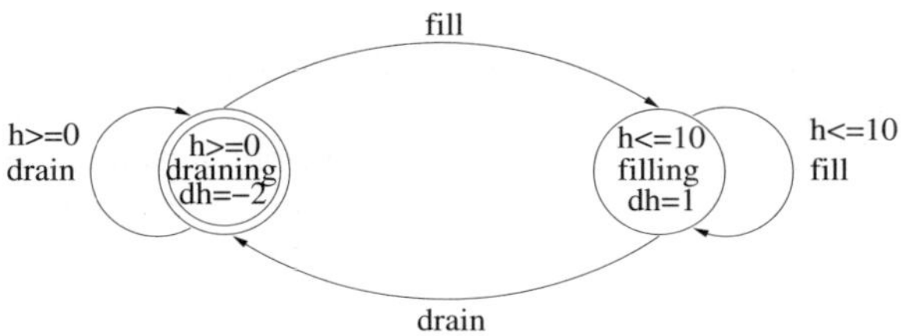

Fig. 3. Hybrid automaton for a tank model

A hybrid model for the tank example is shown in Fig. 3. This model describes the tank level h in a filling and draining process. Draining is two times faster than filling. Although possible, there are no guards or resets on the transitions, but invariants in the locations determine when exactly control is allowed to stay there. Note that there also exist different timed automaton models with invariants, deadlines and urgency. Mostly this has little effect on the expressiveness, cf. (Stahl, 1998), but allows more or less convenient notations.

2.2 Computational Model

In the previous section, we described very briefly how to derive a behavior from each description model. However, the models themselves are merely syntax and in order to formally derive a *semantics*, i.e., the system's behavior, the system description can be mapped to a mathematical abstract representation. This abstract representation is also called *computational model* and represents the semantics.

One way to describe a computational model is a state transition system. It consists of states and and also has the ability to represent the fact that in any given state the system reacts to certain actions and might enter new system states. This pair of system states is then called a *transition*. The semantics of a system is determined by the sequences of all transitions in a system that start from some given initial state. One formal way to describe these state transition systems are *Kripke structures* (Kripke, 1963), named after the logician Saul A. Kripke who used transition systems to define the semantics of modal logics. Transition systems are graphs consisting of states, transitions and a function that maps each state to a set of properties which hold in that state.

Formally, we define a Kripke structure as follows: Given a set of atomic properties P, also called propositions, a Kripke structure $K = (S, S_0, R, \mu)$ contains the following components:

- S is a set of states,
- $S_0 \subseteq S$ is a set of initial states,
- $R \subseteq S \times S$ is a transition relation, which is required to be total, i.e., for every state $s \in S$ there exists an $s' \in S$ such that $(s, s') \in S$, and
- $\mu : S \longrightarrow 2^P$ is a labeling function that assigns a set of propositions to every state.

An execution sequence of a Kripke structure is defined as a possibly infinite sequence $\pi = s_0 s_1 s_2 \ldots$ such that $s_0 \in S_0$ and for every index $i > 0$ in π we have $(s_{i-1}, s_i) \in R$. This means starting from the initial state we go along a path in the graph represented by the Kripke structure. The semantics of a system described by a Kripke structure is the set of all its sequences, i.e., all possible paths from all initial states.

In order to describe the semantics of a system model it is translated into such a computational model first. This means, the system model represents the syntax and the computational model the semantics. For the different types of automata presented above, the computational models are also different. While discrete automata only have to reflect the control location in a state, timed and in particular hybrid systems need to reflect time as well in a state. Since time is dense for both of the latter models, it is not always guaranteed to find a finite representation of these systems. However, using abstract or symbolic state representations, i.e., the clustering of concrete states into equivalence classes, in many cases a finite representation is possible also for timed and so called *linear hybrid automata*. The latter are hybrid automata which only allow fixed (but arbitrary) rates for the continuous variables. A finite representation is important in order to guarantee termination for algorithmic approaches like model checking.

2.3 Temporal Logics

In addition to a formal description of the system, it is also necessary to describe the requirements posed on a system in a formal style. There are different ways to do so. One fundamental issue is to choose between an operational or a declarative way. In this context operational means, e.g., to use automata themselves to specify the desired properties. The advantage is that the same framework is used for modeling the system and specifying the requirements. However, it is often a bit tedious to formulate requirements as automata, and automata are sometimes not as easy to understand as requirements. The declarative way means using logics to specify the requirements.

As mentioned before, we are mainly interested in reactive systems and, therefore, are concerned about the states of a system as well as the transitions between these states. Since basic propositional logic allows to reason only about states, but not sequences of states or transitions, so-called *temporal logic* (Pnueli, 1981) is used in order to remedy this fact. Temporal logic extends propositional logic, i.e., Boolean propositions with connectivities such as logical conjunction, disjunction and negation, with *modal operators*. These are operators like *always* or *eventually* that allow reasoning over execution sequences and can be combined with the usual connectivities.

Let us define propositional logic first. Based on propositions p logical expressions can be constructed by the following rules:

$$\varphi := p \mid \neg\varphi \mid \varphi_1 \wedge \varphi_2$$

Other Boolean connectives like "$\vee$", "$\Rightarrow$", and "$\Leftrightarrow$" can be derived from "$\neg$" and "$\wedge$" as usual.

The semantics is straightforward and therefore not shown here. Next, we present the extension from propositional to temporal logic. In general we can define and distinguish between two main temporal sub-logics, namely, linear time and branching time.

Linear Time Temporal Logic. One way to describe requirements is to define desired sequences in time. Linear Time Temporal Logic (LTL) allows to reason about paths in computational models like Kripke structures. In order to do so, propositional logic is extended by the following basic modal operators:

- $\bigcirc$. This denotes the modality "next" and requires that a property holds in the next state of a path, e.g., a path π in a Kripke structure satisfies $\bigcirc\varphi$ if and only if φ is satisfied in the second state of π.
- $\mathcal{U}$. This denotes the infix modality "until". I.e., a path π in a Kripke structure satisfies the expression $\varphi\mathcal{U}\psi$ if and only if ψ is satisfied in some later state of π, and φ holds in all states in between, including the first state of π. This is meant by the expression "φ until ψ".

LTL is founded on these basic modalities and their free combination with propositional logic. From these the following useful abbreviations can be defined:

- $\Diamond$ means "eventually", and a path π satisfies the expression $\Diamond\varphi$ if and only if there exists a state in π which satisfies φ.
- $\Box$ means "always", and a path π satisfies $\Box\varphi$ if and only if all states in π satisfy φ.

Branching Time Temporal Logic. In contrast to LTL branching time logics do not reason over single paths but over sets of paths, more precise, trees. One logic which does so is called *Computational Tree Logic* (CTL) which is propositional logic extended by *path quantifiers* and *temporal operators*. The temporal operators are the same as in LTL presented above. The path quantifiers are "$\exists$" which requires a single path to exist that satisfies some property and "$\forall$" which requires all paths of the computational model to satisfy some property.

CTL formulas are constructed from propositional logic, temporal operators and path quantifiers in the following way: Every formula starts with a path quantifier, every path quantifier is immediately followed by a temporal operator, and every temporal operator is preceded by a path quantifier.

This allows to build formulas such as

- $\exists\Box\varphi$, which means that there exists a path where always, i.e, for all states, φ holds, and
- $\exists\Diamond\forall\Box\varphi$, which means there exist a path with a certain state from whereon for all paths, i.e., branches, φ is always true.

Remarks. Note that CTL and LTL not only use different means to describe system properties, but in general there are LTL formulas which cannot represented in the CTL framework, and vice versa. Moreover, while linear time appears to be conceptually simpler than branching time, the latter is often computationally more efficient.

For both types of logics there exist real-time extensions. This means the logics provide the possibility to reason about explicit time and distances. We do not go into detail here.

2.4 Tools and Limitations

Returning to the initial task of checking $M \models \varphi$, model checking is, as mentioned, an algorithmic (i.e., automatic) way to decide whether a model M satisfies φ or not. There are several tools supporting model checking. For discrete automata and logics like CTL or LTL there are SMV (McMillan, 2000) and SPIN (Holzmann, 1997) as the most prominent ones. For checking timed automata with real-timed logics there are UPPAAL (Larsen et al., 1997), KRONOS (Olivero and Yovine, 1993) and extensions of SPIN. For linear hybrid systems HyTech (Henzinger et al., 1997) is a tool that enables to check reachability of certain states of the corresponding linear hybrid automaton. Moreover, there are many more tools which are also based on other system description models as well as logics.

For checking reactive systems one of the presented system models and logics is often used. However, due to fundamental limitations not every model and every logics is applicable for model checking. Timed and even more hybrid systems are restricted to certain classes, since a finite state representation in whatever way has to be guaranteed in order to keep model checking possible. Problem classes for which there cannot be any general algorithmic solutions are called *undecidable*.

Despite of these basic fundamental restrictions model checking has also to cope with serious complexity issues which are described in the next section.

3 Complexity Issues

One of the main drawbacks of state-based formal verification methods is the so-called *state explosion problem:* When a large system consists of several smaller components (e.g., automata) running in parallel, the number of global states increases exponentially with the number of components. For instance, consider a system of 20 automata working in parallel, each of which having 10 local states. This amounts to 10^{20} global states. The simple task of enumerating these states on a machine that needs only one nanosecond per state (which is considerably fast at the time of writing) already takes over 3000 years. Building and searching a graph based on these states takes significantly longer and is far beyond today's memory capabilities.

The state explosion problem is inherent in any system having parallel structures and poses a major complexity problem to any verification method based on the exhaustive enumeration of global states. Several techniques have been developed to minimize the impact of this problem on the time and memory consumption of the model checking process. Often a model checking algorithm uses a combination of several such techniques, which are discussed in the following.

Note that although all these methods can result in a significant speedup in practice, they are limited by the worst case complexity inherent to the problem (Lichtenstein and Pnueli, 1985, Vardi and Wolper, 1994, Kupferman et al., 2000). E.g., model checking LTL or CTL properties for Kripke structures is polynomial in $n \log m$ where n is the length of the formula and m is the size of the Kripke structure. When it comes to concurrent programs, i.e, different automata composed in parallel the problem is already PSPACE-complete even for a fixed formula. The same holds for model checking real-time systems in a timed variant of CTL.

3.1 Global vs. Local Strategy

In accordance with the two parameters of the model checking problem, the model M and the requirement φ, there are two basic strategies when designing a model checking algorithm, the "global" and the "local" strategy (Merz, 2001). "Global" means the algorithm operates recursively on the structure of φ and evaluates each sub-formula over the whole M, while the local strategy checks only parts of the state space at a time but for all sub-formulas of φ. The worst-case complexity of both approaches is the same, however, the average behavior can differ significantly in practice. Traditionally, LTL model checking is based on local approaches while for CTL global algorithms are applied.

3.2 "On-The-Fly" Techniques

The classical model checking approach builds a complete state transition graph of the system and performs a search on this graph. But often a large part of the graph is not traversed during the search or is even unreachable from the initial state(s) of the search. Therefore it is often a good idea to construct the graph in an "on-the-fly" fashion (Courcoubetis et al., 1992, Bhat et al., 1995). That is, only the part of the graph that is currently needed is constructed during the search and kept in memory for later reuse, often supported by caching algorithms.

3.3 Efficient Data Structures

A considerable amount of memory can be saved using efficient data structures during the model checking process. One prominent example are *binary decision diagrams* (BDDs) (Bryant, 1986, Bryant, 1992), which are used as a compact representation of Boolean functions. Ken McMillan suggested in his PhD thesis (McMillan, 1992) to use them for model checking, and today BDDs and similar data structures are the key solution for efficient memory usage in many kinds of computation software.

In the field of timed automata the observation that despite their continuous nature, clocks are often compared only to each other and a finite and bounded number of constants, opened the possibility to discretize the state space for model checking. So called *clock regions* are stored in data structures like *difference bounded matrices* (DBMs) (Bellman, 1957, Dill, 1990) and are used in most model checking tools for timed automata like KRONOS (Olivero and Yovine, 1993) and UPPAAL (Larsen et al., 1997).

3.4 Abstraction

Abstraction is a fundamental concept used in all formal verification methods. Abstracting means replacing a concrete object with an abstract one which is more universal, and therefore, often has a simpler structure than before. A well-chosen abstraction simplifies as much as possible, without losing too much information about the concrete object. Abstractions can be used in different ways during the specification and verification process:

- Building the system model: Every translation from a real-life system or an informal system description into a formal model is an abstraction.
- Optimizing the system model: Depending on the property that is to be checked, different abstractions of the system model can be useful, e.g., by abstracting from data, time, or continuous variables to obtain simpler models.
- Reducing the complexity of model checking: Model-checkers often use abstractions to minimize time and space usage, e.g., by introducing symbolic states.

When abstracting a system model, often a so-called *safe abstraction* is chosen: Whenever a property holds for the abstract system, it also holds for the concrete system. The converse, however, does not always hold, due to the over-approximation which occurs in the abstraction process. A positive model checking result on a safe abstraction therefore means that the concrete system also fulfills the property, whereas a negative result can either mean that the concrete system is not correct or that the abstraction is too coarse.

Thus, when getting a negative result, the counterexample provided by the model-checker is examined to see if the error will also occur in the concrete system. If it doesn't, a finer abstraction has to be chosen.

3.5 Compositionality

Another important concept is compositionality. In a compositional approach the system model is split into components. Each component is then specified as a single entity, and its correct behavior can be proved by model checking. The specifications of all components are then combined to get the global property of the system model. A prerequisite for this approach is that the behavior of the components is completely described by its specifications such that the behavior of the global system model only depends on these specifications and not on any additional information about the internal structure of the components.

The advantage of such an approach is obvious. Consider the example at the beginning of this section (20 automata, 10 local states each). A compositional approach yields 20 applications of a model checking algorithm, each of which involving only 10 states, whereas the global approach applies model checking once, but on a set of 10^{20} states. There is, however, some (often significant) overhead for the decomposition of the system model and the construction and the composition of the local specifications.

Section 4 discusses the compositional verification approach which is subject of our research project "Integrierte algorithmische und deduktive Verifikation verteilter Steuerungssysteme für hybride Prozesse" ("Integrated algorithmic and deductive verification of distributed control systems for hybrid processes") in the DFG KONDISK program.

4 Compositional Verification

A compositional approach to verification aims at deducing properties of a system from a local analysis of its constituent parts. Since each subsystem, or module, is dependent on inputs from its environment, this environment must somehow be represented to carry out a local analysis. In the trivial case the module's behavior is unchanged by the environment, whilst in the worst case the interactions might be so intensive that any useful analysis requires a representation of the environment that is equivalent to the composed system.

However, in some domains of application, such as chemical engineering, the modules depend only on a few other modules and only via a few interface channels. In that case, a simplified representation of the environment will enable a less complex local analysis. The problem is how to:

- obtain such a simplified representation and
- ensure that the local analyses do indeed allow deductions about the composed system.

One approach is to compose the environment of a module and then simplify it step by step. This can be referred to as *compositional minimization*. The simplification method must ensure the validity of the deduction, i.e., conserve certain properties with respect to composition.

In the next section some notation is introduced, afterwards the assumption/ commitment methods is presented followed by the formulation of two proof rule paradigms. Finally, the approach is illustrated by an example.

4.1 Groundwork

Modules and Environments. Consider a system S that can be divided into several *modules*, or subsystems, working in parallel:

$$S = S_1 || \dots || S_n. \tag{1}$$

The respective *environment* E_i for each module S_i is the composition of the remaining automata of S:

$$E_i = S_1 || \dots || S_{i-1} || S_{i+1} || \dots || S_n. \tag{2}$$

The behavior of a module can be represented by a discrete or hybrid automaton S. In order to specify that a module fulfills certain requirements, two formalisms exist: properties and abstractions.

Properties. A property of an automaton can be specified in a temporal logic formula. This provides a compact description of a requirement if it concerns only a certain aspect of the behavior of the automaton. However, formulas can become very long and tedious to handle manually.

Abstractions. If a requirement defines the set of desired behavior in an exhaustive manner, it may better be described by an automaton. In practice, the desired behavior of an automaton S_i can be specified as $\hat{S}_i$ by copying the automata while omitting all undesired locations and states. The abstraction is denoted as $S_i \preceq \hat{S}_i$, meaning that any behavior of S_i finds a matching representation within the specification $\hat{S}_i$.

Tableaux and Test Automata. A subclass of temporal logic formulas, sufficiently large for practical applications, can also be represented by automata (Clarke et al., 1999). The automaton T_ϕ representing a formula ϕ can be derived algorithmically by a *tableaux* construction. As a result, a *test automaton* $\hat{S}_i^T$ can be constructed in order to verify an abstraction $S_i \preceq \hat{S}_i$ using model checking. The test automaton contains a *fail* state that is reachable if $S_i \not\preceq \hat{S}_i$ so that

$$S_i || \hat{S}_i^T \models \neg reach(fail) \quad \Rightarrow \quad S_i \preceq \hat{S}_i. \tag{3}$$

4.2 The Assumption/Commitment Paradigm

Consider the behavior of a module S_i. Let

$$S_i \models (a_i, c_i), \tag{4}$$

denote that S_i commits itself to fulfilling the *commitment* c_i under the *assumption* a_i. The pair (a_i, c_i) is called an *assumption/commitment-pair* (a/c-pair). A number of alternative notations can be found in literature, e.g., $\langle a_i \rangle S_i \langle c_i \rangle$ (Pnueli, 1984).

The goal of the compositional analysis is to show that the composed system S fulfills a certain requirement corresponding to a global commitment c. As an a/c-pair, this is written as $S \models (true, c)$. If S is a system that depends on outside input, e.g., human interaction, additional global assumptions a about the unspecified environment of S can be included:

$$S \models (a, c). \tag{5}$$

The a/c-method consists of finding local a/c-pairs (a_i, c_i) for each module S_i such that the combination of the commitments fulfills the assumptions in such a way that the conclusion (5) holds. A major problem results from the fact that if the a/c-pairs combine in a circular way, the conclusion is not valid unless further knowledge is included in the proof. Consider an example system $S = S_1 || S_2$ for which the following holds:

$$\begin{array}{ll} S_1 \models (a_1, c_1), & c_1 \Rightarrow a_2, \\ S_2 \models (a_2, c_2), & c_2 \Rightarrow a_1. \end{array} \tag{6}$$

Since for logical expressions a and b

$$(a \Rightarrow b) \wedge (b \Rightarrow a) \equiv (a \wedge b) \vee (\neg a \wedge \neg b), \tag{7}$$

it can only be deduced from (6) that S_1 and S_2 either both fulfill their commitments or both don't: $S_1 || S_2 \models (true, (c_1 \wedge c_2) \vee (\neg c_1 \wedge \neg c_2))$.

If circularity occurs, it must be broken by including appropriate additional conditions B. *Temporal induction* can be used to solve this problem (Alur and Henzinger, 1999): First, it is shown that in its initial state $S_1 || \ldots || S_n \models a_1, \ldots, a_n$. In the induction step it must be established that given valid commitments c_i no transition occurring in the system can violate any of the a_{i+1}. This relates to (7) as:

$$\big(a_0 \wedge b_0 \wedge \forall k \in \mathbb{N}.(a_k \Rightarrow b_{k+1}) \wedge (b_k \Rightarrow a_{k+1})\big) \Rightarrow \big(\forall k \in \mathbb{N}.a_k \wedge b_k\big)$$

In summary, the aim of the assumption/commitment-paradigm is to combine a/c-pairs (a_i, c_i) with additional conditions B to the following proof rule:

$$\frac{\begin{array}{c} S_1 \models (a_1, c_1) \\ \vdots \\ S_n \models (a_n, c_n) \\ B(a_1, \ldots, a_n, c_1, \ldots, c_n, a, c) \end{array}}{S_1 || S_2 || \ldots || S_n \models (a, c)}. \tag{8}$$

The selection of appropriate a/c-pairs is the creative task of the analyst and difficult to automate. The following section describes how to automate the verification of the individual a/c-pairs. Afterwards, two paradigms are presented that can provide a starting set of a/c-pairs that can then be modified to suit the particular application (Frehse et al., 2002).

Application using Automata. The a/c-pairs (4) can be verified automatically if they are represented by automata. Let A_i be the automaton that represents the behaviors of the environment E_i fulfilling a_i and C_i be the automaton that represents all behaviors of S_i that fulfill c_i:

$$A_i \preceq E_i \wedge A_i \models a_i,$$
$$C_i \preceq S_i \wedge C_i \models c_i. \tag{9}$$

A_i and C_i can be obtained manually from E_i and S_i, or by using the tableau construction $A_i = T_{a_i}, C_i = T_{c_i}$. Then (4) is equivalent to

$$A_i||S_i \preceq A_i||C_i. \tag{10}$$

This inequality can be verified with a model checking tool using a test automaton construction (3).

Chain Proof Rule. In a chain rule form, which was used in the beginnings of a/c reasoning (Pnueli, 1984), the assumption/commitment proof becomes simple and requires no further additional logical conditions or explicit deduction:

$$\frac{\begin{array}{c} S_1 \preceq \hat{S}_1 \\ \hat{S}_1||S_2 \preceq \hat{S}_1||\hat{S}_2 \\ \vdots \\ \hat{S}_1||\hat{S}_2||\ldots||\hat{S}_{n-1}||S_n \preceq \hat{S}_1||\hat{S}_2||\ldots||\hat{S}_n \end{array}}{S_1||S_2||\ldots||S_{n-1}||S_n \preceq \hat{S}_1||\hat{S}_2||\ldots||\hat{S}_n}. \tag{11}$$

It can be interpreted in the following way: $\hat{S}_1$ has to capture the behavior of S_1 for all possible inputs. $\hat{S}_2$ has to simulate S_2 with the inputs from $\hat{S}_1$, which is easier than with all possible inputs. For the last module $\hat{S}_n$, only the behavior occurring under the influence of $\hat{S}_1||\ldots||\hat{S}_{n-1}$ has to be taken into account.

The proof of (11) is straightforward and can be done by iteratively applying the equations to their successors. This rule is simple, but in the following sense, it can't be improved:

- Adding a term $\hat{S}_i$ to both sides of one of the inequalities will destroy the soundness unless further conditions are included.
- Removing a term $\hat{S}_{i+1}$ will lead to a wider range of inputs that S_i will have to cooperate with.

Let A denote an automaton modeling a global assumption as part of the initial conditions. The automata A_i and C_i become:

$$A_1 = A, \quad A_i = \hat{S}_1||\ldots||\hat{S}_{i-1} \text{ for } i > 1,$$
$$C_i \preceq \hat{S}_i. \tag{12}$$

In order to reduce the complexity of the proof steps, the assumption can be widened, i.e., for $j < i$ any $\hat{S}_j$ can be dropped from both sides of (12) at any step. This, however, might lead to an abstraction that is too wide and violates one of the proof steps. If the proof fails because the interactions of the modules cannot be captured by the abstractions in a chain sequence, the assumption should be made more restrictive by adding any $S_j, j > i$, to both sides of (12) at any step. This in turn will increase the complexity.

Circular Proof Rule. The following proof rule, also referred to as Assume/Guarantee rule, has successfully been applied to small real-time and hybrid systems (Henzinger et al., 1998c). In order to verify that $S_i|| \dots ||S_n$ meets the specifications $\hat{S}_i|| \dots ||\hat{S}_n$ the following proof is carried out:

$$\frac{\begin{array}{c} S_1||\hat{S}_2|| \dots ||\hat{S}_{n-1}||\hat{S}_n \preceq \hat{S}_1||\hat{S}_2|| \dots ||\hat{S}_n \\ \hat{S}_1||S_2|| \dots ||\hat{S}_{n-1}||\hat{S}_n \preceq \hat{S}_1||\hat{S}_2|| \dots ||\hat{S}_n \\ \vdots \\ \hat{S}_1||\hat{S}_2|| \dots ||\hat{S}_{n-1}||S_n \preceq \hat{S}_1||\hat{S}_2|| \dots ||\hat{S}_n \\ B(S_1, \dots, S_n, \hat{S}_1, \dots, \hat{S}_n) \end{array}}{S_1||S_2|| \dots ||S_{n-1}||S_n \preceq \hat{S}_1||\hat{S}_2|| \dots ||\hat{S}_n}. \tag{13}$$

Additional conditions B are needed to avoid that the composition of the original modules shows a behavior that can't be met by more than one of the abstractions, in which case the proof would fail. Temporal induction can be applied to accomplish soundness of the proof (Alur and Henzinger, 1999).

With the following definition for A_i and C_i, the constituents of (13) can be obtained from (10):

$$\begin{aligned} A_i &= \hat{S}_1|| \dots |\hat{S}_{i-1}||\hat{S}_{i+1}|| \dots ||\hat{S}_n, \\ C_i &\preceq \hat{S}_i. \end{aligned} \tag{14}$$

4.3 Example

The following example shall illustrate the above methodology. The delivery of raw materials (educts) for a chemical batch process must be in tune with the downstream reactor schedule. In a decentralized control scheme, the delivery schedule can be set within certain limits that guarantee compatibility with the downstream recipe. Once those limits are set, the downstream must in turn consume the delivered raw materials in time.

Such a delivery schedule S_1 can be modeled by a timed automaton as shown in Fig. 4. The delivery takes place at least every 3 min., which in the automaton is represented by the invariant $x < 3$ in the initial state, indicated by the double line. When the educt is stored in a buffer tank, the delivery schedule provides a signal, represented by the label $educt\_ready$, to the controllers and remains in a waiting state in order to give the recipe controller time to drain the buffer tank. The guard $x = 1$ on the transition back to the initial state forces the automaton to wait exactly 1 min. before the next delivery can take place.

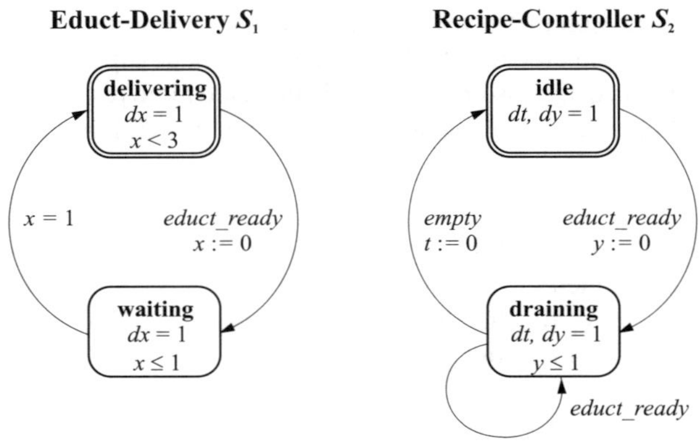

Fig. 4. Automata for the delivery and the recipe controller

A second automaton S_2 in Fig. 4 models the recipe controller. It remains in an $idle$ state until the $educt\_ready$ signal is issued. The controller then goes into the state $draining$ in which the valves are open to drain the educt into the reactors. To avoid including a tank model, the draining process has been abstracted by allowing the controller to stay in the draining state at most 1 min., which is assumed to be the maximum time needed for draining. This is implemented using the clock y that is reset at the transition labeled with $educt\_ready$. The invariant $y \leq 1$ forces the automaton to leave the state $draining$ after 1 min. at the latest and proceed to send the $empty$ signal and return to the initial state. A clock t has been included to measure the time that elapses between two consecutive $empty$ signals.

The requirement to be verified is whether the buffer tank is emptied at least once every 4 min. In temporal logic, this can be represented as a commitment:

$$c_2 = \forall\Box\,(t < 4) \tag{15}$$

The automaton representation C_2 corresponding to c_2 is shown in Fig. 5 together with the test automaton C_2^T that is used to verify $S_1||S_2 \preceq C_2$ using a model checking tool.

In this example, a chain rule deduction is carried out. Starting with the requirement (15) it must be shown that

$$S_1||S_2 \models (true, c_2), \tag{16}$$

From a local analysis of S_2 the assumption a_2 as represented in Fig. 6 was obtained and it was verified that indeed $S_2 \models (a_2, c_2)$ using model checking by showing that $A_2||S_2||C_2^T \models \neg reach(fail)$. Finally it must be shown that S_1 fulfills the assumption a_2. To do so, the commitment of S_1 is set equal to a_2 and it remains to prove $S_1 \models (true, c_1)$. This is checked using the test automaton C_1^T from Fig. 6 to show that $S_1||C_1^T \models \neg reach(fail)$.

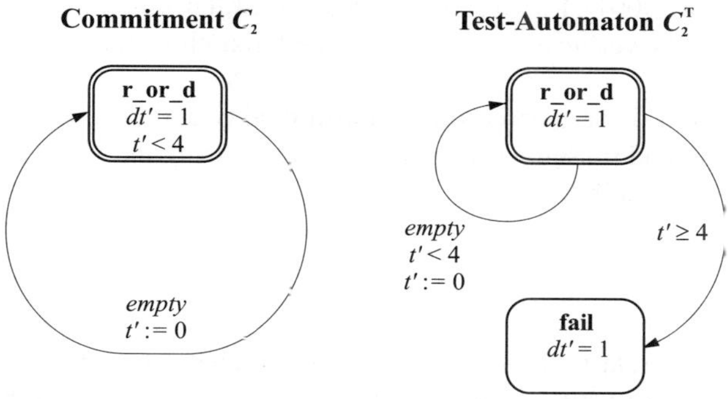

Fig. 5. Automata for representing and for testing the commitment of the recipe controller

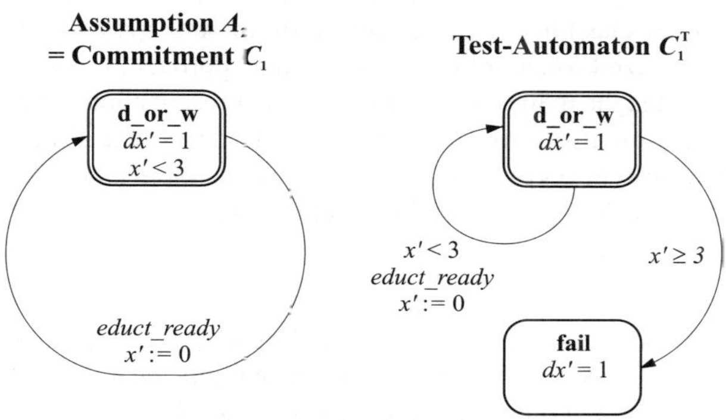

Fig. 6. Automata for representing and for testing the commitment of the delivery schedule

Finally, it can be deduced in either a/c or abstraction representation that the requirement c_2 holds also for the composed system:

$$\begin{array}{c} S_1 \models (true, c_1) \\ S_2 \models (a_2, c_2) \\ c_1 = a_2 \\ \hline S_1 || S_2 \models (true, c_2), \end{array} \qquad \begin{array}{c} S_1 \preceq C_1 \\ A_2 || S_2 \preceq A_2 || C_2 \\ C_1 \preceq A_2 \\ \hline S_1 || S_2 \preceq C_1 || C_2. \end{array} \tag{17}$$

4.4 Related Work

Since it is obvious that noncompositional methods will always be limited by the state explosion problem, there is currently very active research in the field of compositional verification. Thomas Henzinger et al. use an assume-guarantee principle in hierarchical hybrid system design which supports nesting of parallel and serial composition (Henzinger et al., 2001). Rajeev Alur et al. use the modeling language CHARON for the modular design of interacting hybrid systems addressing different aspects of hierarchy (Alur et al., 2001). Ranjit Jhala and Ken McMillan employ compositional model checking for the verification of a processor microarchitecture (Jhala and McMillan, 2001).

Henrik Ejersbo Jensen et al. have worked on compositionality and abstraction in the field of timed automata and presented several case studies using the UPPAAL tool (Jensen, 1999, Jensen et al., 2000).

5 Conclusions

Formal verification completes the classical tools of simulation and analysis of controlled processes by providing a conservative and correct way to analyze the functioning of the system. The drawback of the existing methods lies in their consumption of either manual work or computational resources. Compositional verification methods reduce the costly step of model checking to the composition of relatively simple subsystems. This article summarized some of the compositional approaches to verification in literature and illustrated the methods with an example. Present approaches for model checking rely on the formalisms of timed or linear hybrid automata. Techniques to approximate nonlinear hybrid systems by linear hybrid automata are available, see e.g. (Stursberg, 2000b). An application to a chemical process has shown promising results (Frehse et al., 2001).

Acknowledgements. This investigation was funded by the German Research Council (DFG) through the grant EN152/29-1 and RO1122/7-1. The authors wish to thank the writers of the proposal, Yassine Lakhnech and Stefan Kowalewski.

Part IV

Controller Synthesis

Abstraction Based Supervisory Controller Synthesis for High Order Monotone Continuous Systems

Thomas Moor[1] and Jörg Raisch[2]

[1] Research School of Information Sciences and Engineering
Australian National University, Canberra
[2] Lehrstuhl für Systemtheorie technischer Prozesse
Otto-von-Guericke Universität Magdeburg, and
Systems and Control Theory Group
Max-Planck-Institut für Dynamik komplexer technischer Systeme
Magdeburg, Germany

Abstract. Abstraction based approaches to hybrid control systems synthesis have so far been mostly limited to problems with low-order linear continuous dynamics. In this paper, results from the theory of monotone dynamical systems are used to efficiently compute discrete abstractions for a class of nonlinear models. Furthermore, a situation is investigated where the high-dimensional plant state converges to a low-dimensional manifold; in the proposed approach the computational effort is governed by the dimension of the low-order manifold without neglecting the high-order dynamics. Results are applied to synthesize a discrete event controller for the automatic start-up of a nonlinear distillation column model of order 42.

1 Introduction

The control of a physical, biological or chemical process by a digital computer program often leads to heterogeneous systems that include both continuous and discrete event dynamics. Such systems are referred to as *hybrid control systems* and, in general, exhibit highly complex behaviours. From an engineering perspective, the systematic design of hybrid systems to meet given performance specifications is of particular interest; e.g. (Asarin et al., 2000b, Cury et al., 1998, Davoren and Nerode, 2000, Koutsoukos et al., 2000, Lunze et al., 1997, Lygeros et al., 1999, Zhivoglyadov and Middleton, 1999). When investigating suitable methodologies for particular design objectives and system classes, the development of synthesis procedures that are both efficient and reliable represents a major challenge. The demand for reliability results from safety critical applications, such as air traffic control or the control of chemical processes. As it is only for extremely restrictive conditions that hybrid controller synthesis problems can be solved directly on the basis of a hybrid plant model, a common approach is to resort to approximation based methods; e.g. (Cury et al., 1998, Koutsoukos et al., 2000, Lichtenberg et al., 1999a, Lunze et al., 1997, Moor and Raisch, 1999b, Philips et al., 1999, Raisch and O'Young, 1998). In this context, reliability becomes a nontrivial issue: does a controller which has been synthesized for a particular approximation also achieve the design objectives for the original problem? Roughly speaking, an affirmative answer can be given under the condition that the approximation accounts for all signals on

S. Engell, G. Frehse, E. Schnieder (Eds.): Modelling, Analysis and Design of Hybrid Systems, LNCIS 279, pp. 247–265, 2002.

which the original plant can possibly evolve. An approximation that exhibits this crucial *behavioural inclusion property* is referred to as an *abstraction*.

This paper specifically addresses the synthesis of discrete event controllers for continuous plants with discrete external signals. For this class of synthesis problems we have developed an abstraction scheme of particular convenience, namely *l-complete approximation*. While for all parameter values $l \in \mathbb{N}_0$ the behavioural inclusion property is fulfilled, it can be seen that increasing l corresponds to improving the approximation accuracy. The latter is an essential feature as it allows to systematically adjust the approximation accuracy to the application at hand. By construction, any l-complete approximation can be realized by a finite automaton. Hence, the hybrid synthesis problem is transformed into a purely discrete problem, which can subsequently be solved using methods from DES theory. Some further detail on l-complete approximation in the context of supervisory controller synthesis is given in Sect. 2; for a self contained development of the topic the reader is referred to (Moor and Raisch, 1999b, Moor et al., 2002, Raisch and O'Young, 1998). Our work on abstraction based supervisory control and related topics in hybrid control theory was partly funded by *Deutsche Forschungsgemeinschaft* under the KONDISK-scheme (research grants Ra 516/3-1, Fr 598/6-1, Fr 598/6-2, and Fr 598/6-3). This support is gratefully acknowledged. Beside our core contribution to abstraction based supervisory control (Lichtenberg et al., 1999a, Moor, 1998, Moor, 2000, Moor and Raisch, 1999b, Moor et al., 1998, Moor et al., 2002, Raisch, 1998, Raisch, 2000a, Raisch, 2000b, Raisch and O'Young, 1997, Raisch and O'Young, 1998), we have also addressed hierarchical (Moor et al., 2001b, Raisch et al., 2001, Raisch and Itigin, 2000, Raisch et al., 2000) and modular (Moor et al., 2001a, Moor and Raisch, 2000) extensions to our approach. We have discussed computational procedures for linear dynamics under time-driven and event-driven sampling (Franke et al., 2000, Moor and Raisch, 1999a, Raisch and O'Young, 1997) and applied our results in a number of case studies (Klein et al., 2000, Klein et al., 1998, Klein et al., 1999, Klein and Raisch, 1998, Raisch et al., 1998).

In principle, the computational procedure to generate l-complete approximations is straightforward. There are, however, two major problems that have limited applications to a class of plant models that appears small when compared to the generality of the theoretical framework provided by (Moor and Raisch, 1999b): (i) quantization cells have to be tracked under the progress of time and intersected with other quantization cells. Clearly, this is a difficult problem if the right hand side of the differential or difference equation is nonlinear in the continuous state variable. (ii) Computational effort "explodes" with growing state dimension. Hence, applications have been restricted to fairly low-dimensional plant models. From a more general perspective, both limitations are closely related to the absence of efficient procedures for high-order and/or nonlinear reachability analysis and, in principle, also apply to other approximation based synthesis methods. In this contribution, we derive conditions for high order nonlinear systems that allow efficient computation of l-complete approximations without sacrificing reliability in the subsequent controller design.

The paper is organized as follows: in Sect. 2, we briefly summarize the considered controller synthesis problem and the basic procedure for l-complete approximation. In Sect. 3, we introduce the notion of monotone dynamical systems and show why it is extremely helpful for computing discrete abstractions for a certain class of nonlinear systems. In Sect. 4, we explore a situation that allows treatment of high-dimensional systems. Finally, in Sect. 5, we apply our results to synthesize a DES controller for the start-up of a distillation column that is described by a nonlinear model of order 42.

2 Abstraction Based Supervisory Control

In earlier work (Moor and Raisch, 1999b, Moor et al., 2002, Raisch and O'Young, 1997, Raisch and O'Young, 1998), we combine techniques from J.C. WILLEMS' behavioural systems theory (Willems, 1989, Willems, 1991) and P.J. RAMADGE and W.M. WONHAM's supervisory control theory (Ramadge and Wonham, 1987, Ramadge and Wonham, 1989) to address the problem of supervisory controller synthesis for a fairly general class of hybrid systems and to establish an abstraction based solution procedure. The purpose of this section is to briefly summarize the consequences of (Moor and Raisch, 1999b, Moor et al., 2002) for the more specific case of sampled continuous systems with discrete-valued inputs and outputs; it is this class of plants that we will develop efficient abstraction procedures for in the following sections. While we need to provide an unambiguous framework for the abstraction step, technical aspects of the controller synthesis step are omitted and only an informal outline is given. For a detailed discussion, the reader is referred to (Moor and Raisch, 1999b, Moor et al., 2002).

Plant model. Consider the plant

$$x(k+1) = F(x(k),\, u(k)) \qquad \text{and} \qquad y(k) \in G(x(k))\,, \tag{1}$$

where

- the *input signal* $u\colon \mathbb{N}_0 \to U$ is a sequence of input symbols from the finite input alphabet U, $|U| \in \mathbb{N}$; [1]
- the *state trajectory* $x\colon \mathbb{N}_0 \to \mathbb{R}^n$ is a sequence of real-valued states;
- the *state transition map* $F\colon \mathbb{R}^n \times U \to \mathbb{R}^n$ uniquely determines the successor state from the current state and the current input symbol;
- the *measurement map* $G\colon \mathbb{R}^n \to 2^Y \setminus \{\emptyset\}$ models a quantized measurement of the continuous state; it is not required to be deterministic.
- the *output signal* $y\colon \mathbb{N}_0 \to Y$ is a sequence of measurement symbols from the finite output alphabet Y, $|Y| \in \mathbb{N}$.

[1] We denote the positive integers $\mathbb{N}$ and the non-negative integers $\mathbb{N}_0$. We use $|A| \in \mathbb{N}$ to indicate that A is a nonempty finite set.

This class of plant models includes a scenario of particular application relevance, namely sampled continuous dynamics with a discrete-valued input switching between a finite number of operation modes and a discrete-valued output being generated by measurement quantisation. Assuming a constant sampling period $\Delta > 0$, the state transition map is given by

$$F(\xi, \mu) := \Phi^{\mu}_{\Delta}(\xi)\,, \tag{2}$$

where for each $\mu \in U$ the map $\Phi^{\mu}_{t} \colon \mathbb{R}^n \to \mathbb{R}^n$, $t \in \mathbb{R}^{+}_{0}$, denotes the flow induced by a vector field $f_{\mu} \colon \mathbb{R}^n \to \mathbb{R}^n$; i.e. $z(t) = \Phi^{\mu}_{t}(z_0)$ solves the ODE

$$\dot{z}(t) = f_{\mu}(z(t)) \tag{3}$$

for the initial condition $z(0) = z_0$, and we assume unique existence of such a solution on the entire time axis. Note that if we were concerned with the system behaviour between given sampling instants, we could adopt our set-up to the case of event-driven sampling. In the latter case, the occurrence of events is entirely determined by the system (e.g., by elements of the continuous state vector z crossing certain thresholds) instead of being restricted to a fixed time grid. For linear dynamics, this case has been addressed in (Franke et al., 2000, Moor, 1998, Moor and Raisch, 1999a).

By allowing the measurement map G to be nondeterministic, the quantization cells

$$G^{-1}(\nu_j) \subseteq \mathbb{R}^n, \quad \nu_j \in Y, \quad j = 1, \ldots |Y|,$$

may cover (instead of partition) the continuous state space. This models the practically important case where measurement information is, to a certain extent, ambiguous.

Supervisory control. From the perspective of a potential controller, the system exhibits a discrete event behaviour: at the k-th sampling instant, the supervisor applies an input symbol $u(k) \in U$ and then waits for the next measurement symbol $y(k{+}1) \in Y$. Naturally, for the problem of controller synthesis, this *external behaviour* plays a key role. We formally define the external behaviour $\mathfrak{B}$ induced by (1) as

$$\mathfrak{B} := \{(u,\, y) \colon \mathbb{N}_0 \to U \times Y \,|\, \exists x \colon \mathbb{N}_0 \to \mathbb{R}^n : \text{Eq. (1) holds for all } k \in \mathbb{N}_0 \}\,; \tag{4}$$

i.e. $\mathfrak{B}$ denotes the set of all pairs of input and output signals on which the plant model (1) can possibly evolve. This definition is consistent with J.C. WILLEMS' behavioural systems theory, where a dynamical system is characterized by the set of trajectories that are compatible with the phenomenon it models.

Following the concepts of RAMADGE and WONHAM's supervisory control theory for DESs, the task of a supervisor is to restrict the plant behaviour $\mathfrak{B} \subseteq (U \times Y)^{\mathbb{N}_0}$ such that the closed loop is guaranteed to exhibit only acceptable signals. This specification can be formally represented by the set of acceptable external signals, denoted $\mathfrak{B}_{\text{spec}} \subseteq (U \times Y)^{\mathbb{N}_0}$. Similar to the plant, the supervisor is characterized by a behaviour $\mathfrak{B}_{\text{sup}} \subseteq (U \times Y)^{\mathbb{N}_0}$, which denotes the set of external signals it can evolve on. The closed-loop behaviour is the intersection $\mathfrak{B}_{\text{cl}} = \mathfrak{B} \cap \mathfrak{B}_{\text{sup}}$, i.e. only those pairs of input and output signals "survive closing the loop" that are consistent with both plant and controller dynamics. The supervisor $\mathfrak{B}_{\text{sup}}$ is said to *enforce the specification* $\mathfrak{B}_{\text{spec}}$ if $\mathfrak{B}_{\text{cl}} \subseteq \mathfrak{B}_{\text{spec}}$. However, when interconnecting plant and supervisor one needs to ensure that the supervisor respects the input-output structure of the plant; i.e. the supervisor may enable or disable certain input events at any time but no restrictions must be imposed on the plant outputs. If the latter condition holds, $\mathfrak{B}_{\text{sup}}$ is said to be *admissible* w.r.t. $\mathfrak{B}$; see (Moor and Raisch, 1999b) for a formal definition of admissibility. The problem of supervisory controller synthesis can then be stated as follows:

Given a plant behaviour $\mathfrak{B}$ and a specification $\mathfrak{B}_{\text{spec}}$, a supervisor $\mathfrak{B}_{\text{sup}}$ is said to be a solution to the supervisory control problem $(\mathfrak{B}, \mathfrak{B}_{\text{spec}})$ if (i) $\mathfrak{B}_{\text{sup}}$ is admissible w.r.t. $\mathfrak{B}$, and (ii) $\mathfrak{B}_{\text{sup}}$ enforces the specification.

Discrete abstractions. Suppose both $\mathfrak{B}$ and $\mathfrak{B}_{\text{spec}}$ were realized by finite automata. Not surprisingly, the controller synthesis problem could then be treated by a slightly modified version of known methods from DES theory; e.g. (Ramadge and Wonham, 1989). For this case, efficient procedures are known which either compute a finite automaton realization of a solution $\mathfrak{B}_{\text{sup}}$ or find that no such solution exists. However, the plant (1) is defined on the continuous state space $\mathbb{R}^n$, and a finite automaton realization of $\mathfrak{B}$ can only exist if $\mathbb{R}^n$ can be decomposed by a finite partition into sets of states that are indistinguishable under all external signals. This is a very restrictive condition and, in general, we can *not* assume that $\mathfrak{B}$ is realizable by a finite automaton. A method to overcome this problem is to first construct a finite automaton that *approximates* the hybrid plant and then to solve the synthesis problem for the approximation. Various variants of this approach have been discussed, e.g., in (Cury et al., 1998, Koutsoukos et al., 2000, Lunze et al., 1997, Philips et al., 1999, Raisch and O'Young, 1997, Raisch and O'Young, 1998). In (Moor and Raisch, 1999b, Moor et al., 2002), we justify this approximation based approach by providing a sufficient condition for a solution obtained at the approximation level to remain valid for the actual hybrid plant:

Consider a plant approximation $\mathfrak{B}_{\text{ca}} \subseteq (U \times Y)^{\mathbb{N}_0}$, a specification $\mathfrak{B}_{\text{spec}}$, and a solution $\mathfrak{B}_{\text{sup}}$ of the supervisory control problem $(\mathfrak{B}_{\text{ca}}, \mathfrak{B}_{\text{spec}})$. Assume that each behaviour $\mathfrak{B}_{\text{ca}}$, $\mathfrak{B}_{\text{spec}}$, $\mathfrak{B}_{\text{sup}}$ is realized by a finite automaton. If $\mathfrak{B}_{\text{ca}} \supseteq \mathfrak{B}$, then $\mathfrak{B}_{\text{sup}}$ also solves $(\mathfrak{B}, \mathfrak{B}_{\text{spec}})$, where $\mathfrak{B}$ denotes the external behaviour of the plant model (1). *See (Moor and Raisch, 1999b), Theorem 25, and (Moor et al., 2002), Sect. 6.*

Note that the nontriviality of this result is due to the requirement that any solution $\mathfrak{B}_{\text{sup}}$ respects the input output structure of $\mathfrak{B}$. A plant approximation $\mathfrak{B}_{\text{ca}}$ is said to be a *discrete abstraction* of $\mathfrak{B}$, if (i) the behavioural inclusion $\mathfrak{B}_{\mathrm{ca}} \supseteq \mathfrak{B}$ is fulfilled, and (ii) $\mathfrak{B}_{\mathrm{ca}}$ is realizable by a finite automaton. The supervisory controller synthesis problem has thus been reduced to the construction of a discrete abstraction $\mathfrak{B}_{\mathrm{ca}} \supseteq \mathfrak{B}$ that is sufficiently accurate such that $(\mathfrak{B}_{\mathrm{ca}},\ \mathfrak{B}_{\mathrm{spec}})$ is solvable.

l-Complete approximation. In the case of time invariant systems, a particularly suitable discrete abstraction is the so called *strongest l-complete approximation* $\mathfrak{B}_l \supseteq \mathfrak{B}$, where $l \in \mathbb{N}$ is a parameter. Formally, $\mathfrak{B}_l$ can be characterized by

$$\mathfrak{B}_l := \{(u, y)\colon \mathbb{N}_0 \to U \times Y |\ (u, y)\big|_{[k,k+l]} \in \mathfrak{B}\big|_{[0,l]}\ \forall\, k \in \mathbb{N}_0\}\,, \tag{5}$$

where the restriction operator $(\,\cdot\,)|_{[k,k+l]}\colon (U \times Y)^{\mathbb{N}_0} \to (U \times Y)^{(l+1)}$ picks the finite string ranging from the k-th to the $(k+l)$-th pair of external events and discards its absolute location on the time axis:

$$(u, y)|_{[k,k+l]} := [\,(u(k), y(k)),\ \ldots\ (u(k+l), y(k+l))\,] \in (U \times Y)^{(l+1)}\,. \tag{6}$$

It can be naturally extended to sets of signals:

$$\mathfrak{B}|_{[0,l]} := \{(u, y)\big|_{[0,l]} \in (U \times Y)^{(l+1)} \mid (u, y) \in \mathfrak{B}\}\,. \tag{7}$$

Note that $\mathfrak{B}|_{[0,l]}$ is a finite set as both U and Y are finite. The most relevant features of the strongest l-complete approximation $\mathfrak{B}_l$ are that (i) accuracy is monotone in l, i.e. $\mathfrak{B}_l \supseteq \mathfrak{B}_{l+1} \supseteq \mathfrak{B}$, and that (ii) a finite realization can be easily derived from the restricted plant behaviour $\mathfrak{B}|_{[0,l]}$; see (Moor and Raisch, 1999b), Corollary 11. Hence, there is no need to evaluate (5) in order to construct $\mathfrak{B}_l$. All that remains to be done is the computation of $\mathfrak{B}|_{[0,l]}$ and we recall the following iterative procedure from (Moor and Raisch, 1999b):

Theorem 1. *Let $\mathfrak{B} \subseteq (U \times Y)^{\mathbb{N}_0}$ denote the external behaviour of* (1). *For $(u,\ y) \in (U \times Y)^{\mathbb{N}_0}$ and $l \in \mathbb{N}_0$, iteratively define the sets of states $\mathcal{X}((u, y)|_{[0,l]}) \subseteq \mathbb{R}^n$ that are compatible with the strings $(u, y)|_{[0,l]}$:*

$$\mathcal{X}((u, y)\big|_{[0,0]}) := G^{-1}(y(0))\,, \tag{8a}$$

$$\mathcal{X}((u, y)\big|_{[0,\lambda+1]}) := F(\mathcal{X}((u, y)\big|_{[0,\lambda]}),\ u(\lambda)) \cap G^{-1}(y(\lambda+1))\,, \tag{8b}$$

for $\lambda = 0, \ldots l-1$. Then,

$$(u, y)\big|_{[0,l]} \in \mathfrak{B}\big|_{[0,l]} \quad \iff \quad \mathcal{X}((u, y)|_{[0,l]}) \neq \emptyset\,. \tag{9}$$

According to the above theorem, $\mathfrak{B}|_{[0,l]}$ can be established via a finite iteration of images under F and intersections with the quantization cells G^{-1}. Then, the methods presented in (Moor and Raisch, 1999b, Moor et al., 2002) allow the construction of a discrete abstraction of the hybrid plant and finally the synthesis of a supervisory controller.

While this approach has been successfully applied to a number of examples, there are two major limitations from a practical point of view. First, for nonlinear continuous dynamics, images of sets of states under F can, in general, not be computed efficiently. Roughly speaking, one is often left with the simulation of an exhaustive number of initial conditions $\xi_0 = x(0)$; it is then naively assumed that $\mathcal{X}((u,y)|_{[0,l]}) = \emptyset$ whenever no $\xi_0 \in \mathcal{X}((u,y)|_{[0,l]})$ can be found. Clearly, this implies the risk of omitting a particular string from $\mathfrak{B}|_{[0,l]}$, hence from $\mathfrak{B}_{\mathrm{ca}} = \mathfrak{B}_l$, therefore violating the requirement $\mathfrak{B}_{\mathrm{ca}} \supseteq \mathfrak{B}$. Second, for high dimensional continuous dynamics, a reasonably accurate quantization leads to computationally intractable output alphabets Y. In the following two sections, we identify a broad class of hybrid systems where the above iterative procedure can be refined in order to gain substantial computational efficiency.

3 Discrete Abstractions for Monotone Systems

For *monotone dynamical systems* (see (Smith, 1995) for a comprehensive treatment of the subject), it is possible to efficiently estimate the sets of compatible states $\mathcal{X}((u,y)|_{[0,l]})$ and to derive a discrete abstraction from those estimates. In general, monotonicity is defined with respect to an arbitrary partial order. In this paper, we consider the specific partial order $\preccurlyeq$ which, for $a, b \in \mathbb{R}^n$, is defined by

$$a \preccurlyeq b \quad :\Longleftrightarrow \quad \forall\, i \in \{1, \ldots n\} : \; a_i \leq b_i. \tag{10}$$

Hence, $a \preccurlyeq b$ if and only if $b - a$ lies in the nonnegative convex cone $\mathbb{R}^n_+ := \{\xi \in \mathbb{R} | \xi \geq 0\}^n$.

Definition 1. The map $g : \mathbb{R}^q \to \mathbb{R}^n$ is called *order preserving* if $a \preccurlyeq b$ implies $g(a) \preccurlyeq g(b)$.

Note that a map is order preserving if all its partial derivatives are nonnegative. The image of a *box*

$$Q(a,b) := \{c |\; a \preccurlyeq c \preccurlyeq b\} \tag{11}$$

under an order preserving map g can be efficiently over-approximated via the images of a and b, i.e. $g(Q(a,b)) \subseteq Q(g(a), g(b))$. It is this property that allows efficient approximation of monotone systems.

In the following, we consider dynamical systems

$$\dot{z}(t) = f(z(t)) \tag{12}$$

and assume that, for any initial condition $z(0) = z_0$, there exists a unique solution $\Phi_t(z_0)$ for all $t \geq 0$. The dynamical system (12) is called *monotone*, if ordered states remain ordered under the progress of time:

Definition 2. The dynamical system (12) is *monotone*, if the flow $\Phi_t \colon \mathbb{R}^n \to \mathbb{R}^n$ induced by the vector field $f \colon \mathbb{R}^n \to \mathbb{R}^n$ is order preserving for all $t \geq 0$.

A monotonicity test can be stated in terms of the off-diagonal entries of the Jacobian of f:

Theorem 2. *(see e.g. (Smith, 1995)) The dynamical system $\dot{z} = f(z)$ is monotone if*

$$\frac{\partial f_i}{\partial z_j} \geq 0 \quad \forall i \neq j\,. \tag{13}$$

As an example, consider a linear system $\dot{z}(t) = A\,z(t)$. If all eigenvalues of A lie in $\mathbb{R}$, then there exists a real linear transformation that transforms A in its Jordan normal form. Clearly, the transformed system is monotone by Theorem 2. For further illustration, Fig. 1 shows two state trajectories $z(t)$ and $\hat{z}(t)$ of the monotone system $\dot{z}(t) = \left(\begin{smallmatrix}-1 & 1\\ 0 & -1\end{smallmatrix}\right) z(t)$. For the respective initial conditions $z(0) = (0,\,-1.2)^\mathsf{T}$ and $\hat{z}(0) = (0.2,\,-1)^\mathsf{T}$ we have $z(0) \preccurlyeq \hat{z}(0)$, and hence, by monotonicity, $z(t) \preccurlyeq \hat{z}(t)$ for all $t \geq 0$. This is confirmed by Fig. 1, which also clarifies that monotonicity of a dynamical system must not be confused with monotonicity of individual components of state trajectories: in the example, $z_1(t)$ and $\hat{z}_1(t)$ clearly fail to be monotonously increasing (or decreasing) as functions of t.

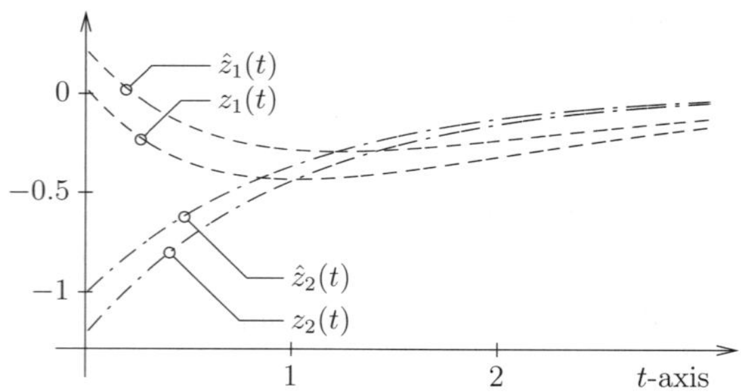

Fig. 1. State trajectories of a linear monotone system.

In consequence, for monotone systems, there is no need to integrate a huge number of states. Instead, the temporal evolution of a box $Q(\zeta_a, \zeta_b)$ can be over-approximated by evaluating the flow for the two points ζ_a, ζ_b only: $\Phi_t(Q(\zeta_a, \zeta_b)) \subseteq Q(\Phi_t(\zeta_a), \Phi_t(\zeta_b))$. Clearly, this is independent of the state dimension.

We now turn to the discrete abstraction of the hybrid plant model (1), with sampled continuous dynamics (3). Obviously, during each sampling interval, the continuous dynamics depends on a fixed control symbol $\mu \in U$. Under the assumption that the continuous system (3) is monotone for each $\mu \in U$, it immediately follows that the transition function $F(\,\cdot\,,\,\mu)$ defined in (2) is order preserving. We further assume that measurement symbols ν_j, $j = 1, \ldots, p$, correspond to bounded boxes in $\mathbb{R}^n$, i.e.

$$G^{-1}(\nu_j) = Q(a_j, b_j), \quad a_j, b_j \in \mathbb{R}^n,\ a_j \preccurlyeq b_j\,. \tag{14}$$

Obviously, a finite number of bounded boxes (14) can not cover the entire $\mathbb{R}^n$. Hence, we need an additional *out of range symbol* ‡ with

$$G^{-1}(\ddagger) = \mathbb{R}^n \setminus \bigcup_{1 \le j \le p} G^{-1}(\nu_j), \quad \text{to give} \quad Y = \{\nu_1, \dots, \nu_p\} \cup \{\ddagger\}. \qquad (15)$$

Based on the iteration (8a), (8b), we are now in a position to provide easily computable *conservative estimates* $\hat{\mathcal{X}}((u,y)|_{[0,l]}) \subseteq \mathbb{R}^n$ for the sets of compatible states. Using

$$\hat{F}(Q(a,b),\, \mu) := Q(F(a,\mu), F(b,\mu))$$

as an over-approximation of the continuous evolution of a box $Q(a,b)$ under the order preserving flow Φ^{μ}_{Δ}, we define:

- if $y(0) = \nu_j \neq \ddagger$ for some j, let

$$\hat{\mathcal{X}}((u,y)\big|_{[0,0]}) := G^{-1}(\nu_j)\,; \qquad (16)$$

- if $y(0) = \ddagger$, let

$$\hat{\mathcal{X}}((u,y)\big|_{[0,0]}) := \mathbb{R}^n\,. \qquad (17)$$

And, for $\lambda = 0, \dots l-1$:

- if $y(\lambda+1) \neq \ddagger$ and $\hat{\mathcal{X}}((u,y)\big|_{[0,\lambda]}) \neq \mathbb{R}^n$, let

$$\hat{\mathcal{X}}((u,y)\big|_{[0,\lambda+1]}) := \hat{F}(\hat{\mathcal{X}}((u,y)\big|_{[0,\lambda]}),\, u(\lambda)) \cap G^{-1}(y(\lambda+1))\,; \qquad (18)$$

- if $y(\lambda+1) \neq \ddagger$ and $\hat{\mathcal{X}}((u,y)\big|_{[0,\lambda]}) = \mathbb{R}^n$, let

$$\hat{\mathcal{X}}((u,y)\big|_{[0,\lambda+1]}) := G^{-1}(y(\lambda+1))\,; \qquad (19)$$

- if $y(\lambda+1) = \ddagger$ and $\hat{\mathcal{X}}((u,y)\big|_{[0,\lambda]}) \neq \mathbb{R}^n$ and $\hat{F}(\hat{\mathcal{X}}((u,y)\big|_{[0,\lambda]}),\, u(\lambda)) \not\subseteq \cup_{1 \le j \le p} G^{-1}(\nu_j)$, let

$$\hat{\mathcal{X}}((u,y)\big|_{[0,\lambda+1]}) := \hat{F}(\hat{\mathcal{X}}((u,y)\big|_{[0,\lambda]}),\, u(\lambda))\,; \qquad (20)$$

- if $y(\lambda+1) = \ddagger$ and $\hat{\mathcal{X}}((u,y)\big|_{[0,\lambda]}) \neq \mathbb{R}^n$ and $\hat{F}(\hat{\mathcal{X}}((u,y)\big|_{[0,\lambda]}),\, u(\lambda)) \subseteq \cup_{1 \le j \le p} G^{-1}(\nu_j)$, let

$$\hat{\mathcal{X}}((u,y)\big|_{[0,\lambda+1]}) := \emptyset\,; \qquad (21)$$

- if $y(\lambda+1) = \ddagger$ and $\hat{\mathcal{X}}((u,y)\big|_{[0,\lambda]}) = \mathbb{R}^n$, let

$$\hat{\mathcal{X}}((u,y)\big|_{[0,\lambda+1]}) := \mathbb{R}^n\,. \qquad (22)$$

Note that (16)–(22) iteratively define the sets $\hat{\mathcal{X}}((u,y)|_{[0,l]})$ for all external signals $(u,y) \in (U \times Y)^{\mathbb{N}_0}$ and for all $l \in \mathbb{N}_0$: (16) and (17) define $\hat{\mathcal{X}}((u,y)|_{[0,0]})$ while (18)–(22) systematically define $\hat{\mathcal{X}}((u,y)|_{[0,\lambda+1]})$ in terms of $\hat{\mathcal{X}}((u,y)|_{[0,\lambda]})$. Note also that $\hat{F}$ is only applied to bounded boxes. By construction, the sets $\hat{\mathcal{X}}((u,y)|_{[0,l]})$ are guaranteed to be supersets of the sets of compatible states $\mathcal{X}((u,y)|_{[0,l]})$. Formally:

Proposition 1. *Assume that for each $\mu \in U$ the state transition map $F(\,\cdot\,,\mu)$ is order preserving and the output map G is defined by* (14)–(15). *Then, for all external signals $(u,y) \in (U \times Y)^{\mathbb{N}_0}$ and for all $l \in \mathbb{N}_0$ the following inclusion holds:*

$$\hat{\mathcal{X}}((u,y)\big|_{[0,l]}) \supseteq \mathcal{X}((u,y)\big|_{[0,l]}) \,. \tag{23}$$

Proof. Pick an arbitrary external signal $(u,y) \in (U \times Y)^{\mathbb{N}_0}$. For $l = 0$ the claim follows immediately from (16) and (17). For $l \neq 0$, the proof is by induction w.r.t. $\lambda = 0, \ldots l-1$: we assume $\hat{\mathcal{X}}((u,y)|_{[0,\lambda]}) \supseteq \mathcal{X}((u,y)|_{[0,\lambda]})$ and show in each of the cases corresponding to (18)–(22) that $\hat{\mathcal{X}}((u,y)|_{[0,\lambda+1]}) \supseteq \mathcal{X}((u,y)|_{[0,\lambda+1]})$. First, observe that for the cases (19) and (22) the inclusion

$$\hat{\mathcal{X}}((u,y)|_{[0,\lambda+1]}) \supseteq \mathcal{X}((u,y)|_{[0,\lambda+1]})$$

follows immediately. For the remaining cases, note that, by monotonicity,

$$\hat{F}(Q(a,b),\mu) \supseteq F(Q(a,b),\mu)$$

holds for any $a,\ b \in \mathbb{R}^n$, $\mu \in U$. Hence,

$$\hat{F}(\hat{\mathcal{X}}((u,y)|_{[0,\lambda]}, u(\lambda)) \supseteq F(\mathcal{X}((u,y)|_{[0,\lambda]}, u(\lambda)).$$

For the case (18) one obtains

$$\hat{\mathcal{X}}((u,y)|_{[0,\lambda+1]}) \supseteq F(\mathcal{X}((u,y)|_{[0,\lambda]}),\ u(\lambda)) \cap G^{-1}(y(\lambda+1)) = \mathcal{X}((u,y)|_{[0,\lambda+1]}).$$

The same argument resolves case (20). Only case (21) remains. From condition $\hat{F}(\hat{\mathcal{X}}((u,y)|_{[0,\lambda]}),\ u(\lambda)) \subseteq \cup_{1 \leq j \leq p} G^{-1}(\nu_j)$ one obtains

$$F(\mathcal{X}((u,y)|_{[0,\lambda]}),\ u(\lambda)) \subseteq \cup_{1 \leq j \leq p} G^{-1}(\nu_j).$$

Together with (15), this implies $F(\mathcal{X}((u,y)|_{[0,\lambda]}),\ u(\lambda)) \cap G^{-1}(\ddagger) = \emptyset$, and, hence,

$$\mathcal{X}((u,y)|_{[0,\lambda+1]}) = \emptyset = \hat{\mathcal{X}}((u,y)|_{[0,\lambda+1]}).$$

Remark 1. The assumption of quantization boxes instead of more general (bounded) quantization sets does not imply any loss of generality: in the latter case, we would simply replace $G^{-1}(\ldots)$ by $Q(\inf G^{-1}(\ldots),\ \sup G^{-1}(\ldots))$ in the above iteration (16)–(22).

As an immediate consequence of Proposition 1, we obtain a finite abstraction $\mathfrak{B}_{\mathrm{ca}}$.

Corollary 1. *Under the same hypothesis as in Proposition 1, the following inclusions hold:*

$$\hat{\mathfrak{B}}\big|_{[0,l]} := \{(u,y)\big|_{[0,l]} \mid \hat{\mathcal{X}}((u,y)\big|_{[0,l]}) \neq \emptyset\} \supseteq \mathfrak{B}\big|_{[0,l]}, \tag{24}$$

$$\mathfrak{B}_{\mathrm{ca}} := \{(u,y) \mid (u,y)\big|_{[k,k+l]} \in \hat{\mathfrak{B}}\big|_{[0,l]} \;\forall\, k \in \mathbb{N}_0\} \supseteq \mathfrak{B}_l \supseteq \mathfrak{B} \tag{25}$$

A finite realization of $\mathfrak{B}_{\mathrm{ca}}$ can now be constructed in the same manner as for $\mathfrak{B}_l$, see (Moor et al., 2002, Moor and Raisch, 1999b), – we merely have to replace $\mathfrak{B}\big|_{[0,l]}$ by $\hat{\mathfrak{B}}\big|_{[0,l]}$. This completes the discrete abstraction procedure for monotone dynamical systems. Note that we do not assume linearity; our results are therefore applicable to nonlinear monotone dynamics.

4 Handling High-Order Dynamics

Many complex technical processes, although intrinsically high-dimensional, converge to a low-dimensional manifold within a short time. Distillation columns are a well-known example: a first principles modelling approach leads to a large number of ODEs describing the temporal evolution of concentrations on each tray of the column. When a column is operated, however, these concentrations stop to be arbitrary and form a concentration profile that can be described by very few parameters.

This particular structure can be exploited in the following way: instead of quantizing the high-dimensional plant state space, only a well defined neighbourhood of the relevant part of the respective manifold is covered by quantization cells and hence provides measurement information; the "rest" of the state space returns the out of range symbol "‡". For a formal treatment of this idea, let

$$h_\mu : \mathbb{R}^q \to \mathbb{R}^n, \; q < n, \tag{26}$$

represent a continuously differentiable parametrization of a q-dimensional manifold $\mathcal{M}_\mu$ in $\mathbb{R}^n$, i.e. $\mathcal{M}_\mu = h_\mu(\mathbb{R}^q)$. Naturally, both the manifold and its parametrization may depend on the control symbol μ. Assume h_μ to be order preserving and $\mathcal{M}_\mu$ to be attractive, i.e.

$$\lim_{t\to\infty} \operatorname{dist}(\mathcal{M}_\mu, \Phi_t^\mu(z_0)) = 0\,, \tag{27}$$

for all initial conditions $z_0 \in \mathbb{R}^n$, where

$$\operatorname{dist}(X,\zeta) := \inf\{\|\zeta - \xi\| \mid \xi \in X\} \tag{28}$$

denotes the distance of a point $\zeta \in \mathbb{R}^n$ to a set $X \subseteq \mathbb{R}^n$ w.r.t. some norm $\|\cdot\|$. Let the bounded subset $P \subset \mathbb{R}^q$ represent the relevant operating range on $\mathcal{M}_\mu$ and, for a given $\delta > 0$,

$$\mathcal{V}_\delta(h_\mu(P)) := \{\zeta \mid \operatorname{dist}(h_\mu(P),\zeta) < \delta\} \tag{29}$$

the neighbourhood of $h_\mu(P)$ that is to be covered by quantization cells.

We give an explicit formula for quantisation cells covering $\mathcal{V}_\delta(h_\mu(P))$ for the case where the operators $\text{dist}(\,\cdot\,)$ and $\mathcal{V}_\delta(\,\cdot\,)$ refer to the so called weighted infinity norm; i.e. $\|\cdot\| := \|\cdot\|_\infty^\beta$ with $\|\xi\|_\infty^\beta := \max_i |\beta_i \xi_i|$ for the weighting vector $\beta = (\beta_1, \ldots \beta_n)^\mathsf{T}$. Subject to the constraints $\beta_i > 0$, $\sum \beta_i/n = 1$, the weights β may be chosen arbitrarily but are assumed to be fixed for the scope of this paper. Note that the closure of a neighbourhood of a bounded box w.r.t. $\|\cdot\|_\infty^\beta$ is again a bounded box:

$$\overline{\mathcal{V}}_\delta(Q(a,b)) = Q(a - \delta\beta^{-1}; \, b + \delta\beta^{-1}), \tag{30}$$

where $\beta^{-1} := (\beta_1^{-1}, \ldots \beta_n^{-1})^\mathsf{T}$, and $\overline{\mathcal{V}}_\delta(X)$ denotes the closure of $\mathcal{V}_\delta(X)$. The *diameter* of a box w.r.t. $\|\cdot\|_\infty^\beta$ is defined by

$$\text{diam}(Q(a,b)) := \sup\{\|\xi - \zeta\|_\infty^\beta \mid \xi, \zeta \in Q(a,b)\} = \|a - b\|_\infty^\beta \,. \tag{31}$$

Given a finite a number of (q-dimensional) boxes covering P – they are referred to as *parameter cells* – we define the (n-dimensional) measurement quantisation cells by

$$P \subseteq \bigcup_{1 \le j \le p_\mu} Q(a_j, b_j) \;=:\; \hat{P} \subset \mathbb{R}^q \,, \quad a_j, b_j \in \mathbb{R}^q \,, \tag{32}$$

$$G^{-1}(\nu_j^\mu) := Q(h_\mu(a_j) - \delta\beta^{-1}; \, h_\mu(b_j) + \delta\beta^{-1}) \subset \mathbb{R}^n \,. \tag{33}$$

This is illustrated in Fig. 2, where, for simplicity, dependence on μ has been omitted and all β_i are equal. Then, as required, the quantisation cells cover $\mathcal{V}_\delta(h_\mu(P))$. Furthermore, referring to a Lipschitz constant of h_μ, the diameter of the parameter cells can be chosen such that the measurement quantisation cells meet a given accuracy requirement, i.e. the measurement cells do not exceed a given maximum diameter. Formally, this can be stated as follows:

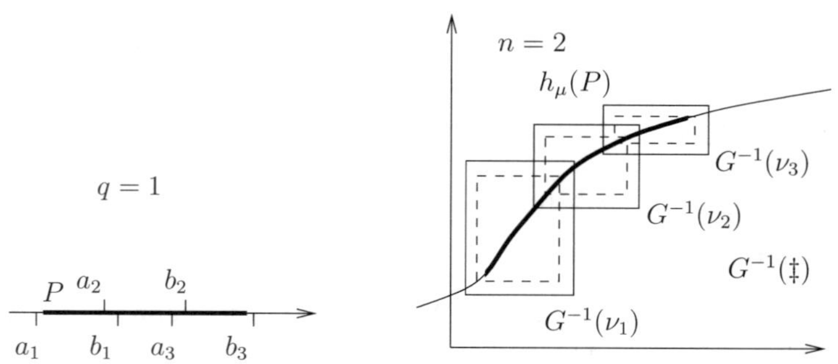

Fig. 2. Quantization of neighbourhood of $h_\mu(P)$.

Proposition 2. *Given the order preserving and continuously differentiable map* $h_\mu : \mathbb{R}^q \to \mathbb{R}^n$, *let* $L > 0$ *denote a Lipschitz constant w.r.t.* $\|\cdot\|_\infty^\beta$ *for* h_μ *on the domain* $\hat{P} \subset \mathbb{R}^q$. *Then* $\mathrm{diam}(G^{-1}(\nu_j^\mu)) \leq L\ \mathrm{diam}(Q(a_j, b_j)) + 2\delta$. *Let* γ *denote the maximum diameter of the parameter cells in the finite cover* (32). *Then*

$$\overline{\mathcal{V}}_{\delta+\gamma L}(h_\mu(\hat{P})) \quad \supseteq \quad \bigcup_{1\leq j\leq p_\mu} G^{-1}(\nu_j^\mu) \quad \supseteq \quad \mathcal{V}_\delta(h_\mu(P))\,. \tag{34}$$

Proof. The existence of a Lipschitz constant L is ensured by continuous differentiability of h_μ and boundedness of $\hat{P}$. As an immediate consequence, observe $\mathrm{diam}(\,Q(h_\mu(a_j),\,h_\mu(b_j))\,) \leq L\ \mathrm{diam}(Q(a_j, b_j))$. By the triangle inequality, we obtain

$$\mathrm{diam}(G^{-1}(\nu_j^\mu)) \leq L\ \mathrm{diam}(Q(a_j, b_j)) + 2\delta.$$

To show the first of the two inclusions in (34), pick an arbitrary point

$$\xi \in \cup_{1\leq j\leq p_\mu} G^{-1}(\nu_j^\mu)$$

and an integer j such that $\xi \in \overline{\mathcal{V}}_\delta(\,Q(h_\mu(a_j),\,h_\mu(b_j))\,)$. Hence, there exists a point $\zeta \in Q(h_\mu(a_j),\,h_\mu(b_j))$ with $\|\xi - \zeta\|_\infty^\beta \leq \delta$. Obviously, $Q(h_\mu(a_j),\,h_\mu(b_j))$ has a nonempty intersection with $h_\mu(\hat{P})$, and therefore

$$\mathrm{dist}(h_\mu(\hat{P}),\,\zeta) \leq \mathrm{diam}(\,Q(h_\mu(a_j),\,h_\mu(b_j))\,) \leq \gamma L.$$

This implies $\mathrm{dist}(h_\mu(P),\,\xi) \leq \delta + \gamma L$. Hence, $\xi \in \overline{\mathcal{V}}_{\delta+L\gamma}(h_\mu(P))$, completing the proof of the first inclusion in (34). To show the second inclusion, take any $\zeta \in \mathcal{V}_\delta(h_\mu(P))$. Then there exists a $p \in P$, $\xi := h_\mu(p)$, such that $\|\xi - \zeta\|_\infty^\beta < \delta$. By (32), we can find a j such that $p \in Q(a_j, b_j)$. As h_μ is order preserving, this implies $\xi = h_\mu(p) \in Q(h_\mu(a_j),\,h_\mu(b_j))$. Hence, $\zeta \in \mathcal{V}_\delta(Q(h_\mu(a_j),\,h_\mu(b_j))$, and, by (30), $\zeta \in G^{-1}(\nu_j^\mu)$. This proves the second inclusion in (34).

The part of $\mathbb{R}^n$ not covered by any of the cells $G^{-1}(\nu_j^\mu)$, $j = 1, \ldots p_\mu$, $\mu \in U$, again returns the out of range symbol ‡, i.e.

$$G^{-1}(\ddagger) := \mathbb{R}^n \setminus \bigcup_{1\leq j\leq p_\mu,\ \mu\in U} G^{-1}(\nu_j^\mu)\,, \tag{35}$$

such that the set of measurement symbols is given by

$$Y := \bigcup_{\mu\in U} \{\nu_1^\mu, \ldots, \nu_{p_\mu}^\mu\} \cup \{\ddagger\}\,. \tag{36}$$

This concludes the construction of a measurement quantization based on lower dimensional attractive manifolds. The reduction of the number of required quantization cells is quite significant. If, for example, one was to cover a bounded subset of $\mathbb{R}^n$ by cells not exceeding a certain diameter $\varrho > 0$, the number of required cells would be of the order $\mathcal{O}(1/\varrho^n)$. By the above method, only $\mathcal{O}(|U|/\varrho^q)$ cells are necessary to cover the corresponding portion of the manifolds $\mathcal{M}_\mu$, $\mu \in U$.

A discrete abstraction can again be obtained via Theorem 1 or, assuming monotonicity of the system dynamics, by Corollary 1, and a supervisor that is synthesized for the abstraction is guaranteed to enforce the specification for the original hybrid plant. While we have significantly reduced the number of cells, the dimension of each individual cell $G^{-1}(\nu_j^\mu)$ is not affected and the propagation over time of each such cell is with respect to the full-dimensional dynamics. As indicated, the manifold $\mathcal{M}_\mu$ may very well depend on the input symbol μ and Theorem 1 (or Corollary 1) still ensures the crucial inclusion $\mathfrak{B}_{\mathrm{ca}} \supseteq \mathfrak{B}$. Note that neither Theorem 1 nor Corollary 1 refer to the attractiveness of $\mathcal{M}_\mu$ and therefore the respective statements remain true even if $\mathcal{M}_\mu$ fails to be attractive. From the construction of the measurement quantization, however, the discrete abstraction $\mathfrak{B}_{\mathrm{ca}}$ can only be expected to be reasonably accurate if changes in the input signal only occur when the state trajectory evolves within $\mathcal{V}_\delta(h_\mu(P))$. If the state trajectory does *not* approach $\mathcal{V}_\delta(h_\mu(P))$, the resulting abstraction will not purvey sufficient information on the underlying plant dynamics and we can not expect that a nontrivial specification can be enforced for the abstraction.

Given a continuous system (3), a constructive proof for the existence of an attractive manifold $\mathcal{M}_\mu$, in general, is a nontrivial problem. However, in contrast to hybrid controller synthesis, non-linear stability analysis refers exclusively to continuous dynamics and has been discussed in depth for many application relevant ODEs. In Sect. 5, we give an example of a chemical process that demonstrates how our hyrid controller synthesis framework benefits from a rich knowledge base regarding the non-linear process dynamics. A class of hybrid control problems for which an attractive manifold $\mathcal{M}_\mu$ is readily known to exist occurs in hierarchical control architectures, in which a continuous plant is subject to a number of alternative *low-level* continuous controllers; see (Moor et al., 2001b). In this configuration, a *high-level* discrete input symbol $\nu \in U$ implements the activation of the respective low-level controller. In particular, for each μ the system (3) represents a continuous closed-loop model, which in many applications is required to exhibit stable state components by any resonable design objective. Again, the enforcement of such low-level design objectives refers to continuous dynamics only and for the solution of these control problems one can draw from the literature on non-linear control.

5 Start-Up of a Distillation Column

We consider a distillation column in pilot plant scale which is operated at the Institut für Systemdynamik und Regelungstechnik in Stuttgart. It is about 10 m high, and consists of 40 bubble cap trays (consecutively numbered by $i = 2, \ldots, 41$ from bottom to top), a reboiler ($i = 1$) and a condenser ($i = 42$), see Fig. 3. Feed is supplied on tray 21. Our application example is the separation of methanol and propanol.

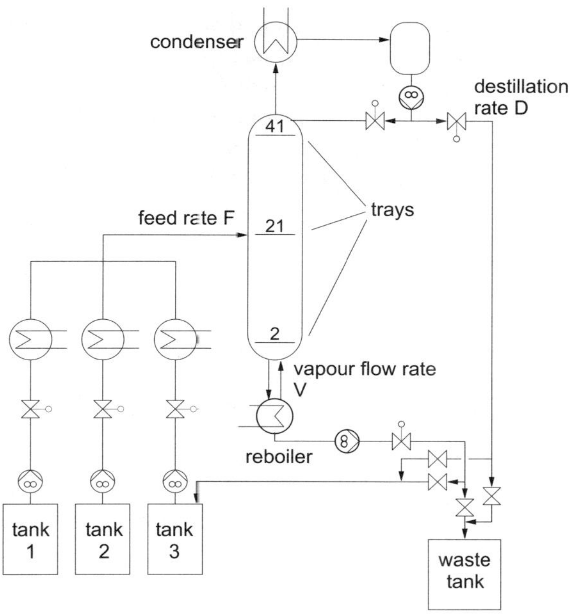

Fig. 3. Distillation column

The following steps can be distinguished during conventional column start-up: initially, the column trays are partially filled with liquid mixture from the previous experimental run. Further feed is added, and the column is heated up until boiling point conditions are established in the whole column. During this start-up step, the column is operated at total reflux and reboil. At the end of this step, a single concentration front is established. The position of this front depends on the initial concentration and varies from experiment to experiment. In a second step, the feed F, and the control inputs (distillate and vapour flow rate, D and V) are adjusted to their desired steady state values, and the initial front splits into two fronts. Then, in a third step, the two fronts move *very* slowly towards their steady state. We try to speed up the third step of the start-up procedure by introducing a suitable supervisory control strategy. The starting point for our approximation based controller synthesis is a continuous distillation column plant model which incorporates the following assumptions, which are well justified during the third step of the start-up: constant molar overflows, constant molar liquid holdups, negligible vapour holdups, total condenser, constant relative volatilities, a tray efficiency of one. Therefore, the model is based on material balances only and consists of one nonlinear first-order ODE for each tray, the reboiler, and the condenser (Klein et al., 1999):

$$n_L^i \dot{x}_i = F_L^{i+1} x_{i+1} - F_L^i x_i + F_V^{i-1} y_{i-1} - F_V^i y_i + \begin{cases} F x^F & \text{if } i = 21\,, \\ 0 & \text{else,} \end{cases} \tag{37a}$$

$$y_i = x_i \frac{\alpha}{1 + x_i(\alpha - 1)}\,, \tag{37b}$$

where x_i and y_i are the methanol mole fractions in the liquid and in the vapour on the i-th tray ($i = 2, \dots, 41$), in the condenser ($i = 42$) and the reboiler ($i = 1$); $\alpha = 2.867$ is the relative volatility; $x^F = 0.32$ is the methanol mole fraction in the feed; F_L^i denotes the liquid molar flow rate, F_V^i the vapour flow rate and n_L^i the molar liquid holdup. Numerical values for the latter are given in Table 1. The table also states how F_L^i and F_V^i depend on F, D and V (feed, distillate and vapour flow rate).

Table 1. Flow rates and liquid holdups

	i	F_L^{i+1}	F_L^i	F_V^{i-1}	F_V^i	n_L^i [mol]
condenser	42	0	V	V	0	1.922
stripping	22-41	$V-D$	$V-D$	V	V	1.922
feed tray	21	$V-D$	$F+V-D$	V	V	1.922
rectifying	2-20	$F+V-D$	$F+V-D$	V	V	1.922
reboiler	1	$F+V-D$	$F-D$	0	V	135

The feed flow rate is considered to be constant at $F = 220.0$ mol/h, while D and V are control inputs. For any constant D and V, the system (37a), (37b) has an attractive equilibrium $x^*(D,V)$, which, for the nominal inputs $D_0 = 70.4$ mol/h and $V_0 = 188.2$ mol/h, corresponds to the desired operating point $x_0^* := x^*(D_0, V_0)$ of the distillation column. To speed up the process of approaching x_0^*, we look for a controller that switches between a finite number of constant input values. Considering only values $V > 0$, $D > 0$ such that $F + V - D \geq 0$, monotonicity of (37a), (37b) follows from the criterion given in Theorem 2.

The construction of lower dimensional manifolds $\mathcal{M}_\mu$, which is vital for approximation based discrete control, is based on *wave propagation theory* (Kienle, 2000); it considers particular concentration profiles as waves and discusses their propagation in time and space. Each wave is of the form

$$x_i = p_1 + \frac{p_2 - p_1}{1 + e^{\varrho(i-s)}}\,, \tag{38}$$

where p_1 and p_2 are the asymptotic values of the methanol mole fraction at the bottom and at the top of the wave, s is the so called wave position (point of inflexion) and ϱ is the slope at s. The aspect of wave propagation theory most relevant to our discussion is that during the third startup step, the concentration profile can be represented by two waves of the type (38), one each in the stripping ($1 \leq i \leq 21$) and the rectifying section ($21 < i \leq 42$). Their slopes can be approximated reasonably well by the slopes corresponding to the equilibrium $x^*(D,V)$. For the nominal inputs D_0 and V_0, the slopes turn out to be $\varrho_s = 0.465$ and $\varrho_r = 0.572$ for the stripping section and the rectifying section, respectively. Neglecting the effect of different inputs to the slopes, the lower dimensional manifold under construction becomes independent of the input symbol. If we further assume constant methanol mole fractions in the reboiler and condenser, $x_1 = 0$ and $x_{42} = 1$, the asymptotic values in (38) are uniquely determined by the feed concentration x_{21} and the wave positions $\hat{s}_s$ and $\hat{s}_r$ for the stripping and rectifying section, respectively. [2] Consequently, the wave fronts of interest are parametrized by a map $h\colon \mathbb{R}^3 \to \mathbb{R}^{42}$ mapping parameter triples $(x_{21}, \hat{s}_s, \hat{s}_r)$ to concentration profiles in the high dimensional state space. The i-th component h_i of h evaluates to

$$h_i(x_{21}, \hat{s}_s, \hat{s}_r) := x_{21} \left[(1 - e^{(i-1)\varrho_s}) (1 + e^{(\hat{s}_s - 1)\varrho_s}) \right] \times \left[(1 - e^{20\,\varrho_s}) (1 + e^{(i-22+\hat{s}_s)\varrho_s}) \right]^{-1} \quad (39)$$

for $1 \leq i \leq 21$, and

$$\begin{aligned} h_i(x_{21}, \hat{s}_s, \hat{s}_r) := \big[\, & x_{21} (e^{21\varrho_r} - e^{(i-63+\hat{s}_r)\varrho_r}) \\ & + (1 - x_{21}) (e^{(\hat{s}_r - 21)\varrho_r} - e^{(i-21)\varrho_r}) + e^{(i-42+\hat{s}_r)\varrho_r} - 1 \,\big] \\ & \times \left[(e^{21\varrho_r} - 1) (e^{(i-63+\hat{s}_r)\varrho_r} + 1) \right]^{-1} \end{aligned} \quad (40)$$

for $22 \leq i \leq 42$. Note that all partial derivatives of h are non-negative. Hence, h is order preserving. This completes the construction of $\mathcal{M} \equiv \mathcal{M}_\mu := h(\mathbb{R}^3)$.

We now specify the operating range of the supervisor. For our particular setting, the equilibrium x_5^* corresponds to the parameter triple $x_{21} \approx 0.318$, $\hat{s}_s \approx 10.7$, $\hat{s}_r \approx 28.7$. The bounded box of parameters $P = [0.300,\ 0.340] \times [4.0,\ 20.0] \times [23.0,\ 37.0]$ is considered a reasonably large operation range, which we partition by $p = 139$ parameter cells $Q(a_j, b_j)$, $1 \leq j \leq p$. The high dimensional measurement quantization cells are then constructed by (33) with $\delta = 0.002$. Input symbols $U = \{\mu_1, \ldots \mu_9\}$ are chosen according to Table 2; see (Klein et al., 1999) for a detailed motivation of the particular numerical values.

Table 2. Control symbols

symbol	μ_1	μ_2	μ_3	μ_4	μ_5
D [mol/h]	35.8070	59.3318	82.8566	46.8782	70.4030
V [mol/h]	188.2433	158.6412	129.0391	217.8455	188.2433
symbol	μ_6	μ_7	μ_8	μ_9	
D [mol/h]	93.9278	57.9494	81.4742	104.999	
V [mol/h]	158.6412	247.4476	217.8455	188.2433	

[2] We use the substitutions $22 - s \to \hat{s}_s$ and $63 - s \to \hat{s}_r$ for the wave positions in order to end up with an order preserving map h.

For each input symbol $\mu \in U$, the system (37a), (37b) exhibits a unique solution and hence induces a flow Φ_t^{μ}. With the choice of a particular sampling interval ($\Delta = 10$ min), a hybrid plant model according to Sec. 2 is completely determined.

As a specification, we require the supervisor to drive any initial state within $X_0 = \mathcal{V}_\delta(h(P))$ into the target region $X_f = \overline{\mathcal{V}}_\delta(h(P_f))$ within no more than 20 min, where $P_f = [0.316,\ 0.320] \times [8.5,\ 11.5] \times [27.5,\ 31.0] \subset P$. Choosing one of the quantization cells equal to X_f, this specification can be formalized by the behaviour $\mathfrak{B}_{\text{spec}}\{(u,y)|\ y(k) = \nu_f\ \forall\ k \geq 2\}$, where $G^{-1}(\nu_f) = X_f$ for some $\nu_f \in Y$. Controller synthesis is then successfully carried out based on the estimate sets $\hat{\mathcal{X}}((u,y)|_{[0,l]})$ for $l = 2$. A simulation of the closed loop (consisting of 42nd order continuous plant model and DES controller) is shown in Fig. 4. For each sampling instant, one concentration profile is plotted, the arrows indicate forward evolution in time and the intervals per tray indicate the target region X_f. As the sampling intervals in the closed-loop configuration are chosen to be 10 min, the target region is seen to be reached within 20 min. In contrast, Fig. 5 shows an open-loop simulation for the nominal input V_0 and D_0. Here, one profile every 5 h is plotted, and it takes an overall time of 20 h to reach the target region.

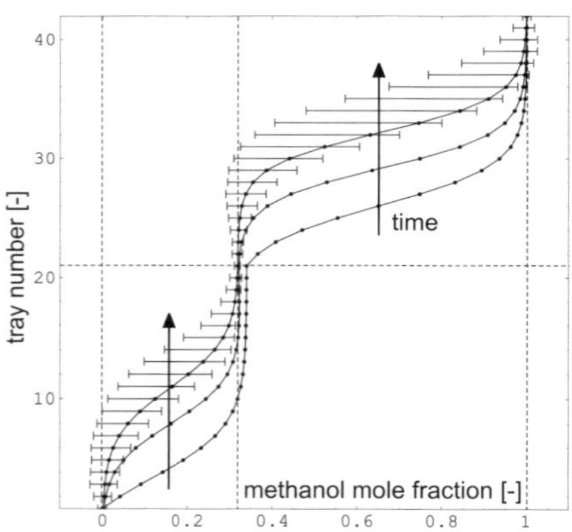

Fig. 4. Closed-loop (Δ=10min)

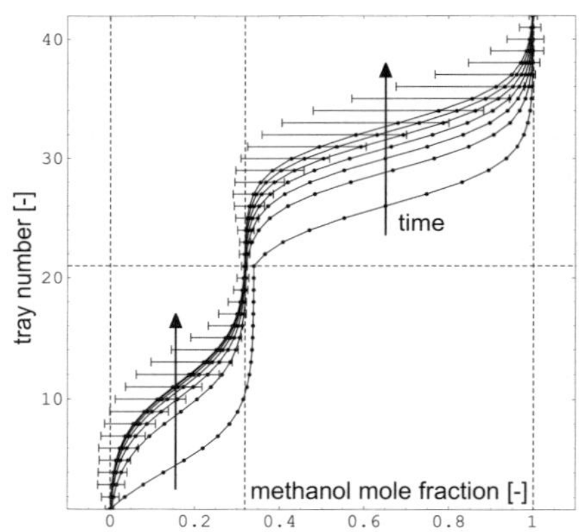

Fig. 5. Open-loop (Δ=5h)

Remark 2. The properties employed for the construction of $\mathcal{M}$ are well motivated by wave propagation theory and also have been validated by simulations and experiments. It follows from the successful completion of the controller synthesis procedure, that our discrete abstraction is accurate enough for the particular purpose. While the insight from the process engineering perspective has been an essential guidance, it is important to note that the reliability of our controller does not depend on the various claims and assumptions regarding the process model: the only relevant requirement is the inclusion $\mathfrak{B}_{\text{ca}} \supseteq \mathfrak{B}$, and this follows purely from the monotonicity of f as discussed in Sect. 3, see Corollary 1.

On a decent workstation, the overall time required for the computation of both the discrete approximation and the supervisory controller is about 10 min. This is a significant performance increase when compared with earlier work (Klein et al., 2000, Klein et al., 1998, Klein et al., 1999) on the very same scenario, but based on exhaustive simulation: there, computations took many hours. Note also the different quality of reliability: while our new approach guarantees the approximation to be conservative, exhaustive simulation may – in principle – overlook critical states.

6 Conclusions

In this paper, we have shown how a general method for the abstraction based synthesis of discrete event controllers can be applied to a class of nonlinear high-order continuous systems, characterised by a monotonicity condition and an attractive low-dimensional manifold. In the presence of strict reliability requirements, abstraction based controller synthesis methods have been mostly restricted to low-order linear plant models and in this sense our contribution constitutes a considerable extension to the range of potential applications. Using monotonicity, the temporal evolution of quantization cells can be conveniently over-approximated even for nonlinear dynamics. This allows for the economical construction of a discrete abstraction for the nonlinear plant dynamics under investigation. Under the assumption that the plant state approaches a low-dimensional manifold, we construct an abstraction that in terms of computational effort depends only on the dimension of the attractive manifold rather than the full order of the plant dynamics. Note that both of our conditions lie completely within the domain of continuous dynamics: whether or not a plant is monotone and whether or not it exhibits an attractive manifold can be assessed by means of the classical theories. One might argue that our conditions are too restrictive for our results to be of practical relevance. This is not true, and we present a real-world example to support our claim to the contrary: based on a 42nd order nonlinear model of a pilot plant scale distillation column, we synthesize a discrete controller that speeds up the column start-up procedure. A comparison with earlier work underlines the achieved computational benefits.

Acknowledgement. We'd like to thank D. Flockerzi for valuable discussions on monotone dynamical systems and A. Kienle, A. Itigin, and E. Klein for their help with the the distillation column scenario.

Hybrid Reconfigurable Control

Jan Lunze and Thomas Steffen

Institute of Automation and Computer Control
Ruhr University Bochum
Universitätsstraße 150
D-44780 Bochum
phone: +49 234 32 28071
http://www.ruhr-uni-bochum.de/atp

Abstract. A severe fault renders a system inoperable by breaking the control loops. The task of control reconfiguration is to change the control structure in response to the detected fault. A two-level approach for control reconfiguration is presented. Firstly, a discrete-event model of the faulty system is used to design a discrete controller that brings the faulty system towards its new equilibrium state. Secondly, by using a linear model valid around this equilibrium, a linear extension to the original control loop is designed that allows the stabilisation of the faulty system.

1 Control Reconfiguration

Any technical system is liable to the occurrence of faults, where typically a fault in a single component affects the whole system. In case of severe faults, the system cannot be held in operation without a severe change in the control algorithm. This paper concerns the problem of reconfiguring the control after the occurrence of faults.

The reconfiguration problem consists of finding and implementing a new control structure in response to the occurrence of a severe fault. Severe faults such as the complete failure of actuators or sensors break the control loops brought about by the nominal controller. It is, therefore, necessary to use a different set of inputs or output for the control task. Once the new control configuration is selected, new controller parameters have to be found. The goal of the reconfiguration is to stabilise the faulty process and to keep it operational. The reconfiguration task is similar to a controller design problem for the faulty process, but it has to be carried out completely automatically during the operation of the system and it can build on the existence of the nominal control loop.

Summary of the Approach Presented here. As reconfiguring the control structure requires discrete decisions such as the choice of new actuators, sensors or set-points, the first part of the reconfiguration problem (reaching the equilibrium) is formulated as a discrete optimisation problem (Sects. 3 and 4). The task of stabilising the faulty system once the new equilibrium is reached will be solved by means of a linear model of the faulty plant (Sect. 5). If both solutions are taken together, a hybrid approach to control reconfiguration emerges. This allows the treatment of non-linear or hybrid systems and produces a linear extension of the control structure to stabilise the system without using pre-designed controllers or manual intervention (Sect. 6).

S. Engell, G. Frehse, E. Schnieder (Eds.): Modelling, Analysis and Design of Hybrid Systems, LNCIS 279, pp. 267–284, 2002.

Related Literature. While *fault detection and isolation* (FDI) have been a subject of intensive research for many years, there are only few approaches to control reconfiguration. The typical way for building a fault-tolerant controller is to manually design controllers for every fault case and to switch among them. A formalisation of this is given in (Blanke et al., 2000a, Blanke et al., 2000b).

Adaptive methods have been used to handle faults that bring about minor parameter deviations (Ioannou, 1996), but they fail to treat major faults which change the structure of the process (not just the parameters). The same is true for the pseudo-inverse method (Gao and Antsaklis, 1991). Model predictive control has been successfully used with respect to severe faults that can be described by additional restrictions on the plant behaviour (Maciejowski, 2002). This approach becomes too complex for larger systems and it does not answer basic questions about the reconfigurability and stability of the system.

A survey of the state of the art of control reconfiguration is given in (Lunze, 2001). Several reconfiguration options are applied to a common example in (Lunze et al., 2000). The approach presented here incorporates results of discrete-event systems theory, as described in (Moody and Antsaklis, 1998). The specific solution given in Sect. 4 uses dynamic programming (Bellman, 1957).

2 The Reconfiguration Problem

The reconfiguration problem consists of finding a new control configuration and a new control algorithm so that the closed-loop system satisfies the given requirements. It is described in detail in this section.

2.1 The Plant Model

The plant model depends on the fault f. It is given in state space form

$$\dot{\mathbf{x}}(t) = \mathbf{g}(\mathbf{x}(t), \mathbf{u}(t), f) \tag{1}$$
$$\mathbf{y}(t) = \mathbf{h}(\mathbf{x}(t)) \tag{2}$$
$$\mathbf{x}(0) = \mathbf{x}_0 \tag{3}$$

where $\mathbf{x} \in \mathbb{R}^n$ is the process state, $\mathbf{u} \in \mathbb{R}^m$ the input, $\mathbf{y} \in \mathbb{R}^r$ the output and $f \in \mathcal{F}$ the fault, where $\mathcal{F}$ is the set of modelled fault cases. The nominal plant is described by (1)–(3) for $f = 0$. The fault f is assumed to be constant during the reconfiguration. It is detected by an FDI module. The system function $\mathbf{g}$ may be piecewise defined like that for switching systems, i.e. the system may show hybrid behaviour.

For the nominal plant there exists a nominal controller. The resulting control loop consisting of the nominal plan and nominal controller is assumed to be stable. Furthermore, it satisfies all requirements concerning disturbance rejection and input tracking.

2.2 Reconfiguration Goal

The faulty process is defined by the same set of equations but with a different value of the fault symbol f. The behaviour may show major differences because the faults include the loss of actuator or sensor functions. Therefore the controller cannot be adapted to the faulty system by simply changing its parameters, but a new control configuration has to be found. Hence, the faults severely change the plant behaviour. The diagnostic task is not considered here and f is assumed to be known.

The reconfiguration goal is to design a new controller for the faulty process. This controller may use alternative sensors and actuators not used by the original controller. In fact, the selection of sensors and actuators is an important part of the reconfiguration problem. The resulting control loop has to be stable and it should show a behaviour that is as similar as possible to the behaviour of the nominal control loop.

2.3 Steps of the Proposed Control Reconfiguration

In the reconfiguration loop the following tasks have to be carried out (Fig. 1):

1. Detect and identify the fault (diagnosis)
2. Find the model of the faulty system
3. Find a sequence of control actions to bring the process to a new equilibrium state (state transition problem)
4. Extend the controller to stabilise the faulty plant at the new equilibrium (linear reconfiguration)

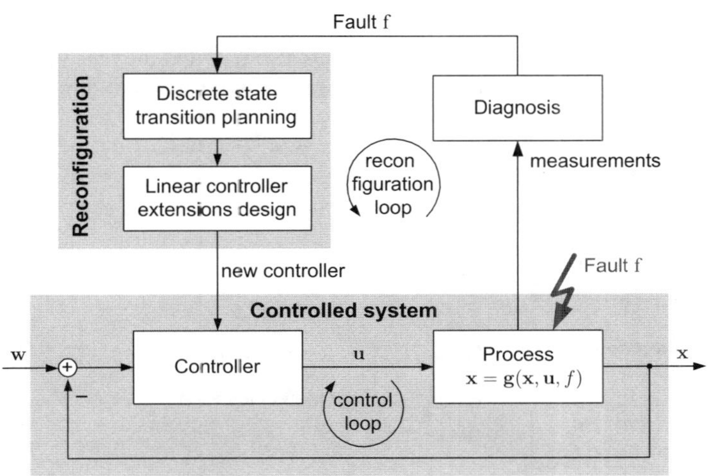

Fig. 1. Controller reconfiguration

This approach focus on steps 2 to 4, since it is assumed that step 1 has been solved (Lunze and Steffen, 2000). Step 3 is necessary because the state that the plant assumes when the fault is identified may differ significantly from the new equilibrium and both may differ from the nominal equilibrium of the faultless process. For step 3, a discrete-event approach is proposed here, where the hybrid plant is treated as a purely discrete-event system (Sect. 4). The reason for this is that the controller in step 4 is designed to deal with small disturbances (Sect. 5), and therefore it may not be able to deal with the non-linear and hybrid effects encountered here. Hence, step 2 has to end up with a discrete-event model that describes the global behaviour of the faulty plant and a linear model describing the faulty plant in the surroundings of the new equilibrium.

3 The Discrete-Event Model of the Faulty Plant

In step 3 a discrete controller is used to solve the state transition problem. For the design of this controller, a discrete model of the process is required.

3.1 Discrete Control Loop

The control loop considered when solving the state transition problem is shown in Fig. 2. It consists of the faulty process

$$\dot{\mathbf{x}} = \mathbf{g}(\mathbf{x}, \mathbf{u}, f) \tag{4}$$

and a discrete static feedback function

$$v = R(z) \tag{5}$$

which has to be found.

The connection between the discrete controller output $v \in \mathcal{V}$ and the real-valued process input $\mathbf{u} \in \mathbb{R}$ is handled by the injector, which assigns every discrete controller output a continuous-variable process input:

$$\mathbf{u} = \mathbf{q}_u(v) \ , \ \mathbf{q}_u : \mathcal{V} \to \mathbb{R}^m \ .$$

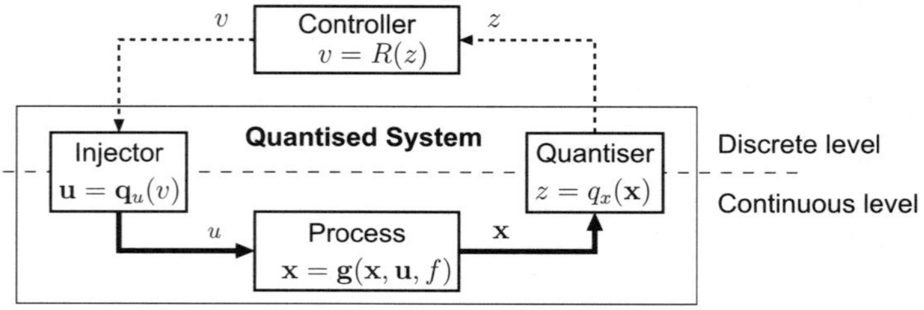

Fig. 2. Hybrid control loop

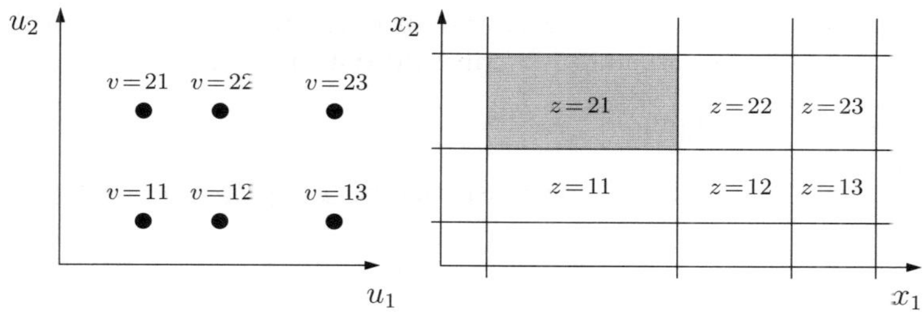

Fig. 3. Input and output discretisation

This function is shown in the left part of Fig. 3: every dot in the input space of $\mathbf{u}$ corresponds to a different discrete value of v.

The quantiser does the opposite conversion from the process state $\mathbf{x} \in \mathbb{R}^{n_x}$ to the discrete-valued controller input $z \in \mathcal{Z}$. It defines a state space partition where every area in the state space is assigned a specific controller input z (see the right part of Fig. 3):

$$z = q_x(\mathbf{x}) \ , \ q_x : \mathbb{R}^n \rightarrow \mathcal{Z} \ .$$

Changes of the quantised state are called events, which occur at times

$$t_1, t_2, \ldots, t_n \ .$$

Because of the static feedback function used, the input v will only change as a consequence of a change of the state z (synchronous events). The equations defining the injector and the quantiser can be rewritten in the form:

$$\begin{aligned} \mathbf{u}(t) &= \mathbf{q}_u(v_k) \quad \text{for } t_k < t < t_{k+1} \\ z_k &= q_x(\mathbf{x}(t)) \ \text{for } t_k < t < t_{k+1} \\ z_k &\neq z_{k+1} \end{aligned}$$

3.2 Abstraction of the Quantised Process Model

In the implementation step, the discrete controller together with the quantiser and the injector have to be applied to the faulty system. However, in the design step the plant is combined with the quantiser and the injector to form the quantised system, which is seen as a discrete-event system with a sequence $v_1, \ldots, v_n$ of input symbols going in and a sequence of output symbols $z_1, \ldots, z_n$ coming out (cf. Fig. 2).

A timed stochastic automaton

$$TSA(\mathcal{Z}, \mathcal{V}, \text{Prob}(z_{k+1}|z_k, v_k), T(z_{k+1}|z_k, v_k))$$

is used as a discrete-event model of the quantised system with input $v \in \mathcal{V}$ and output $z \in \mathcal{Z}$. $\mathcal{Z}$ is the set of state (and output) symbols, $\mathcal{V}$ is the set of input symbols. The third argument Prob$(z_{k+1}|z_k, v_k)$ is the state transition probability from state z_k with input v_k to state z_{k+1}. The last argument describes the time interval

$$T(z_{k+1}|z_k, v_k) = [T_{\text{min}}, T_{\text{max}}]$$

which gives upper and lower limits for the transition time $t_{k+1} - t_k$ dependent upon the input.

3.3 Completeness of the Model

It seems reasonable to claim that the stochastic model TSA behaves exactly like the quantised process. That is, for a given sequence of inputs $v_1, \ldots, v_n$ and a given initial state z_1, the model TSA generates the same set $\mathbb{B}_{\text{Model}}$ of trajectories $z_2, \ldots, z_n$ as the actual system, whose trajectory set is denoted by $\mathbb{B}_{\text{System}}$:

$$\mathbb{B}_{\text{Model}} = \mathbb{B}_{\text{System}} \quad .$$

However, exact discrete-event models typically have an infinite size, so it is necessary to use an approximation (Lunze and Raisch, 2002). For the controller design it is important that the model contains all trajectories of the quantised system which leads to the completeness requirement:

$$\mathbb{B}_{\text{Model}} \supseteq \mathbb{B}_{\text{System}} \quad .$$

In other words: for a given input sequence, the automaton TSA can generate all output sequences of the quantised system.

Several algorithms have been designed to determine the abstracted process model (Lunze and Raisch, 2002). They differ in the way the completeness is achieved and in the processing power required. For the on-line abstraction, a fast and simple algorithm is proposed in (Lunze and Steffen, 2000). The following considerations assume that the discrete model TSA has been found by such an algorithm for the plant subject to the current fault f.

4 Discrete Controller Design

The state transition task is to move the state of the faulty system from where it is when the fault is detected into a specified region around the new equilibrium, where it can be stabilised. The linear controller is not able to perform the state transition, because it cannot account for hybrid effects, nonlinear behaviour or input limitations. A discrete controller is able to cope with these effects if they are reflected by the discrete-event model.

4.1 Control Task

This section shows how the state transition problem can be formulated as a separable optimisation problem. Then, by using Bellman's optimality principle (Bellman,

1957), an efficient solution of the problem is possible. The state transition problem can be interpreted as a staged process. The process inputs are the choices to be made by the controller, and the process state is the system state in the sense of the optimality principle. Every event starts a new stage and only a finite number of events h is considered. An optimisation criterion

$$I(z) = \text{Prob}(\exists j \in 1 \ldots h : \, z_j \in \mathbf{G} \,|\, z_0 = z) \tag{6}$$

for reaching the goal set of states $\mathbf{G}$ is given, and the goal is to find the optimal sequence of decisions (the optimal feedback controller).

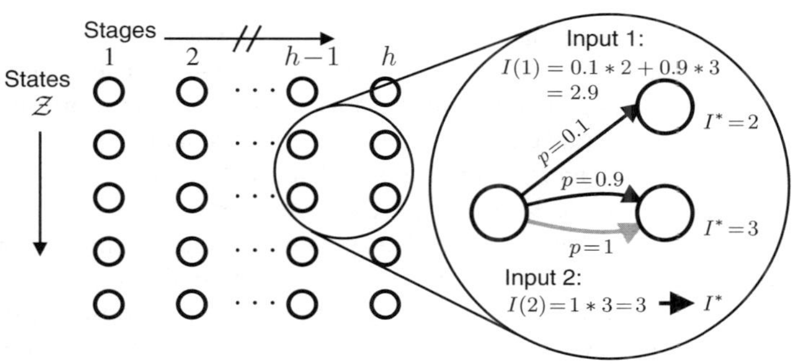

Fig. 4. Solving the state transition problem using Bellman's optimality principle

Bellman's optimality principle states that the optimal control decision at a certain stage can be found without knowledge about earlier stages. This allows the efficient determination of the optimal controller for the given objective function and process model.

4.2 Basic Optimisation Problem

One way of constructing a separable criterion is to use the expected probability of reaching a state within a given set $\mathbf{G}$ of acceptable equilibrium states (depending on the purpose of the process). The main objective is obviously to reach this set $\mathbf{G}$ within the considered number of stages h. The corresponding criterion (6) can be rewritten according to Bellman's optimality principle:

$$I_k(z_k, v_k) = \sum_{z_{k+1} \in \mathcal{Z}} \text{Prob}(z_{k+1}|z_k, v_k) \begin{cases} I^*_{k+1}(z_{k+1}) & \text{for } z_{k+1} \notin \mathbf{G} \\ 1 & \text{otherwise} \end{cases} \tag{7}$$

where the probability of a state transition is given by the stochastic process model. $I_k(z_k, v_k)$ is the objective value for selecting the input v_k when the system is in the state z_k and at stage k. The evaluation of the further development up to stage h is included, but earlier stages don't influence the equation other than through z_k and k.

The optimisation problem can then be written in an iterative form as:

$$v_k^*(z_k) : I_k^*(z_k) = \max_{v_k \in \mathcal{V}} I_k(z_k, v_k) \tag{8}$$

This means: for a given stochastic model Prob$(z_{k+1}|z_k, v_k)$ and a goal set $\mathbf{G}$, find the optimal discrete control $v_k^*(z_k)$ with respect to the optimisation criterion $I_k(z_k, v_k)$.

4.3 Enhanced Optimisation Problem

The basic criterion is not sufficient to describe all the restrictions to be met by the controller, because it does not allow the definition of forbidden states and the resulting state transition may be very slow.

An extended criterion can be defined to include the requirement of avoiding a given set $\mathbf{F}$ of forbidden states. The probability of reaching a state in $\mathbf{F}$ has to be minimised. Since both criteria (reaching $\mathbf{G}$ and avoiding $\mathbf{F}$) may conflict, a priority has to be set. This is done using a ranked multi-value criterion. The first priority is to minimise the probability of entering $\mathbf{F}$. Only if this probability of entering $\mathbf{F}$ is the same for several possible choices (e.g. exactly 0), the second criterion of reaching $\mathbf{G}$ is considered for selecting the better choice. This guarantees that the solution of the optimisation problem avoids the forbidden states if this is at all possible.

The transition time is used as a third criterion. It is impossible to impose a hard time limit, since this would lead to a non-separable criterion (the remaining amount of time depends on the previous stages). Instead, the third criterion is defined as the maximum time necessary to reach the set $\mathbf{G}$ according to the discrete process model. If this criterion is minimised and the resulting value is greater than the available time-span, a state transition within the required time cannot be guaranteed.

The complete ranked criterion is a 3-tuple defined by

$$\mathbf{I}_k(z_k, v_k) = \begin{pmatrix} \sum_{z_{k+1} \in \mathcal{Z}} \text{Prob}(z_{k+1}|z_k, v_k) I_{1,k+1}^*(z_{k+1}) \\ \sum_{z_{k+1} \in \mathcal{Z}} \text{Prob}(z_{k+1}|z_k, v_k) I_{2,k+1}^*(z_{k+1}) \\ -T_{\max}(z_{k+1}|z_k, v_k) + \min_{z_{k+1} \in \mathcal{Z}} I_{3,k+1}^*(z_{k+1}) \end{pmatrix} \tag{9}$$

for $z_k \notin \mathbf{G}$ and $z_k \notin \mathbf{F}$. If z_k is either within $\mathbf{G}$ or $\mathbf{F}$, later states are not considered and a fixed value criterion is used instead:

$$\mathbf{I}_k(z_k, v_k) = (0, +1, 0)^T \text{ for } z_k \in \mathbf{G}$$
$$\mathbf{I}_k(z_k, v_k) = (-1, 0, 0)^T \text{ for } z_k \in \mathbf{F} \ .$$

The order relation on the 3-tuple $\mathbf{I}$ is defined as follows:

$$\begin{aligned} \mathbf{I}(a) < \mathbf{I}(b) \text{ iff } & I_1(a) < I_2(a) \\ & \text{or } I_1(a) = I_1(b) \wedge I_2(a) < I_2(b) \\ & \text{or } I_1(a) = I_1(b) \wedge I_2(a) = I_2(b) \wedge I_3(a) < I_3(b) \ . \end{aligned}$$

This relation guarantees that the three parts of the criterion are always considered in the stated order of priorities. Apart from the criterion, the optimisation problem (8) remains unchanged.

4.4 Algorithm

Because the optimisation problem is separable, the following iterative algorithm can be used to solve it within constant computation time. The time complexity of this algorithm is $O(h\,|\mathcal{Z}|\,|\mathcal{V}|)$, which can be reduced by applying appropriate heuristics.

Given: Prob$(z_{k+1}|z_k, v_k)$, **G**, **F**, $\mathcal{Z}$, $\mathcal{V}$, h
Initialisation: $\forall z \in \mathcal{Z} : \mathbf{I}^*_h(z) = 0$
Loop over k from $h-1$ down to 0:
  Loop over $z \in \mathcal{Z}$:
    Find the optimal input $v^*_k(z_k)$ in response to state z_k:
      1 Calculate $\mathbf{I}(z_k, v_k)$ for every input value $v_k \in \mathcal{V}$.
      2 Select the highest value

$$v^*_k(z_k) : \mathbf{I}^*_k(z_k) = \max_{v_k \in \mathcal{V}} \mathbf{I}(z_k, v_k) \tag{10}$$

End-loops
Result: $R(z_k) = v^*_0(z_k)$ is the optimal discrete controller for (5).

5 Reconfiguration of the Linear Controller

5.1 Way of Solution

Once the faulty process has reached its new equilibrium state, it has to be stabilised there. Because only small disturbances have to be considered, a linear model is sufficient for this task. This problem could be solved by designing a new controller for the linear model of the faulty process.

However, the complete re-design of the controller has two disadvantages. Firstly, controller design is an iterative process that cannot be done fully automatically during the operation of the system. Secondly, a re-design does not utilise the knowledge about the system which is obtained during the design cycle the nominal controller. Note that due to the tests the controller contains knowledge about the actual process that is not be present in the model. It is known that the nominal controller is able to control the actual process. Therefore, the new controller should include the nominal controller and adapt it to the faulty plant. This way of solution is described below.

5.2 Linear Reconfiguration Problem

It is assumed that a linear model of the faulty process is known, which is valid for $q_x(\mathbf{x}) \in \mathbf{G}$. A suitable equilibrium state shall exist and the model has been linearised around it. The same linearisation has been applied to the nominal model. It is further assumed that the nominal controller is applicable to the linearised model of the nominal process for the new equilibrium, which may differ slightly from the linearisation around the original equilibrium.

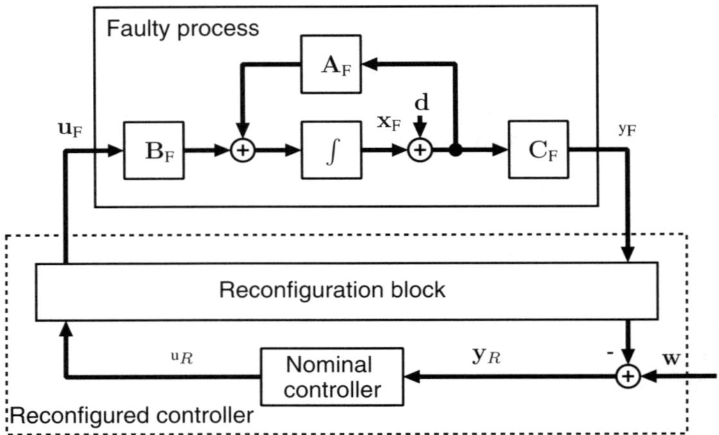

Fig. 5. Generic reconfiguration approach

Nominal and faulty processes are both given in state space form:

$$\dot{\mathbf{x}}_\mathrm{N} = \mathbf{A}_\mathrm{N}\mathbf{x}_\mathrm{N} + \mathbf{B}_\mathrm{N}\mathbf{u}_\mathrm{N} \tag{11}$$

$$\mathbf{y}_\mathrm{N} = \mathbf{C}_\mathrm{N}\mathbf{x}_\mathrm{N} \tag{12}$$

$$\mathbf{x}_\mathrm{N}(0) = \mathbf{x}_0 \tag{13}$$

$$\dot{\mathbf{x}}_\mathrm{F} = \mathbf{A}_\mathrm{F}\mathbf{x}_\mathrm{F} + \mathbf{B}_\mathrm{F}\mathbf{u}_\mathrm{F} \tag{14}$$

$$\mathbf{y}_\mathrm{F} = \mathbf{C}_\mathrm{F}\mathbf{x}_\mathrm{F} \tag{15}$$

$$\mathbf{x}_\mathrm{F}(0) = \mathbf{x}_0 \tag{16}$$

It is assumed that the nominal process can be stabilised (satisfying all requirements) by using the given proportional[1] feedback controller

$$\mathbf{u}_\mathrm{N} = \mathbf{K}_\mathrm{N}\mathbf{y}_\mathrm{N} \ . \tag{17}$$

The goal is to extend the control loop with a generic 4-port interface as shown in Fig. 5 so that the behaviour is as close to the nominal loop as possible. Hence, the reconfiguration task for the linear controller is solved by constructing the reconfiguration block that adapts the nominal controller to the faulty plant.

5.3 Reconfiguration in Case of Sensor Faults

This section explains the reconfiguration of sensor faults, where the nominal and the faulty processes differ in the matrix $\mathbf{C}$ only. The idea is to determine the difference

[1] For the presented approach, the class of controllers is not restricted to static or linear controllers. This assumption is made because it greatly simplifies the analysis of the reconfigured control loop.

$(\mathbf{C}_\mathrm{N} - \mathbf{C}_\mathrm{F})\mathbf{x}$ by means of an observer and to add the difference to the sensor output to restore normal operation (see Fig. 6). Hence, the faulty plant together with the reconfiguration block behaves, from the point of view of the controller, exactly as the nominal plant did.

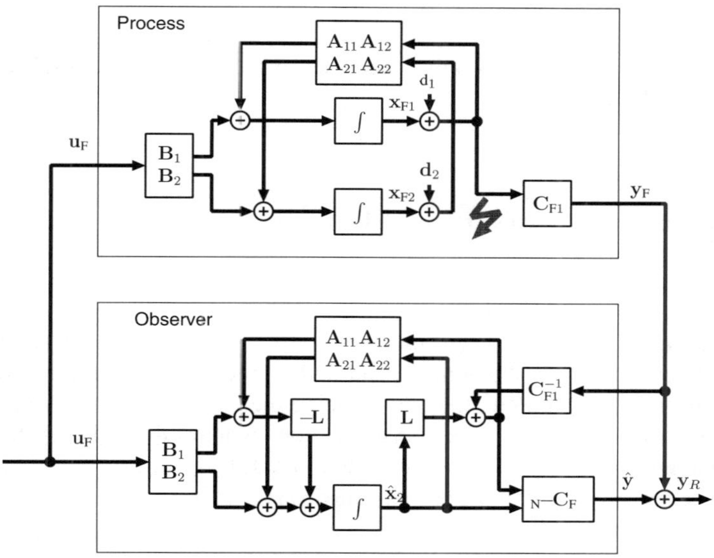

Fig. 6. Reconfiguration for sensor faults using a reduced observer

The design of a reduced-order observer is a standard method of control theory (Lunze, 2002). It is worth noting that the state space is divided into a measurable part $\mathbf{x}_\mathrm{F1}$ (which can be calculated from the output $\mathbf{C}_\mathrm{F}\mathbf{x}_\mathrm{F1}$) and a non-measurable part $\mathbf{x}_\mathrm{F2}$ (which has to be observed). With the introduction of the observation error $\mathbf{e} = \hat{\mathbf{x}}_2 - \mathbf{x}_\mathrm{F2}$, a compact model of the closed loop can be derived

$$\begin{pmatrix} \dot{\mathbf{x}}_\mathrm{F} \\ \dot{\mathbf{e}} \end{pmatrix} = \begin{pmatrix} \mathbf{A} - \mathbf{B}\mathbf{K}_\mathrm{N}\mathbf{C}_\mathrm{N} & -\mathbf{B}\mathbf{K}_\mathrm{N}\mathbf{C}_\mathrm{N2} \\ \mathbf{O} & \mathbf{A}_{22} - \mathbf{L}\mathbf{A}_{12} \end{pmatrix} \begin{pmatrix} \mathbf{x}_\mathrm{F} \\ \mathbf{e} \end{pmatrix} + \begin{pmatrix} \mathbf{B}\mathbf{K}_\mathrm{N} \\ \mathbf{O} \end{pmatrix} \mathbf{w} \tag{18}$$

$$\mathbf{y}_R = \mathbf{C}_\mathrm{N}\mathbf{x}_\mathrm{F} + \mathbf{C}_\mathrm{N2}\mathbf{e} \quad , \tag{19}$$

where the matrix $\mathbf{L}$ has to be chosen appropriately. Two properties are noteworthy:

1. The separation principle states that the observer poles $(\mathbf{A}_{22} - \mathbf{L}\mathbf{A}_{12})$ and the controller poles $(\mathbf{A} - \mathbf{B}\mathbf{K}_\mathrm{N}\mathbf{C}_\mathrm{F})$ can be chosen separately. Therefore the reconfiguration task can be solved by the right choice of $\mathbf{L}$.
2. The observer poles are not controllable through the reference signal $\mathbf{w}$. Therefore, they do not appear in the input-output behaviour. This means that the tracking behaviour (static and dynamic) of the reconfigured control loop is identical to that of the nominal control loop.

The reduced-order observer can be designed on-line to solve the reconfiguration problem. The initial state of the observer should correspond to the plant state. If it is unknown, the observer state should be initialised to the equilibrium.

5.4 Reconfiguration in Case of Actuator Faults

An actuator fault affects the matrix $\mathbf{B}$ only. The exact reproduction of the original trajectory $\mathbf{x}_{\mathrm{N}}(t) = \mathbf{x}_{\mathrm{F}}(t)$ requires

$$\mathbf{B}_{\mathrm{N}}\mathbf{u}_{\mathrm{N}} = \mathbf{B}_{\mathrm{F}}\mathbf{u}_{\mathrm{F}}$$

which is typically impossible because $\mathbf{B}_{\mathrm{N}}$ has a lower rank than $\mathbf{B}_{\mathrm{F}}$ in the case of a severe fault.

The Virtual Actuator. The idea of the virtual actuator is to use a parallel model that simulates the difference between the nominal and faulty processes. The difference is subtracted from the process output, thus reproducing the input/output-behaviour of the nominal process. A series of state transformations reduces the order of the parallel model and guarantees internal stability (Steffen, 2001).

The state space has to be split into a part $\mathbf{x}_{\mathrm{F1}}$ that can be affected directly, and a part $\mathbf{x}_{\mathrm{F2}}$ that is not affected by $\mathbf{B}_{\mathrm{F}}\mathbf{u}_{\mathrm{F}}$. The goal is to bring the $\mathbf{B}$-matrices into the following form

$$\mathbf{B}_{\mathrm{N}} = \begin{pmatrix} \mathbf{B}_{\mathrm{N1}} \\ \mathbf{B}_{\mathrm{N2}} \end{pmatrix} \quad \mathbf{B}_{\mathrm{F}} = \begin{pmatrix} \mathbf{B}_{\mathrm{F1}} \\ \mathbf{O} \end{pmatrix} \quad , \tag{20}$$

where $\mathbf{B}_{\mathrm{F1}}$ is invertible. This transformation is always possible.

The virtual actuator is given by the following equations, where $\hat{\mathbf{x}}_2$ is the state of the virtual actuator and $\mathbf{M}$ a parameter matrix:

$$\dot{\hat{\mathbf{x}}}_2 = (\mathbf{A}_{22} - \mathbf{A}_{21}\mathbf{M})\hat{\mathbf{x}}_2 + \mathbf{B}_{\mathrm{N2}}\mathbf{u}_R \tag{21}$$

$$\hat{\mathbf{u}} = \mathbf{B}_{\mathrm{F1}}^{-1}(\mathbf{I} \;\; \mathbf{M})\mathbf{A}\begin{pmatrix} -\mathbf{M} \\ \mathbf{I} \end{pmatrix}\hat{\mathbf{x}}_2 + \mathbf{B}_{\mathrm{F1}}^{-1}(\mathbf{I} \;\; \mathbf{M})(\mathbf{B}_{\mathrm{N}} - \mathbf{B}_{\mathrm{F}})\mathbf{u}_R \tag{22}$$

$$\hat{\mathbf{y}} = (\mathbf{C}_2 - \mathbf{C}_1\mathbf{M})\hat{\mathbf{x}}_2 \tag{23}$$

The function of the virtual actuator can be explained in two parts:

1. The effects of the faulty actuators $(\mathbf{B}_{\mathrm{N}} - \mathbf{B}_{\mathrm{F}})$ are replaced by a the use of couplings within the process $(\mathbf{A}_{12})$. These couplings are slower, but this is partly compensated for by a differential behaviour of the virtual actuator.
2. Remaining differences are simulated by the virtual actuator and deducted from the process output so that the nominal input/output behaviour is restored.

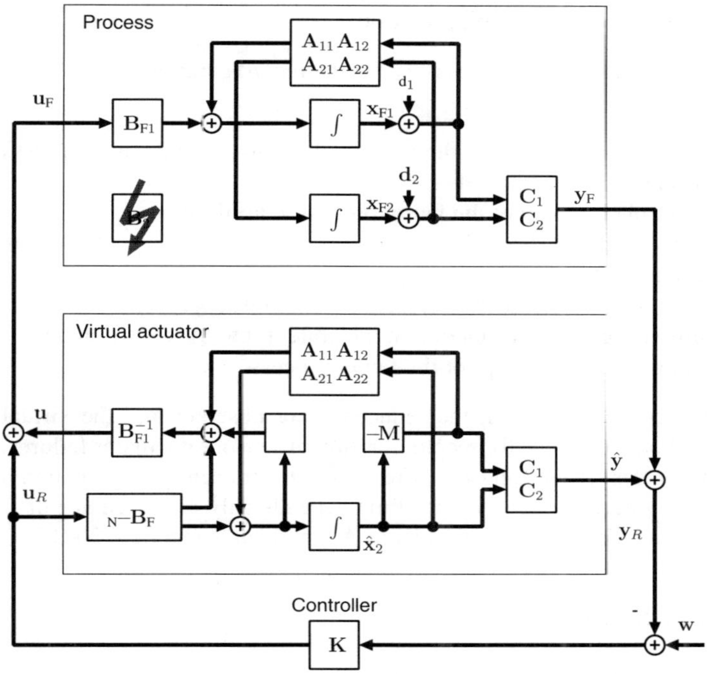

Fig. 7. Reconfigured control loop with virtual actuator

Behaviour of the Reconfigured Control Loop. After applying a state transformation, the behaviour of the reconfigured control loop can be given in a compact equation:

$$\begin{pmatrix} \dot{\tilde{\mathbf{x}}}_\mathrm{F} \\ \dot{\hat{\mathbf{x}}}_2 \end{pmatrix} = \begin{pmatrix} \mathbf{A} - \mathbf{B}_\mathrm{N}\mathbf{K}_\mathrm{N}\mathbf{C} & \mathbf{O} \\ -\mathbf{B}_\mathrm{N2}\mathbf{K}_\mathrm{N}\mathbf{C} & \mathbf{A}_{22} - \mathbf{A}_{21}\mathbf{M} \end{pmatrix} \begin{pmatrix} \tilde{\mathbf{x}}_\mathrm{F} \\ \hat{\mathbf{x}}_2 \end{pmatrix} - \begin{pmatrix} \mathbf{B}_\mathrm{N} \\ \mathbf{B}_\mathrm{N2} \end{pmatrix} \mathbf{K}_\mathrm{N}\mathbf{w} \tag{24}$$

$$\mathbf{y}_\mathrm{F} = \mathbf{C}\tilde{\mathbf{x}}_\mathrm{F} + \mathbf{C}\begin{pmatrix} \mathbf{M} \\ -\mathbf{I} \end{pmatrix} \hat{\mathbf{x}}_2 \tag{25}$$

$$\mathbf{y}_R = \mathbf{C}\tilde{\mathbf{x}}_\mathrm{F} \ . \tag{26}$$

It is obvious that the states of the virtual actuator are not observable via $\mathbf{y}_R$, which means that the input/output behaviour from $\mathbf{u}_R$ to $\mathbf{y}_R$ is completely restored (equal to that of the nominal plant). The poles of the virtual actuator $(\mathbf{A}_{22} - \mathbf{A}_{21}\mathbf{M})$ can be assigned independently of the poles of the controller $(\mathbf{A} - \mathbf{B}_\mathrm{N}\mathbf{K}_\mathrm{N}\mathbf{C})$. The virtual actuator poles do influence $\mathbf{x}_\mathrm{F}$ and $\mathbf{y}_\mathrm{F}$, but if fast poles are assigned via $\mathbf{M}$, the deviation from the nominal trajectory rapidly diminishes. The speed is limited only by actuator and state constraints.

The virtual actuator can be designed and added to the control structure on-line. To avoid jumps in the process input, the virtual actuator state should be initialised to zero.

5.5 Applicability for Control Reconfiguration

In summary, the following steps are necessary for the reconfiguration on the linear level.

Requirements: state $\mathbf{x}_F$ has to be near the new equilibrium and the faulty process has to be controllable and observable
Given: nominal controller, model of the faulty process, equilibrium
1 Calculate the linear model of the faulty process
2 Divide the state space into $\mathbf{x}_1$ and $\mathbf{x}_2$
3 Assign $\mathbf{M}$ or $\mathbf{L}$ using available knowledge, e.g. by an LQR approach
4 Initialise the observer or virtual actuator and integrate it into the control loop
Result: the reconfigured control loop for the faulty process

Due to the properties mentioned, the reduced order observer and the virtual actuator are the methods of choice for reconfiguration of sensor or actuator failures. The equilibrium $\mathbf{x}_F = 0$ is reached and any difference to the trajectory of the nominal system is temporary, where the convergence time depends only on the choice of $\mathbf{L}$ or $\mathbf{M}$. More details and the treatment of faults in $\mathbf{A}$ are given in (Steffen, 2001).

6 Application Example

6.1 Plant Model and Reconfiguration Problem

The reconfiguration approach has been experimentally tested at a titration plant with three tanks (see Fig. 8). The control objective is to maintain a constant level and a constant temperature in the reactor tank $B1$, leading to a constant outflow. To achieve this, hot and cold water can be brought into the reactor from tank $B2$ and $B5$. Heating and cooling are also available.

In the nominal case the level is controlled by adjusting the cold water inflow from tank $B5$, and the temperature is controlled via a pulse-width modulator acting on the heating.

The model of the plant contains three states: the reactor content V_{B1}, the reactor temperature ϑ_{B1} and the content of the cold water tank V_{B5}:

$$\begin{aligned}
\dot{V}_{B1} &= q_{21} + q_{51} - q_{1out} \\
\dot{\vartheta}_{B1} &= (\vartheta_{B2} - \vartheta_{B1})\frac{q_{21}}{V_{B1}} + (\vartheta_{B5} - \vartheta_{B1})\frac{q_{51}}{V_{B1}} + \frac{u_{heat}k_{heat}}{V_{B1}} \\
\dot{V}_{B5} &= k_{P2}u_{P2} - q_{51}
\end{aligned}$$

with

$$\begin{aligned}
q_{21} &= k_{P1}u_{P1} \\
q_{51} &= k_{V1}\,124.5^{u_{V1}}\sqrt{h_{B5} + 1.07} \\
q_{1out} &= k_{V2}\sqrt{\frac{V_{B1}}{A_{B1}} + 1.4} \quad ,
\end{aligned}$$

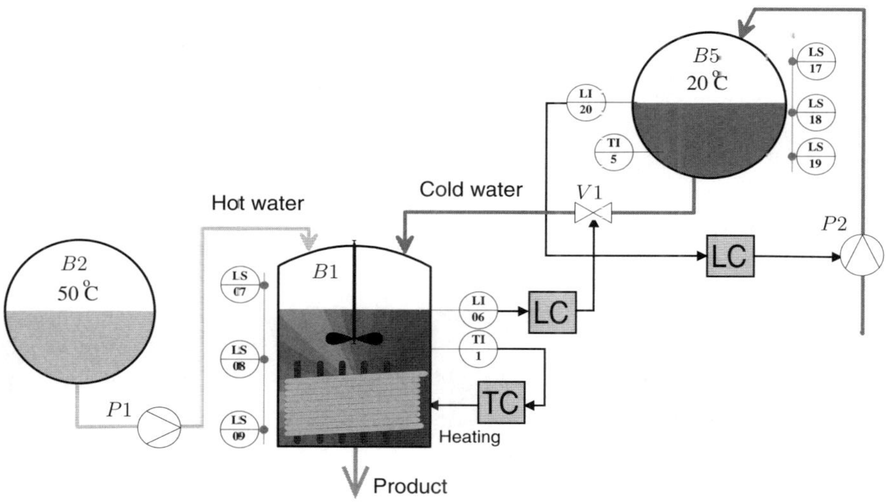

Fig. 8. Relevant part of the titration plant

where h_{B5} is the level in the spherical tank $B5$. Several limits, security interlocks, preconditions and constants apply, which are not given here.

The original proportional controllers are defined by:

$$u'_{V1} = -0.5\, V_{B1} \;,\; u'_{heat} = -0.5\, \vartheta_{B1} \;,\; u_{P2} = -1\, V_{B5}$$

where u'_{V1} is an input linearised variant of u_{V1} and u'_{heat} controls the discrete heating via a pulses-width modulation (PWM).

6.2 Fault Cases

Several different faults have been tested:

1. Fault in the heating
2. Blockage of valve u_{V1} near nominal position
3. Heating cannot be switched off
4. Clogging of valve u_{V1}
5. Increased temperature in $B5$

The second case will be presented here, because it is the most difficult fault requiring a thorough treatment on both levels.

6.3 Solution to the State Transition Problem

Based on the discretisation for states and inputs shown in Tables 1 and 2, an optimisation criterion is defined that is used for all fault cases. The goal region **G** is the

Table 1. Discretisation for states

State	1	2	3	4	5
V_{B1} in dm^3	$12-14$	$14-18$	$18-22$	$22-26$	$26-30$
ϑ_{B1} in °C	$20-36$	$36-44$	$44-60$		
V_{B5} in dm^3	$6-8$	$8-9.8$	$9.8-10.8$	$10.8-14$	

Table 2. Discretisation for inputs

Input	1	2	3	4	5
u_{P1}	0	1			
u_{P2}	0	0.18	0.185	0.20	1
u_{V1}	0	0.7	0.95	1	
u_{heat} in kW	0	1.5	3	6	

nominal state $(3, 2, 1)$, and the important forbidden areas are that $B1$ or $B5$ become empty. The optimisation horizon is $h = 10$. The resulting trajectories of the system with the optimal discrete controller are shown in Fig. 9. The inputs used for u_{P2} by the discrete controller $v = R(z)$ to achieve these trajectories are written next to the states.

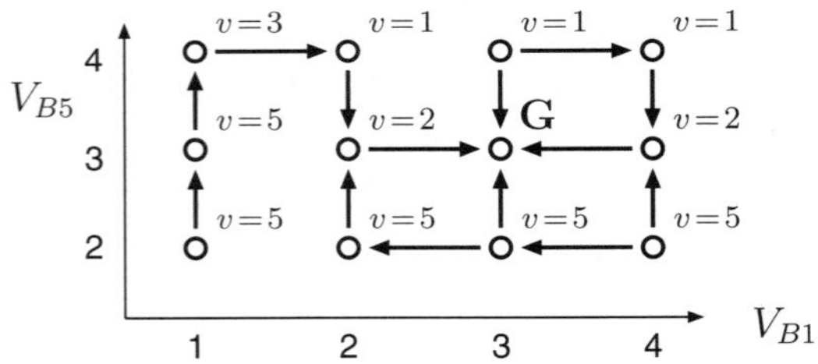

Fig. 9. Discrete controlled trajectories in the state space z

It should be noted that the reachability probability is exactly 1, which means that the transition is guaranteed to be successful. The highlighted trajectory shows that the level of tank $B5$ is temporarily increased. This reduces the transition time significantly compared to the alternative of leaving $B5$ in the nominal interval.

6.4 Linear Controller Reconfiguration

The loss of the actuator u_{V1} does not effect the equilibrium, but it breaks the level control loop for V_{B1}. A common approach would be to design a multi-variable

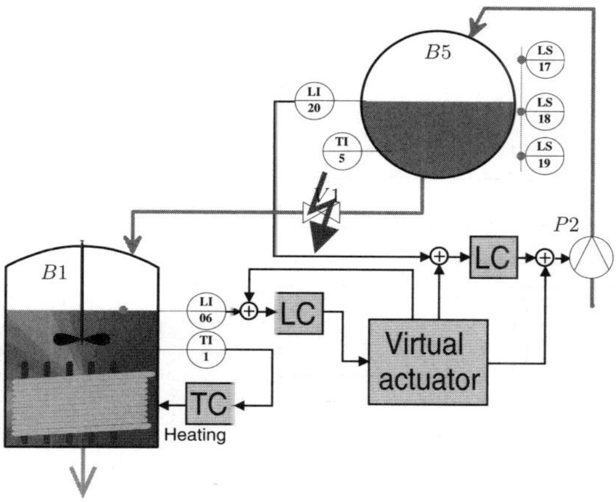

Fig. 10. Use of the virtual actuator to replace u_{V1}

controller for V_{B1} and V_{B5} using u_{P2}, which allows independent pole placement. However, the use of a virtual actuator instead of u_{V1} allows keeping the nominal controller while changing the control structure as little as possible.

The directly influenceable part $\mathbf{x}_{\mathrm{F1}}$ of the plant state is defined by V_{B5} and ϑ_{B1}, while $\mathbf{x}_{\mathrm{F2}}$ is V_{B1}, cf. (20). The parameter $\mathbf{M}$ (a 1×2 matrix) is determined by pole placement. The element of $\mathbf{M}$ that is acting on ϑ_{B1} has no influence on the actuator pole and is therefore set to 0. The other value is set so that the actuator pole is moved to -0.004, which is the limit of acceptable performance in the fault case. Applying equations (21)–(23) to this example leads to

$$\dot{\hat{x}}_2 = -0.004\,\hat{x}_2 + 0.0229\,u_{V2,R} \tag{27}$$

$$\hat{\mathbf{u}} = \begin{pmatrix} 0.015 \\ -0.318 \\ 0 \end{pmatrix} \hat{x}_2 + \begin{pmatrix} -0.107 \\ 1.78 \\ 0 \end{pmatrix} u_{V2,R} \tag{28}$$

$$\hat{\mathbf{y}} = \begin{pmatrix} -8 \\ 0 \\ 1 \end{pmatrix} \hat{x}_2 \ . \tag{29}$$

The function of the virtual actuator can be described as follows (cf. Fig. 10). The input u_{V1} is not available to control the inflow into the main reactor, but this inflow also depends on the level V_{B5}. In order to reach the same effect that the broken actuator had, V_{B5} is increased or decreased by using u_{P2} as necessary. V_{B5} cannot be changed instantaneously. So, this "replacement action" is slower than the nominal control loop, leading to an addition pole ($\hat{\mathbf{x}}_2$). The difference in behaviour is determined by the virtual actuator and deducted from the measurement V_{B1} and V_{B5}. This way, the additional pole remains hidden from the controller.

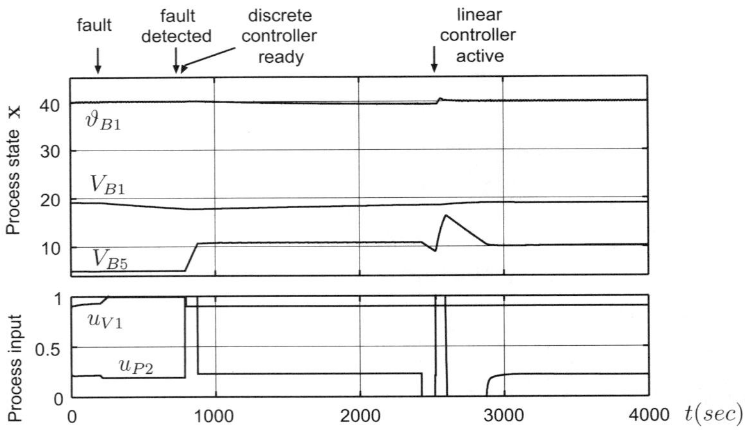

Fig. 11. Reconfiguration simulation

6.5 Results

The simulation results for the reconfiguration of the titration plant shown in Fig. 11 consists of 5 different phases of the controller reconfiguration. After the process has been operating normally for 200 seconds, the actuator u_{V1} gets stuck. The effect on the state is rather small. Hence the fault is not detected before $t = 780\,\text{s}$. It is assumed that 10 seconds later, the discrete controller design is completed and the new controller is activated. The level V_{B5} is increased, so that more water flows into $B1$. The nominal partition is reached at about 2500 seconds and the virtual actuator activated. The remaining state deviation is eliminated in a short time, after which the plant is stabilised at the new equilibrium. In this example, the operation of the plant can be fully restored with only a minor performance degradation.

7 Conclusion

A hybrid approach to control reconfiguration has been described. On the discrete level, the transition of the faulty system state to the new equilibrium state was found as the solution of a discrete optimisation problem. It can be solved in an efficient way, and it allows the treatment of hybrid or strongly non-linear systems.

On the continuous level, a minimal extension of the control structure is proposed for treating the fault. The virtual actuator provides a reconfiguration option of actuator faults similar to the use of a reduced order observer for sensor faults. Due to their advantageous properties, the reduced observer and the virtual actuator seem to be the methods of choice for linear control reconfiguration.

Taken together, the two methods allow the treatment of a detected fault automatically without manual intervention and without having pre-designed controllers. Therefore, this approach helps to increase the availability of a plant with little design effort.

Automatic Design of Controllers for Hybrid Systems Using Genetic Algorithms

Stefan Wegele[1], Eckehard Schnieder[1], and Mourad Chouikha[2]

[1] Institute for control and automation engineering,
TU Braunschweig, Langer Kamp 8,
38106 Braunschweig

[2] Extessy AG,
Hagenring 59,
38106 Braunschweig

Abstract. In this article a design method for robust control of hybrid systems is presented. It uses optimisation algorithms to choose and parametrise an appropriate control algorithm from a formalised library of control algorithms. To prove the robustness of the control chosen a verification technique based on optimisation is used. Both techniques – verification and design – are demonstrated using two examples.

1 Introduction

The automated control design is a very challenging task because of a great variety of hybrid systems. A universal control algorithm for all kinds of hybrid systems that would fulfil all possible quality conditions does not exist. In this article we show how well-known existing algorithms could be used for automated control design.

For this purpose we need a definition of hybrid systems. Any hybrid system could be described by two mathematical models: continuous and discrete. There are many description means for such models (Henzinger, 1996). In this article we model hybrid systems using hybrid automata. For every discrete state of the automaton a system of differential equations is defined (Fig. 1).

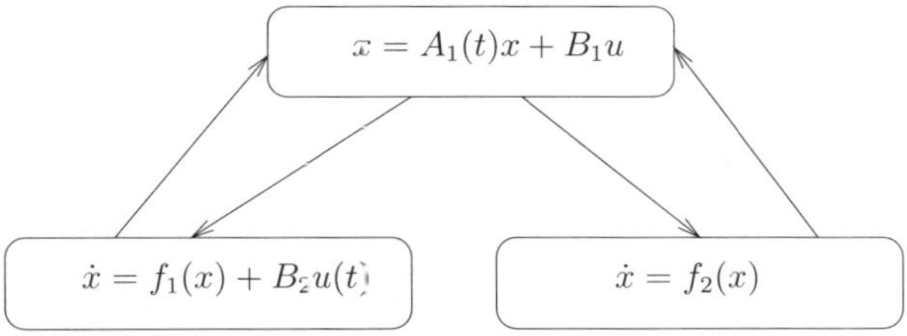

Fig. 1. Example of a hybrid system modelled by means of hybrid automaton

S. Engell, G. Frehse, E. Schnieder (Eds.): Modelling, Analysis and Design of Hybrid Systems, LNCIS 279, pp. 285–294, 2002.

2 Control Design Approach

At the beginning of the control design the following information is given:

- Model of a hybrid system with known inputs and outputs
- design specification. It consists of
  1. parametrised design aim: $f_0 \to \min$
  2. parametrised forbidden area:

     $$F(x) < b$$

     where $x = (x_1, x_2, \ldots, x_n)^T$ - state vector , F - vector of functions, and $b = (b_1, b_2, \ldots, b_m)^T$. They define m conditions for a forbidden area.
- Library with formalised control design methods (e.g. H_∞, PID)
- Verification module
- Optimisation module

First the design aim is defined by the functional

$$f := f_0 \to \min .$$

For the given inputs and outputs some design methods are chosen from the control design library. Using the model of the hybrid system and the functional the design method creates some control algorithm. This algorithm is coupled with the hybrid system and the controlled system is verified by the verification module. If a forbidden area can be reached the initial conditions, disturbances, etc. are transported to the optimisation method and a penalty function f_1 is added to the previous functional:

$$f := f + f_1 \to \min$$

The optimisation method minimise the functional by changing the design parameters of the design method (e.g. the parameters of PID controller or weighting functions in H_∞-design). If the forbidden area can not be avoided (there are such disturbances, that "move" the system into forbidden area), the chosen design method is not robust and another design method must be chosen from the library.

The following sections describe the modules mentioned above.

3 Verification as a Control Design Tool

An analysis of control for linear systems examines several parts:

- stability of the modelled closed loop
- performance of defined kinds of disturbances in some ranges
- robustness of defined ranges of the system parameters
- avoidance of forbidden system states

For well modelled linear systems with nearly constant parameters it is relatively simple to verify some of these points: disturbances influence the system in a linear way. It's sufficient to show the robustness of the system for only one disturbance to know the system behaviour for all disturbances of the same kind. For hybrid systems it's totally different. For a chosen control algorithm you have to test almost the whole state space of the system to be sure that it does not come into any forbidden area. It is not possible because of enormous state space of discrete and continuous subsystems. The common way to handle this problem is to test some relevant system states and to hope that the controlled system is robust enough in the other possible states. Normally one decides which states should be tested and if a problem occurs, one modifies the control algorithm (e.g. by changing some parameters). This task could be automated using automatic verification. A verification algorithm would search in the state space for forbidden areas. The areas found would be communicated to another module. This would modify the control algorithm, until the system remains in the allowed states. So the following modules are needed (Fig. 2). To be able to

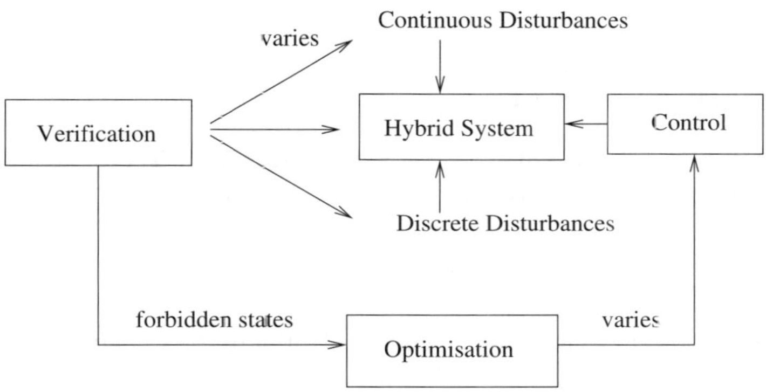

Fig. 2. Structure of the automated control design

automatically verify a hybrid system all the uncertainties of the system and input-signals must be parameterised. A search through all the parameter ranges answers the verification question: if the system enters a forbidden area. In this formulation (search) the verification can be substituted by optimisation (Chouikha, 1999) with a functional, that describes the distance of the system to the forbidden area:

$$f = \|x(t) - S(t)\| \longrightarrow \min_{\Omega,t}$$

with x - current system state
S - forbidden area of the system state space
Ω - uncertainties of the system (disturbances, parameter ranges etc.)
t - time

The verification process can be shown as in Fig. 3. The optimisation algorithm tries to "move" the system into the forbidden area by means of the system uncertainties. The constraints can be evaluated for the solution of the optimisation problem. If they are violated - the forbidden area is reachable. In this case the system parameters which led the system to the forbidden state can be transported to the control optimisation module.

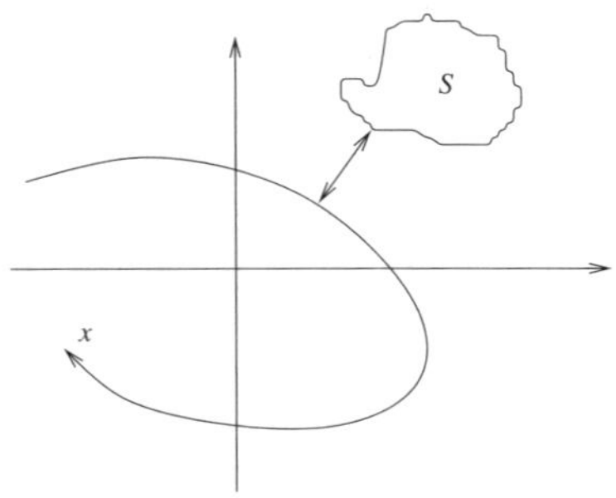

Fig. 3. Minimisation of the distance between the current system state and the forbidden space

3.1 Example of an Engine Verification

As an example consider a control loop of a car diesel engine (Fig. 4). This system is hybrid because it has continuous behaviour:

$$\frac{K_1 s}{a_2 s^2 + a_1 s + a_0} + \frac{K_2}{T_1 s^2 + s}$$

and discrete behaviour (gear level, switching control algorithm). As a control algorithm a PI-controller is used with DT1-Block which would be used during a short period immediately after a jump in the control input. There are big uncertainties in the model parameters because of fabrication tolerances, wear and tear, different

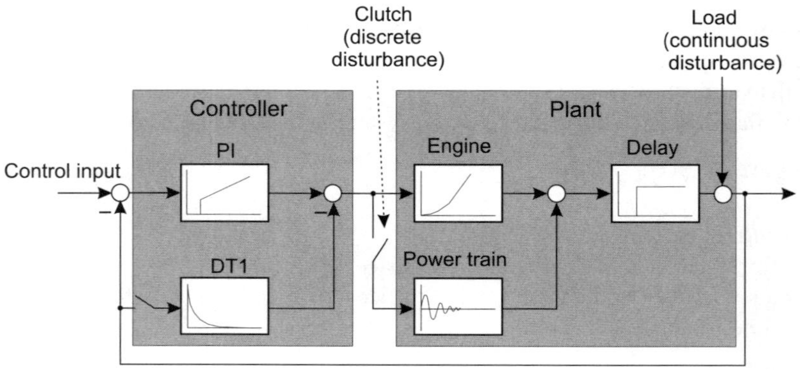

Fig. 4. Closed loop control of the engine

temperatures etc. The system was modelled by a system of differential equations of the order 7. The uncertainties could be defined by simple parameter ranges e.g.

$$K_i = 250 \pm 20\%.$$

with K_i - engine parameter. There are two kinds of disturbances:

- jump $s(T)$ by the air condition at some point of time T_{jump}
- sinus $A\sin(\omega t + \varphi)$ by e.g. indicator

The model parameters with disturbances are defined in the same way:

$$\begin{aligned} T_{Jump} &= [0, 1000] \\ \omega &= 1 \pm 20\% \\ A &= [-45, 45] \end{aligned}$$

In this example discrete events occur as the gears are changed. They can be modelled by a vector of pairs {event, time interval}, e.g. Fig 9. In this way, the verification routine has all the degrees of freedom to find out the worst case. There are several forbidden scenarios for an engine:

- engine stall
- oscillation of rotational speed

For every scenario a forbidden area of the state space can be defined:

$$\begin{aligned} \text{engine stall} &\Longleftrightarrow \omega = 0 \\ \text{oscillation of rotational speed} &\Longleftrightarrow a_i \neq 0, \ \text{for some } i \geq 1 \end{aligned}$$

where a_i denotes any coefficient of the Fourier series:

$$\omega(t) = \sum_{i=0}^{n} a_i \sin(ift + \varphi_i)$$

The definition of the verification by means of optimisation is:

$$\omega \rightarrow \min_{\Omega, t}$$

or

$$a_i \rightarrow \max_{\Omega}$$

If at the global optimum the system remains in the allowed state area - the controller is robust. The question is: what optimisation algorithm is able to verify many kinds of hybrid systems? In practical applications the state space of a system is too large to be tested by brute force method. A good solution in this case could be an evolutionary algorithm e.g. a genetic algorithm (Schöneburg et al., 1996, Michalewicz and Fogel,

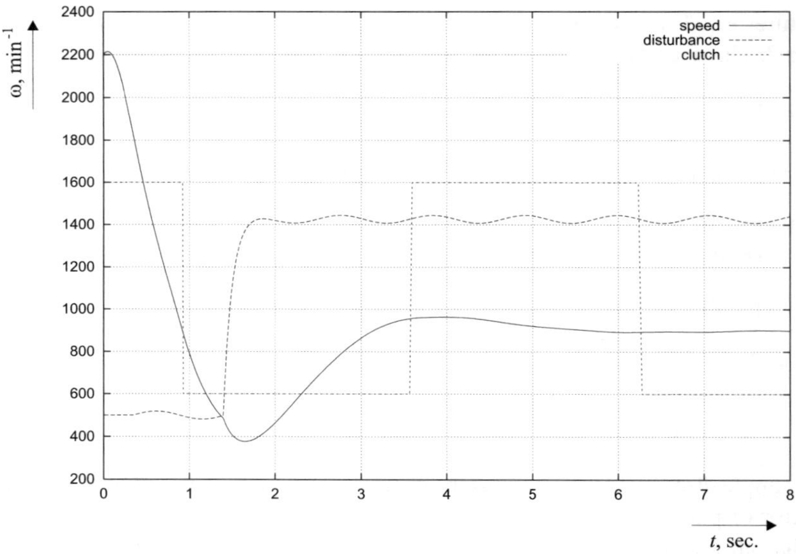

Fig. 5. Verification result of the controlled diesel engine

2000). In the case of the above example the following result was produced (Fig. 5). In this example it was verified if the engine can be stalled by different disturbances and system parameters (efficiency factor, temperature etc.). The results are obvious because the system is relative simple. The worst case can be seen in the diagram:

- The air conditioning switches on immediately before the lowest speed level is reached
- Discrete state is "idle"= clutch is off (because the system is very quick in this state)

The forbidden area begins at 600 rotations/min and during the verification 370 is reached. Hence the control algorithm is not robust and must be modified.

4 Design of Control

4.1 Interface of Control Algorithms

As shown in Fig. 6 an optimisation algorithm receives the information about the violations of allowed system state space and modifies the control algorithm to avoid such violations. To be able to solve this task automatically a formalisation of the transported information is needed. It is simple to formalise the information about the worst-case. It consists of

- initial conditions
- system parameter
- disturbances.

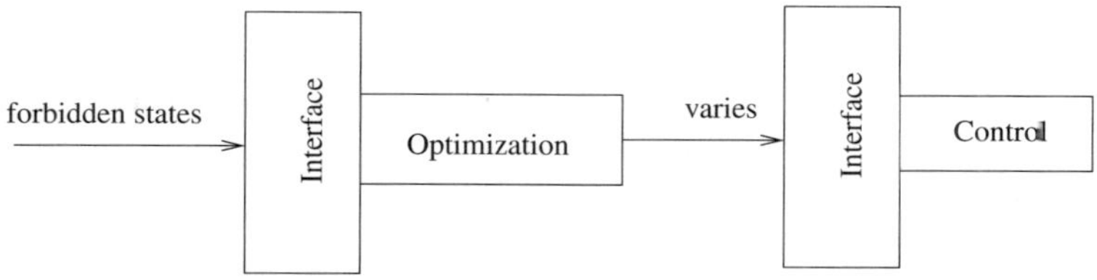

Fig. 6. Structure of the automated control design

To enable modifications all control algorithms must share the same modification interface. Some examples:

- *PID-control* has three parameters, which can be modified independently
- H_∞ *and* H_2*-control* use a linear model of the system, parameter ranges and weighting functions (Müller, 1996).
- *Linear state controller* can be defined by poles of the characteristic polynomial.

The interface of any control algorithm can be defined by a number of parameter within some ranges. It's possible to set any achievable control behaviour if there is enough time to optimise these parameters.

4.2 Simple Discrete Open-Loop Control

In this section a design method of open-loop control for hybrid systems is shown. The main idea here is to create a basis control for an undisturbed precisely modelled system and during verification to improve this algorithm by some simple control algorithms (e.g. P or PID). The strategy is the same as for the system verification. It is assumed that the initial state and the target state of the system are known. The task is to find such signals which would lead the system from the initial state into target state. This search can be substituted by optimisation with a following functional:

$$f = \|x - T\| \longrightarrow \min_{\Psi}$$

T - target state space
x - current system state
Ψ - set of the allowed control possibilities

All the control outputs must be parameterised. In this case the degrees of freedom are the same as for verification. These parameter are varied by the optimisation algorithm. As an instance of such control design an example of hybrid system provided by the Institute for Control at the TU Hamburg-Harburg (Nixdorf and Lunze, 2000b) should be considered (Fig.7).

At the beginning there are four metal discs in the stack. A crane moves the discs to the heating and to the water tank. The aim is to increase the water temperature by

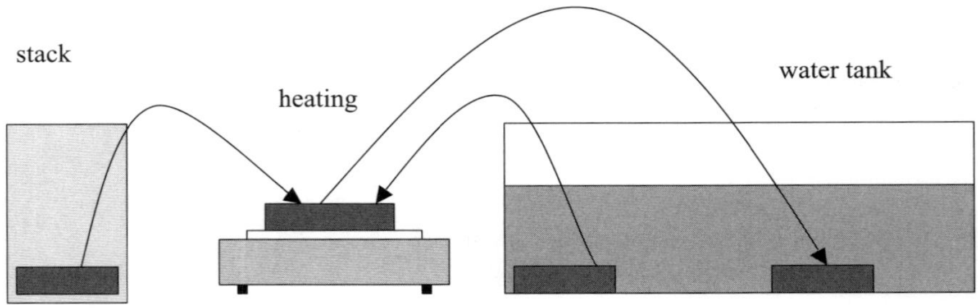

Fig. 7. The hybrid system to be controlled

four degrees in exactly one hour. The system can be influenced by the crane only. Hence the control algorithm is discrete. It should send discrete signals to the crane at the right points of time to achieve the aim. At this point it is assumed that there are no disturbances and the system parameter are exactly known. The formulation of the optimisation aim is:

$$f = |t_{end} - 4| \longrightarrow \min$$

with t_{end} - increment of the water temperature after one hour. There are nine discrete signals at the controller's disposal to enforce the minimum of the functional:

- 1: move disc from stack to heating
- 2-5: move disc from heating to the water tank at the position 1..4
- 6-9: move disc from water tank at position 1..4 to the heating

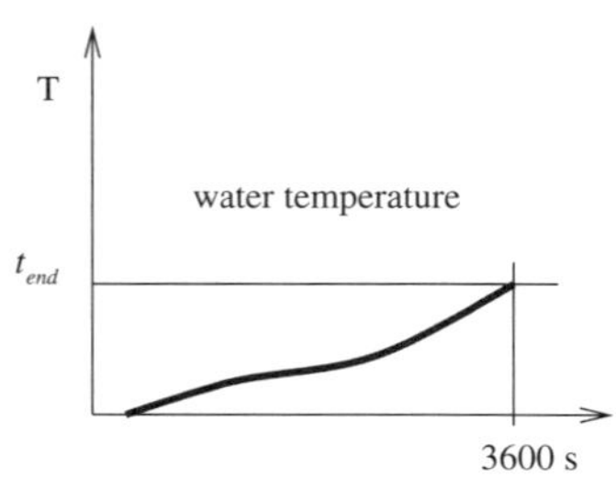

Fig. 8. Definition of t_{end}

$O_1 = [1..9]$	$t_1 = [0..3600]$	$O_2 = [1..9]$	$t_2 = [0..3600]$	...

Fig. 9. State space definition as a vector

The controller should send this instructions to the crane at the right point of time. Therefore the state space formulation is a vector (Fig.9). This vector is some kind of a cookery book: crane interprets the instruction O_i and then it waits t_i seconds until it goes to the next instruction. Defined in this way the optimisation problem can be solved by genetic algorithm. The temperature functions for every of the four discs are shown in the Fig. 10.

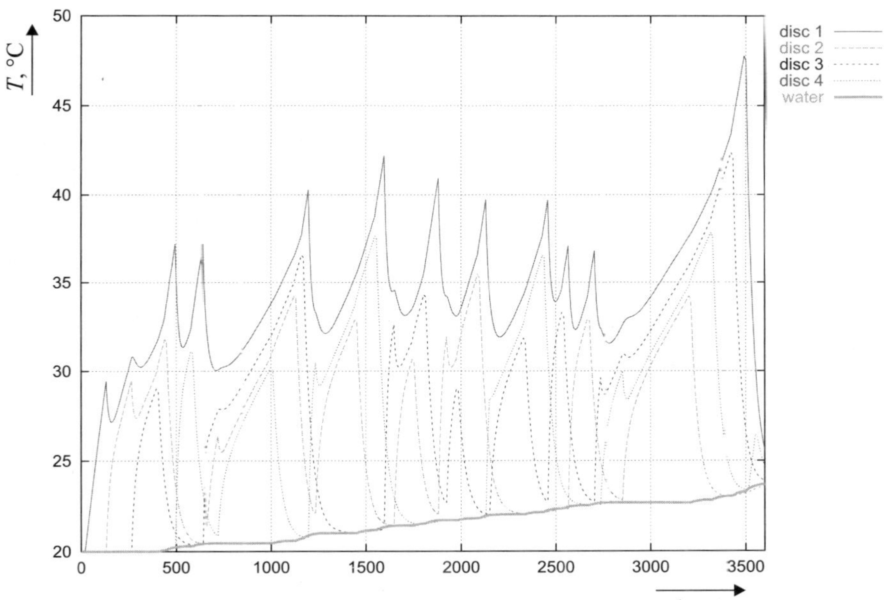

Fig. 10. Structure of the automated control design

The required water temperature is reached at the end of an hour. The whole optimisation takes about 3 minutes on 400 MHz. Pentium computer. The length of the vector (Fig.9) is set to 300 value pairs. But only a small part of it is used because of the time range of one hour. The genetic algorithm is a stochastic optimisation method, that's why it can produce some illegal (senseless) instructions. If there was a senseless instruction in the vector (e.g. move a disc from an empty place) it was ignored. This strategy is similar to recessive/dominant genes in the nature: senseless instructions are recessive and useful ones are dominant genes.

The modification interface of this design method can be defined by means of functional. To enforce the avoidance of a forbidden area a penalty function can be added to the functional

$$f_1 = f_0 + kf(\|x - S\|) \longrightarrow \min$$

where f_0 is the old functional, S - forbidden subspace, x - current system state. The forbidden subspace can be the max allowed temperature of the discs e.g.: $t_i \leq 100$. The verification module has to find the violated forbidden states and to enforce the correction by changing of the weighting factor k.

In this design method the same optimisation algorithm is used for both tasks: verification and control design. In this case both tasks could be integrated into one functional as a sum of control aim and weighted penalty functions for violations of forbidden areas.

5 Conclusions

In this paper we show how to design robust controllers for hybrid systems without creating any new control algorithms. Every control algorithm is robust for particular kinds of systems, and the task is to find and set up such an algorithm from the library of control algorithms. To ensure its robustness, a verification method is used. It tests if the controlled system can reach any forbidden area of the state space. An iterative improvement of the control algorithm is achieved by variation of design parameters. For this purpose the design parameters must be formalised and stored together with the design algorithms in a library.

Synthesis of a Discrete Control for Hybrid Systems by Means of a Petri-Net-State-Model

Christian Müller[1], Philipp Orth[2], Dirk Abel[2], and Heinrich Rake[2]

[1] ABB Corporate Research, Ladenburg, Germany
[2] Institute of Automatic Control, RWTH Aachen University, Aachen, Germany

Abstract. The design of discrete control systems for technical processes leads to hybrid systems. These can be modelled by a combination of several Petri Nets and extended state space models with appropriate interfaces between them. For the analysis of the whole system, the introduced evolution graph is used to describe hybrid reachability, including generally infinite converging cycles by a covered graph. The analysis results are utilised to synthesise minimal extensions to the control system to assure desired system behaviour like liveness or reversibility. The procedure is illustrated by a small example.

1 Introduction

This contribution focusses on discretely controlled hybrid systems. A combination of continuous technical processes and discrete control is found in most cases, e.g. sequential control. The aim is to support the design of such a discrete control with formal methods.

A suitable model of the hybrid system is required for control design, able to describe the process itself and also the control system. In the *Petri-Net-State-Model* Petri Nets are used for modelling the discrete parts of the hybrid system while switched differential equation systems as extended state space models represent the continuous parts of the hybrid system. These two parts can be coupled via appropriate interfaces. That way, large systems with several continuous and discrete parts can be modelled. Each partial model can interact with every other model, continuous or discrete, using the interfaces.

To analyse the behaviour of the system, the evolution graph is used for the Petri-Net-State-Model. This graph is the *hybrid* equivalent to the *reachability graph* of discrete systems. For this reason the classical analysis methods of the graph theory can be applied to hybrid systems, too. The evolution graph provides information e.g. about liveness, reversibility and dead transitions. If some of these properties are lacking in the hybrid system – or at least in the relevant parts of the system describing normal operation conditions – the desired system properties may not be achieved in all situations.

A precondition for the applicability of the analysis procedure is the finiteness of the corresponding evolution graph. In the case of interactions between continuous and discrete parts which are leading to convergence into a stable or limited working cycle, a covered evolution graph can be used. This graph needs only a small amount

S. Engell, G. Frehse, E. Schnieder (Eds.): Modelling, Analysis and Design of Hybrid Systems, LNCIS 279, pp. 295–310, 2002.

of nodes to describe infinitely long transient behaviour. That way generally *unlimited systems* in the classic sense of reachability can be dealt with.

Based on a method to synthesise control corrections (Seiche, 1991, Seiche and Abel, 1993), which is applicable only on pure discrete systems, a synthesis procedure for hybrid systems is proposed. This procedure uses the results of a system analysis to attain the desired system behaviour by minimal extensions to the control. The application of the procedure will be demonstrated by an example of a simple manufacture.

2 Hybrid Model

There are many approaches to model hybrid systems (Antsaklis and Koutsoukos, 1998, Labinaz et al., 1996, David and Alla, 1994, Krebs and Schnieder, 2000). The models differ with respect to complexity of the considered continuous and discrete dynamics and with respect to emphasis on simulation, analysis or synthesis. For modelling it is useful – if not even necessary – to differentiate clearly between the process on one side and the control on the other side, if emphasis lies on the design of a discrete control for a hybrid system. Thus the discretely controlled hybrid system is divided into – possibly several – continuous and discrete event subsystems for modelling, building together the Petri-Net-State-Model.

The discrete event subsystems are modelled by Place/Transition nets (P/T nets) (Abel, 1990, David and Alla, 1994) while the continuous subsystems are represented by switched differential equations (Champagnat et al., 1996, Champagnat et al., 1998). The subsystems are connected via simple binary signals connecting the interfaces integrated into the subsystems. Thus any continuous and any discrete submodel can be coupled with each other and modular and efficient modelling of large systems is possible (Müller and Rake, 1999). A simple Petri-Net-State-Model consisting of one discrete event and one continuous subsystem is depicted in Fig. 1.

The internal dynamics of the discrete system are modelled by a common P/T net $\boldsymbol{N}$. This net is extended by input and output places. To treat the binary input signals in the input vector $\boldsymbol{v_I}$, the net is extended by the input place set $\boldsymbol{I} = \{i_1, i_2, ..., i_n\}$ with a marking always representing the corresponding binary input signal. These input places can only be connected to the net by self-loops since they represent external and *read-only* firing conditions. For the sake of clearness, the output places in the figures are redundant to places of $\boldsymbol{N}$, as the marking of $\boldsymbol{N}$ describes the state of the Petri Net completely. To create the output signals, combined in the output vector $\boldsymbol{v_O}$, the net is extended analogously to the input by the output place set $\boldsymbol{O}$. Every output place has to be 1-safe to represent the signals 1 and 0.

To model the internal continuous dynamics of the continuous subsystems, switched differential equations, represented via an extended state space model, are used:

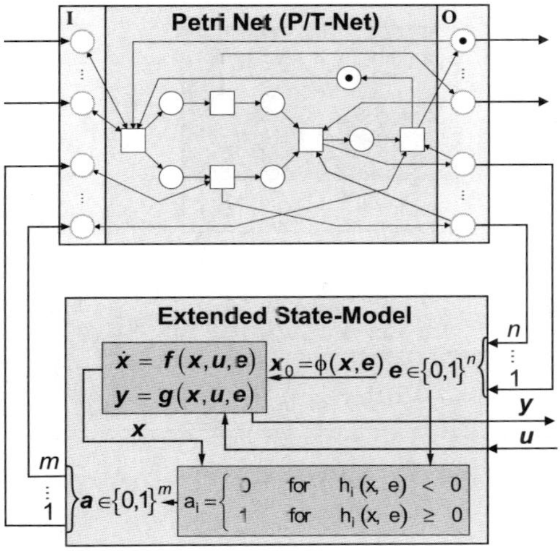

Fig. 1. Petri-Net-State-Model

$$x_0 = \Phi(x(t),\, e(t)) \,, \tag{1a}$$

$$\dot{x} = f(x(t),\, u(t),\, e(t)) \,, \tag{1b}$$

$$y = g(x(t),\, u(t),\, e(t)) \,, \tag{1c}$$

$$a = h(x(t),\, e(t)) \,. \tag{1d}$$

The vector x represents the continuous dynamic states, u is the continuous input vector and y the continuous output vector. A change of the discrete input signals e causes the system dynamics to vary including a possible change of the dimension of the state vector x. Discontinuities of the state variables can be modelled by the mapping Φ by reinitialising the system to the initial state x_0. For communication with other, especially discrete submodels, the binary output signals a are obtained by the binary threshold function h.

The *hybrid state vector* x_h of the system can be represented by the discrete marking M consisting of the markings of every single place and (here separated by a vertical bar) the continuous state vector x

$$\begin{aligned} M &= \left(M(p_1)\; M(p_2)\; \ldots\right)^T , \\ x &= \left(x_1\; x_2\; \ldots\right)^T , \\ x_h &= \left(M^T \mid x^T\right)^T . \end{aligned} \tag{2}$$

A simple example for the modelling of a hybrid system with the Petri-Net-State-Model is shown in Fig. 2. A valve at the outlet switches from one tank to another if a certain level is reached. This can be modelled with a net of two alternating places where the momentarily opened outlet is represented by the place with a marking of

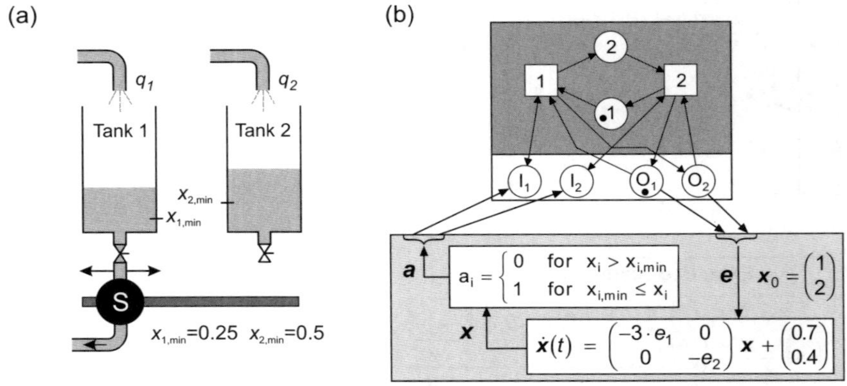

Fig. 2. Example of Two Tanks – Scheme (**a**) and Petri-Net-State-Model (**b**)

one. The continuous dynamics are characterised by variable outflow and constant inflow. Starting at $\boldsymbol{x}_0$ there are no discontinuities in the trajectories, as there are no reinitialisations caused by a function $\boldsymbol{\Phi}$. When a tank level reaches the minimum level, the corresponding output signal a_i equals one and the Petri Net 'wakes up'. Then the enabled transition in the Petri Net fires, the marking of the net changes and the outlet switches to the other tank. Afterwards there is no enabled transition left in the net as the components of the vector $\boldsymbol{a}$ cannot equal one simultaneously, so the treatment of the Petri Net is finished. The whole discrete process in the Petri Net happens instantly. The continuous dynamics have changed because of the altered vector $\boldsymbol{e}$ and the signal a_i is reset to zero, since the tank level is now above the minimum level.

In classical Petri Nets the moment an enabled transition fires is not determined. An enabled transition may fire but does not have to. To guarantee *deterministic behaviour* of the whole system, every transition has to fire as soon as it is enabled – this agreement was already used for the explanations concerning the example above. The duration of firing is assumed to be zero, so the continuous modules encounter changes in $\boldsymbol{v_O}$ after all – possibly sequentially – enabled transitions have fired. This means that only after a change of the input vector $\boldsymbol{v_I}$ to the net – which is equivalent to a change in the corresponding binary threshold functions $\boldsymbol{h}$ – a transition may be enabled again. In other words, the marking of the Petri Net $\boldsymbol{N}$ may change only when a new input signal v_i occurs. These changes then take place immediately.

This leads to the definition of *Invariant Behaviour States* (IB-states). Beginning with the firing of all enabled transitions in a so-called maximal step (Müller, 2002), a change in the discrete parts of the system results in constant output vectors $\boldsymbol{v_O}$ of the Petri Nets and constant inputs $\boldsymbol{e}$ to the continuous parts, at least for a while. As the switching of the switched differential equations $\boldsymbol{f}$ and $\boldsymbol{g}$ is dependent on the input vector $\boldsymbol{e}$, the differential equations remain unaltered only for constant input. Their behaviour remains constant until a change in the threshold functions $\boldsymbol{h}$ results in a change of the discrete marking of the Petri Net. The altered marking leads to a new input vector $\boldsymbol{e}$ to the continuous system. The invariant behaviour of the dynamics in the continuous part is related to the momentary discrete marking. An IB-state is characterised by the discrete marking and the initial continuous state, giving information about the whole succeeding behaviour of the state variables which is determinable because the differential equations are known. The duration Δt of an IB-state and the changes of the binary outputs $\boldsymbol{a}$ at the end of an IB-state are therefore also known. The term *state transition* will denote the transition between two IB-states in the following.

3 Analysis of the Petri-Net-State-Model

3.1 Evolution Graph

All IB-states which can be reached from a given initial hybrid state $\boldsymbol{x}_{h_0}$ form the hybrid reachability set. The reachability set can be represented by a directed graph, the evolution graph E_h, where the nodes correspond to the reachable IB-states and the arcs represent the transitions between the IB-states (David and Alla, 1994). A node of the evolution graph is shown in Fig. 3. The lighter left side of the node (IB-state) signifies the corresponding marking $\boldsymbol{M}$ of the Petri Net during an IB-state, while the darker right side contains the continuous state $\boldsymbol{x}$ at the beginning of an IB-state. The arcs are labelled with the duration Δt of the preceding IB-state, the signals a_i causing the state transition and all Petri Net transitions t_j firing consequently in a maximal step.

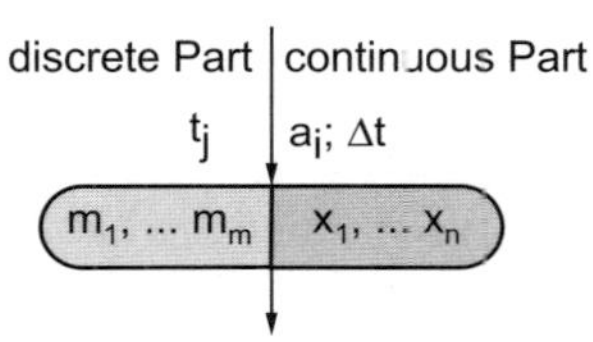

Fig. 3. Example for a Node of the Evolution Graph

The evolution graph is the hybrid equivalent to the reachability graph for discrete event systems. Therefore graph-theoretical methods for the analysis of Petri Nets can be principally applied to hybrid systems, too. Because of the integration of time into the evolution graph it is also possible to consider the continuous behaviour. Dead transitions of the Petri Net, dead output signals a_i and total deadlocks of the hybrid system can be detected immediately in the evolution graph (Müller and Rake, 1999, Müller and Rake, 2000). Branches in the evolution graph arise only from classical conflicts of several transitions in the discrete event system.

3.2 Covered Evolution Graph

The reachability analysis of a hybrid system and the synthesis procedure described later on is based on the finiteness of the evolution graph. In general this is not the case.

But often and especially technical systems show transient behaviour converging into stable or finite cycles. An example for this is the two tank system of Fig. 2. The state plot of its two states in Fig. 4 indicates such a finite cycle for the chosen initial values $x_1 = 1$ and $x_2 = 2$.

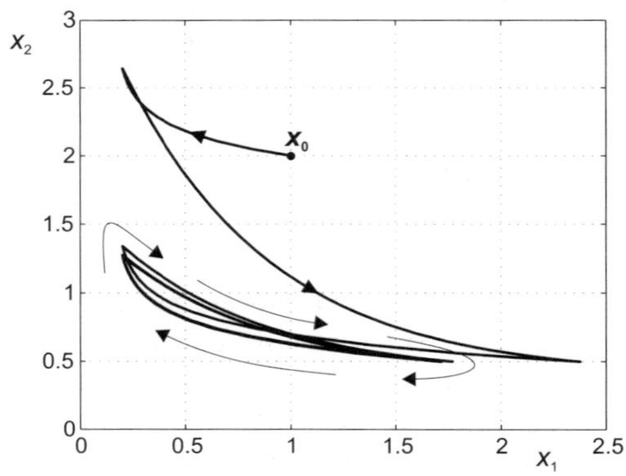

Fig. 4. Example – State Plot

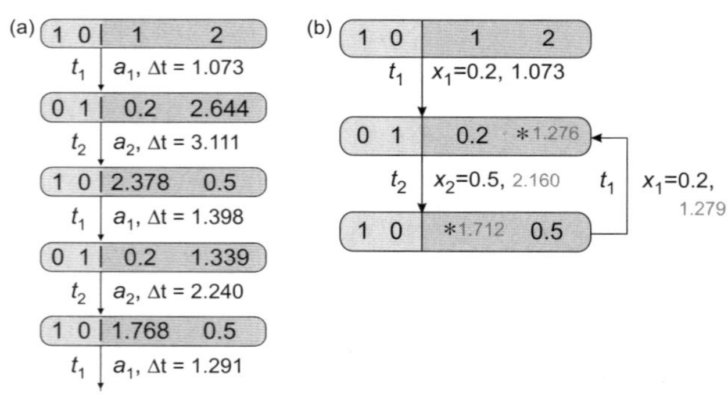

Fig. 5. Example – Evolution Graph (**a**) and Covered Evolution Graph (**b**)

The beginning of the corresponding infinite evolution graph is shown in Fig. 5a – cut after some nodes and free of branches, because the belonging Petri Net contains only two alternating states and is therefore free of conflicts. Since there is no new information in the repeating sequences of the cycle, the infinite evolution graph can be represented by the compact – and finite – *covered evolution graph* (Fig. 5b). In the covered evolution graph all IB-states in a cycle are reduced to nodes with the same discrete marking and partially different continuous states (Müller, 2002). The differing states are replaced by asterisks and the limit values. The arcs are labelled with grayed limit values for the time differences Δt in the same manner, while the threshold values remain constant in the example and can be specified exactly.

How to detect cycles during the construction of the evolution graph and the general construction of the covered evolution graph for several cycles is out of the scope of this paper. For details see (Müller, 2002). Proofs under which initial and general conditions such a cycle or several cycles in a system converge to stable or finite cycles can be found in (Müller and Rake, 1999, Müller, 2002, Branicky, 1998, Petterson, 1999, Müller et al., 2001, He and Lemmon, 1998, McMillan, 1995, Johannson and Rantzer, 1998).

3.3 Condensation of an Evolution Graph

A subgraph E_s of a graph E is called a strong component of E if and only if E_s is strongly connected, i. e. between every pair of nodes in E exists a directed path in both directions. A node of a graph E which is not strongly connected with any other node of the graph E constitutes the only node of the corresponding strong component E_s. The condensation E^k of a graph E is a reduced graph where the nodes represent the *strong components* of E.

The property of liveness is assigned to a strong component of an evolution graph if all discrete transitions t_j fire at least once inside of the strong component and if all output signals a_i of the continuous system change inside of the strong component. In other words, at least one arc of the subgraph E_s is labelled with t_j for every transition of the discrete parts of the system and at least one arc of the subgraph E_s is labelled with a_i for every output signal of the continuous parts of the system. If no transition t_j and no output signal a_i fires or changes respectively inside of the strong component, it is called dead. All strong components consisting of a single node are therefore dead.

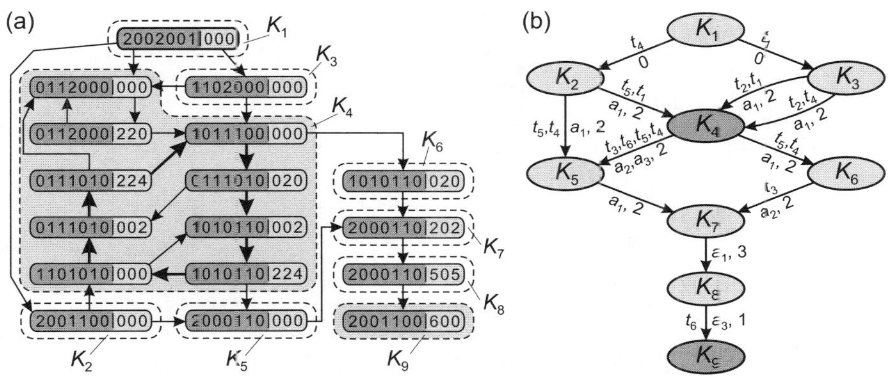

Fig. 6. Example of a (simplified) Evolution Graph (**a**) and its Condensation (**b**)

The *condensation* of an evolution graph (Müller and Rake, 2000) reveals some elementary properties very easily. E. g. a hybrid system is only reversible if all nodes of the evolution graph are strongly connected, resulting in a condensation consisting of only one node. On the other hand, if any drain in the condensation – with all arcs ending in the node and no arc beginning at it – is dead, the hybrid system may run into a deadlock, as the comparison between the example of an evolution graph (with labelling of the arcs omitted) and its condensation in Fig. 6 shows for node K_9 in the condensation.

4 Manufacturing Example

A simple example will be used for demonstration purposes in the following. It consists of two furnaces for thermal treatment of the workpieces and two parallel production units (Fig. 7). Each unit has one machine to process workpieces. Both production units are coupled by a manipulator which secures the supply with workpieces. This constitutes the critical part of the process, as the handling time of the workpieces is limited: the momentary temperature of the workpieces cooling down may not fall below a minimum temperature. Otherwise they cannot be processed in the production units. If this handling time passes without the corresponding workpiece being processed, the manufacturing process stops. The connections from the production units to the environment are designed to fit the needs of the system so that there are always enough workpieces at the entry and enough free places at the exit.

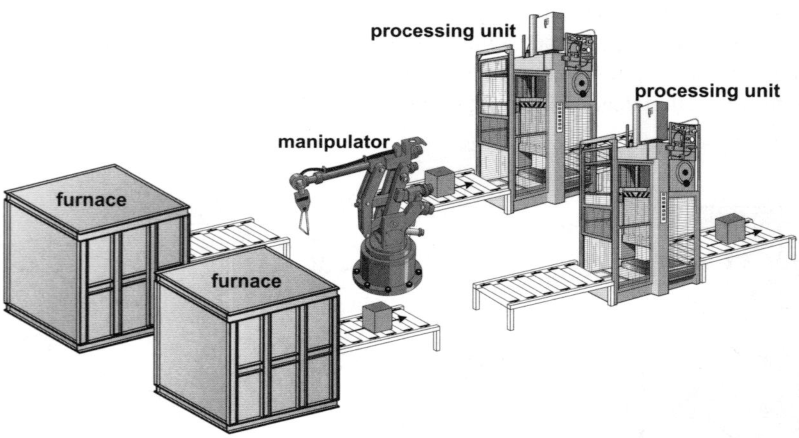

Fig. 7. Manufacturing Example – Scheme

Figure 8 shows the Petri-Net-State-Model of the manufacturing process. The upper half of the figure shows the discretely modelled part of the system. The place p_7 represents the manipulator and the places p_3 and p_6 correspond to the processing in the production units. Dynamics in the continuous part are reduced for simplification to the integration of time with constant time lengths as thresholds (t_T: minimum handling time needed by manipulator; t_A: maximum handling time allowed; t_{Pi}

processing time on unit i). The hybrid state vector consists of the discrete marking $\boldsymbol{M}$ and the continuous states x, x_{P1} and x_{P2}

$$\begin{aligned} \boldsymbol{x_h} &= \left(\boldsymbol{M}^T \mid \boldsymbol{x}^T\right)^T , \\ \boldsymbol{M} &= \left(M(p_1)\ M(p_2)\ M(p_3)\ M(p_4)\ M(p_5)\ M(p_6)\ M(p_7)\right)^T , \\ \boldsymbol{x} &= \left(x\ \ x_{P1}\ \ x_{P2}\right)^T . \end{aligned} \tag{3}$$

The solution of the conflict of the two production units in accessing the manipulator (transitions t_1 and t_4) must be provided by the control. The strategy to solve the conflict is of course dependent e.g. on the given time lengths.

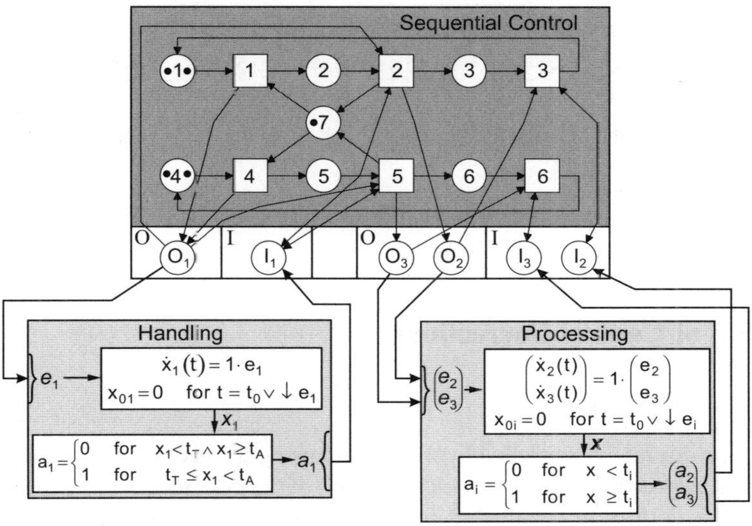

Fig. 8. Manufacturing Example – Petri-Net-State-Model

The evolution graph in Fig. 9 – which is valid only for the given time constants – serves to analyse the behaviour of the manufacturing process. Obviously there is a dead IB-state/deadlock in the lower right of the figure but there are no dead transitions. The deadlock is reached if two conditions are fulfilled: first the remaining processing time of a workpiece on a production unit is still greater than the maximum handling time allowed; secondly the manipulator nevertheless tries to serve this production unit with a new workpiece taken out of a furnace.

The condensation of the evolution graph contains besides some components with only one IB-state the component K_4 and the deadlock in K_9 (see also Fig. 6). K_4 is live with respect to both the outgoing signals and the transitions. The component K_4 contains also the desired behaviour of the manufacturing process: there are several possible cycles in K_4 one of which is emphasised in Fig. 9. These cycles are the

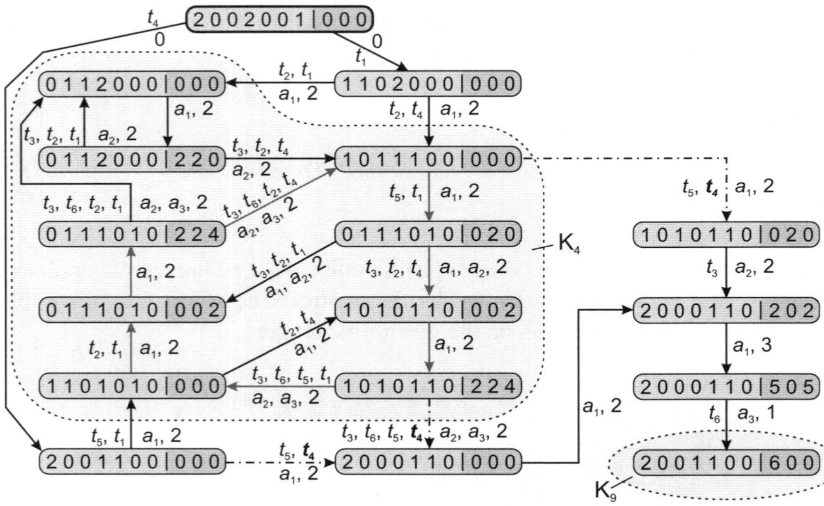

Fig. 9. Evolution Graph of the Manufacturing with $t_T = 2, t_A = 5, t_{P1} = 4$ and $t_{P2} = 6$

hybrid equivalent to T-invariants in classical Petri Nets since both reproduce the initial hybrid state or marking respectively.

5 Synthesis of Control Corrections

5.1 Objective and Conditions

The analysis of a discretely controlled hybrid system using the evolution graph and its condensation can reveal several relevant properties of the system. Very important properties are freedom from deadlocks, reversibility and liveness with respect to transitions in the discrete part of the system or with respect to the output signals of the continuous part. These properties are possibly interesting only for some parts of a system. An emergency shutdown will e.g. often lead to a system with a deadlock as intended behaviour which is not reversible. Apart from this the system should potentially be free from deadlocks. Also the starting behaviour is generally not to be reproduced – like in Fig. 9 the transition from the thick bordered initial IB-state via two alternative paths to the condensation's component K_4. For normal production operation (the node K_4 and all succeeding nodes in the condensation) reversibility may be an objective.

An even more important property of hybrid systems is liveness. If it is lacking, this is a hint that in certain states of the system some of the modelled events can no longer occur and certain parts of the modelled system behaviour are out of reach. This may be desired in case the modelled behaviour of the controlled system is off specifications, but not for normal operation of the controlled system with working control. The control has to ensure that the states which are not desired will not be reached. The aim of a first correction of the control in the example must be the prevention of the transitions into the deadlock in K_9 (the dashed lines in Fig. 9) by introduction of additional conditions. Properly designed conditions which prevent only the undesired transitions – *minimal extensions* to the control – shorten the system's degrees of freedom minimally, leaving most possibilities for further objectives like optimality (Seiche, 1991).

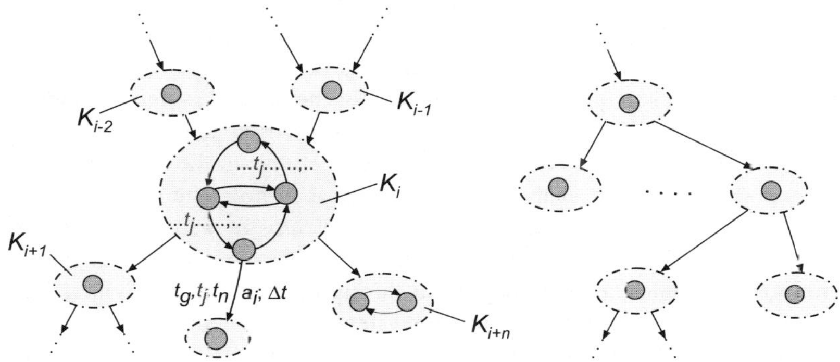

Fig. 10. Correctable (**a**) and not correctable (**b**) condensation of an evolution graph

A condition for a successful synthesis of control corrections is a condensation with a structure generally comparable to Fig. 10a: at least one component K_i of the condensation must contain the desired system behaviour. State transitions leaving the component may be prevented if the firing of the triggering transition is controllable, which applies to all transitions inside of the discrete control but not necessarily to discrete parts of the controlled system.

A hybrid system with a condensation similar to Fig. 10b is not correctable, as the transitions between different states may be prevented but leads to new dead components, because the synthesis cannot introduce cyclic runs. If this is the case, the restrictions by the already existing control are too strong or the process itself does not allow the desired behaviour. Therefore the control or the process – or their modelling – have to be reworked.

5.2 Determination of Additional Firing Conditions

The condensation E_h^k of the evolution graph E_h of a hybrid system is the basis of the synthesis as the interesting and generally desired properties are included. First, the components containing the desired system behaviour as depicted have to be chosen. They build up the modified condensation $\hat{E}_h^k$ (in the left of Fig. 10 the components K_1 to K_i). With the condensation $\hat{E}_h^k$ the corresponding evolution graph $\hat{E}_h$ is also known. The Petri-Net-State-Model has to be modified in such a way that its graph-theoretical analysis results in the evolution graph $\hat{E}_h$ instead of E_h and in the corresponding condensation $\hat{E}_h^k$ instead of E_h^k.

The state transitions to be prevented can be determined by comparing all state transitions with a starting node in $\hat{E}_h$. If the end node of a transition lies also in $\hat{E}_h$, the system behaviour may not be changed concerning this transition. If it lies outside (only in E_h, but not in $\hat{E}_h$), it has to be prevented. A transition between IB-states consists of several transitions in the discrete system and changes in the output signals of the continuous system. The *critical transition* out of this sequence of discrete transition has to be determined which is responsible for the state transition. E. g. in Fig. 10 this is the transition t_j triggering the state transition from component K_i to K_{i+2}. But each critical transition not only has at least one reachable state, when it should not fire but when it is activated. In general, other states will enable the transition and firing is required, as it is needed for the desired system behaviour. As disabling of the transition in all cases would be too restrictive, the synthesis has to distinguish between states in which the transition belongs to the desired system behaviour and should remain enabled and in which it is not. The corresponding hybrid states in the evolution graph are defined as *activator* $\boldsymbol{a_i^j}$ and *deactivator* $\boldsymbol{d_i^j}$ respectively, if transition t_j is considered in IB-state i. The activators and deactivators are generally not IB-states, as they may occur during the firing of a maximal step. Only if t_j is the first transition fired in a maximal step, the corresponding activator or deactivator can be found in the evolution graph. Otherwise they have to be calculated from the last IB-state applying the firing sequence of the maximal step up to t_j. The activators and deactivators can be expressed by

$$\boldsymbol{x_h} = \boldsymbol{a_1^j} \lor \boldsymbol{x_h} = \boldsymbol{a_2^j} \lor \ldots \Rightarrow \text{enable } t_j \;, \tag{4a}$$

$$\boldsymbol{x_h} = \boldsymbol{d_1^j} \lor \boldsymbol{x_h} = \boldsymbol{d_2^j} \lor \ldots \Rightarrow \text{disable } t_j \;. \tag{4b}$$

For the example of the manufacturing process the component K_4 in Fig. 9 comprises the desired system behaviour. Together with the three nodes for starting behaviour (K_1, K_2 and K_3 in Fig. 6) the modified condensation $\hat{E}_h^k$ and the corresponding modified evolution graph $\hat{E}_h$ is given. A comparison between original and modification shows the three dashed state transitions as critical ones. On closer inspection all state transitions are caused by the critical transition t_4. There are two activators and two deactivators which lead to new firing conditions for the transition t_4.

$$\boldsymbol{a_1^4} = (1\;0\;1\;1\;0\;1\;1|0\;0\;2)^T \tag{5a}$$

$$\boldsymbol{a_2^4} = (1\;0\;1\;2\;0\;0\;1|0\;0\;0)^T \tag{5b}$$

$$\boldsymbol{d_1^4} = (1\;0\;1\;1\;0\;1\;1|0\;2\;0)^T \tag{6a}$$

$$\boldsymbol{d_2^4} = (2\;0\;0\;1\;0\;1\;1|0\;0\;0)^T \tag{6b}$$

5.3 Determination of Significant Firing Conditions

A close look at (4a) reveals that the whole hybrid state vector (3) has to be checked for the treatment of critical transitions. This leads to complex and large extensions to the hybrid model – e.g. only the activator in (5a) needs ten new firing conditions for the critical transition t_4 and the same holds for every other activator or deactivator:

$$\begin{aligned} M(p_1) &= 1 \wedge M(p_2) = 0 \wedge M(p_3) = 1 \wedge M(p_4) = 1 \wedge M(p_5) = 0 \\ \wedge M(p_6) &= 1 \wedge M(p_7) = 1 \wedge x = 0 \wedge x_{P1} = 0 \wedge x_{P2} = 2 \\ &\Rightarrow \text{activate } t_4 \ . \end{aligned} \tag{7}$$

Comparing the activators and deactivators in the equations above it is obvious that not all elements of the hybrid state vector are needed to determine whether a transition has to be enabled or disabled. Evidently some elements of the state vector are even equal for all activators and deactivators. A direct comparison of them shows that there are some *significant places and states*. This is the case because often only local parts of a system are responsible for undesired behaviour. The comparison between activators and deactivators can be formalised by introducing the difference matrix $\boldsymbol{\Delta_j}$. This matrix contains the vectorial differences between all activators and deactivators of the critical transition t_j

$$\boldsymbol{\Delta}^j = \left(\boldsymbol{d}_1^j - \boldsymbol{a}_1^j; \ \ldots \ ; \boldsymbol{d}_1^j - \boldsymbol{a}_q^j; \boldsymbol{d}_2^j - \boldsymbol{a}_1^j; \ \ldots \ ; \boldsymbol{d}_p^j - \boldsymbol{a}_q^j\right) \ .$$

Elements not equal to zero in a row of the matrix show that the corresponding activator and deactivator differ concerning the marking of the corresponding place or in the value of the corresponding continuous state respectively. The element of the matrix is consequently suitable to differentiate the corresponding activator and deactivator. If a whole line is different from zero, the corresponding element of the hybrid state vector alone is significant to differentiate between all activators and deactivators. In general all lines will contain zero elements, meaning that a linear combination of – generally as few as possible – significant states/lines has to be chosen. The difference matrix of the example reads as follows:

$$\begin{aligned} \boldsymbol{\Delta}^4 &= (\boldsymbol{d}_1^4 - \boldsymbol{a}_1^4; \boldsymbol{d}_2^4 - \boldsymbol{a}_1^4; \boldsymbol{d}_1^4 - \boldsymbol{a}_2^4; \boldsymbol{d}_2^4 - \boldsymbol{a}_2^4) \\ &= \begin{pmatrix} 1-1 & 2-1 & 1-1 & 2-1 \\ 0-0 & 0-0 & 0-0 & 0-0 \\ 1-1 & 0-1 & 1-1 & 0-1 \\ 1-1 & 1-1 & 1-2 & 1-2 \\ 0-0 & 0-0 & 0-0 & 0-0 \\ 1-1 & 1-1 & 1-0 & 1-0 \\ 1-1 & 1-1 & 1-1 & 1-1 \\ \hline 0-0 & 0-0 & 0-0 & 0-0 \\ 2-0 & 0-0 & 2-0 & 0-0 \\ 0-2 & 0-2 & 0-0 & 0-0 \end{pmatrix} = \begin{pmatrix} 0 & 1 & 0 & 1 \\ 0 & 0 & 0 & 0 \\ 0 & -1 & 0 & -1 \\ 0 & 0 & -1 & -1 \\ 0 & 0 & 0 & 0 \\ 0 & 0 & 1 & 1 \\ 0 & 0 & 0 & 0 \\ \hline 0 & 0 & 0 & 0 \\ 2 & 0 & 2 & 0 \\ -2 & -2 & 0 & 0 \end{pmatrix} . \end{aligned} \tag{8}$$

This difference matrix contains no line with no zero entries, meaning that at least two places and/or states have to be used for correction of the control. The difference between activator $\boldsymbol{a}_1^4$ and the deactivator $\boldsymbol{d}_1^4$ is zero for all elements of the discrete part (see the part above the horizontal line in the first row of $\boldsymbol{\Delta^4}$ which contains only zeros). That means at least one continuous state has to be used for correction of the control. One possible combination would be the significant place p_6 with the significant state x_{P2} to the following instruction

$$M(p_6) = 0 \vee x_{P2} \geq 2 \Leftrightarrow \text{enable}\, t_4 \ . \tag{9}$$

So the transition t_4 may fire if and only if production unit 2 is free or if two units of time have passed during processing of a workpiece. The example can be easily solved manually by finding a suitable linear combination, but larger examples and difference matrices require a mathematical formulation of the problem. All possible solutions $\boldsymbol{s_i}$ with combinations of significant places and states are included in the diophantine matrix equation

$$\begin{gathered}\left(\boldsymbol{D^j}\right)^T \cdot \boldsymbol{s} > \boldsymbol{0}\ , \\ D_{kl}^j = |signum(\Delta_{kl}^j)| = \begin{cases} 0 & \text{for} \quad \Delta_{kl}^j = 0 \\ 1 & \text{for} \quad \Delta_{kl}^j \neq 0 \end{cases} \qquad \text{and} \quad s_i \in \{0,\, 1\}\ .\end{gathered} \tag{10}$$

The inequality relation can be solved with the same algorithm that is used to determine the place and transition invariants in a Petri Net. Every solution $\boldsymbol{s}$ describes a linear combination of lines in $\boldsymbol{D^j}$ so that the result is free of zeros. The elements s_i of $\boldsymbol{s}$ equal to zero represent places or states that are not necessary to correct the control. Therefore especially those solutions of (10) are interesting which have a maximum of zero elements s_i, meaning that a minimum of new firing conditions has to be used and the corrected model stays clearly arranged. For the example one obtains the diophantine matrix equation

$$\begin{pmatrix} 1 & 0 & 1 & 1 & 0 & 1 & 0 & | & 0 & 0 & 0 \\ 0 & 0 & 0 & 1 & 0 & 1 & 0 & | & 0 & 2 & 0 \\ 1 & 0 & 1 & 0 & 0 & 0 & 0 & | & 0 & 0 & 2 \\ 0 & 0 & 0 & 0 & 0 & 0 & 0 & | & 0 & 2 & 2 \end{pmatrix} \cdot \boldsymbol{s} > \begin{pmatrix} 0 \\ 0 \\ 0 \\ 0 \end{pmatrix} . \tag{11}$$

Among others the following three solution vectors $\boldsymbol{s_i}$ are found:

$$\boldsymbol{s_1} = (0\,0\,0\,0\,0\,1\,0|0\,0\,1)^T , \tag{12a}$$

$$\boldsymbol{s_2} = (0\,0\,1\,0\,0\,0\,0|0\,1\,0)^T , \tag{12b}$$

$$\boldsymbol{s_3} = (0\,0\,0\,1\,0\,0\,0|0\,0\,1)^T . \tag{12c}$$

I. e. the combination of the significant place p_6 and the significant state x_{P2} in (12a), p_3 and x_{P1} in (12b) or p_4 and x_{P2} in (12c) can be used to correct the control. In all cases the objective to reduce the system behaviour in the evolution graph from E_h to $\hat{E}_h$ is reached.

5.4 Realisation of the Control Corrections

The last step in the synthesis procedure is to realise the gained new firing conditions for the critical transitions in the model. As there are both discrete and continuous *conditions* which have to influence the system only in case of a deactivator state but which must not change the system behaviour in other cases, there is the need of a suitable construction in the Petri Net. A *self-loop* is such a construction which only interrogates the marking of a place without changing it, having only an influence on the activation and therefore on the firing of a transition. Self-loops have already been used for checks of the inputs from continuous systems. If the synthesis leads to new conditions for the continuous part these lead to additional threshold functions evaluating state space variables. For each new threshold function a new output port is introduced into the continuous system and also a corresponding input place in the discrete part. This input place is connected to the critical transition as described above via self-loops.

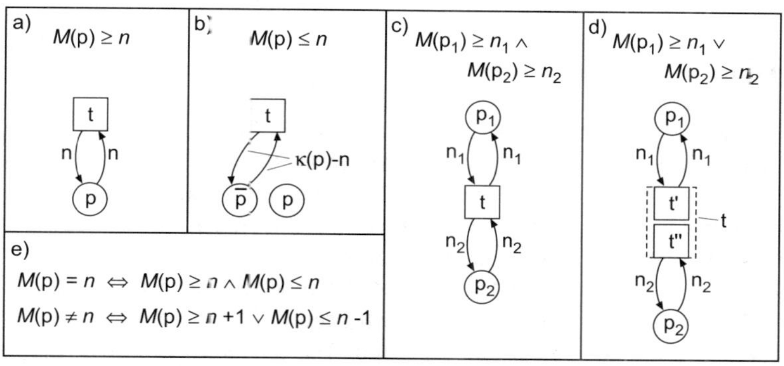

Fig. 11. Basic construction methods to apply the additional conditions for places

The basic elements to realise (in-) equations and their logical combinations are shown in Fig. 11 (Seiche, 1991). The inequation $M(p) \leq n$ in case b) needs the introduction of a *complementary place* $\bar{p}$ – a place with the same capacity κ as p, but with the complementary marking $M(\bar{p}) = \kappa - M(p)$, so the sum of both markings is always the capacity. This is realised by connecting $\bar{p}$ with the same arcs with identic weights as p but with changed directions of the arcs. In the case of a Boolean *OR* operation parallel *congruent* copies of a transition are used like in case d).

Realising the control corrections (9) and (12a) respectively in the example of the manufacturing process needs the split of transition t_4 into two congruent conditions t_{4a} and t_{4b} (Fig. 12). The continuous condition leads to the new input place I_4. The discrete comparison $M(p_6) = 0$ leads in accordance with Fig. 12 to two inequations but only one self-loop with the complementary place $p_8 = \bar{p}_6$ for $M(p_6) \leq 0$. The second resulting inequation $M(p_6) \geq 0$ is not needed, because the marking of a place is always greater or equal to zero and therefore the self loop with p_6 would have a weight of zero. In consequence the self-loop can be omitted. Redoing an analysis of the corrected hybrid system shows that the desired behaviour for the manufacturing process now is realised.

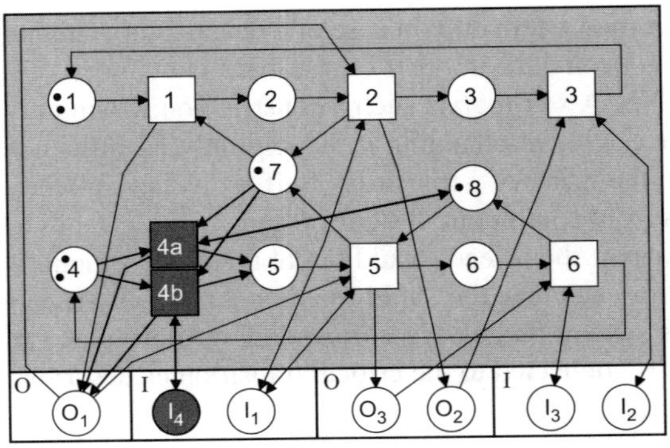

Fig. 12. Corrected Control for the Manufacturing Example

6 Conclusion

In this article, a model for the description of hybrid systems, like technical processes combined with a discrete control, is introduced. The Petri-Net-State-Model involves modules of Petri Nets for the discrete partial models and extended state space models to describe the continuous parts, coupled via binary interfaces. The chosen modular structure allows efficient modelling of large systems.

The sequence of hybrid states (IB-states) can be represented in the evolution graph, a hybrid reachability graph. Systems with transient converging behaviour can be described with the covered evolution graph. The (covered) evolution graph and its condensation serve to analyse the behaviour of the hybrid system. It gives – especially in the case of the design of a discrete control – very valuable hints on conflicts between transitions in the discrete part of the system. These may be critical in the sense that the firing of a transition in a certain system state prevents the desired system behaviour.

A procedure to synthesise corrections for the discrete control in a hybrid system – solving the depicted critical conflicts – is presented, based on the results of the reachability analysis by aid of the evolution graph. This procedure expands the control of the hybrid system by minimal additional firing conditions to prevent *only* undesired state transitions which are declared during or after analysis. The results of the synthesis procedure are hybrid, as both discrete and continuous states are referred to determine activation of transitions. The procedure is demonstrated in the example of a manufacturing process.

Nonlinear Hybrid Dynamical Systems: Modeling, Optimal Control, and Applications

Martin Buss[1], Markus Glocker[2], Michael Hardt[2], Oskar von Stryk[2], Roland Bulirsch[3], and Günther Schmidt[4]

[1] Control Systems Group, Technische Universität Berlin, Berlin, Germany
[2] Simulation and Systems Optimization Group, Technische Universität Darmstadt, Darmstadt, Germany
[3] Zentrum Mathematik, Technische Universität München, München, Germany
[4] Institute of Automatic Control Engineering, Technische Universität München, München, Germany

Abstract. Nonlinear hybrid dynamical systems are the main focus of this paper. A modeling framework is proposed, feedback control strategies and numerical solution methods for optimal control problems in this setting are introduced, and their implementation with various illustrative applications are presented. Hybrid dynamical systems are characterized by discrete event and continuous dynamics which have an interconnected structure and can thus represent an extremely wide range of systems of practical interest. Consequently, many modeling and control methods have surfaced for these problems. This work is particularly focused on systems for which the degree of discrete/continuous interconnection is comparatively strong and the continuous portion of the dynamics may be highly nonlinear and of high dimension. The hybrid optimal control problem is defined and two solution techniques for obtaining sub-optimal solutions are presented (both based on numerical direct collocation for continuous dynamic optimization): one fixes interior point constraints on a grid, another uses branch-and-bound. These are applied to a robotic multi-arm transport task, an underactuated robot arm, and a benchmark motorized traveling salesman problem.

1 Introduction

The recent interest in nonlinear hybrid dynamical systems has forced the merger of two very different modeling and control methodologies, namely those for discrete and for continuous systems. The investigation of hybrid systems attempts to effectively unite these two formalisms in order to model, investigate, and design these systems with analytical and numerical tools. The attempt to provide a unified hybrid modeling scheme well-suited to the study of hybrid dynamical systems has inspired many researchers (Back et al., 1993, Branicky et al., 1998, Brockett, 1993, Engell, 1997, Labinaz et al., 1996, Nenninger et al., 1999, Nerode and Kohn, 1993, Tavernini, 1987), including the hybrid modeling approach presented here which is based on previous work in (Buss, 2000). The characteristic behavior of hybrid systems is discussed and illustrated using this modeling scheme. In particular, the multiple potential dynamical events that may occur due to the strong interconnection of discrete and continuous elements are highlighted.

S. Engell, G. Frehse, E. Schnieder (Eds.): Modelling, Analysis and Design of Hybrid Systems, LNCIS 279, pp. 311–335, 2002.

Theoretical work on controllability properties of nonlinear hybrid dynamical systems is still in its early stages and to date only several problems of low state and control dimension can be thoroughly understood (van der Schaft and Schumacher, 2000). Nevertheless, there has been a strong interest in numerical methods for determining controllers for these systems, inspired from the success of such approaches in conventional nonlinear optimal control problems. Nonlinear optimal control plays a key role in modern mechatronics and robotics, in particular in the area of path, trajectory, and action planning. To mention some of the many applications: walking pattern and trajectory planning (Hardt et al., 2000), mobile robot path planning (Kondak and Hommel, 2001), optimal payload (weight) lifting, and acrobatics (Albro and Bobrow, 2001, Martin and Bobrow, 1997), etc. Numerical algorithms designed for hybrid optimal control problems (HOCPs) with variable structure, nonlinear differential equations have recently been published (Branicky et al., 1999, Buss et al., 2000a, Hedlund and Rantzer, 1999, Tomlin, 1999). These efforts were applied to low-dimensional illustrative problems, yet the results presented here demonstrate that numerical methods do exist which are promising for dealing with realistic, higher-dimensional system models.

The key to numerically solving HOCPs seems to be the combination of efficient numerical solvers – such as direct collocation – for optimal control problems together with (heuristical) approaches to reduce the combinatorial complexity of the discrete event aspect in HOCPs (Buss et al., 2000a, von Stryk and Glocker, 2000, von Stryk and Glocker, 2001, von Stryk, 2000). This paper presents numerical solution techniques for HOCPs with applications in mechatronics and robotics. An example problem of three robotic arms cooperatively transporting an object from an initial to a goal position is solved suboptimally by fixing interior point times and state constraints to fixed values on a grid. The trajectory planning problem of an underactuated robot with an unactuated joint equipped with a holding brake in the passive joint is solved by branch-and-bound to obtain optimal hybrid trajectories, in particular, the optimal number of switches for the holding brake. Finally the solution for the benchmark motorized traveling salesman problem is presented which is a problem that is easily scalable to higher dimensions.

The solution approaches presented here rely on the efficient numerical tool DIRCOL, which implements a direct collocation method to approximately solve nonlinear optimal control problems by advanced nonlinear programming methods (von Stryk, 2001), see also (Betts, 1998, Hardt et al., 2000, von Stryk and Bulirsch, 1992). The organization of the paper is as follows: Sect. 2 proposes the *Hybrid State Model* HSM as a general hybrid modeling framework. Hybrid feedback control architectures are introduced in Sect. 3. In Sect. 4 a broad class of HOCPs is defined. In Sect. 4.2 numerical solution strategies to obtain suboptimal solutions on interior point constraints on grids and a branch-and-bound strategy are proposed. The solution of three illustrative hybrid problems in robotics are presented in Sect. 5 followed by a discussion of more realistic, higher dimensional problems currently being investigated.

2 Modeling of Hybrid Dynamical Systems

A conventional continuous dynamical system is described by the velocity vector field $\boldsymbol{f}(\boldsymbol{x}, \boldsymbol{u}, t)$, which depends on the continuous state $\boldsymbol{x}$, the continuous control input $\boldsymbol{u}$, and time t; the continuous output $\boldsymbol{y}_x$ is generated by the output function $\boldsymbol{h}_x(\boldsymbol{x}, \boldsymbol{u}, t)$. The dynamics of a lumped parameter continuous time systems are thus defined by a set of ordinary differential (algebraic) equations. Systems with purely discrete state dynamics are often modeled by a finite state automaton or a Petri-Net. Interconnections of these very different system descriptions are denoted as hybrid dynamical systems and a variety of modeling paradigms have been proposed for which we refer to (Branicky et al., 1998, Engell, 1997, Labinaz et al., 1996, Nenninger et al., 1999, Schlegl et al., 2000). The hybrid modeling approach presented here is rooted in the theory of continuous dynamical systems and includes discrete system elements such as discontinuous nonlinearities and switching actions as extensions to these systems. This leads to a general hybrid system model for the class of systems denoted as hybrid dynamical systems (HDS).

A HDS consists of, in addition to continuous dynamical system aspects, a *discrete (symbolic) state* $\boldsymbol{q} \in \mathbb{N}^l$, a *discrete (symbolic) control input* $\boldsymbol{v} \in \mathbb{N}^k$, a *discrete (symbolic) system output* $\boldsymbol{y}_q$, *discrete event* generating functions $\boldsymbol{s}_j$, and *discrete dynamics* $\boldsymbol{\phi}_j$, see Fig. 1. The continuous dynamical behavior is the result of the velocity vector field $\boldsymbol{f}(\cdot)$. Discrete events are caused by the discontinuity indicator functions s_j and hybrid successor states are specified by transition (jump) maps ϕ_j, $j = 1, \ldots, n_s$. Hence, the hybrid dynamics are specified by the three components

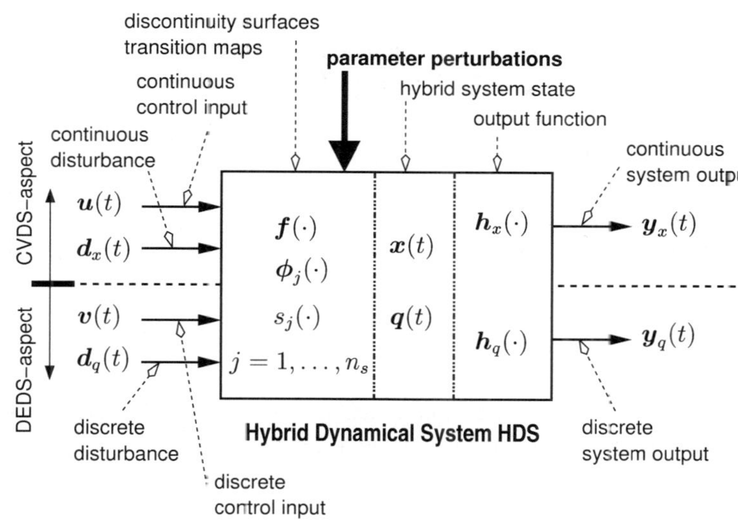

Fig. 1. Hybrid dynamical system (HDS) with continuous variable (CVDS) and discrete-event (DEDS) aspects composed of input, output, and state vectors, discontinuity surfaces and jump maps

$\boldsymbol{f}(\cdot)$, $s_j(\cdot)$, $\boldsymbol{\phi}_j(\cdot)$, see left part of Fig. 1. Inputs to the hybrid dynamical system are the continuous control input $\boldsymbol{u}(t)$, the discrete control input $\boldsymbol{v}(t)$, the continuous disturbance $\boldsymbol{d}_x(t)$, and the discrete disturbance signals $\boldsymbol{d}_q(t)$. The hybrid output $\boldsymbol{y}(t) = [\boldsymbol{y}_x(t)^T \boldsymbol{y}_q(t)^T]^T$ is produced by the output functions $\boldsymbol{h}(\cdot) = (\boldsymbol{h}_x(\cdot),\ \boldsymbol{h}_q(\cdot))$.

2.1 The Hybrid State Model

In this section the *hybrid state model* (HSM) is proposed for the modeling of a fairly general class of nonlinear hybrid dynamical systems. The model is related to the Branicky-Borkar-Mitter BBM model, see (Branicky, 1993, Branicky, 1994a, Branicky, 1994b, Branicky, 1994c, Branicky, 1995, Branicky, 1996, Branicky, 1998, Branicky et al., 1998). The main difference lies in the use of discontinuity surfaces defined by switching functions instead of jump sets used in the BBM model. A benefit of the HSM model is that switching functions have close ties to variable structure control; another advantage is that simulation and implementation of the HSM is straightforward.

Definition 1 (HSM). A hybrid dynamical system (HDS) is defined by its hybrid state model (HSM) as follows:

$$\dot{\boldsymbol{x}} = \boldsymbol{f}(\boldsymbol{x}, \boldsymbol{u}, \boldsymbol{q}, t) \qquad \text{if } s_j(\boldsymbol{x}, \boldsymbol{u}, \boldsymbol{q}, \boldsymbol{v}, t) \neq 0,\ j = 1, \dots, n_s \tag{1}$$

$$\begin{bmatrix} \boldsymbol{x}(t^+) \\ \boldsymbol{q}(t^+) \end{bmatrix} = \boldsymbol{\phi}_j(\boldsymbol{x}, \boldsymbol{u}, \boldsymbol{q}, \boldsymbol{v}, t^-) \text{ if } s_j(\boldsymbol{x}, \boldsymbol{u}, \boldsymbol{q}, \boldsymbol{v}, t) = 0,\ j \in \{1, \dots, n_s\} \tag{2}$$

$$\boldsymbol{y} = \boldsymbol{h}(\boldsymbol{x}, \boldsymbol{u}, \boldsymbol{q}, \boldsymbol{v}, t)\ , \tag{3}$$

where (1), (2) describe the continuous and discrete dynamic behavior, respectively; the notation $\boldsymbol{x}(t^+)$ denotes the successor state (limit from the right) of $\boldsymbol{x}$ at time t. The hybrid output $\boldsymbol{y}$ is generated by (3). The continuous state vector $\boldsymbol{x}(t) \in \mathcal{X} \subseteq \mathbb{R}^n$ and the discrete state vector $\boldsymbol{q}(t) \in \mathcal{Q} \subseteq \mathbb{N}^l$ together form the hybrid state vector

$$\zeta(t) = \begin{bmatrix} \boldsymbol{x}(t) \\ \boldsymbol{q}(t) \end{bmatrix} \in \mathcal{X} \times \mathcal{Q} \subseteq \mathbb{R}^n \times \mathbb{N}^l\ .$$

The continuous control input $\boldsymbol{u}(t) \in \mathcal{U} \subseteq \mathbb{R}^m$ belongs to the set $\mathcal{U}$ of permissible controls. The discrete (symbolic) control input vector is $\boldsymbol{v}(t) \in \mathcal{V} \subseteq \mathbb{N}^k$. The hybrid output vector

$$\boldsymbol{y}(t) = \begin{bmatrix} \boldsymbol{y}_x \\ \boldsymbol{y}_q \end{bmatrix} \in \mathcal{Y} \subseteq \mathbb{R}^p \times \mathbb{N}^r$$

combines a p-dimensional continuous output $\boldsymbol{y}_x$ and a r-dimensional discrete (symbolic) output $\boldsymbol{y}_q$; $\boldsymbol{y}$ is generated by the hybrid output function

$$\boldsymbol{h} : \mathcal{X} \times \mathcal{U} \times \mathcal{Q} \times \mathcal{V} \times \mathbb{R} \to \mathbb{R}^p \times \mathbb{N}^r\ . \tag{4}$$

The continuous behavior of the HDS is given by the vector field

$$\boldsymbol{f} : \mathcal{X} \times \mathcal{U} \times \mathcal{Q} \times \mathbb{R} \to \mathbb{R}^n \tag{5}$$

Discontinuous behavior of the HDS is caused by events occurring when the hybrid state intersects discontinuity surfaces

$$s_j : \mathcal{X} \times \mathcal{U} \times \mathcal{Q} \times \mathcal{V} \times \mathbb{R} \to \mathbb{R} , \tag{6}$$

for $j = 1, \dots , n_s$. Note, that the discontinuity surfaces may depend on the continuous and/or the discrete control input $\boldsymbol{u}(t)$, $\boldsymbol{v}(t)$. The hybrid successor state

$$\zeta(t_1^+) = \begin{bmatrix} \boldsymbol{x}(t_1^+) \\ \boldsymbol{q}(t_1^+) \end{bmatrix} , \tag{7}$$

after discrete events is given by the transition (jump) maps

$$\phi_j : \mathcal{X} \times \mathcal{U} \times \mathcal{Q} \times \mathcal{V} \times \mathbb{R} \to \mathcal{X} \times \mathcal{Q} , \tag{8}$$

see also (2). As long as all discontinuity surface functions $s_j(\boldsymbol{x}, \boldsymbol{u}, \boldsymbol{q}, \boldsymbol{v}, t) \neq 0$, for $j = 1, \dots , n_s$, the system trajectory evolves continuously according to (1).

Remark 1. A sliding-mode condition (Schlegl et al., 1997) also fits into the model from Definition 1 when it is permitted that infinitely many discrete transitions occur in a finite time period. Results describing such cases may be found in (Schlegl et al., 1997, Doğruel and Özgüner, 1995, Doğruel et al., 1996).

Remark 2. It has been shown that the BBM model incorporates alternative modeling formalisms such as the Tavernini TAV model (Tavernini, 1987), the Back-Guckenheimer-Myers BGM model (Back et al., 1993), the Nerode-Kohn NK model (Nerode and Kohn, 1993) and the Brockett BRO model (Brockett, 1993). This applies here as well to the proposed HSM defined in Definition 1, which also includes further modeling paradigms such as (Nenninger et al., 1999), see (Buss, 2000) for a detailed discussion.

2.2 Characterization of Hybrid Dynamic Behavior

The dynamic behavior of a HDS is strongly influenced by discontinuities in its system trajectories. Discontinuities include *state resets* (SR) resulting in *state jumps*, *vector field switches* (VFS) resulting in a switch of the velocity vector field, and their combination (SRVFS). These may be triggered by a *time event* (TE) occurring at a certain time or by a *state event* (SE) if the system state reaches a certain value. Further events include *control events* (CE) caused by the introduction of a hybrid control action into the discrete control input or *disturbance events* (DE) caused by discrete disturbance inputs. These events may be interdependent as, for example, a SE may either be induced externally (*controlled*) as a result of a CE or DE or induced internally (*autonomous*) (van der Schaft and Schumacher, 2000). Other dynamic effects of HDS include chaotic behavior, see e.g. (Bühler and Koditschek, 1993, Engell et al., 1997), or sliding mode, see e.g. (Schlegl et al., 1997, Utkin, 1992). Further discussion of hybrid dynamic characteristics may be found in (Buss, 2000).

In Fig. 2 an example of a typical path for a hybrid trajectory is plotted. The HDS starts with the discrete state $\boldsymbol{q} = \boldsymbol{q}_1$ and continuous state $\boldsymbol{x}(0) \in \mathcal{X}_1 \subseteq \mathbb{R}^n$ and evolves within the portion of state space open from the left on the left-hand side of Fig. 2. As soon as the discontinuity surface $s_1 = 0$ is reached, the hybrid state is reinitialized with a state reset (SR) after which the system trajectory continues in the discrete state $\boldsymbol{q} = \boldsymbol{q}_2$ corresponding to the continuous portion of state space $\mathcal{X}_2 \subseteq \mathbb{R}^n$. The trajectory then enters into a CE region when the discontinuity surface s_2 is crossed. The CE in this case must first be triggered by a discrete control input $\boldsymbol{v} = \boldsymbol{v}_2$ which occurs upon reaching approximately the center of the CE region. The resulting SR causes the system to make the transition into the discrete state $\boldsymbol{q} = \boldsymbol{q}_3$ and its respective portion of state space $\mathcal{X}_3 \subseteq \mathbb{R}^n$. There a TE occurs in combination with a SR whereby the discrete state does not change after the TE. The portions of state space $\mathcal{X}_3, \mathcal{X}_4 \subseteq \mathbb{R}^n$ corresponding to the discrete states $\boldsymbol{q}_3, \boldsymbol{q}_4$ are separated by a discontinuity surface s_3 from one another. The system trajectory reaches this discontinuity surface and enters with its fulfillment of the necessary sliding-mode conditions into a sliding state along the discontinuity surface $s_3 = 0$. Finally the existence conditions for the sliding-mode are no longer fulfilled resulting in the system evolution in the discrete state $\boldsymbol{q} = \boldsymbol{q}_4$ in the state space region $\mathcal{X}_4$ until the SE $s_4 = 0$.

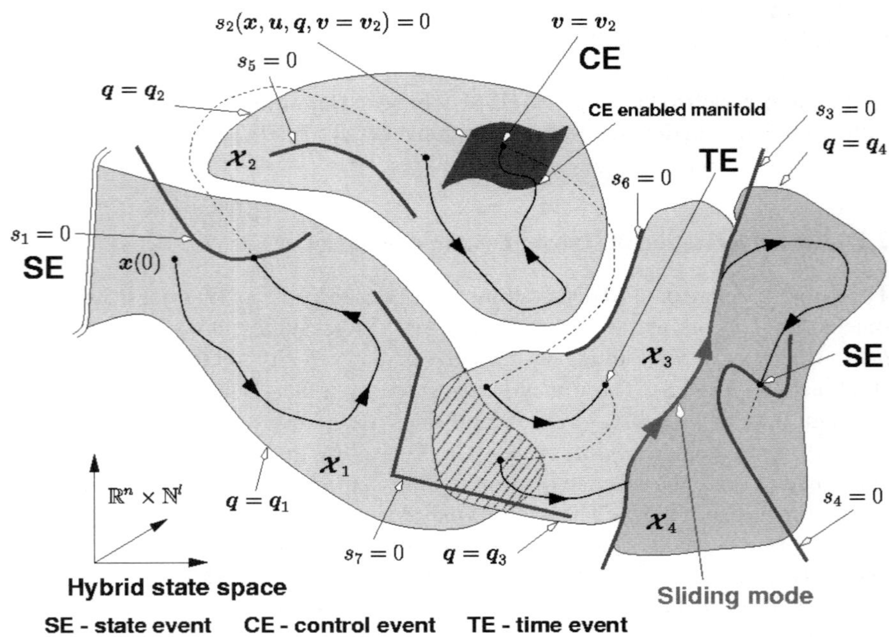

Fig. 2. An example of the evolution of a typical hybrid system trajectory in a hybrid state space

In Fig. 2 further examples are displayed of discontinuity surfaces s_5, s_6, s_7 that are irrelevant for the example trajectory. Furthermore it is shown how state space regions corresponding to certain discrete states, e.g. $\mathcal{X}_3$ and $\mathcal{X}_4$, can overlap. The allowable region $\mathcal{X}_1$ corresponding to the discrete state $\boldsymbol{q}_1$ continues unbounded into infinity in Fig. 2. The portrayal of the hybrid state space in Fig. 2 is planar, usually it will be of much higher dimension.

3 Hybrid Feedback Control

In this investigation of hybrid dynamical systems, a general hybrid control architecture is proposed consisting of three main parts, see Fig. 3: (i) the hybrid process model, cf. Sect. 2; (ii) the hybrid controller (HC) controlling this process to be discussed in this section; and (iii) the hybrid reference trajectory generator (HRG). The synthesis of reference trajectories implemented in the HRG as solutions to hybrid optimal control problems will be discussed in Sect. 4.

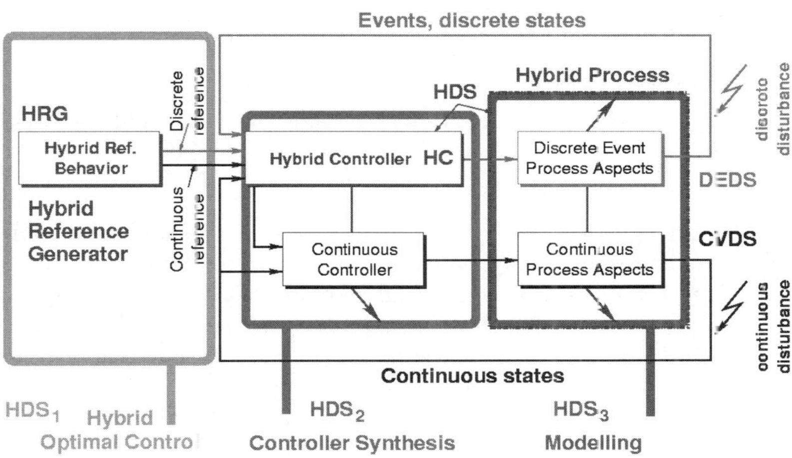

Fig. 3. General hybrid control architecture

Hybrid Control and Error Compensation. Taking the HSM of Definition 1 as the basis for modeling a HDS and keeping in mind the control architecture described above, it is possible to generalize classical control concepts such as output-following control to the hybrid case. The resulting hybrid output control (HOC) block diagram with hybrid control signals is depicted in Fig. 4. The hybrid output controller in Fig. 4 compares hybrid reference values with actual output values and produces hybrid control signals such that the output tracks the reference value with small error. Calculating the error between discrete reference value and the actual discrete output is an important question which has received little attention. A discussion can be found in (Buss, 2000). An obvious way, for example, to define the discrete comparison operator ⊎ would be to perform the arithmetic difference of two discrete values resulting in an integer-valued discrete error. In principle, the goal of a hybrid

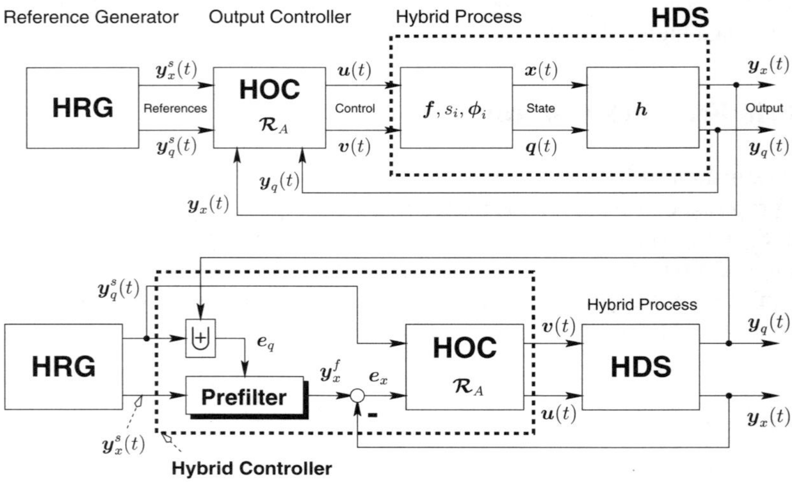

Fig. 4. Hybrid output control (top) and hybrid error compensation by means of a continuous prefilter (bottom)

controller is to eventually make the hybrid control error small. In case of a discrete error, this may not be easy as the hybrid process may be in contact with a moving system other than that assumed by the hybrid controller. One solution to hybrid error compensation is shown in Fig. 4, where a discrete error activates a continuous prefilter to modify the continuous reference $\boldsymbol{y}_x^s \to \boldsymbol{y}_x^f$ in such a way that both the discrete as well as the continuous control error eventually vanish. Similar concepts are a discrete prefilter, more complicated discrete dynamics in the compensation controller, or a combined reference generator adaptation scheme, see (Buss, 2000) for details.

4 Hybrid Optimal Control

The discrete-continuous process model of a hybrid optimal control problem (HOCP) consists of a set of ordinary differential or differential-algebraic equations of variable structure and variable constraint equations. The system structure varies among a (finite) discrete set of system descriptions each of which is associated with a specific discrete state of the considered hybrid system. The challenging aspect of this model is that the value of the discrete variable can determine the sequence, type and number of phase dynamics. Thus, the dynamics in a phase and even the dimension or number of constraints may be completely different for different values of the discrete variable.

4.1 Hybrid Optimal Control Problem

The HOCP is to find optimal hybrid (i.e., continuous $\boldsymbol{u}$ and discrete $\boldsymbol{v}$) control trajectories such that an integral cost index, typically an integral of a function of the hybrid system state and control input, is minimized subject to the system dynamics, initial, terminal, and further equality or inequality constraints.

Definition 2. The HOCP is defined as the minimization of the real valued, hybrid cost index J

$$\min_{\boldsymbol{u},\,\boldsymbol{v}} J(\boldsymbol{u},\boldsymbol{v}) = \Theta + \int_{t_0}^{t_f} \psi(\boldsymbol{x},\boldsymbol{u},\boldsymbol{q},\boldsymbol{v},t)\,\mathrm{d}t\,, \tag{9}$$

subject to

$$\dot{\boldsymbol{x}} = \boldsymbol{f}(\boldsymbol{x},\boldsymbol{u},\boldsymbol{q},\boldsymbol{v},t) \qquad \text{if } s_j(\boldsymbol{x},\boldsymbol{u},\boldsymbol{q},\boldsymbol{v},t) \neq 0 \quad j = 1,\dots,n_s \tag{10}$$

$$\begin{bmatrix} \boldsymbol{x}(t_i^+) \\ \boldsymbol{q}(t_i^+) \end{bmatrix} = \boldsymbol{\phi}_j(\boldsymbol{x},\boldsymbol{u},\boldsymbol{q},\boldsymbol{v},t_i^-) \text{ if } s_j(\boldsymbol{x},\boldsymbol{u},\boldsymbol{q},\boldsymbol{v},t_i^-) = 0 \quad j \in \{1,\dots,n_s\} \tag{11}$$

$$\boldsymbol{u}(t) \in \mathcal{U} \subset \mathbb{R}^{n_u},\ \boldsymbol{v}(t) \in \mathcal{V} \subset \mathbb{Z}^{n_v},$$

$$\boldsymbol{x}(t) \in \mathcal{X} \subset \mathbb{R}^{n_x},\ \boldsymbol{q}(t) \in \mathcal{Q} \subset \mathbb{Z}^{n_q}, \quad \forall t \in [t_0,t_f] \tag{12}$$

$$0 \leq \boldsymbol{g}(\boldsymbol{x},\boldsymbol{u},\boldsymbol{q},\boldsymbol{v},t), t \in [t_0,t_f] \text{ inequality constraints,} \tag{13}$$

$$\boldsymbol{x}(t_0) = \boldsymbol{x}_0,\ \boldsymbol{q}(t_0) = \boldsymbol{q}_0 \text{ initial conditions,} \tag{14}$$

$$\boldsymbol{x}(t_f) = \boldsymbol{x}_f,\ \boldsymbol{q}(t_f) = \boldsymbol{q}_f \text{ terminal conditions,} \tag{15}$$

where the initial and final times, written as t_0, t_f, are free or fixed, s_j are the n_s switching functions and ϕ_j denotes the explicit phase transition conditions (jump maps) occurring at the zeros of one of the switching functions. The Mayer type part Θ of the performance index is a general function of the phase transition times (events) t_i, $i = 0,\dots,N$ and of the continuous $\boldsymbol{x}(t_i^-)$, $\boldsymbol{x}(t_i^+)$ and discrete states $\boldsymbol{q}(t_i^-)$, $\boldsymbol{q}(t_i^+)$ just before and just after the $N-1$ interior transition events and at the beginning and final times respectively written as

$$\Theta := \Theta[\,\boldsymbol{x}(t_0^+),\dots,\boldsymbol{x}(t_N^-); \boldsymbol{q}(t_0^+),\dots,\boldsymbol{q}(t_N^-); t_0,\dots,t_N] \in \mathbb{R}\,.$$

Here, $t_f = t_N$ is assumed while the number of phases N may be given or free. The integrand ψ is a real-valued function of the continuous/discrete state and control variables and of time.

The minimization of (9) is subject to the initial and terminal conditions (14), (15), admissible values for the continuous/discrete control variables (12), and inequality constraints (13). Obviously, valid hybrid optimal trajectories must obey the differential equations (10) and the discrete-based phase transition equations (11). The optimization parameters to be determined are the continuous $\boldsymbol{u}(t)$ and discrete control input trajectories $\boldsymbol{v}(t)$ and all, some, or none of the phase transition times.

The solutions to the HOCPs described in Definition 2 are deterministic *open-loop* trajectories. Like in conventional optimal control this problem class can be generalized to a stochastic setting or to treat issues like optimal closed-loop feedback control. The numerical solution of closed-loop hybrid feedback control problems, however, is at even a much earlier stage and the primarily finite-element based solution strategies that have been presented for their solution (Branicky et al., 1999, Hedlund and Rantzer, 1999, Tomlin, 1999) cannot readily handle nonlinear systems of more than three dimensions due to the well-known curse of dimensionality (Hardt et al., 2000).

A framework for modeling and (optimally) controlling mixed logical dynamical systems described by linear dynamic equations subject to linear inequalities involving real and integer variables has been proposed by (Bemporad and Morari, 1999a). The on-line optimization problems resulting from a predictive control scheme are solved numerically by application of a mixed-integer quadratic programming branch-and-bound method. However, the approach is not applicable to our class of HOCPs with *nonlinear* dynamics equations subject to *nonlinear* constraints.

4.2 Numerical Solution Strategies

A set of several different numerical strategies is presented here for the approximation of the solution to the HOCP. The basis for the suboptimal solution strategies is the highly efficient direct collocation method implemented in the software package DIRCOL (von Stryk, 2001) to approximately solve optimal control problems using solutions to (sparse) nonlinear programs. DIRCOL was primarily designed for the solution of optimal control problems related to piecewise continuous, nonlinear dynamical systems though it handles well important discrete system components such as unknown interior time events (TE) when state resets (SR) or vector field switches (VFS) may occur. Other discrete state aspects it cannot handle directly such as the number of interior SR or VFS events. These aspects must be specified in advance. For this reason, the proposed solution strategy is to use DIRCOL in the inner optimization iteration and other strategies to solve for the combinatorial aspect of the discrete-event in an outer level optimization. The key to cope with the possibly overwhelming combinatorial complexity of HOCPs is to reduce the number of candidates to be evaluated in the outer iteration.

After providing some insights into the method DIRCOL, two alternatives HOCP solution strategies will be shown: (i) suboptimal solution with interior event time and state constraints fixed on a grid combined with graph search, and (ii) transformation to a mixed-binary-optimal control problem and its subsequent solution using a branch-and-bound algorithm.

Sparse Direct Collocation

The numerical method of sparse direct collocation implemented in DIRCOL can efficiently solve multi-phase optimal control problems with a fixed discrete state trajectory. The state $\boldsymbol{x}$ is approximated by cubic Hermite polynomials $\tilde{\boldsymbol{x}}(t) = \sum_j \alpha_j \hat{\boldsymbol{x}}_j(t)$

and the control vector $\boldsymbol{u}$ by piecewise linear functions $\tilde{\boldsymbol{u}}(t) = \sum_k \alpha_k \hat{\boldsymbol{x}}_k(t)$ on a discretization grid $t_i^c = t_1^{(i)} < t_2^{(i)} < \ldots < t^{(i)}_{n_t^{(i)}} = t_{i-1}^c$ in each phase. The state differential equations (10) are pointwise fulfilled at the grid points and grid midpoints, resulting in a set of nonlinear NLP equality constraints $\boldsymbol{a}(\boldsymbol{y}) = 0$. The control or state inequality constraints are to be satisfied at the grid points resulting in a set of nonlinear NLP inequality constraints $\boldsymbol{b}(\boldsymbol{y}) \geq 0$. The vector $\boldsymbol{y}$ contains the n_y parameters $\boldsymbol{y} = (\alpha_1, \alpha_2, \ldots, \beta_1, \beta_2, \ldots, p, t_1, \ldots, t_{N-1}, t_f)^T$ where $p_i \in [0,1]$, $i = 1, \ldots, n_p$ denotes the set of relaxed binary variables. With ϕ as the parameterized cost index (18), the nonlinearly constrained optimization problem may be written as the nonlinear program (NLP)

$$\min_{\boldsymbol{y}} \phi(\boldsymbol{y}) \quad \text{subject to} \quad \boldsymbol{a}(\boldsymbol{y}) = 0,\ \boldsymbol{b}(\boldsymbol{y}) \geq 0\,. \tag{16}$$

The transcription of the optimal control problem to an NLP is made by DIRCOL (von Stryk, 2001), the NLP is solved efficiently with the advanced SQP-based sparse nonlinear program solver SNOPT (Gill et al., 1997), and subsequently DIRCOL processes the solution to provide state and control trajectories, error estimates and output that may be used to verify the optimality of the solution.

Important features of the method are:

- As the grid becomes finer, the discretized solution converges to a solution of the Euler-Lagrange differential equations (EL-DEQs) according to the Maximum Principle.
- Reliable estimates of the adjoint variable trajectories $\tilde{\lambda}$ along the discretization grid may be derived from the Lagrange multipliers of the NLP. They enable a verification of the optimality conditions of the discretized solution without solving explicitly the EL-DEQs.
- Local optimality error estimates can be derived which enable efficient strategies for successively refining a first solution on a coarse grid.
- The NLP Jacobians $(\boldsymbol{\nabla a}(\boldsymbol{y}), \boldsymbol{\nabla b}(\boldsymbol{y}))$ are sparse and structured, permitting the use of sparse solvers.
- Computation is fast because ODE simulation and control optimization are performed simultaneously (unlike shooting methods).
- In extension of (10), the method is also applicable to systems described by differential-algebraic equations of differential index 1. In this case, the algebraic state variables are discretized analogously to the control variables by piecewise linear functions.

Suboptimal Solution Technique

Suboptimal solutions may be obtained by fixing interior point times and states to fixed values on a (fine) grid. Between all these grid points standard optimal control problems with fixed boundary conditions are solved. Finally, the suboptimal solution to the HOCP is obtained by a graph search with each grid point forming nodes and the optimal cost weighing the vertices of this graph. This solution strategy is applied to solve the cooperative multi-arm transport problem in Sect. 5.1, see also (Buss et al., 2000a, Buss, 2000, Denk, 1999). Disadvantages of this approach are the possibly high number of multi-point boundary value problems to be solved and the inherent suboptimality of the obtained solution. On the other hand, an appealing advantage is that by problem understanding one often has good insight as to how the grids need to be specified, and that useful solutions usually can be obtained easily.

Branch-and-Bound

The solution method for mixed-binary optimal control problems (MBOCP) using a combination of sparse direct collocation and branch-and-bound was first presented in (von Stryk, 2000) and further investigated in (Buss et al., 2000a, von Stryk and Glocker, 2000, von Stryk and Glocker, 2001). Given certain assumptions, the HOCP may be transformed into a MBOCP with a simple transformation of its discrete variables. For this we assume:

(A1) The number $N-1 \geq 0$ of event times t_i and, thus, the number N of phases are finite and known (this assumption may be circumvented with yet another "outer" iteration to vary N).

(A2) The discrete state variable $\boldsymbol{q}$ and the discrete control variable $\boldsymbol{v}$ are constant in each phase and may only change at an event t_i.

Each discrete variable $q_k(t)$ (or $v_l(t)$), $0 \leq t \leq t_f$, is described by an integer variable $\boldsymbol{z}_k \in \mathbb{Z}^{n_c+1}$ with $q_k(t) = z_{k,i}$ in the i-th phase. A scalar, integer variable z_1 with given lower and upper bounds $z_1 \in [z_{1,\min}, z_{1,\max}] \subset \mathbb{Z}$ can be transformed into a binary variable $\boldsymbol{\omega} \in \{0,1\}^{n_\omega}$ of dimension n_{z_1} by

$$z_1 = z_{1,\min} + \omega_1 + 2^1\omega_2 + \ldots + 2^{n_\omega - 1}\omega_{n_\omega}, \tag{17}$$

with $n_\omega = 1 + \text{INT}\{\log(z_{1,\max} - z_{1,\min})/\log 2\}$. In this manner, a binary control vector $\boldsymbol{\omega}$ may be used to represent both the unknown discrete state $\boldsymbol{q}$ in each phase and the discrete control variable $\boldsymbol{v}$ which controls the order and types of phase transitions.

The MBOCP is to minimize the real-valued, hybrid performance index

$$J[\boldsymbol{u}, \boldsymbol{\omega}] = \Theta + \sum_{i=1}^{N} \int_{t_{i-1}}^{t_i} \psi(\boldsymbol{x}(t), \boldsymbol{u}(t), \boldsymbol{\omega}, t)\, \mathrm{d}\, t \tag{18}$$

subject to (10)-(15) with the discrete variables $\boldsymbol{q}$ and $\boldsymbol{v}$ substituted by the binary control vector $\boldsymbol{\omega} \in \{0,1\}^{n_\omega}$ in both Θ and ψ. The solutions of the MBOCP are the optimal (open loop) trajectories of $\boldsymbol{x}^*(t)$, $\boldsymbol{u}^*(t)$, $0 \leq t \leq t_f$, the optimal phase transition times $t_i^{c\,*}$, the possibly free final time t_f^* , and the optimal binary control vector $\boldsymbol{\omega}^*$.

Remark 3. The nature of the binary control vector $\boldsymbol{\omega}$ appearing in the MBOCP is twofold. On the one hand it represents the discrete control variable $\boldsymbol{v}$ that controls the order and types of phase transitions, on the other hand it also represents the discrete state $\boldsymbol{q}$ in each phase.

To avoid solving all $\{0,1\}^{n_w}$ MBOCPs, a branch-and-bound strategy in combination with a binary search tree is employed: The subproblems solved by DIRCOL provide approximate upper and lower bounds to the MBOCP performance index. If the lower bound at a node is greater than the global upper bound, that branch is discarded. The comparison of subproblem solutions is additionally aided by the use of the optimality error estimate (confidence interval) computed by DIRCOL (von Stryk, 2001). A subproblem is constructed by either fixing a component of the binary control vector $\boldsymbol{\omega}_i$ to 0 or 1 or relaxing it $0 \leq \boldsymbol{\omega}_i \leq 1$, $i \in \{1, 2, \ldots, n_\omega\}$. The MBOCP is thus reduced to a "continuous" multi-phase optimal control problem.

Remark 4. The B&B procedure on the binary control vector requires existence of solutions to relaxed MBOCPs, or more precisely, the existence of continuous relaxations to the MBOCP. For some MBOCPs, numerical solutions may not exist for their relaxations. When they exist, the relaxed binary variables may not necessarily have any physical meaning with respect to the underlying application. This however does not present any numerical difficulties. The solution of subproblems in the B&B is analagous to the application of the interior-point solution method to linear programming problems. The iterative procedure normally first delivers a well-defined solution at termination of the algorithm. Usually additional modeling effort will be required in defining suitable "meta"-MBOCPs allowing useful relaxations analogously to the definition of superstructures for mixed-integer nonlinear programming problems (Adjiman et al., 1998).

Remark 5. As it must be expected that some modeling effort for the MBOCP is required before applying numerical methods, it has been suggested to derive suitably simplified and problem specific "screening models" (Allgor and Barton, 1997). A screening model can be solved to simultaneously guarantee global optimality and to yield a rigorous lower bound on the solution of the MBOCP, thus avoiding the need for dealing with relaxed MBOCPs. An application for a simple batch process development has successfully been investigated in (Allgor and Barton, 1997). Although in principle the idea seems to be applicable to a wide class of problems, there is no constructive way to obtain a screening model for a concrete MBOCP.

Remark 6. The challenge in solving relaxed MBOCPs during the binary tree search cannot be underestimated. There is no numerical method available that solves optimal control problems with nonlinear dynamics defined in multiple phases and subject to nonlinear constraints and with phase transitions at unknown times *guaranteeing* the global optimum or that even guarantees a locally optimal solution in general at all. However, not only the global optimum is of interest. For many types of MBOCPs, even a "good" solution obtained by the proposed approach that significantly improves the initial guess will be highly appreciated.

The branch-and-bound procedure is outlined as follows:

1. Find a global upper bound. Make an initial guess for $\boldsymbol{\omega}$ and solve the resulting control problem with $\boldsymbol{\omega}$ fixed;
2. At the root node, relax all binary variables ($0 \leq \omega_i \leq 1, i \in \{1, 2, \ldots, n_\omega\}$) and solve to obtain a lower bound to the solution;
3. Select the branching variable ω_i and solve both subproblems with that component set to 0 and 1 thereby creating two offspring to the current node;
4. Select the next node where to continue the branching process by either: Breadth First Search (node with minimal performance out of those with the least amount of fixed components), Depth First Search (node with minimal performance out of those with the maximum amount of fixed components), Minimum Bound Strategy (node with minimal performance);
5. If the lower bound in a node is greater than the current best upper bound of the whole search tree, then all subsequent branches from this node are trimmed.

Depending on the problem, this approach may get caught in local minima which can be avoided by perturbations for the relaxed problems. It is also hard to guarantee that trimmed branches do not contain the true global minimum. A positive note is that useful suboptimal solutions are readily computable.

5 Applications

5.1 Multi-arm Transportation Task

Figure 5 shows a cooperative multi-arm transport task. The square object is initially on the right and is to be transported to the elevated goal position on the left. This is to be accomplished by picking up the object with transport arm 1, handing it over to arm 2, then to arm 3, and finally placing it in the goal position. Each transport arm j has two rotational joints $\theta_{j,i}$ driven by control input torques $u_{j,i}$, $j = 1, 2, 3$, $i = 1, 2$. The effector of each transport arm can be opened/closed to grasp/release the object by a discrete control input v_j. The transportation task should be performed such that the cost index of quadratic power consumption is minimized

$$\min_{u_{j,i}(t), v_j(t)} \quad J = \int_0^{t_f} \sum_{j=1}^{3} \sum_{i=1}^{2} (u_{j,i}\, \dot{\theta}_{j,i})^2 \, dt \, .$$

To solve this HOCP we need to determine the optimal hybrid control trajectories $u^*_{j,i}(t)$, $v^*_j(t)$, the positions, velocities and times of object handover. The physical parameters of the multi-arm system are assumed as: mass $m_1 = m_2 = 5$, length $l_1 = l_2 = 1$ of link 1, 2, respectively, object mass $m_o = 10$, ground distance from arm mount point $x_g = 1.5$. The distance between two arms is $d = 1.5$, the grid points for possible handovers of arm 1 are at $y_{1,ho} = -0.75$, $x_{1,ho} = 1.5/x_{1,ho} = 1$ (ground/air), and likewise for the other arms.

For each arm $i = 1, 2, 3$ the hybrid model has 4 discrete states $q_i = 1, 2, 3, 4$ as follows: $q_i = 1$: arm has no contact with environment, effector open; $q_i = 2$: arm

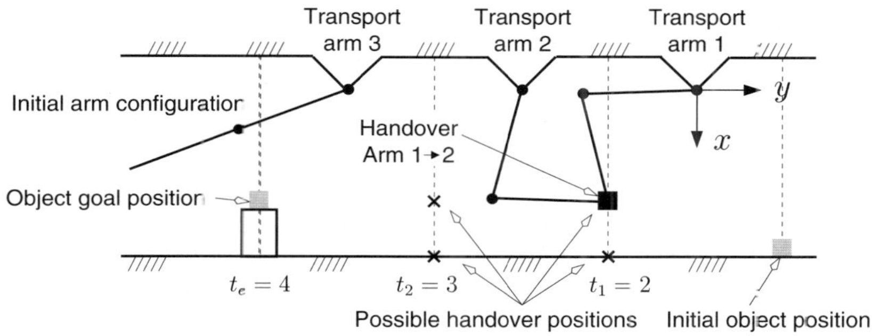

Fig. 5. Cooperative multi-arm transport task

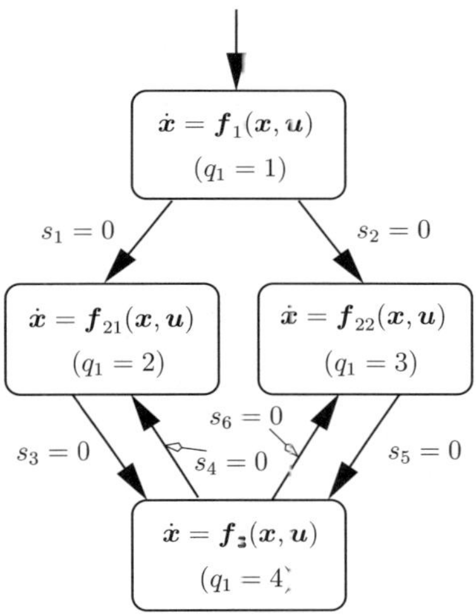

Fig. 6. Hybrid model for a single arm

holds object in configuration 1 (elbow right) object has contact to ground; $q_i = 3$: arm holds object in configuration 2 (elbow left) object has contact to ground; $q_i = 4$: arm holds object in the air, no contact with environment. The variable structure q_i dependent motion differential equation for arm i then are:

$$\dot{\boldsymbol{x}}_i = \boldsymbol{f}(\boldsymbol{x}_i, \boldsymbol{u}_i, q_i) = \begin{cases} \boldsymbol{f}_1(\boldsymbol{x}_i, \boldsymbol{u}_i) & \text{if } q_i = 1 \\ \boldsymbol{f}_{21}(\boldsymbol{x}_i, \boldsymbol{u}_i) & \text{if } q_i = 2 \\ \boldsymbol{f}_{22}(\boldsymbol{x}_i, \boldsymbol{u}_i) & \text{if } q_i = 3 \\ \boldsymbol{f}_3(\boldsymbol{x}_i, \boldsymbol{u}_i) & \text{if } q_i = 4 \end{cases} \tag{19}$$

Note that if $q_i = 2, 3$ the arm is also subject to a kinematic equality constraint as ground contact needs to be maintained. Environment forces must also be taken into account during such phases. The complete hybrid model of a single arm is shown in Fig. 6.

Applying the suboptimal solution strategy outlined in Sect. 4.2, the coupling of the optimal control problems is first eliminated for each of the transport arms by

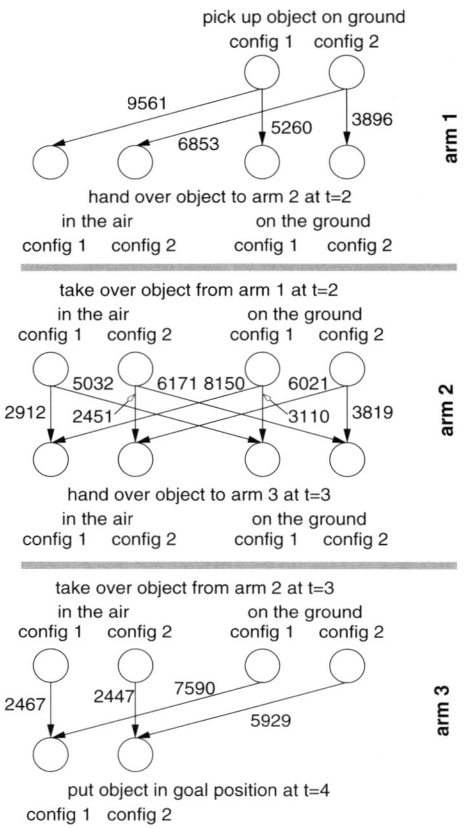

Fig. 7. Feasible handover TPBVPs for each arm

fixing the possible times and states of handover to constant values on a grid, see Fig. 5. The object handover time from arm 1 to 2 is fixed to $t_1 = 2$ and only two possible handover positions (on the ground and in the air and at zero velocity) are considered. Some of the handover possibilities can be excluded because of internal arm collision problems, e.g. handover in the air between arms 1, 2 with configuration 2, 1, respectively.

All remaining feasible handover TPBVPs (Two Point Boundary Value Problems) and the cost of the optimal solutions obtained by DIRCOL are shown in Fig. 7. The three subgraphs are then combined into the complete graph in Fig. 8, in which the best suboptimal solution is obtained by minimum path search; also marked in Fig. 8. The best suboptimal solution to the transport task is to pick up the object by arm 1 and hand it over to arms 2/3 in the air at the fixed positions and times as shown in Fig. 5. Figure 9 shows some snapshots of the suboptimal coordinated transportation task.[1].

[1] An animated movie of the suboptimal solution to the multi-arm transportation task is available at `http://www.rs.tu-berlin.de/videos`

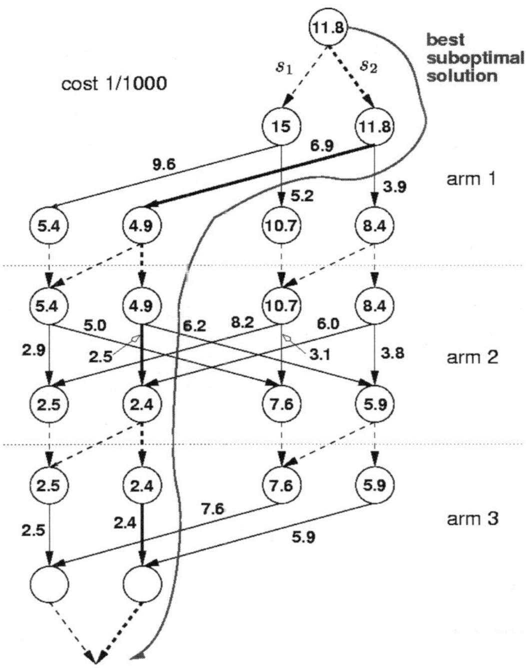

Fig. 8. Graph connecting all feasible discrete sequence candidates

5.2 Underactuated Two Degree-of-Freedom Robot Arm

The trajectory planning example application is considered for a 2-link SCARA robotic arm with two rotational degrees-of-freedom, yet only one actuated (R2D1). In the first joint a torque u_1 may be applied while the second joint may be influenced only by a holding brake controlled by $v_1(t) \in \{0, 1\}$, see Fig. 10 and (Mareczek et al., 1999, Mareczek et al., 1998). The brake can only be set when the second joint has reached a zero relative velocity. A discrete control action can switch back and forth between the passive and locked modes for the second joint while a continuous control force is applied to the first joint actuator. We are interested in finding not only the optimal continuous state and control trajectories, but also the optimal discrete strategy composed of the optimal *number* and *times* of the switches necessary to move the R2D1 from a given initial state to a goal state.

The following $\mathcal{H}_2$ performance index is considered

$$J[u_1, v_1] = \int_0^{t_f} (\boldsymbol{x}(t) - \boldsymbol{x}_f)^T \boldsymbol{W} (\boldsymbol{x}(t) - \boldsymbol{x}_f) + \alpha (u_1(t) - u_{1,f})^2 \, \mathrm{d}t \qquad (20)$$

where $\boldsymbol{W} \in \mathbb{R}^{4\times 4}$, $\boldsymbol{W} \geq 0$, and $\alpha > 0$. Here, we use $\boldsymbol{W} = \boldsymbol{I}$ and $\alpha = 1$. Furthermore, $\boldsymbol{x}_f \in \mathbb{R}^4$ denotes a desired final state, and $u_{1,f}$ is the control value for which the system is at equilibrium at $\boldsymbol{x}_f$. The final time is constrained, e. g., by

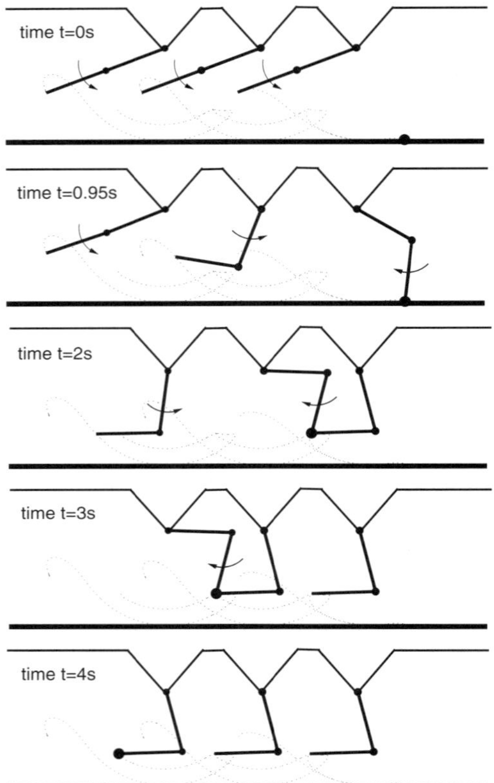

Fig. 9. Snapshot sequence of suboptimal transport solution

$t_f \leq 10\,\text{s}$. The HOCP is to minimize J subject to the robot dynamics

$$\begin{aligned} \ddot{\boldsymbol{\theta}} &= \begin{pmatrix} u_1 \\ 0 \end{pmatrix} - v_1(t)\,\boldsymbol{F}_1(\boldsymbol{\theta}(t),\dot{\boldsymbol{\theta}}(t)) - (1-v_1(t))\,\boldsymbol{F}_2(\boldsymbol{\theta}(t),\dot{\boldsymbol{\theta}}(t)) \\ \boldsymbol{F}_i(\boldsymbol{\theta},\dot{\boldsymbol{\theta}}) &= \boldsymbol{M}_i^{-1}(\boldsymbol{\theta})\big(\boldsymbol{C}_i(\boldsymbol{\theta},\dot{\boldsymbol{\theta}}) + \boldsymbol{g}_i(\boldsymbol{\theta}) + \boldsymbol{r}_i(\dot{\boldsymbol{\theta}})\big), \quad i = 1,2 \end{aligned} \tag{21}$$

$$\begin{array}{llll} \boldsymbol{x}(t) & = (\theta_1(t),\dot{\theta}_1(t),\theta_2(t),\dot{\theta}_2(t)) & \boldsymbol{u}(t) \in \mathcal{U} = \mathbb{R} \\ \boldsymbol{x}(0) & = \boldsymbol{x}_0 = (1.2,\ 0,\ 0.8,\ 0)^T & \boldsymbol{x}(t) \in \mathcal{X} = SO(1) \times SO(1) \times \mathbb{R}^2 \\ \boldsymbol{x}(t_f) & = \boldsymbol{x}_f = (\pi/2,\ 0,\ -\pi/2,\ 0)^T & \boldsymbol{v}(t) \in \mathcal{V} = \{0,1\} \\ v_1(t_f) = 1 & \text{(brake on)} & \boldsymbol{q}(t) \in \mathcal{Q} = \emptyset \end{array} \tag{22}$$

where $\boldsymbol{M}_i$ are the mass-inertia matrices for each dynamical configuration, $\boldsymbol{C}_i$ are the vectors of Coriolis and centrifugal forces, $\boldsymbol{g}_i$ are the vectors of gravitational forces, and $\boldsymbol{r}_i$ are the friction forces. The physical parameters in standard units are: $l_1 = 0.300$, $l_{c1} = 0.206$, $l_{c2} = 0.092$, $I_1 = 0.430$, $I_2 = 0.127$, $m_1 = 10.2$, $m_2 = 5.75$.

The optimal control problem for R2D1 is formulated as a MBOCP, and the numerical approach discussed in Sect. 4.2 is applied. The time $t_f \leq 10$ is initially divided

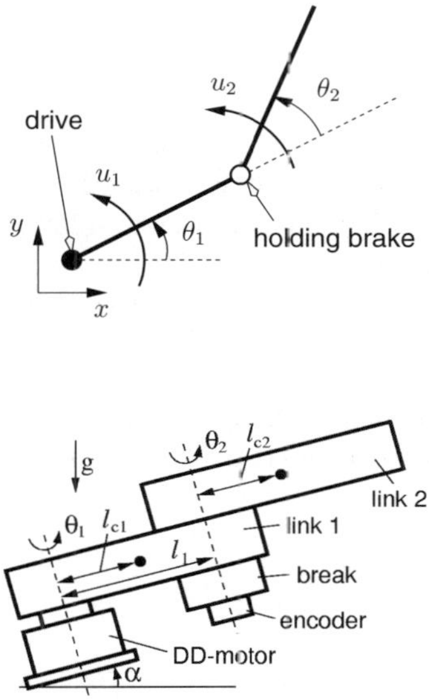

Fig. 10. Kinematic structure of R2D1 (Mareczek et al., 1998, Mareczek et al., 1999)

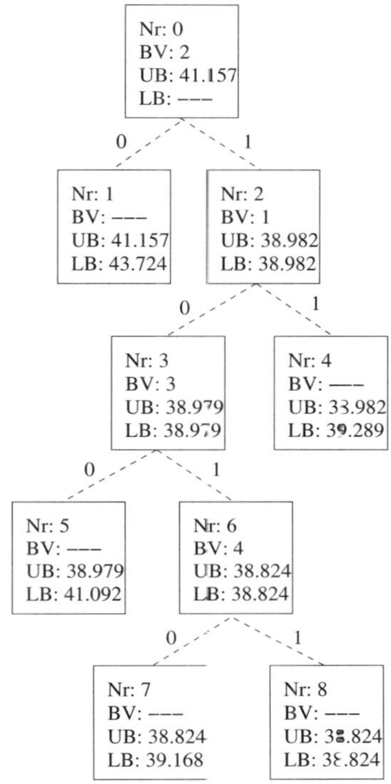

Fig. 11. Branch-and-bound search using *minimum bound* strategy. *Nr* – node number from search order, *BV* – branching variable, *UB* – global upper bound, *LB* – lower bound for branch

into a fixed number $m = 8$ of phases, though the intermediate times corresponding to the phase transitions may vary freely. Included in the problem formulation are a set of constant, unknown binary parameters $p_i \in \{0, 1\}, i = \{1, \dots, n_p\}$ which are related to the unknown binary variables ω_i. They determine the total number of switches and indicate at which of the pre-defined phase transitions a switch occurs. The first component p_1 indicates in which discrete state the system starts, $\{p_1 = 0$, brake off; $p_1 = 1$, brake on$\}$. The remaining components of $\boldsymbol{p}$ are a binary representation of the total number of switches taking place during the time interval. For example, if five switches occur beginning with the brake off, then $p = [p_1\ p_2\ p_3\ p_4] = [0\ 1\ 0\ 1]$ and the switches are assigned to the predefined phase transitions using the scheme: $p_k = 1 \ \Rightarrow\ 2^{(n_p-k)}$ switches with one every 2^{k-1} phase transitions beginning with number $2^{(k-2)th} + 1$. Fig. 12 depicts the phase transitions over which the binary parameter p_k exerts an influence.

The branch-and-bound search strategy was used together with a minimum-bound node selection strategy. Figure 11 displays the complete binary search path for the problem. An initial solution with $\boldsymbol{p}$ fixed at $[0\ 1\ 0\ 0]$ (4 switches) is first calculated

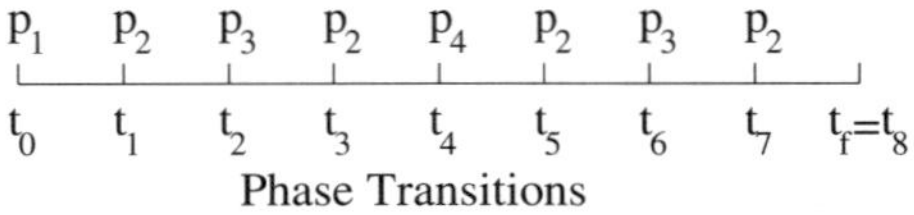

Fig. 12. Phase transitions influenced by binary parameters p_k

to obtain an upper bound of $J^* = 41.157$. Lower bounds were first calculated for the children of the root node, and the second binary variable is arbitrarily first selected as the branching variable. The final optimal solution has a discrete solution of $\boldsymbol{p}^* = [0\ 1\ 1\ 1]$ corresponding to 7 switches starting with the brake off and an objective value of $J^* = 38.824$. As is normally the case in a branch-and-bound search, the search procedure ends if an integer solution obtained from a relaxed problem is the new best lower bound. In this case, our optimal solution was obtained already at node 2, after the third optimization run. The search though was continued here to verify the solution and ensure that it did not correspond to a local minimum.

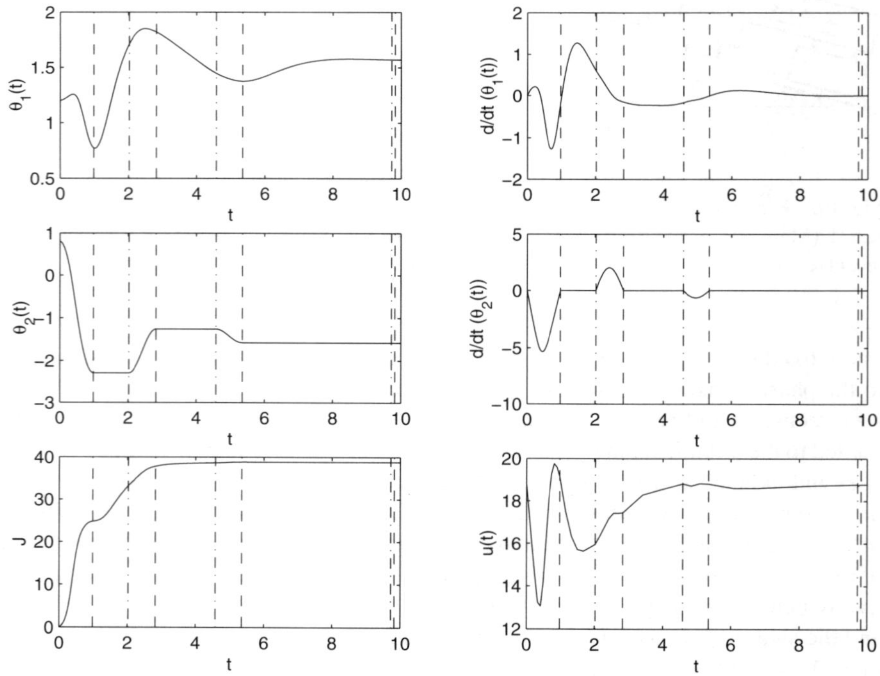

Fig. 13. Final optimal hybrid switching solution with 7 switches

In order to avoid convergence to a local minimum, at intermediate steps all relaxed binary parameters in the optimization are initialized to 0.5 to perturb the system away from its starting values and therewith avoid local minima. The final solution[2] as displayed in Fig. 13 has an optimality error of $\tilde{w} = 0.567$ (von Stryk, 2001). The incremental difference in the objective decreases rapidly with an increasing number of switches such that the solution with 5 or 6 switches lie within the error margin for the optimal solution with 7 switches. The optimality tolerance (Gill et al., 1997, von Stryk, 2001) set at 10^{-4} may then be reduced to obtain more accurate solutions in order to correctly distinguish between them. It is also possible at this point to lengthen the search by reinitializing the binary search with more predefined phase transitions thereby allowing for more switches to take place. The average computational time by DIRCOL for each optimal control problem (the solution at a given node) was 19.6 seconds on a Pentium III 500 MHz computer, the average grid size $\sum_{i=1}^{N} n_t^{(i)}$ was 56.3, and the average NLP dimension was $n_y = 278$, $n_a = 230$.

5.3 The Motorized Traveling Salesman

We consider the hybrid dynamical extension of one of the most popular combinatorial optimization problems: *A motorized salesman is on his way to visit n_c cities at most one time. He is not allowed to stop in the cities, instead he should drive through them on a smooth curve. He starts at the origin and returns there after his journey. How should he steer and accelerate and in which order should he pass through the cities to minimize the overall traveling time?*

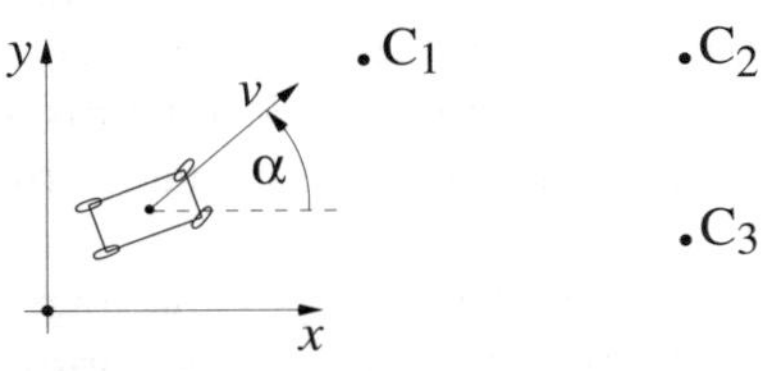

Fig. 14. Motorized traveling salesman problem (MTSP)

In the standard setting as a combinatorial optimization problem, the interconnections between two cities are independent of each other. In the problem setting here, the salesman has to travel on a smooth curve and the performance in between two cities depends on the overall selection of the continuous (steering wheel, gas and brake pedal) and discrete (order of cities) controls. This benchmark hybrid optimal control problem serves to demonstrate the strong interaction of continuous and discrete dynamics that may occur for even low dimensional systems.

[2] An animated movie of the final solution for the R2D1 robot control is available at `http://www.sim.informatik.tu-darmstadt.de/videos`

The motorized traveling salesman (MTSP) can be described by a simplified kinematical model describing a point mass moving in a (x, y)-plane

$$\begin{aligned} \dot{x}(t) &= v_x(t), & x(0) &= 0 = x(t_f), \\ \dot{y}(t) &= v_y(t), & y(0) &= 0 = y(t_f), \\ \dot{v}_x(t) &= a_x(t), & v_x(0) &= 0 = v_x(t_f), \\ \dot{v}_y(t) &= a_y(t), & v_y(0) &= 0 = v_y(t_f), \\ a_x^2 + a_y^2 &\leq 7 \quad . \end{aligned} \tag{23}$$

Hereby v_x and v_y denote the velocity and a_x, a_y the acceleration or braking of the car in x respectively in y direction, i.e., the continuous state and control variables. The MTSP is formulated as an MBOCP according to Section 4.2 by $\boldsymbol{u} = (a_x, a_y)$, $\boldsymbol{x} = (x, y, v_x, v_y)$ and

$$\min_{\boldsymbol{u},\, \boldsymbol{\omega}} J[\boldsymbol{u}, \boldsymbol{\omega}] := t_f + 0.002 \int_0^{t_f} (u_1^2 + u_2^2)\, \mathrm{dt} \tag{24}$$

$$\boldsymbol{r}_{(i)}(\boldsymbol{x}(t_i^-), \boldsymbol{x}(t_i^+), \boldsymbol{\omega}, t_i) := \begin{pmatrix} x(t_i^-) \\ y(t_i^-) \end{pmatrix} - \sum_{k=1}^{N-1} \omega_{i,k} \begin{pmatrix} x_k \\ y_k \end{pmatrix} \tag{25}$$

$$\boldsymbol{x}(t_i^+) = \boldsymbol{x}(t_i^-) \tag{26}$$

$$\sum_{i=1}^{N-1} \omega_{i,k} = 1, \quad \sum_{k=1}^{N-1} \omega_{i,k} = 1, \quad 0 \leq \omega_{i,k} \leq 1 \tag{27}$$

At the end of each phase the salesman must visit one of the $(N-1)$ cities $(x_k, y_k)^T$. This is ensured by (25). The linear constraints make sure, that each city is visited exactly once. Thus the final matrix $\Omega = (\omega_{i,k})_{i,k \in \{1,\ldots,N-1\}} \in \mathbb{R}^{(N-1)\times(N-1)}$ has in each column and each row exactly one entry equal to 1. The other values are equal to 0. If $\omega_{i,k} = 1$, the k-th city is visited at the end of the i-th phase. Each tour is a permutation of the $(N-1)$ cities. Thus each feasible matrix Ω can be obtained by a permutation of the columns of the identity matrix.

If the salesman has to visit $(N-1)$ cities, then there are $(N-1)!$ possible tours, including the symmetric ones. Figure 15 shows solutions to three possible scenarios. In the present formulation $(N-1)^2$ binary values are used resulting in a branch & bound tree with a depth of $(N-1)^2$ and a breadth of $2^{(N-1)^2}$ nodes. The tree has $(2^{(N-1)^2+1} - 1)$ nodes; most of them are infeasible though with respect to the linear constraints (27).

If a tree search is performed beginning at the root of the tree without the knowledge of an upper bound for the problem, at least $(N-1)^2$ nodes have to be analyzed to obtain an initial upper bound. In our numerical experiments, however, even more steps are usually needed to reach the leaves. Thus, the search for a optimum should begin at the leaves of the search tree until an initial upper bound is provided. The branch & bound algorithm starts afterwards to prove whether this bound is optimal (in convex cases) or to find a better one.

For each of the tours the continuous controls and switching times were optimized using the direct collocation method of Sect. 4.2 with respect to the terminal time t_f

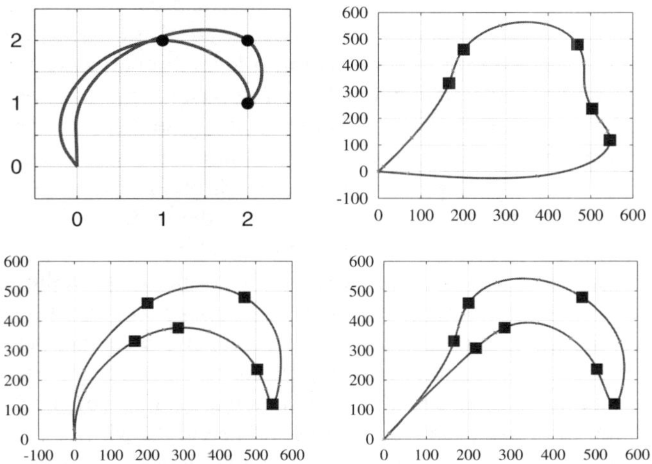

Fig. 15. Solutions for the MTSP for 3, 5, 6, and 7 cities

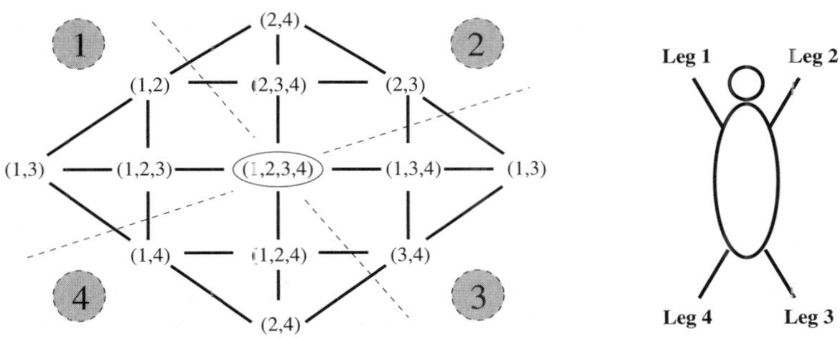

Fig. 16. Hybrid automaton for the quadruped. The nodes represent the different discrete states; the numbers in parentheses refer to the numbers of the support legs. Edges indicate discrete transitions (a leg has either broken ground contact or just entered a contact condition)

for a given discrete variable, i. e., order of cities, i. e., sequence of phases. To start the iterative direct collocation method, initial guesses for the switching points consisting of $t_{i,\text{estimate}} = i$, $i = 1, \ldots, N$, are used. A linear interpolation of the coordinates of the cities is applied as an initial guess for x and y, whereas v and a were initially set to zero. Computational times for obtaining a final solution can vary between a few minutes (for 5 cities) and several hours (for 7 cities) on a Pent. III, 900 MHz PC.

6 Other Problems

The robotic applications presented in this work serve primarily as illustrative examples to demonstrate the complexity existing in the optimal control of strongly interconnected discrete–continuous systems. A more realistic and challenging problem however that is currently being investigated using these approaches is the gait generation problem for four-legged robots. Quadrupeds are ideal for many applications due to their increased dexterity in comparison to legged robots with more legs and its increased stability compared to a biped. An unsolved problem, however, remains the determination of the optimal gait for moving at a given velocity where the order of leg movement and ground contact conditions at each moment in time are discrete characteristics of the problem. Preliminary work on this problem may be found in (Hardt and von Stryk, 2000). Fig. 16 displays the hybrid automaton for quadruped legged locomotion. Each node represents a different discrete state, where a different combination of legs are supporting the quadruped. A periodic gait is characterized, apart from the periodicity of its continuous states, by the discrete condition that each leg has exactly one period of ground contact and another period without contact during the gait. As a result, periodic gaits are represented by periodic paths which must visit all four quadrants in the hybrid automaton (Fig. 16) and then return to its starting point; thus, this problem is closely related to the MTSP.

The underlying HOCP for step sequence planning in humanoid walking is also an open challenge; see (Lorch et al., 2000) for preliminary results combining step sequences from pre-calculated suboptimal step primitives. Another important robotic problem within this context is manipulation using multi-fingered dextrous robotic hands (Schlegl et al., 2002b).

7 Conclusions

A methodology for the modeling and control of hybrid nonlinear dynamical systems is presented. The dynamical model, feedback solutions, and the numerical methods presented for the solution of hybrid optimal control problems are all geared towards the analysis of hybrid problems where the degree of discrete–continuous interconnection is strong, and the continuous dynamics may be highly nonlinear and of high dimension. In particular, the hybrid optimal control problem (HOCP) is defined and two approaches are described for its solution. The first approach decouples HOCPs by fixing interior point time and state constraints to a grid of possible values. Then, solutions to the decoupled TPBVPs are obtained, their optimal cost assigned to a graph with nodes representing the grid points and vertices the optimal cost. In this graph the best suboptimal solution is found by minimum path search. Alternatively, a branch-and-bound strategy is proposed based on the decomposition of HOCPs into MBOCPs. Binary variables are successively relaxed to obtain upper and lower bounds on the solutions. The search in the resulting solution tree is performed by branch-and-bound. The solutions to three hybrid control problems in robotics illustrate the effectiveness and scalability of the numerical methods presented here.

Acronyms

B&B	Branch and Bound
CE	Control Event
DE	Disturbance Event
EL-DEQ	Euler-Lagrange Differential Equation
HDS	Hybrid Dynamical System
HOCP	Hybrid Optimal Control Problem
HSM	Hybrid State Model
MBOCP	Mixed-Binary Optimal Control Problem
MTSP	Motorized Traveling Salesman Problem
NLP	Nonlinear Program
SQP	Sequential Quadratic Programming
SR	State Reset
SRVFS	State Reset and Vector Field Switch
TPBVP	Two Point Boundary Value Problem
TE	Time Event
VFS	Vector Field Switch

Generation of Optimal Control Policies for Systems with Switched Hybrid Dynamics

Olaf Stursberg[1,2], Sebastian Panek[1], Jochen Till[1], and Sebastian Engell[1]

[1] Process Control Laboratory (CT-AST), University of Dortmund, 44221 Dortmund (Germany)
[2] currently: Carnegie Mellon University, Dept. Electrical and Computer Engineering, Pittsburgh, PA 15213 (USA); olaf@ece.cmu.edu

Abstract. This contribution presents an approach to synthesize controllers for hybrid dynamic systems such that a given set of formal specifications is satisfied. The considered dynamics is represented by sets of possibly nonlinear ODEs among which is switched according to control inputs or autonomous events. The specifications define target states that have to be reached from an initial region, and sets of 'forbidden' states that must be avoided during the system evolution. In order to solve the control task, the nonlinear dynamics is approximated by linear models which are iteratively updated along the computed state trajectory. The computation is performed by solving a series of mixed-integer linear programming problems in a moving horizon fashion. The mixed integer programs are based on a new disjunctive formulation which allows an efficient solution. In order to select a specific control policy from the set of feasible solutions, a performance criterion is chosen that combines costs for the distance to the target, for the transition times, and the variation of the control inputs. The paper describes the modeling and synthesis procedure, and illustrates the result for a non-trivial processing system example.

1 Discrete Controllers for Hybrid Dynamic Systems

In the last decade, intensive research on model-based design of discrete controllers for hybrid systems has led to several approaches, most of which can be characterized by one of the two design paths shown in Fig. 1: The first one, usually referred to as *verification* approach (solid line), starts with the manual design of a model or implementation of the controller. The term *discrete controller* is herein understood as a logic which responds to certain *events* (occurring in the plant behavior) by an appropriate signal which is chosen from a finite discrete set. The designer tries to create the logic such that a given set of requirements is fulfilled for the supposed behavior of the plant. For the composition of the controller model and a formal representation of the hybrid dynamic behavior of the plant, the step of formal analysis then reveals if the controller satisfies the requirements for the actual plant behavior. If the requirements are verified the design can be accepted, otherwise the designer has to introduce appropriate modifications and to run the analysis again. Examples for approaches that follow this idea can be found in, e. g., (Asarin et al., 2000a, Chutinan and Krogh, 1999b, Kowalewski et al., 1999). The alternative approach of *controller synthesis* (dashed line in Fig. 1) combines the hybrid dynamic model and the requirement specification in an algorithmic procedure to directly get a controller

S. Engell, G. Frehse, E. Schnieder (Eds.): Modelling, Analysis and Design of Hybrid Systems, LNCIS 279, pp. 337–352, 2002.

that establishes the desired properties of the plant. Examples for approaches that follow this path are published in (Asarin et al., 2000b, Chouikha and Schnieder, 1999, Gokbayrak and Cassandras, 2000, Moor et al., 2001a, Trontis and Spathopoulos, 2001). Some of these methods aim at computing all admissible control inputs that fulfill the requirements for specific scenarios. The result of these procedures, a set of controllers that provably ensure the required properties of the plant, is comparable to that of the verification approach. So far, no general results exist on the comparison of the verification and the synthesis approach with respect to the computational costs. But the synthesis seems to be preferable with regard to the facts that it does not contain a manual design step, and that it avoids the possibly iterative loop of analysis and controller modification if a desired property is repeatedly refuted.

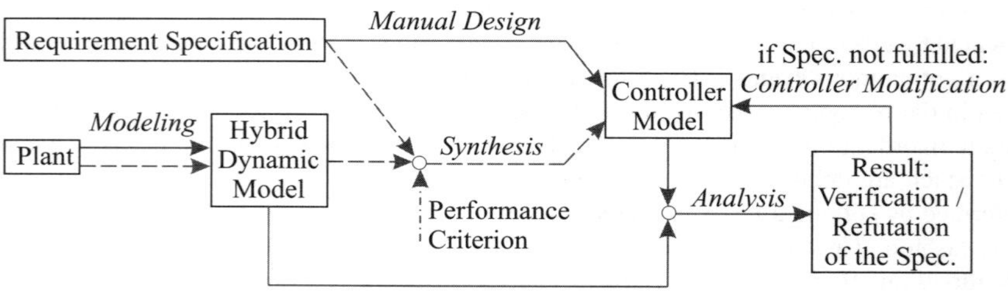

Fig. 1. Design of Discrete Controllers for Hybrid Plants

Rather than computing all admissible control inputs for a specific situation and the given requirements, recent research on synthesis methods has focussed on considering performance criteria additionally. These criteria select the one control strategy with the best performance from the set of admissible options. The idea of such an optimal control setting has been described in several publications, see e.g. (Branicky et al., 1998, Buss et al., 2002, Broucke et al., 2000, Bemporad et al., 2002, Dimitriadis et al., 1996b, Hedlund and Rantzer, 1999, Lincoln and Rantzer, 2001, De Schutter, 1999, Sussmann, 1999, Stursberg and Engell, 2002, Tomlin et al., 2000, Xu and Antsaklis, 2001, Zhang and Cassandras, 2001). The approaches differ in the formulation of the performance criterion (or 'cost functional'), the type of constraints, and the techniques to solve the optimization. With respect to the latter, the use of mixed integer programming is one possibility: The discrete and continuous control options are mapped into integer or continuous variables respectively, and the system dynamics is considered as optimization constraints formulated over these variables. The optimization routine returns those values of the control variables which maximize the performance criterion.

In this contribution, we adopt the scheme of using mixed integer programming to synthesize a controller: The plant dynamics is initially modelled by a hybrid automaton with switched continuous-time nonlinear dynamics (Sect. 2), and then approximated by linear discrete-time dynamics to allow a more efficient computation. The linearization is updated in each time-step for the current state of the nonlinear system. The controller synthesis is formulated as an optimization problem in which the discrete part of the dynamics is encoded by constraints for continuous and binary variables (Sect. 3). The solution by mixed-integer programming leads to control trajectories that obey specifications of desired and forbidden states, and optimize the chosen performance criterion, as illustrated for the example of a controlled chemical reactor in Sect. 4.

With respect to the methods described in literature, the largest similarities are apparently given to the approaches in (Buss et al., 2002, Buss et al., 2000b) and (Bemporad and Morari, 1999a, Bemporad et al., 1999). Buss et al. use a two-level solution scheme in which one level searches over the discrete options while the other accounts for the optimization of the nonlinear dynamics by using direct collocation techniques. The difference to our approach is that we avoid solving a complex nonlinear problem but rather use updated linear approximations and a moving horizon setting to reduce the complexity. The approach introduced by Bemporad et al. models hybrid systems with linear dynamics as so-called Mixed-Logical Dynamical Systems. The optimal control problem for these systems is then solved by mixed integer linear or multi-parametric quadratic programming. In difference to this method, we use so-called *disjunctive formulations* to represent the constraints and the performance criterion, and we employ a moving horizon scheme with variable time steps.

2 Modeling by Hybrid Automata

The plant dynamics considered in this paper combines continuous behavior with two types of discrete phenomena: switching between different continuous dynamics depending on the current state and switching triggered by discrete input variables[1]. The latter are appropriate to model controls that can only be altered between countably many (usually only a few) discrete options, while additional continuous control variables are defined on continuous sets.

2.1 Nonlinear Continuous-Time Dynamics

The initial format to model the plant behavior is that of hybrid automata with nonlinear continuous-time dynamics according to (Stursberg and Engell, 2002):

Definition 1 *Hybrid Automaton* $A_{N,C}$

***Syntax**: A hybrid automaton with switched nonlinear continuous time dynamics is given by:*

$$A_{N,C} = (\boldsymbol{X}, \boldsymbol{U}, \boldsymbol{V}, E, Z, \boldsymbol{f}, \phi) \tag{1}$$

with the following components: The state vector $\boldsymbol{x}(t)$ *is defined on the convex state space* $\boldsymbol{X} := \{\boldsymbol{x} \in X \mid \boldsymbol{C} \cdot \boldsymbol{x} \leq \boldsymbol{d}, \boldsymbol{C} \in \mathbb{R}^{q \times n}, \boldsymbol{d} \in \mathbb{R}^{q \times 1}, q \in \mathbb{N}\}$*. The vector of continuous inputs* $\mathbf{u}(t)$ *is defined on* $\boldsymbol{U} = [u_1^-, u_1^+] \times \ldots \times [u_{m_u}^-, u_{m_u}^+]$ *with* $u_j^-, u_j^+ \in \mathbb{R}$.

[1] The extension to linear jump functions that discontinuously reset the continuous state depending on the state and/ or on external inputs is possible, but not included in this report.

The discrete input $\boldsymbol{v}(t) \in \boldsymbol{V} = \{\boldsymbol{v}_1, \ldots, \boldsymbol{v}_{n_v}\}$, $dim(\boldsymbol{v}) = m_v$ *switches between finitely many options and only finitely often at times* t_k *in a time interval* $[t_0, t_f]$*. A set* $E = \{E_1, \ldots, E_{n_E}\}$ *of hyperplanes* $E_j := \{\boldsymbol{x} \in \boldsymbol{X} \mid \boldsymbol{c}_j \cdot \boldsymbol{x} = d_j, \boldsymbol{c}_j \in \mathbb{R}^{1\times n}, d_j \in \mathbb{R}\}$ *partitions the state space into a set* $R = \{R_1, \ldots, R_{n_R}\}$ *of convex and disjunct regions* $R_i := \{\boldsymbol{x} \in \boldsymbol{X}, E_i \in E \mid J = \{1, 2, \ldots, n_E\}, H \subseteq J, \forall h \in H : \boldsymbol{c}_h \cdot \boldsymbol{x} \sim_h d_h, \sim_h \in \{<, \leq\}\}$ *such that* $\bigcup_{i=1}^{n_R} R_i = \boldsymbol{X}$*. The set of discrete states* $Z = \{z_1, \ldots, z_{n_Z}\}$ *is formed by a mapping* $\rho : R \rightarrow Z$ *that assigns one discrete state* $z_i \in Z$ *to each region* R_i*.*

The continuous dynamics is given such that $\boldsymbol{f} : \boldsymbol{X} \times \boldsymbol{U} \times \boldsymbol{V} \times Z \rightarrow \mathbb{R}^n$ *defines* $\boldsymbol{x}(t)$ *as the unique solution of an ODE* $\dot{\boldsymbol{x}}(t) = \boldsymbol{f}_z(\boldsymbol{x}(t), \boldsymbol{u}(t), \boldsymbol{v}(t))$ *for a time interval* $[t_k, t_{k+1}[$ *between two switching events. At each time* t*, the dynamics* $\boldsymbol{f}_{z_i}$ *is valid which is assigned to the discrete state* z_i *with* $R_i = \rho^{-1}(z_i)$*,* $\boldsymbol{x}(t) \in R_i$*. According to the transition function* $\phi = Z \times \boldsymbol{X} \times \boldsymbol{X} \times R \rightarrow Z$*, a transition* $z_1 \rightarrow z_2$ *with* $z_1, z_2 \in Z$ *occurs at* t_k*, if:* $\boldsymbol{x}(t_k^-) \in R_1$*,* $\rho(R_1) = z_1$ *and* $\boldsymbol{x}(t_k) \in R_2$*,* $\boldsymbol{x}(t_k) \notin R_1$*,* $\rho(R_2) = z_2$*. Hence, if* R_1 *is left across* E_j*, the transition guard is:* $\boldsymbol{c}_j \cdot \boldsymbol{x}(t_k^-) = d_j \wedge \boldsymbol{c}_j \cdot \boldsymbol{x}(t_k) > d_j$ *(if* E_j *belongs to* R_1*), or* $\boldsymbol{c}_j \cdot \boldsymbol{x}(t_k^-) < d_j \wedge \boldsymbol{c}_j \cdot \boldsymbol{x}(t_k) = d_j$ *(if* E_j *is assigned to* R_2*) respectively. It is required that* $\boldsymbol{x}(t)$*,* $t \in [t_0, t_f]$ *fulfills the continuity condition* $\boldsymbol{x}^+ = \boldsymbol{x}(t^*)$ *for all discrete transitions and switching events at a time* t^**. (*$\boldsymbol{x}^+$ *is the time successor of* $\boldsymbol{x}(t^*)$*.)*

Semantics*: Let* $T = \{t_0, t_1, t_2, \ldots, t_N, t_f\}$ *contain the initial time* t_0*, the final time* t_f *and all points of time* $t_k \in]t_0, t_f[$*, at which a transition according to* ϕ *or a switching in* $\boldsymbol{v}(t)$ *occurs. A valid run* $r : T \rightarrow Z \times \boldsymbol{X}$ *of* $A_{N,C}$ *is then the finite sequence* $r(t_0), r(t_1), r(t_2), \ldots, r(t_N), r(t_f)$ *of hybrid states* $r(t_k) = (z(t_k), \boldsymbol{x}(t_k))$ *such that:*

(a) $A_{N,C}$ *is initialized to:*
$r(t_0) = (z(t_0), \boldsymbol{x}_0)$ *with* $z(t_0) \in Z$*,* $\boldsymbol{x}_0 = \boldsymbol{x}(t_0) \in \boldsymbol{X}$ *with* $\boldsymbol{x}_0 \in R^* \in R$*,* $\rho(R^*) = z(t_0)$*.*

(b) and evolves according to:
$r(t_{k+1}) = (z(t_{k+1}), \boldsymbol{x}(t_{k+1}))$ *following from the assignments:*
- $z(t_{k+1}) = \phi(z(t_k), \boldsymbol{x}(t_{k+1}^-), \boldsymbol{x}(t_{k+1}), R_j)$ *with* $\boldsymbol{x}(t_{k+1}^-) \in R_j$*,* $\rho(R_j) = z(t_k)$*, and*
- $\boldsymbol{x}(t_{k+1}) = \int_{t_k}^{t_{k+1}^-} \boldsymbol{f}_{z_k}(\boldsymbol{x}(t), \boldsymbol{u}(t), \boldsymbol{v}(t))dt$ *with* $\boldsymbol{v}(t) = \boldsymbol{v}_j \in \boldsymbol{V}$ *for* $t \in [t_k, t_{k+1}[$*.* ◇

The run of $A_{N,C}$ can informally be understood as follows: Starting from the initial hybrid state $r(t_0)$ the system evolves according to the continuous dynamics that is determined by the discrete state z_0, the constant discrete input vector $\boldsymbol{v}(t_0)$, and the (possibly) altering continuous input $\boldsymbol{u}(t)$ until the time t_1 is reached. At this time, either the discrete input $\boldsymbol{v}(t_1)$ is changed (effecting the continuous dynamics in $t \in [t_1, t_2[$) or the discrete state is altered if $\boldsymbol{x}(t)$ enters into a new region. The updated hybrid state $r(t_1)$ is the starting point for the next phase of continuous evolution, and so on.

While this type of model captures the behaviors of several systems at a level that is suitable for controller design (e.g., for certain transition procedures in processing systems, see (Stursberg and Engell, 2001)), the use within a synthesis algorithm is involved: In the first place, the fact that the control inputs $\boldsymbol{v}(t)$ can be switched arbitrarily often (as far as finitely often) in the interval $[t_0, t_f]$ imposes 'too much' freedom on the search for the desired controller. Secondly, the use of arbitrary nonlinear dynamics makes the efficient solution of the synthesis task intractable in general. In fact, decidability results are so far only existent for a very limited class of hybrid systems (Vidal et al., 2001).

2.2 Linear Discrete-Time Dynamics

For these reasons, we resort to a linear and discrete-time formulation of the hybrid dynamics for optimization, and use $A_{N,C}$ to remove the approximation error periodically over the considered time horizon (see Sect. 3.3). In order to obtain simpler dynamics the following transformations are carried out: Assume that linearization points $\boldsymbol{x}_i^c$ are chosen for each region R_i of $A_{N,C}$, as well as for the interval of each continuous input ($\boldsymbol{u}^c$). Then, the linearization of the dynamics $\boldsymbol{f}$ of $A_{N,C}$ for each combination of $\boldsymbol{x}_i^c$ and a discrete input vector $\boldsymbol{v}_k := \boldsymbol{v}(t_k) \in \boldsymbol{V}$ is denoted by: $\dot{\boldsymbol{x}}(t) = \boldsymbol{A}^l_{z,\boldsymbol{v}_k}\boldsymbol{x}^l(t) + \boldsymbol{B}^l_{z,\boldsymbol{v}_k}\boldsymbol{u}^l(t) + \boldsymbol{L}^l_{z,\boldsymbol{v}_k}$ with $\boldsymbol{x}^l(t) = \boldsymbol{x}(t) - \boldsymbol{x}_i^c$, $\boldsymbol{u}^l(t) = \boldsymbol{u}(t) - \boldsymbol{u}^c$.

Assume furthermore that a time span $\triangle t_k = t_{k+1} - t_k$ is given over which the discrete input $\boldsymbol{v}(t) = \boldsymbol{v}_k$ and the continuous input $\boldsymbol{u}(t) = \boldsymbol{u}_k := \boldsymbol{u}(t_k)$ are hold constant ($t \in [t_k, t_{k+1}[$). The solution of the linearized continuous dynamics at time t_{k+1} can then be written as:

$$\begin{aligned}\boldsymbol{x}_{k+1} &= e^{\boldsymbol{A}^l_{z,\boldsymbol{v}_k}\cdot\triangle t_k}\cdot\boldsymbol{x}^l_k + \int_{t_k}^{t_{k+1}} e^{\boldsymbol{A}^l_{z,\boldsymbol{v}_k}\cdot t_{k+1}-\tau}d\tau\cdot(\boldsymbol{B}^l_{z,\boldsymbol{v}_k}\cdot\boldsymbol{u}^l_k + \boldsymbol{L}^l_{z,\boldsymbol{v}_k})\\ &=: \boldsymbol{A}_{z,\boldsymbol{v}_k,\triangle t_k}\cdot\boldsymbol{x}^l_k + \boldsymbol{B}_{z,\boldsymbol{v}_k,\triangle t_k}\cdot\boldsymbol{u}^l_k + \boldsymbol{L}_{z,\boldsymbol{v}_k,\triangle t_k},\end{aligned} \tag{2}$$

with matrices $\boldsymbol{A}_{z,\boldsymbol{v}_k,\triangle t_k}$, $\boldsymbol{B}_{z,\boldsymbol{v}_k,\triangle t_k}$, and $\boldsymbol{L}_{z,\boldsymbol{v}_k,\triangle t_k}$ which depend on the region (denoted by z), the discrete input, and the time span. A hybrid automaton with continuous dynamics according to (2) can now be defined as follows:

Definition 2 *Hybrid Automaton* $A_{L,D}$

***Syntax**: A hybrid automaton with switched linear discrete time dynamics:*

$$A_{L,D} = (\boldsymbol{X}, \boldsymbol{U}, \boldsymbol{V}, E, Z, \boldsymbol{f}^D, \phi^D, T) \tag{3}$$

consists of the state space $\boldsymbol{X}$, the continuous input space $\boldsymbol{U}$, and the set of discrete inputs $\boldsymbol{V}$ as for $A_{N,C}$. The partitioning of $\boldsymbol{X}$ into a set of polyhedral regions $R = \{R_1, \ldots, R_{n_R}\}$ by the set E of switching planes is also the same. The trajectories $\boldsymbol{x}(t)$, $\boldsymbol{u}(t)$, and $\boldsymbol{v}(t)$ are now defined on a discrete time domain $t_k \in T = \{t_0, t_1, \ldots, t_f\}$, i.e., the variables are constant on each time interval $\triangle t_k := [t_k, t_{k-1}[$. The ordered set of time intervals is denoted by $\triangle T = (\triangle t_0, \ldots, \triangle t_{f-1})$ The discrete state set $Z = \{z_1, \ldots, z_{n_z}\}$ again results from an assignment $\rho : R \to Z$ of one discrete state to each region $R_i \in R$.

The continuous state transfer function $\boldsymbol{f}^D : \boldsymbol{X} \times \boldsymbol{U} \times \boldsymbol{V} \times \boldsymbol{Z} \times \triangle T \to \mathbb{R}^n$ determines a new continuous state according to the linear, discrete-time equation

$$\boldsymbol{x}_{k+1} := \boldsymbol{x}(t_{k+1}) = \boldsymbol{A}_{z,\boldsymbol{v}_k,\triangle t_k}\cdot\boldsymbol{x}(t_k) + \boldsymbol{B}_{z,\boldsymbol{v}_k,\triangle t_k}\cdot\boldsymbol{u}(t_k) + \boldsymbol{L}_{z,\boldsymbol{v}_k,\triangle t}$$

with matrices $\boldsymbol{A}_{z,\boldsymbol{v}_k,\triangle t_k} \in \mathbb{R}^{n\times n}$, $\boldsymbol{B}_{z,\boldsymbol{v}_k,\triangle t_k} \in \mathbb{R}^{n\times p}$, *and* $\boldsymbol{L}_{z,\boldsymbol{v}_k,\triangle t_k} \in \mathbb{R}^{n\times 1}$ *that are determined by* $z_k = z(t_k) \in Z$, $\boldsymbol{v}_k$, *and the current time span* $\triangle t_k$.

The transition function $\phi^D : Z \times \boldsymbol{X} \times \boldsymbol{X} \times R \to Z$ *specifies the current discrete state: For two regions* $R_a, R_b \in R$, *a transition* $z_k \to z_{k+1}$ *occurs at* t_{k+1}, *if:* $\boldsymbol{x}_k \in R_a$, $\rho(R_a) = z_k$ *and* $\boldsymbol{x}_{k+1} \in R_b$, $\rho(R_b) = z_{k+1}$, $\boldsymbol{x}_{k+1} \notin R_a$. *If* R_a *is left across* E_j *the transition guard is* $(\boldsymbol{c}_j \cdot \boldsymbol{x}_k \leq d_j) \wedge (\boldsymbol{c}_j \cdot \boldsymbol{x}_{k+1} > d_j)$ *if* E_j *belongs to* R_a, *or* $(\boldsymbol{c}_j \cdot \boldsymbol{x}_k < d_j) \wedge (\boldsymbol{c}_j \cdot \boldsymbol{x}_{k+1} \geq d_j)$ *if* E_j *is assigned to* R_b.

***Semantics**: Transitions and changes in* $\boldsymbol{v}(t)$ *and* $\boldsymbol{u}(t)$ *are possible at the points of time in* T. *A valid run of* $A_{L,D}$ *is defined by* $r : T \to \boldsymbol{X} \times Z$ *as the sequence* $r(t_0), r(t_1), \ldots, r(t_f)$ *of hybrid states* $r(t) = (\boldsymbol{x}(t), z(t))$ *such that:*

(*a*) $A_{L,D}$ *is initialized to:* $r(t_0) = (\boldsymbol{x}_0, z(t_0))$ *with* $\boldsymbol{x}_0 = \boldsymbol{x}(t_0) \in R^* \in R$, $\rho(R^*) = z(t_0) \in Z$.
(*b*) *and it evolves with:* $r(t_k) = (\boldsymbol{x}_k, z_k)$ *for* $t_k \in T \setminus t_0$ *according to:*
1. *continuous evolution:*
   $\boldsymbol{x}_{k+1} = \boldsymbol{f}^D(\boldsymbol{x}_k, \boldsymbol{u}_k, \boldsymbol{v}_k, z_k, \triangle t_k)$;
2. *discrete transitions:*
   $z_{k+1} = \phi(z_k, \boldsymbol{x}_k, \boldsymbol{x}_{k+1}, R_j)$ *with* $x_k \in R_j$, $x_{k+1} \notin R_j$. ◇

The main difference in the evolution of $A_{N,C}$ and $A_{L,D}$ is the fact that discrete changes occur only at points of time that are contained in T, i.e., transitions are taken not exactly on the boundaries of regions but at the first $t_k \in T$ encountered after a hyperplane is crossed.

3 Controller Generation Based on Optimization

3.1 Formulation as Optimization Problem

The objective of the control synthesis is to determine the trajectories $\boldsymbol{v}(t)$ and $\boldsymbol{u}(t)$ for $t \in [t_0, t_f]$ for which $A_{N,C}$ fulfills a set of given specifications. Since we consider those specifications over $\boldsymbol{x}(t)$ which refer to specific scenarios, particularly the transitions from an initial state into a target set, the control trajectories to be determined are called *control policies*:

Definition 3 *Optimal Control Policies*

Those control trajectories $\boldsymbol{v}^\star(t) = (\boldsymbol{v}_0^\star, \ldots, \boldsymbol{v}_k^\star, \ldots, \boldsymbol{v}_{t_f-1}^\star)$ *and* $\boldsymbol{u}^\star(t) = (\boldsymbol{u}_0^\star, \ldots, \boldsymbol{u}_k^\star, \ldots, \boldsymbol{u}_{t_f-1}^\star)$ *with* $\boldsymbol{v}_k^\star \in \boldsymbol{V}$, $\boldsymbol{u}_k^\star \in \boldsymbol{U}$, $k \in K = \{0, 1, \ldots, t_f - 1\}$ *which lead to the state trajectory* $\boldsymbol{x}^\star(t) = (\boldsymbol{x}_0^\star, \ldots, \boldsymbol{x}_k^\star, \ldots, \boldsymbol{x}_{t_f}^\star)$ *of* $A_{N,C}$ *such that:*

- $\boldsymbol{x}_0^\star = \boldsymbol{x}(t_0)$ *is the specified initial state,*
- $\boldsymbol{x}_{t_f}^\star \in \boldsymbol{X}_T$ *is contained in a given target region* $\boldsymbol{X}_T = [x_{T,1}^-, x_{T,1}^+] \times \ldots \times [x_{T,n}^-, x_{T,n}^+] \subset \boldsymbol{X}$ *with* $x_{T,j}^-, x_{T,j}^+ \in \mathbb{R}$ *and* $\boldsymbol{x}_0 \notin \boldsymbol{X}_T$,

- *$\boldsymbol{x}_k^\star \notin \boldsymbol{X}_{F,i}$ for all k and all forbidden regions $\boldsymbol{X}_{F,i} := \{\boldsymbol{x} \in \boldsymbol{X}_{F,i} \mid \boldsymbol{C}_F \cdot \boldsymbol{x} \leq \boldsymbol{d}_F, \boldsymbol{C}_F \in \mathbb{R}^{q_F \times n}, \boldsymbol{d}_F \in \mathbb{R}^{q_F \times 1}\}$ from a set $\boldsymbol{X}_F = \{\boldsymbol{X}_{F,1}, \ldots, \boldsymbol{X}_{F,\eta_F}\}$, where $\boldsymbol{X}_{F,i} \subset \boldsymbol{X}$ and $\boldsymbol{X}_{F,i} \cap \boldsymbol{X}_T = \emptyset$,*
- *and such that a performance criterion $\Omega(\boldsymbol{x}(t), \boldsymbol{u}(t), \boldsymbol{v}(t), t)$ is minimized.*

The corresponding optimal control policies of $\boldsymbol{A}_{L,D}$ with respect to a discrete-time formulation of the performance criterion $\Omega(\boldsymbol{x}_k, \boldsymbol{u}_k, \boldsymbol{v}_k, t_k)$ are denoted by: $\hat{\boldsymbol{v}}^\star(t)$, $\hat{\boldsymbol{u}}^\star(t)$, and the corresponding optimal state trajectory by $\hat{\boldsymbol{x}}^\star(t)$.

◇

Note that the determination of the policies $\hat{\boldsymbol{v}}^\star(t)$, $\hat{\boldsymbol{u}}^\star(t)$ corresponds to the synthesis of discrete controllers as referred to in Sect. 1 if $\boldsymbol{u}(t)$ is not considered (i.e., the system has either no continuous inputs, $m_u = 0$, or existing continuous inputs are set to fixed values). Equivalently, the case $n_v = 0$ or $n_v = 1$ refers to completely continuously controlled automata $A_{N,C}$ and $A_{L,D}$.

The polyhedral forbidden regions can be considered to specify parts of the state space which must not be reached during the complete evolution, e.g., for safety reasons.

If control strategies have to be computed for complete manufacturing or production processes, which include a set of target regions that must be reached sequentially, the strategies can be obtained as concatenations of optimal control policies: $(\boldsymbol{x}_{0,I}^\star, \ldots, \boldsymbol{x}_{t_{f,I},I}^\star =: \boldsymbol{x}_{0,II}^\star \in \boldsymbol{X}_{T,I}, \boldsymbol{x}_{1,II}^\star, \ldots, \boldsymbol{x}_{t_{f,II},II}^\star =: \boldsymbol{x}_{0,III}^\star \in \boldsymbol{X}_{T,II}, \ldots)$ if $i \in \{I, II, \ldots\}$ in $\boldsymbol{x}_{k,i}^\star$ denotes the number of the policy.

The determination of the optimal control policies $\hat{\boldsymbol{v}}^\star(t)$, $\hat{\boldsymbol{u}}^\star(t)$ for $A_{L,D}$ is formulated as the following optimization problem:

$$\begin{aligned}
(\hat{\boldsymbol{x}}^\star(t), \hat{\boldsymbol{v}}^\star(t), \hat{\boldsymbol{u}}^\star(t)) = \min_{\boldsymbol{v}_k, \boldsymbol{u}_k} \ & \Omega(\boldsymbol{x}(t_k), \boldsymbol{u}(t_k), \boldsymbol{v}(t_k), t_k) \\
\text{s.t.} \ & \boldsymbol{x}_0 = \boldsymbol{x}(t_0) \\
& \boldsymbol{x}_k \in \boldsymbol{X}, \boldsymbol{x}_k \notin \boldsymbol{X}_{F,i} \ \forall \boldsymbol{X}_{F,i} \in \boldsymbol{X}_F \ \forall t_k \in \{t_0, \ldots, t_f\} \\
& \boldsymbol{x}(t_f) \in \boldsymbol{X}_T \\
& \text{and subject to the dynamics of } A_{L,D}.
\end{aligned} \tag{4}$$

The discrete-time formulation of the performance criterion in (4) is given by:

$$\Omega = \sum_{k=0}^{|K|+1} (\alpha(t_k, \boldsymbol{x}_k) + \delta(t_k)) + \sum_{k=0}^{|K|} \beta(t_k, \boldsymbol{u}_k) + \sum_{k=1}^{|K|} \gamma(t_k, \boldsymbol{v}_k) \tag{5}$$

with the following terms:

- $\alpha(t_k, \boldsymbol{x}_k) = \mu_1(t_k) \cdot \|\boldsymbol{w}_1 \cdot (\boldsymbol{x}_k - \boldsymbol{X}_T)\|_1$ describes the distance between the current state and (the nearest boundary) of the target region (weighted over the state components by $\boldsymbol{w}_1$);
- $\beta(t_k, \boldsymbol{u}_k) = \mu_2(t_k) \cdot \|\boldsymbol{w}_2 \cdot (\boldsymbol{u}_k - \boldsymbol{u}_S)\|_1$ contains the deviation of $\boldsymbol{u}_k$ from a reference vector $\boldsymbol{u}_s$ (with weights $\boldsymbol{w}_2$);

- $\gamma(t_k, \boldsymbol{v}_k) = \mu_3(t_k) \cdot \left\{ \begin{matrix} w_3: & \text{if } \boldsymbol{v}(t_{k-1}) \neq \boldsymbol{v}_k \\ 0: & \text{else} \end{matrix} \right\}$ adds the amount w_3 to Ω if the input $\boldsymbol{v}(t)$ switches;
- $\delta(t_k) = \left\{ \begin{matrix} \mu_4(t_k): & \text{if } \boldsymbol{x}_k \neq \boldsymbol{X}_T \\ 0: & \text{else} \end{matrix} \right\}$ increases the costs until the target region is reached.

The factors $\mu_1(t_k), \ldots, \mu_4(t_k)$ are appropriate weights for the contributions of these four terms.

3.2 Transformation of $A_{L,D}$ into Optimization Constraints

In order to be able to solve (4) by optimization, the logical part of the dynamics of $A_{L,D}$ and of the performance criterion (5) have to be transformed into constraints in equation-based form. One possibility for this transformation is the so-called *M-approach* (Glover, 1975, Williams, 1978). Its principle is to introduce binary auxiliary variables and to express the validity of specific constraints by sets of inequalities. Referring to the logical decisions involved in (4), this means that, e.g., the equivalence $(\boldsymbol{c}_j \cdot \boldsymbol{x}_k \leq d_j) \Leftrightarrow (b = 1)$ is expressed by:

$$\boldsymbol{c}_j \cdot \boldsymbol{x}_k - d_j \leq M^+ \cdot (1 - b), \quad \boldsymbol{c}_j \cdot \boldsymbol{x}_k - d_j < M^+ \cdot (1 - b) \tag{6}$$

where b is a binary auxiliary variable and M^-, M^+ are two constants that limit the value of the left-hand side of the inequalities. Using such constructions, the transitions of $A_{L,D}$ are formulated as the change of values of binary variables. This concept has been used in (Stursberg and Engell, 2001, Stursberg and Engell, 2002), and in the context of Mixed-Logical Dynamical Systems (Bemporad and Morari, 1999a) to transfer discrete dynamics into optimization constraints.

The problem of the M-approach for our purpose is that it requires a large number of binary auxiliary variables as well as it produces a large number of inequalities – both factors are crucial for the computational complexity of the solution of (4). In (Stursberg and Panek, 2002), the use of *disjunctive formulations* is proposed as an alternative which requires a considerably smaller number of binary variables. The idea is to replace a disjunctive expression (as, e.q., the decision if an automaton is in one of two states) by its convex relaxation (using only continuous auxiliary variables), and to force the relaxation variables to zero or one by appropriate constraints. Since continuous variables usually have a smaller impact on the solution performance, it has been shown that this formulation can be solved considerably more efficient (Stursberg and Panek, 2002). The basic idea of the convex hull relaxation has been introduced in (Balas, 1985) and extended in (Lee and Grossmann, 2000, Vecchietti and Grossmann, 1999): Assume that a set of constraints $\boldsymbol{g}_j(\boldsymbol{x}) \leq \boldsymbol{0}$, $j \in J$ is given, only one of which can be valid at one time. The convex hull of $\bigvee_{j \in J} \boldsymbol{g}_j(\boldsymbol{x}) \leq \boldsymbol{0}$ can then be written as:

$$\boldsymbol{x} = \sum_{j \in J} \boldsymbol{w}_j, \quad \boldsymbol{0} \leq \boldsymbol{w}_j \leq \lambda_j \cdot \boldsymbol{x}^+$$

$$\lambda_j \cdot \boldsymbol{g}_j(\boldsymbol{w}_j/\lambda_j) \leq 0, \quad \sum_{j \in J} \lambda_j = 1, \quad 0 \leq \lambda_j \leq 1. \tag{7}$$

The vectors $\boldsymbol{w}_j$ contain continuous auxiliary variables, $\boldsymbol{x}^+$ is the vector of upper bounds of $\boldsymbol{x}$, and the scalars λ_j determine the weight of the constraints $\boldsymbol{g}_j$. If $\lambda_j = 1$ applies, the corresponding disjunctive term evaluates to be true. As shown in (Lee and Grossmann, 2000), the advantage of this formulation is not only that binary variables are not required but also that it leads to smaller solution spaces during the optimization as if the M-approach is used.

Interpreting the logical decisions in (4) as disjunctions is obvious: At a time t_k, the current continuous state $\boldsymbol{x}_k$ can be contained only in one region and, of course, only one discrete input $\boldsymbol{v} \in \boldsymbol{V}$ can be applied. Hence, the dynamics of $A_{L,D}$ is expressed in disjunctive form as:

$$\bigvee_{i=1}^{n_R} \bigvee_{l=1}^{n_V} \begin{bmatrix} \boldsymbol{C}_i \cdot \boldsymbol{x}_k \sim \boldsymbol{d}_i \\ \boldsymbol{v}(t_k) = \boldsymbol{v}_l \\ \boldsymbol{x}_{k+1} = \boldsymbol{A}_{z_i,\boldsymbol{v}_l,\triangle t_k} \cdot \boldsymbol{x}_k + \boldsymbol{B}_{z_i,\boldsymbol{v}_l,\triangle t_k} \cdot \boldsymbol{u}_k + \boldsymbol{L}_{z_i,\boldsymbol{v}_l,\triangle t_k} \end{bmatrix} \tag{8}$$

with $\sim \in \{<, \leq\}$. This formulation applies if the points of time in T (and with that the time intervals $\triangle t_k \in \triangle T$) are fixed in advance. If the switching times are optimization parameters itself, the disjunction has to be extended to these options: If $\triangle\tilde{T} = \{\triangle\tilde{t}_1, \ldots, \triangle\tilde{t}_{n_{\triangle\tilde{T}}}\}$ denotes the specified set of time steps from which a particular value is chosen for $\triangle t_k$, the disjunction becomes:

$$\bigvee_{i=1}^{n_R} \bigvee_{l=1}^{n_V} \bigvee_{p=1}^{n_{\triangle\tilde{T}}} \begin{bmatrix} \boldsymbol{C}_i \cdot \boldsymbol{x}_k \sim \boldsymbol{d}_i \\ \boldsymbol{v}(t_k) = \boldsymbol{v}_l \\ \triangle t_k = \triangle\tilde{t}_p \\ \boldsymbol{x}_{k+1} = \boldsymbol{A}_{z_i,\boldsymbol{v}_l,\triangle\tilde{t}_p} \cdot \boldsymbol{x}_k + \boldsymbol{B}_{z_i,\boldsymbol{v}_l,\triangle\tilde{t}_p} \cdot \boldsymbol{u}_k + \boldsymbol{L}_{z_i,\boldsymbol{v}_l,\triangle\tilde{t}_p} \end{bmatrix} \tag{9}$$

In a similar way the logical decisions in (5) can be understood as disjunctive terms, i.e., the current state $\boldsymbol{x}_k$ is either below, in, or above the target region $\boldsymbol{X}_T$, and the current input $\boldsymbol{v}_k = \boldsymbol{v}(t_k)$ is either equal to or different from $\boldsymbol{v}_{k-1}$.

In order to express these disjunctive expressions in equation-based form, we use the convex hull relaxation (7). For example, the test whether the current state $\boldsymbol{x}_k$ is contained in a specific region $(\boldsymbol{C}_i \cdot \boldsymbol{x}_k \leq \boldsymbol{d}_i) \leftrightarrow (\eta_{k,i} = 1)$ is replaced by the inequalities:

$$\boldsymbol{C}_i \cdot \boldsymbol{w}_{k,i} \leq \eta_{k,i} \cdot \boldsymbol{d}_i, \quad \boldsymbol{0} \leq \boldsymbol{w}_{k,i} \leq \eta_{k,i} \cdot \boldsymbol{x}^+, \quad \boldsymbol{0} \leq \boldsymbol{x}_k \leq \boldsymbol{x}^+$$

$$\sum_{i=1}^{n_R} \boldsymbol{w}_{k,i} = \boldsymbol{x}_k, \quad \sum_{i=1}^{n_R} \eta_{k,i} = 1, \tag{10}$$

where $\eta_{k,i} \in [0,1]$. To enforce that the variables $\eta_{k,i}$ assume only the values zero or one, binary variables $b_{k,q} \in \{0,1\}$ with $q \in Q = \{1, \ldots, \lceil log_2(n_R) \rceil\}$ are introduced. The values 0 or 1 are assigned to $\eta_{k,i}$ by formulating linear equations of the type:

$$\eta_{k,i} \sim c + \sum_{q \in Q'} \pm b_{k,q} \tag{11}$$

with $c \in \mathbb{Z}$, $\sim \in \{\leq, \geq\}$, and $Q' \subseteq Q$.

The extension of (10) to the case with negative continuous variables x_k is straightforward by simply shifting the origin.

The rigorous application of (7) leads to the following equation-based form of the disjunctive expressions in (8):

$$\sum_{i=1}^{n_R} \eta^I_{k,i} = 1, \quad \sum_{l=1}^{n_V} \eta^{II}_{k,l} = 1, \quad \sum_{p=1}^{n_{\triangle\tilde{T}}} \eta^{III}_{k,p} = 1 \tag{12}$$

$$\forall\, i \in \{1, \ldots, n_R\}: \quad \eta^I_{k,i} = \sum_{l=1}^{n_V} \sum_{p=1}^{n_{\triangle\tilde{T}}} \chi_{k,i,l,p}$$

$$\boldsymbol{C}_i \cdot \boldsymbol{w}_{k,i} \leq \eta^I_{k,i} \cdot \boldsymbol{d}_i, \quad \boldsymbol{w}_{k,i} = \sum_{l=1}^{n_V} \sum_{p=1}^{n_{\triangle\tilde{T}}} \boldsymbol{\xi}_{k,i,l,p} \tag{13}$$

$$\forall\, l \in \{1, \ldots, n_V\}: \quad \eta^{II}_{k,l} = \sum_{i=1}^{n_R} \sum_{p=1}^{n_{\triangle\tilde{T}}} \chi_{k,i,l,p} \tag{14}$$

$$\forall\, p \in \{1, \ldots, n_{\triangle\tilde{T}}\}: \quad \eta^{III}_{k,p} = \sum_{i=1}^{n_R} \sum_{l=1}^{n_V} \chi_{k,i,l,p} \tag{15}$$

$$\sum_{i=1}^{n_R} \sum_{l=1}^{n_V} \sum_{p=1}^{n_{\triangle\tilde{T}}} \chi_{k-1,i,l,p} = 1 \tag{16}$$

$$\sum_{i=1}^{n_R} \sum_{l=1}^{n_V} \sum_{p=1}^{n_{\triangle\tilde{T}}} \boldsymbol{\xi}_{k-1,i,l,p} = \boldsymbol{x}_{k-1}, \quad \boldsymbol{0} \leq \boldsymbol{\xi}_{k-1,i,l,p} \leq \chi_{k-1,i,l,p} \cdot \boldsymbol{x}^+$$

$$\sum_{i=1}^{n_R} \sum_{l=1}^{n_V} \sum_{p=1}^{n_{\triangle\tilde{T}}} \boldsymbol{\pi}_{k-1,i,l,p} = \boldsymbol{u}_{k-1}, \quad \boldsymbol{0} \leq \boldsymbol{\pi}_{k-1,i,l,p} \leq \chi_{k-1,i,l,p} \cdot \boldsymbol{u}^+ \tag{17}$$

$$\boldsymbol{x}_k = \sum_{i=1}^{n_R} \sum_{l=1}^{n_V} \sum_{p=1}^{n_{\triangle\tilde{T}}} \boldsymbol{A}_{z_i,\boldsymbol{v}_l,\triangle\tilde{t}_p} \cdot \boldsymbol{\xi}_{k-1,i,l,p} + \boldsymbol{B}_{z_i,\boldsymbol{v}_l,\triangle\tilde{t}_p} \cdot \boldsymbol{\pi}_{k-1,i,l,p} + \chi_{k-1,i,l,p} \cdot \boldsymbol{L}_{z_i,\boldsymbol{v}_l,\triangle\tilde{t}_p} \tag{18}$$

for $k \in K \cup \{t_f\} \setminus t_0$. These equations should be understood as follows: First, three sets of continuous auxiliary variables are introduced for each point of time t_k, such that $\eta^I_{k,i} \in [0,1]$, $\eta^{II}_{k,l} \in [0,1]$, $\eta^{III}_{k,p} \in [0,1]$. These variables continuously encode which region $\eta^I_{k,i}$, discrete input $\eta^{II}_{k,p}$, and time step $\eta^{III}_{k,p}$ respectively is valid in t_k. The variables are forced to the values zero or one by equations of the type (11) which are defined for binary variables b_{k,q^I}, $b_{k,q^{II}}$, and $b_{k,q^{III}}$. This requires overall $\lceil \log_2(n_R) \rceil + \lceil \log_2(n_V) \rceil + \lceil \log_2(n_{\triangle \tilde{T}}) \rceil$ binary variables for every point of time t_k.

The equations (12) enforce that for t_k only one variable $\eta^I_{k,i}$, $\eta^{II}_{k,l}$, and $\eta^{III}_{k,p}$ can assume the value one while the others are forced to zero. By using the formulation introduced with (10), the equations (13) establish the relation between $\eta^I_{k,i}$ and the region assigned to the discrete state z_i: If $\eta^I_{k,i} = 1$, then the continuous state $\boldsymbol{x}_k$ lies within the region defined by $\boldsymbol{C}_i$ and $\boldsymbol{d}_i$. The continuous variables $\boldsymbol{w}_{k,i}$ play the same role as in (10), but their values are here determined by additional continuous variables $\boldsymbol{\xi}_{k-1,i,l,p}$. The latter represent the product $\boldsymbol{x}_k \cdot \chi_{k,i,l,p}$, where $\chi_{k,i,l,p} \in [0,1]$ combines the information about the current region, input, and time step. The equations (12-16) ensure implicitly that $\chi_{k,i,l,p}$ can only assume the values zero or one, and that only one of these variables is one for a given k. The first line of (13) and (14,15) force those variables $\chi_{k,i,l,p}$ to zero, for which the corresponding variables $\eta^I_{k,i}$, $\eta^{II}_{k,l}$, and $\eta^{III}_{k,p}$ are zero as well.

The equations (17) then determine the values of $\boldsymbol{x}_{k-1}$ and $\boldsymbol{u}_{k-1}$ depending on $\boldsymbol{\xi}_{k-1,i,l,p}$ and on products $\boldsymbol{\pi}_{k-1,i,l.p} := \boldsymbol{u}_{k-1} \cdot \chi_{k,i,l,p}$. The vectors $\boldsymbol{x}^+$ and $\boldsymbol{u}^+$ again denote an upper bound of the values of the state and continuous input variables (the origin of which is shifted if necessary to obtain only positive values). Finally, the variables $\boldsymbol{\xi}_{k-1,i,l,p}$, $\boldsymbol{\pi}_{k-1,i,l,p}$, and $\chi_{k,i,l,p}$ determine the relevant entries of the matrices $\boldsymbol{A}_{z_i,\boldsymbol{v}_l,\triangle \tilde{t}_p}$, $\boldsymbol{B}_{z_i,\boldsymbol{v}_l,\triangle \tilde{t}_p}$, and $\boldsymbol{L}_{z_i,\boldsymbol{v}_l,\triangle \tilde{t}_p}$, and with that the new continuous state $\boldsymbol{x}_k$.

The formulation becomes simpler if the time steps $\triangle t_k$ are fixed to one specific value ($n_{\triangle \tilde{T}} = 1$). In this case, the variables $\eta^{III}_{k,p}$, the index p, and all sums over this index can be omitted. The number of required binary variables then reduces to $\lceil \log_2(n_R) \rceil + \lceil \log_2(n_V) \rceil$ per t_k.

The equations (12-18) together with (in)equalities that formulate the state space, the initial state, the target region, and the forbidden regions form the set of constraints of the optimization problem (see (4)). The exclusion of forbidden states ($\boldsymbol{x}_k \notin \boldsymbol{X}_{F,i}\ \forall\ t_k$) is modelled by using the formulation in (6): The equivalence $(\boldsymbol{x}_k \in \boldsymbol{X}_{F,i}) \Leftrightarrow (b_{F,i} = 1)$ is expressed with equations of the type (6), and the constraint $b_{F,i} = 0$ explicitly excludes that a combination of inputs and time step is chosen that leads to $\boldsymbol{x}_k \in \boldsymbol{X}_{F,i}$ (if existing). Alternatively a disjunctive formulation similar to (10) can be used such that a variable $\eta^I_{k,i}$ (i referring to the forbidden region in this case) is set to zero for all k.

In order to obtain a problem that is completely given in equation-based form, the logical decisions in the performance criterion have to be reformulated also. The disjunctive formulation according to (10) is suitable for this purpose as well.

3.3 Complexity Reduction by Using Limited Look-Ahead Horizons

The disjunctive formulation presented in the previous section requires a relatively small number of binary variables – however, due to the combination of binary variables b_{k,q^I}, $b_{k,q^{II}}$, $b_{k,q^{III}}$ and the continuous variables contained in (12-18), mixed-integer programming is used to solve the optimization. Since the formulation is completely linear we employ mixed-integer linear programming, notably Branch-and-Bound algorithms, for the solution.

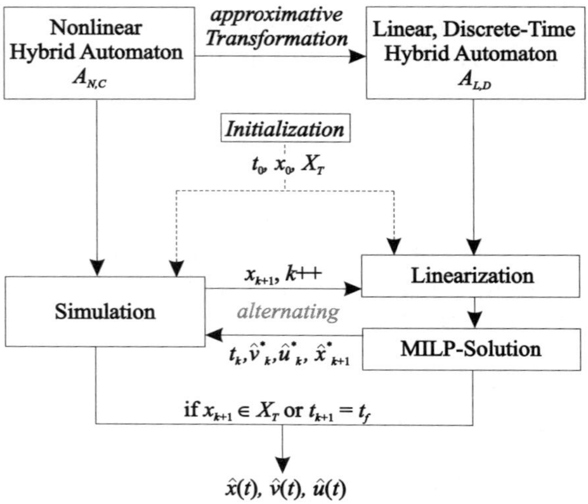

Fig. 2. The solution with moving horizons

An inherent drawback of all discrete-time optimization problems with $\omega := \lceil \log_2(n_R) \rceil + \lceil \log_2(n_V) \rceil + \lceil \log_2(n_{\triangle \tilde{T}}) \rceil$ different binary options at a time t_k is, that the number of binary variables grows linearly with $|K|$, and hence the number of possible solutions over the complete horizon with $\omega^{|K|}$. The optimization is therefore limited to 'short' horizons. As a remedy, we use a moving horizon strategy as it is known from model predictive control (see Fig. 2): For each $k \in K$, the optimization in (4) with the performance criterion (5) and a formulation of the dynamics of $A_{L,D}$ according to (12-18) is solved by MILP. This optimization is performed for a horizon $P = \{1, 2 \ldots, k_P\}$ with usually $k_P << |K|$. The first control vectors $\hat{\boldsymbol{u}}_1^\star$ and $\hat{\boldsymbol{v}}_1^\star$ of the solution are applied to the original nonlinear automaton, and the latter is simulated for the time-period $\triangle t_k$, giving the new hybrid state $(\boldsymbol{x}_{k+1}, z_{k+1})$. The dynamics of the nonlinear automaton is linearized for this new state, i.e., the matrices $\boldsymbol{A}_{z_i,\boldsymbol{v}_k,\triangle t_k}$, $\boldsymbol{B}_{z_i,\boldsymbol{v}_k,\triangle t_k}$, and $\boldsymbol{L}_{z_i,\boldsymbol{v}_k,\triangle t}$ are updated to improve the model accuracy[2]. In the next cycle, the horizon P is shifted one time step ahead, and the MILP is solved starting from the new state. The alternating use of the linear and nonlinear automaton corrects the differences between the behavior of both models in each step.

[2] If the current region is left within the prediction horizon, those matrices are used which are obtained from the linearization of $A_{N,C}$ for the center point of the new region.

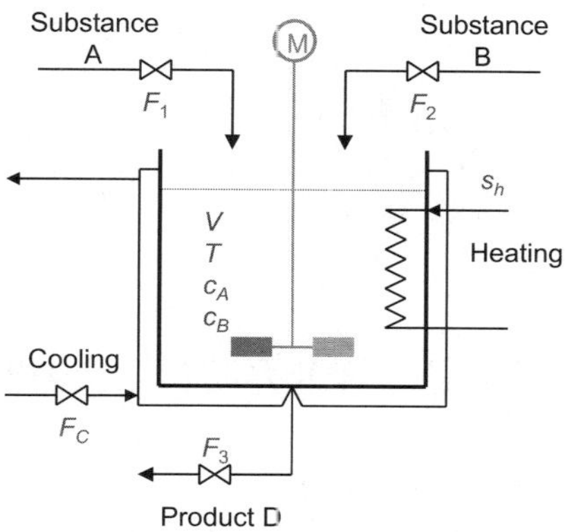

Fig. 3. Chemical reactor

This procedure is repeated until either $\boldsymbol{X}_T$ is reached or until a specified limit of time steps is exceeded.

The complexity of this procedure grows only linearly with $|K|$. There is, however, a trade-off between this gain with respect to the complexity, and the result of the optimization: The obtained solution can deviate from the optimal control policies $\boldsymbol{v}^\star(t)$ and $\boldsymbol{u}^\star(t)$ of $A_{N,C}$ due to the limited look-ahead horizons. As described in (Stursberg and Engell, 2002), this deviation can be reduced if increasing time steps $\triangle t_k$ are used over the horizon P. This allows covering a relatively large time span in each cycle even though a small number of time steps is considered.

4 Control of the Startup Procedure of a Chemical Reactor

The approach is illustrated by computing the control policies for the start-up procedure of a chemical reactor. Within this reactor two dissolved substances A and B react exothermically to form a product D (Fig. 4). A and B are fed into the reactor by two streams F_1 and F_2 with temperatures T_1 and T_2, and the outlet stream (for which a high concentration of D is desirable) is denoted by F_{out}. The reactor is equipped with a device to heat the content up to a required reaction temperature, and a cooling jacket allows to remove heat from the system once the reaction has started to produce an excess of heat. The streams F_1 and F_2 as well as the heater ($s_H \in \{0,1\}$) can each be switched between two modes, i.e., $\boldsymbol{v} = (F_1, F_2, s_H)^{\mathrm{T}}$ can assume 8 different values. The stream F_{out} and the flow of cooling liquid F_C can be continuously controlled within given intervals, i. e., they form the continuous input $\boldsymbol{u} = (F_{out}, F_C)^{\mathrm{T}}$. The state vector $\boldsymbol{x} = (V, T_R, c_A, c_B)^{\mathrm{T}}$ contains the volume V, the temperature T_R, and the concentrations c_A and c_B of the reactor content. The reactor dynamics $\dot{\boldsymbol{x}} = \boldsymbol{f}(\boldsymbol{x}, \boldsymbol{u}, \boldsymbol{v})$ is specified by:

$$V \in [0.1, 0.8]:$$
$$\boldsymbol{f}^I = \begin{pmatrix} F_1 + F_2 - F_{out} \\ (F_1 \cdot (T_1 - T_R) + F_2 \cdot (T_2 - T_R))/V \\ \quad + F_C \cdot k_1 \cdot (T_C - T_R) \cdot (k_2/V + k_3) - k_4 \cdot r \\ (F_1 \cdot c_{A,1} - c_A \cdot (F_1 + F_2))/V + k_9 \cdot r \\ (F_2 \cdot c_{B,2} - c_B \cdot (F_1 + F_2))/V + k_{10} \cdot r \end{pmatrix}$$
$$V \in \,]0.8, 2.2]:$$
$$\boldsymbol{f}^{II} = \left(f_1^I,\ f_2^I + s_H \cdot k_6 \cdot (T_H - T_R) \cdot (k_7 - \tfrac{k_8}{V}),\ f_3^I,\ f_4^I \right)^{\mathrm{T}} \tag{19}$$

with $r = c_A \cdot c_B^2 \cdot \exp(-k_5/T_R)$. The valid set of equations depends on V in order to model that the heater is effective only above a certain liquid level. The temperature behavior switches if this level is crossed. Since $\boldsymbol{f}^I$ and $\boldsymbol{f}^{II}$ are strongly nonlinear, we further partition the state space into two temperature intervals (at $T_R = 320$ K) such that four different regions exist overall.

The control objective for this system is to determine control policies $\boldsymbol{u}(t)$ and $\boldsymbol{v}(t)$ that drive the reactor from an initial state $\boldsymbol{x}_1 = (0.1, 300, 0, 0)$ into the target region

$$\boldsymbol{X}_T = [1.49, 1.51] \times [343, 347] \times [0.13, 0.17] \times [0.28, 0.32].$$

This region corresponds to a mode of stable production, in which the reactor is filled up to a desired level, a required reaction temperature is reached and the product concentration is sufficiently high. Throughout the operation the safety constraints $T_R \leq 360$ and $V \leq 1.55$ have to be satisfied (i.e., parts of the state space which violate these constraints correspond to forbidden regions).

For simplicity, we consider only the first term of the objective function (the distance to $\boldsymbol{X}_T$) and a constant time step. The optimization is carried out with $|P| = 3$ and $\triangle t_k = 20$ using the solver CPLEX (Brooke et al., 1998). The optimization results obtained are shown in Fig. 4: The trajectories show that F_1, F_2 start with high values in order to fill up the reactor quickly, and then assume lower values in order to maintain a throughput with the desired yield. A high priority for filling in the initial phase has been established by choosing the weighting factors $\mu_1(t_k)$ and $\boldsymbol{w}_1$ accordingly. When the volume has crossed the value at which the heating becomes effective, s_H assumes the value one in order to heat up the content. After 700 sec the combination of heating and cooling maintains a temperature within the target interval. The concentrations of A and B first rise during the filling procedure and are then driven into the desired intervals by increasing the temperature and decreasing the throughput.

The control policies were computed by using the moving horizon setting described in Sect. 3.3, and the state trajectory $\boldsymbol{x}(t)$ reaches $\boldsymbol{X}_T$ after 60 cycles. This result was obtained in 13 minutes of computation time.

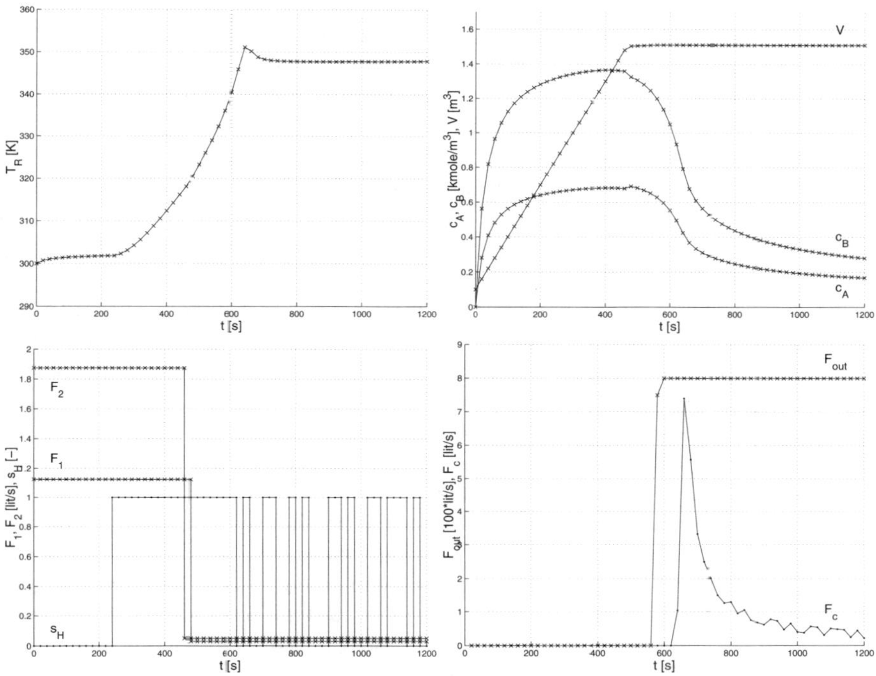

Fig. 4. Optimized state trajectories $\boldsymbol{x}(t) = (V, T_R, c_A, c_B)$ and control trajectories $\boldsymbol{v}(t) = (F_1, F_2, s_H)$, $\boldsymbol{u}(t) = (F_{out}, F_C)$ of the nonlinear hybrid model.

5 Conclusions

We have reported on an approach to determine discrete and continuous control policies for hybrid systems with switched dynamics. The characterizing steps of the method are:

- an initial modeling of the system as a hybrid automaton with possibly nonlinear continuous time dynamics,
- the approximation by a series of automata with discrete time and linear dynamics such that the automata are repeatedly updated along the computed state trajectory,
- the formulation of the synthesis task as a mixed integer linear programming (MILP) problem in which the transition dynamics of the automaton and the logical decisions within the performance criterion are encoded by specific disjunctive formulations, and
- the computation of the control policies by solving a series of MILP problems in a moving horizon setting to allow the application to larger horizons.

In order to assess the approach, three important issues have to be considered, namely the use of approximating linear models, the efficiency of the disjunctive formulation, and the solution on moving horizons: Using linearized models for the optimization can, of course, lead to a deviation of the computed policy $\hat{\boldsymbol{u}}^\star(t)$, $\hat{\boldsymbol{x}}^\star(t)$ from the optimal solution $\boldsymbol{u}^\star(t), \boldsymbol{x}^\star(t)$ for the nonlinear hybrid automaton. But since the linear model is updated for each state $\boldsymbol{x}_k$ for which the next inputs $\boldsymbol{v}_k$, $\boldsymbol{u}_k$ are computed, the deviations should be small if the original continuous dynamics is only slightly nonlinear. For strong nonlinearities in the dynamics $\boldsymbol{f}$ of $A_{N,C}$ however, the deviation may be significant. It remains a topic of future research to quantify this deviation and compare the results to methods that use the nonlinear model for optimization directly as, e.g., (Buss et al., 2002).

As discussed in (Stursberg and Panek, 2002), the use of disjunctive formulations to represent the MILP problems has the following advantages: The number of binary variables, as an important factor for the complexity of the optimization problem, can be significantly smaller as if the same task is expressed by using formulations according to (6). Even more important is the fact that the convex hull relaxation given in (7) leads to smaller solution spaces as if a relaxation of (6) is used. This is shown in (Balas, 1985, Lee and Grossmann, 2000).

With respect to the use of moving horizons, it first has to be noted that a limited look-ahead horizon can of course add to a deviation of $\hat{\boldsymbol{u}}^\star(t)$, $\hat{\boldsymbol{x}}^\star(t)$ from $\boldsymbol{u}^\star(t)$, $\boldsymbol{x}^\star(t)$. On the other hand, the application of the method to large horizons (as, e.g., 60 steps for a system of the size of the reactor) would not be possible with a single optimization (for the example, more than $5^{|K|} \approx 1.7 \cdot 10^{41}$ options). It must hence be the objective to use a horizon with a relatively small number of sampling points $|P|$ while a large time horizon is covered to improve the prediction. In (Stursberg and Engell, 2002) we propose a modification in which increasing time steps $\triangle t_k$ over P enable prediction horizons that cover the complete path into $\boldsymbol{X}_T$.

Whereas the simplifications introduced in Sect. 2.2 make the method applicable to systems of the size of the reactor example, the extension to industrial-scale systems, which may consist of several interacting components, is still an unresolved problem. Our current research focusses on extending the applicability by considering techniques of decomposition and abstraction, and developing solution techniques that are tailor-made for the problem setting investigated in this paper.

Acknowledgements. This investigation was funded by the German Research Council (DFG) through the grant EN152/24-1.

Part V

Applications

Definition of a Type of Continuous-Discrete High-Level Petri Nets and Its Application to the Performance Analysis of Train Protection Systems

Gebhard Decknatel, Roman Slovák, Eckehard Schnieder

Institute of Control and Automation Engineering, Technical University of Braunschweig, Langer Kamp 8, 38106 Braunschweig, Germany
E-mail:{decknatel | slovak | schnieder}@ifra.ing.tu-bs.de

Abstract. We present a hybrid (continuous-discrete) type of high-level Petri Nets, and describe how it can be applied to the study of train control systems. In addition to the elements of discrete high-level Petri Nets, the nets used here have continuous transitions, which are able to manipulate real-valued attributes of tokens. We outline two methods for simulating the hybrid models with a standard Petri Net tool. The simulation is then used for the performance analysis of moving and fixed block train protection systems.

1 Introduction

The purpose of the project described here is to study the possibilities of modelling hybrid systems, i.e. mixed continuous-discrete systems. More specifically, the aim is to examine a form of integrated modelling using a language which can be used to describe continuous as well as discrete processes and their interaction, and to test that language by applying it to hybrid aspects of railway systems.

The formalism used here is a hybrid variant of high-level Petri Nets. The modelling, simulation and analysis of these hybrid nets is supported by a tool for high-level Petri Nets, which is extended with functions for continuous transitions. Two different approaches for the simulation were implemented, based on numerical integration and on discrete events.

The simulation model represents the movement of trains across a given topology including points and signals. It includes the position and length of each train at any given time as well as the length of the track ahead of the train which is required to come to a halt. This allows to model different types of train protection and to investigate their properties with regard to performance.

The performance of a railway system is usually measured in terms of two key parameters: train speed and train line capacity. The train speed is important because it directly affects travel times and the line capacity measures the maximal volume of traffic carried by a line. The subject for the performance analysis here is the improvement which can be achieved by application of moving block in comparison to conventional fixed block train protection.

S. Engell, G. Frehse, E. Schnieder (Eds.): Modelling, Analysis and Design of Hybrid Systems, LNCIS 279, pp. 355–367, 2002.

2 Language

Languages for mixed continuous-discrete systems can be built by extending a formalism for discrete systems in a way which allows the additional description of continuous processes. One class of description means for discrete systems which can form a basis for such an approach are Petri Nets.

Petri Nets can be used to model, analyse and simulate discrete systems. There are several approaches to extend Petri Nets by continuous features such that they can be applied to hybrid systems as well. In the project described here, one such approach was used on the basis of Coloured Petri Nets.

2.1 Hybrid Petri Nets

In the last 15 years, a large number of formalisms for mixed systems on the basis of Petri Nets have been presented, which can be summarized under the term hybrid Petri Nets, see also the bibliography by A. Giua and A. Piccaluga (Giua and Piccaluga, 2002). There are different types of (discrete) Petri Nets and there are different ways to introduce continuous processes into them, such that there are many different types of hybrid Petri Nets. The main criterion for the classification of those different types of nets is the question whether they use anonymous (black) tokens or individual (structured) tokens and thus belong to the class of high-level nets.

The first definitions of continuous and hybrid Petri Nets were based on Place/-Transition nets (David and Alla, 1987). In Place/Transition nets, each place may carry an arbitrary natural number of indistinguishable tokens, which are added or subtracted by the discrete firing of transitions. In continuous nets such as described by R. David and H. Alla in (David and Alla, 1987), the number of tokens on the places can be any positive real number, and the transitions can move these tokens continuously with a constant or varying velocity. From these continuous nets, Hybrid Petri Nets are derived as nets containing both the discrete net elements from Place/Transition nets and the continuous places and transitions from continuous nets (David and Alla, 1992, Le Bail et al., 1991).

The interaction between discrete and continuous parts of a model is represented by the arcs which connect discrete and continuous nodes. An arc from a continuous place to a discrete transition can be used to model the fact that a discrete event occurs when a continuous variable crosses a threshold. An arc in the opposite direction signifies a discrete jump in the continuous variable. Arcs between continuous transitions and discrete places can be used to switch the continuous token flow through the transition. In this case, the connections have to be symmetrical such that the continuous firing of the continuous transition may depend on the marking of the discrete place but does not change it.

The anonymity of the tokens in Place/Transition Nets and in the hybrid nets based on them imposes a certain limitation on the ability to model complex systems. This has lead to the definition of high-level nets (Jensen and Rozenberg, 1991), where the tokens are distinguishable and may contain structured data. This allows to build more succinct and manageable models than can be obtained using nets with anonymous tokens. Two basic types of high-level nets are Predicate/Transition Nets (Genrich, 1987) and Coloured Petri Nets (Jensen, 1997).

Several approaches exist to extend high-level nets for the modeling of mixed processes. One of them was introduced by A. Giua and E. Usai in (Giua and Usai, 1996) as High-Level Hybrid Petri Nets (HLHPN), see also (Giua and Usai, 1998). They represent a combination of high-level nets with the concepts of continuous nets with anonymous tokens. There are again discrete as well as continuous places and transitions. The marking of a continuous place is a vector of real numbers. The length of that vector is a fixed property of the place, whereas the values of the elements of the vector may be changed by continuous transitions. One continuous place of a HLHPN is thus equivalent to a vector of continuous places in a hybrid Petri Net with anonymous tokens. A net always possesses a fixed number of continuous variables, which are identified by their position within the net.

A different approach for the integration of continuous processes into high-level nets was presented by R. Wieting in (Wieting, 1996). The goal of this approach is to combine high-level Petri Nets with differential algebraic equations and object-oriented concepts. Instead of introducing special continuous places, the continuous variables are defined as attributes of structured (individual) tokens. The resulting hybrid Nets are called HyNets. The discrete basis for HyNets is a type of high-level nets called THOR-Nets (Timed Hierarchical Object-Related Nets). These nets possess two mechanisms for timed transitions and a hierarchy concept where subnets trigger each other through special call- and stop-transitions. The language for inscriptions and type definitions is C++, tokens are called objects. In addition to the usual arcs, THOR-Nets contain consuming arcs and two types of test arcs. Transitions may have a firing capacity (maximum number of parallel occurrences).

The extension of THOR-Nets into HyNets mainly consists of the introduction of continuous transitions, which are connected to places by undirected continuous arcs (and test arcs). An extension of the inscription language allows the notation of differential equations. When a continuous transition fires, it changes the attribute values of tokens on those places to which it is connected by continuous arcs. The continuous manipulation is indicated by a firing action which may contain algebraic and differential equations. The concept of the evolution graph, which was introduced for hybrid nets with anonymous tokens in (Le Bail et al., 1991) is extended to HyNets.

2.2 An Extension of Coloured Petri Nets

The background for the approach presented here is the application of a hybrid formalism to the domain of railway systems. It is based on high-level nets since they offer the ability to describe individual trains moving across a topology which is configured from generic elements and interacting with many individual signals using just one generic train model and one model for each type of topology element or signal.

Among the definitions for high-level nets the Coloured Petri Nets (CPN) by K. Jensen (Jensen, 1997) were selected, being one of earliest and still most important

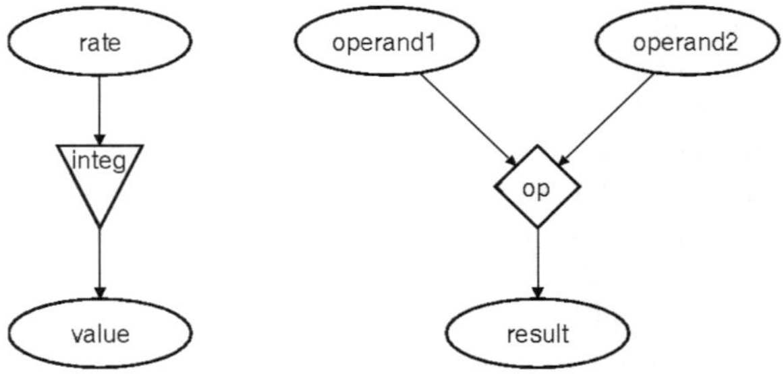

Fig. 1. Two forms of continuous transitions

definitions of high-level nets, which is also supported by a widely distributed standard tool (Design/CPN, 2002). The extension consists of the introduction of two forms of continuous transitions, namely integrating and algebraic continuous transitions. As the rules for the syntax and behaviour of the discrete net are left unchanged, there are no special places or other new features. Figure 1 shows the graphical representation of the continuous transitions. We chose triangles for integrating and tilted squares for the algebraic transitions to distinguish them from discrete transitions (rectangles) and places (circles, ovals).

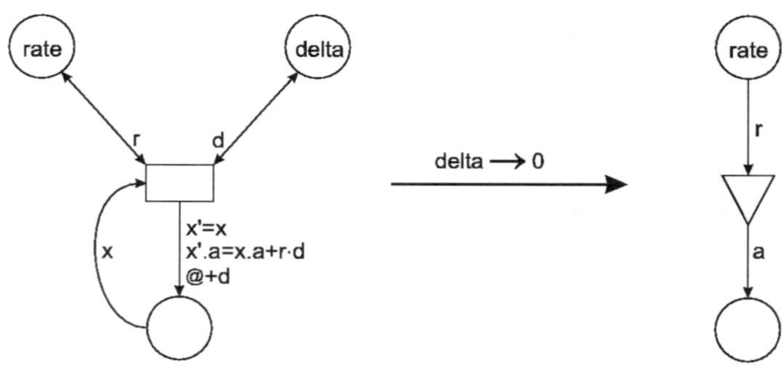

Fig. 2. From discrete to continuous transitions

The integrating continuous transitions represent the most basic form of continuous interaction in the net where the value of a given attribute of one token indicates the rate of change of the value of an attribute of another token. Together with the discrete transitions, this construct allows to model hybrid systems with piecewise constant rates. Figure 2 shows the relationship between continuous transitions and discrete transitions: the continuous transitions are simply timed discrete transitions with infinitely small time increments (Decknatel and Schnieder, 1998).

The algebraic transitions were introduced in order to be able to model relationships between continuous variables. The type of relationship is indicated by an element from a given set of operators.

A non-hierarchical timed CP-net is a tuple CPN = $(\Sigma, P, T, A, N, C, G, E, I, R, r_0)$, where the elements of the tuple are the set of colour sets (Σ), places (P), transitions (T), and arcs (A) plus the node (N), colour (C), guard (G), arc expression (E) and initialisation (I) functions, the set of time values (R) and the start time (r_0) (Jensen, 1992, Jensen, 1997). This definition is extended to hybrid nets by adding a function H, which partitions the set T of transitions into the set of discrete, integrating and algebraic transitions.

The definitions of the behaviour of (timed) CP-nets given in (Jensen, 1992, Jensen, 1997) also apply to these hybrid Petri Nets. Just as for regular transitions, a binding $b(\delta)$ of a continuous transition δ maps each variable of δ to a value, respecting the type of the variable as well as the guard of the transition. A continuous binding element is a pair $(\delta, b(\delta))$, and a continuous step is a multiset of continuous binding elements. A discrete step is enabled if there are enough tokens on the input places for all transitions (i.e. binding elements) in the step to fire concurrently, where all of the tokens involved have time stamps less than or equal to the current model time. For continuous transitions this is slightly different, since they may share a common rate and also manipulate a particular token (goal) together. Hence, enabling a continuous step depends on the individual binding elements and not on their sum; the continuous bindings are enabled independently of each other. Whenever a continuous binding element is enabled, it becomes active, i.e. it fires continuously, changing the value of the goal attribute as indicated by the rate.

From the continuous perspective, a discrete net with time possesses only piecewise constant functions. With the continuous transitions it is possible to construct linear functions with piecewise constant slope, or even functions of higher degrees by cascading the continuous transitions.

3 Models

CPN models are structured in pages, which are separate (non-hierarchical) nets. Pages can be connected by transition refinement and fusion sets. A refined transition serves as a placeholder for another page, which is called the subpage of the transition. A fusion set is a collection of places which behave as one single place, i.e. all places within the set always carry the same tokens. This can be used in order to simplify the graphical layout of a net by creating duplicates of a place which is connected to several transitions, and it can be used to connect separate pages. The behaviour of a hierarchical net using transition refinement and fusion places is defined as the behaviour of the corresponding non-hierarchical net, which is constructed by replacing each refined transition by its subpage and merging the places of each fusion set.

The models used in this project consist of about thirty pages. There is always one central page which contains the track topology. This allows to quickly adapt the model to different scenarios by simply replacing the topology page. Figure 3 shows a part from a model of a single line with a small station.

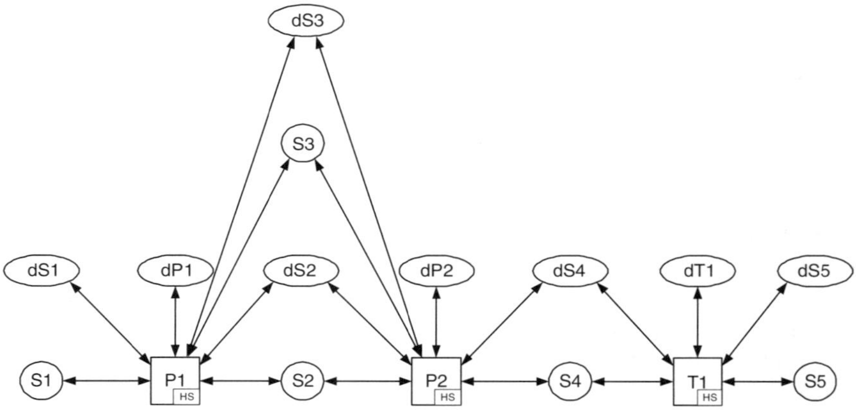

Fig. 3. Model of the track topology: Sections, borders and points

The places S1 to S5 represent track sections, the transitions P1 and P2 stand for points (switches), and the transition T1 is the border between the sections S4 and S5, which can be used as the position of a signal or the end of a block. The small ovals are places which carry specific attributes of their associated place or transition: the place dP1 for example contains data about the point P1. All transitions have subpages, which is indicated by the small rectangles marked HS. There are two subpages in total, one for section borders and one for points, such that all transitions within such a net are refined by one of those two subpages.

The actual net in Design/CPN contains some further annotations which are omitted here, such as the type (color set) of the places, the mapping between the places around a transition and the places on its subpage, initial markings and arc inscriptions.

As an example for an integrating transition, Fig. 4 shows the page which connects acceleration and velocity, with the integrating transition accel. It has a subpage which is responsible for the realization of the (pseudo-) continuous behaviour. The two places belong to fusion sets, as indicated by the boxes marked FG. These fusion sets connect this page to the other pages in the model. The colour sets (types) of the places are indicated next to the places (ACCELERATION and VELOCITY).

In order to control the movement of the trains, their braking distance is added to the model. The far end of the braking curve, i.e. the position at which the train would come to a halt if a given deceleration was applied, is moving ahead of the train. The velocity of that point depends on the current velocity and acceleration of the train, and its braking ability (assumed deceleration). Since acceleration and velocity of the train change dynamically, the velocity of the front end of the braking curve is calculated using continuous algebraic transitions as shown in Fig. 5.

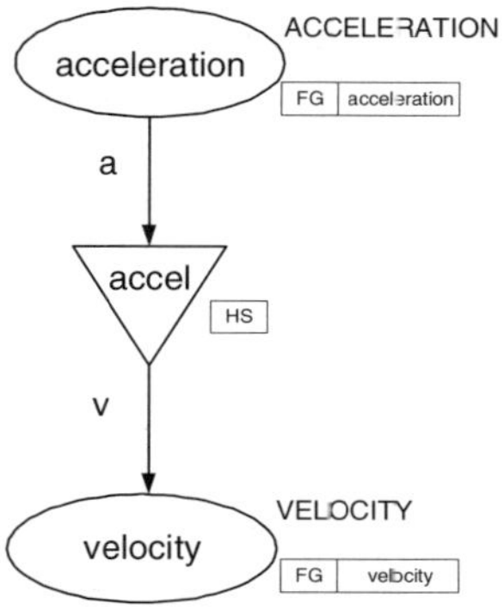

Fig. 4. Integrating transition linking acceleration and velocity

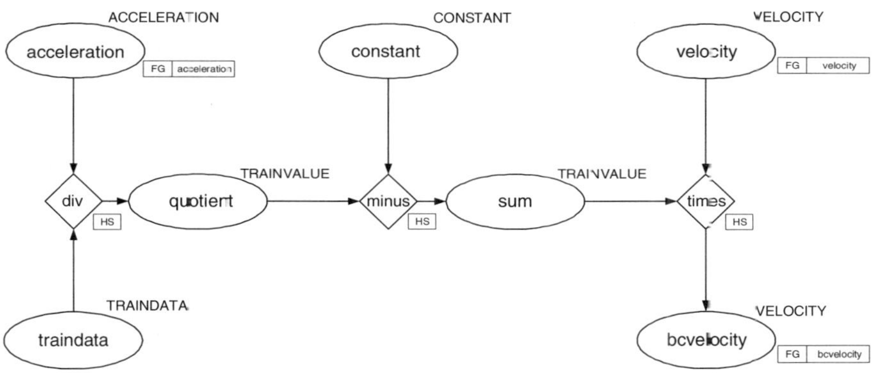

Fig. 5. Continuous calculation of the velocity of the front

4 Simulation

In order to be able to use the model described above for simulation, the behaviour of the continuous processes as described by the continuous transitions has to be simulated. One possibility would be to build an interface to a tool for the continuous simulation. In order to stay within the language of Petri Nets and to use the simulator which is part of the tool Design/CPN, a translation of the continuous behaviour into discrete terms is required.

4.1 Numeric Integration

The first solution is based on the idea to hide the discrete approximation of the continuous processes as a subpage below the continuous transitions. We added a function to the tool which constructs such a subpage automatically on the basis of a few parameters which are supplied by the user as comments to the continuous transition. One advantage of this approach is that it is flexible and that the modeller does not have to deal with the discretization in detail. A disadvantage is the inaccuracy of the approximation and the reduction in performance as soon as several continuous transitions with small steps sizes are active.

4.2 Discrete-Event Simulation

In order to overcome these disadvantages, a second simulation approach using discrete events was implemented. The main difference is that, instead of continuously accessing the tokens, the (pseudo-) continuous transitions fire only at event times, setting the continuous evolution of the attribute values by parameters. The tokens thus not only contain the current value of such an attribute, but also parameters which describe the continuous behaviour of that value. In our example, we used quadratic polynomials, such that the function describing the behaviour of an attribute value is given by three parameters (coefficients).

For each transition which has a firing condition which depends on the value of such an attribute, such as the crossing of a threshold, there is one auxiliary transition which calculates the event time when the condition with respect to that attribute will become true. This auxiliary transition fires when the parameters of the attribute change. The simulation clock is automatically set to the smallest event time in the future as soon as no more transitions are enabled.

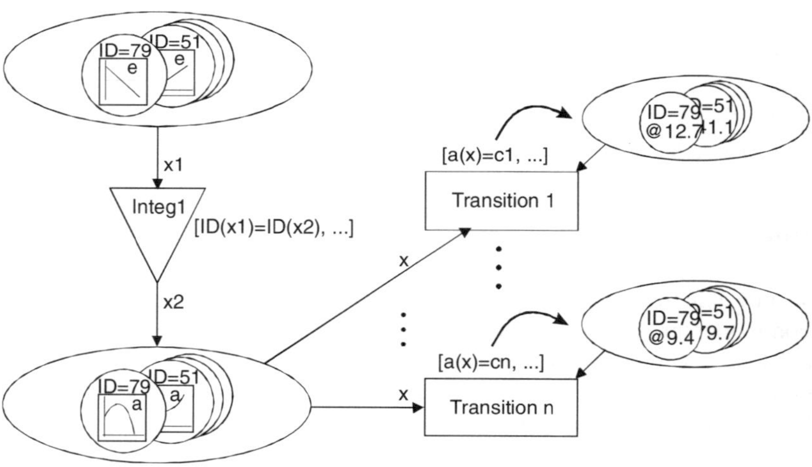

Fig. 6. Principle of discrete-event simulation

Figure 6 illustrates the idea of the discrete event simulation: The large ovals are places, the circles within them represent tokens, the rectangles are discrete transitions and the triangle is an integrating continuous transition. Inside the tokens, the continuous evolution of attribute values is symbolized by a small graph. The integrating transition “Integ1” sets the attribute “a” within the tokens on its output place at the bottom by integrating the value of the attribute "e" within the tokens on its input place. For each binding element the tokens have to agree on the attribute "ID", as specified by the guard expression next to the transition.

The guards of the transitions on the right refer to the variable "a", which is translated into event times by an auxiliary transition (not shown). The places on the right hold one token for each continuous binding element. These tokens have time stamps such that the transitions cannot fire before the global clock reaches the specified time. A continuous step is thus represented by a step of the global clock. Technically, the marking of the net does not change during such a step, since the continuous evolution of the attributes is already (statically) included in the tokens.

This form of simulation requires the calculation of event times, which represents a limitation in terms of modelling. The advantage here is that there is no numerical approximation, leading to higher accuracy and less computational overhead. Another advantage of the discrete-event simulation is that it allows the automatic computation of the reachability graph (occurrence graph, OG) using the OG tool which is part of Design/CPN.

5 Performance Analysis

Contemporary railway signalling systems are based on dividing the track into track sections known as blocks and ensuring that trains are separated by a suitable number of blocks. Such systems are called fixed-block train protection systems (Pachl, 1999). In a fixed-block signalling system the rails are divided into blocks by track circuits or by axle counters. The signalling system knows the position of a train only by the relatively coarse measure of block occupancy. The indication of the occupancy is realised by fixed signal aspects. The length of a block is usually given by the maximal length of the braking distance of a train running with the maximal track speed.

Moving block is a protection policy which has neither blocks nor signal aspects (Lee, 2000, Ning, 1998). The system is based on a continuous or frequent calculation of the free headway, i.e. the safe distance ahead of each train, and then relaying the appropriate speed, braking or acceleration rate to each train. A precise knowledge of the train's position, speed and length and fixed details of the line are necessary for the calculation of the next stopping position of each train. The position of this virtual target point is based on the normal braking distance of the train plus a safety distance which considers the accuracy of the positioning system, the length of the overlap and the reaction time of the braking system.

In order to test the developed modelling language and modelling approach a performance analysis of a free line with no stop and a line with a single station stop was chosen. To allow an easier model verification the overlap was neglected for simulation.

5.1 Performance Analysis of a Free Track

For the purpose of performance analysis a track of 10 km length was modelled. The middle part of the line can be divided into 8 fixed blocks of length 1 km. This restricts the maximal speed using the fixed block signalling system to 160 km/h. In the construction of the model ten instances of the discrete movement across block borders were used.

The performance analysis was carried out using discrete-event simulation by running three trains on the modelled line. The objective of the simulations performed for a specific signalling system is to determine the headway between trains. The length of the trains was considered 250 m and the acceleration and braking rate of $0.076\ m/s^2$. A typical result of a simulation is presented in Fig. 7. In this case the initial velocity of the trains was set at 90 km/h and in the first half of the line four blocks were modelled.

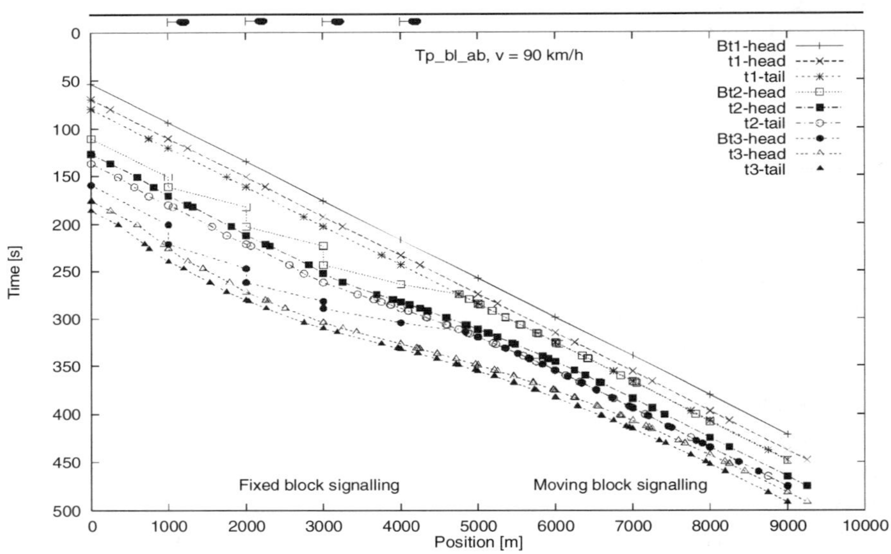

Fig. 7. Characteristic run over consecutive areas with fixed block and moving block signalling

Figure 7 shows the run of three trains over time. In addition to the movement of front and rear ends of the trains (*t-head, t-tail*), the positions of the far ends of the braking curves (*Bt-head*) are included as well. One can see a significant time expansion of the run in the fixed-block area in comparison to the run under the moving block signalling system. The long headway has a negative influence on the performance of the line.

The dependence of the maximal line performance, expressed in trains per hour, on the velocity of the trains with their considered length and ideal operation conditions is shown in Fig. 8. The analysis results shows the performance improvement and the most effective operating area of the moving block signalling system. The maximal increase of factor 2.8 was reached with the speed of 71.6 km/h and is numerically conform with similar investigations (Six, 1996, Uebel, 2000). The maximal speed of the line by the fixed block protection corresponds to the braking distance of the train, which cannot exceed the length of a block section. However, the moving block (see also Fig. 8) does not imply such a systematic restriction which is desirable mainly on lines with mixed kinds of transport.

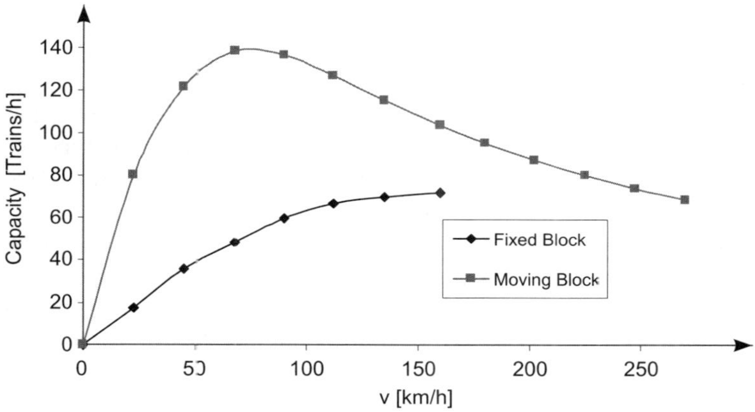

Fig. 8. Track capacity depending on velocity

Another possibility is to show the dependence of the performance on the train density, expressed in trains per kilometer, in form of the so called fundamental diagram, see Fig. 9.

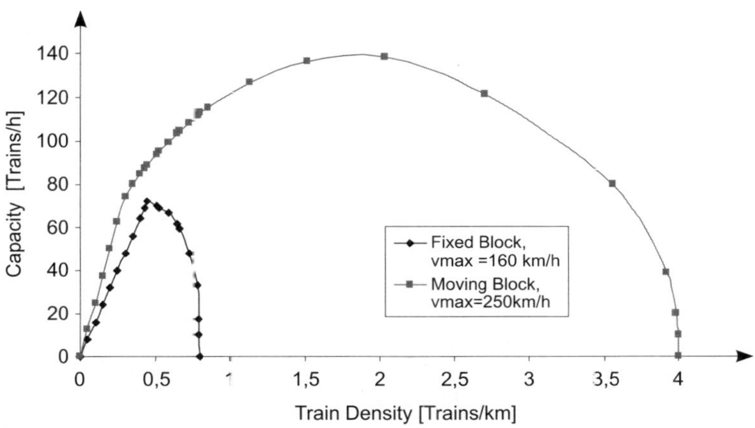

Fig. 9. Fundamental diagram: Track capacity depending on train density

The gradient of the secant gives the value of the mean velocity and the maximum of the curve shows the critical train density. The further increase of train density leads to decrease of performance. One can see a considerable flexibility reached by the use of the moving block signalling system.

5.2 Performance Analysis of a Track with a Station Stop

A particular performance improvement is expected by the use of moving block in rapid transit systems on highly frequented urban lines. This was the reason to include a station stop in the track model. A station stop can by modelled easily by including special signals which are connected with a clock, which counts down the time according to a given rate (here -1). The stop dwell time was set for each train to 32 s. After expiration of the dwell time the train is started automatically.

The charts in Fig. 10 show a typical run of three trains over time. In this case the maximal track speed of 135 km/h was considered. On the left the simulation result using fixed block signalling can be seen and on the right the result for moving block. The charts especially show the small loss of time of the run using moving block signalling system which increases the track performance considerably.

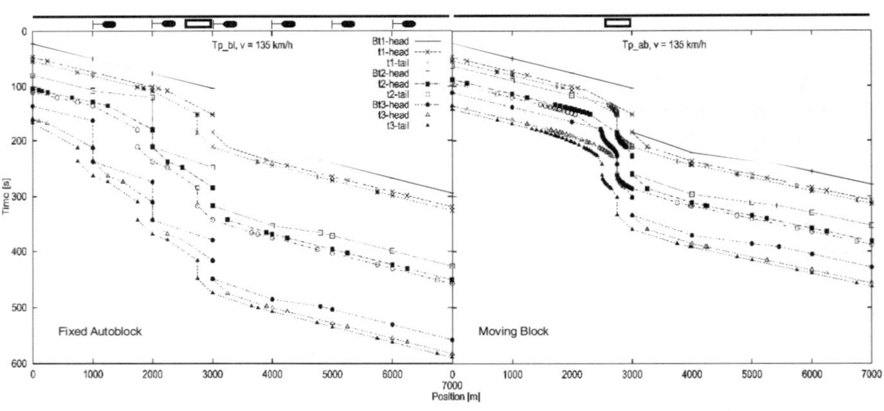

Fig. 10. Characteristic run with station stop using fixed block and moving block signalling

The stop dwell time causes that the performance, measured on exit of the test track, remains for different velocities more or less constant ($v > 40\text{km/h}$). Assuming ideal operating conditions the moving block performance has reached 48.26 trains per hour in comparison to 27.32 trains per hour using fixed block signalling which represents an improvement of 76.6%.

6 Summary and Outlook

A particular variant of hybrid high-level nets on the basis of Coloured Petri Nets was developed and used to model railway systems as hybrid systems. This allows to combine models of complex (discrete) control systems with continuous processes such as the train motion. Using a discrete-event approach we were able to achieve a reasonably efficient simulation and to employ it for specific problems of railway traffic. In addition it is possible to automatically construct the reachability graph for any given scenario.

We showed how performance analysis of a track can be carried out using this hybrid railway model. The results of simulations show the large potentials of moving block signalling to increase the track performance. The advantages are significant for trains with non-stop scenarios as well as for included station stops. Furthermore, the real operating condition can be considered including inaccuracies of the positioning system, system reaction time as well as stochastic disturbances (Zhu, 2001).

Simulation for Analysis of Aircraft Elevator Feedback and Redundancy Control

Pieter J. Mosterman[1], Manuel A. Pereira Remelhe[2], Sebastian Engell[2], and Martin Otter[1]

[1] Institute of Robotics and Mechatronics, DLR Oberpfaffenhofen, P.O.Box 1116, D-82230 Wessling, Germany
[2] Process Control Laboratory, Department of Chemical Engineering, University of Dortmund, 44221 Dortmund, Germany

Abstract. Safety critical systems such as aircraft require functional and hardware redundancy to achieve prescribed safety levels. Discrete event control is applied to ensure that a safe system configuration is available at all times. Since, at present, formal verification techniques are restricted to models with few continuous states, in this paper, simulation is used to verify that the overall system operates according to the requirements when an actuator failure occurs. The feasibility study to modelling and simulation of complex controlled systems presented here is characterised by (i) a complex object-oriented model of aircraft dynamics, including gravity, aerodynamics, etc., (ii) the specification of the discrete event redundancy control by a domain specific formalism that includes statecharts, (iii) the usage of energy based hybrid bond graphs to model the dynamics of the hydraulic actuators, (iv) model integration on the model level as well as on the data level, (v) support of DAEs with dynamically changing index and (vi) illustrative simulation results.

1 Introduction

Redundancy is one of the most important techniques to achieve the desired level of safety in systems such as aircraft, nuclear plants, chemical plants, and other safety critical applications. Its basic premise is to include redundant functionality into a system that can be activated when failures of the normal operating components occur and to validate and select normal behaviour (e.g., voting procedures).

1.1 Aircraft Attitude Control

To illustrate the concept, consider the primary (attitude) control surfaces of an aircraft as shown in Fig. 1. The ailerons are used to control roll, the elevators control pitch, and the rudder controls yaw motion. This paper concentrates on the pitch control, performed by the elevators. Each of the elevators is positioned by one of two actuators, the other one operates as a passive load. Discrete-event control embedded on two primary flight control units (PFCU) selects the controlling actuator and ensures that the redundant actuator is loading. Each PFCU controls one actuator per elevator, so that both elevators can be controlled, even if one PFCU fails completely. The PFCUs also generate the position control signals for the four actuators. Feedback control is

S. Engell, G. Frehse, E. Schnieder (Eds.): Modelling, Analysis and Design of Hybrid Systems, LNCIS 279 pp. 369–390, 2002.

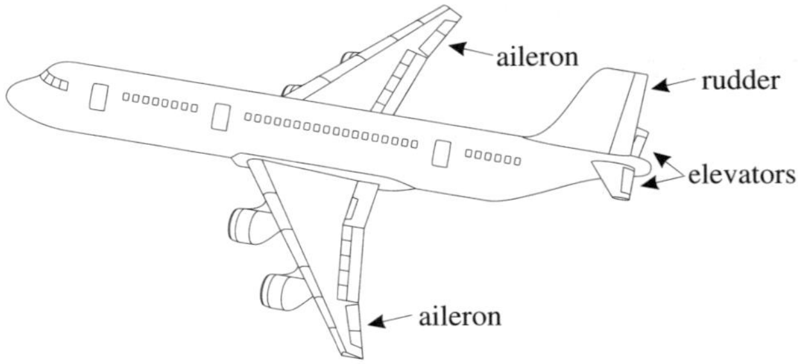

Fig. 1. Primary control surface of an airplane

used for normal operation whereas so-called direct link control is applied to single actuators in the case of certain failures. To ensure minimal transient disturbances caused by actuator switching, the loading actuator should be shadowing the control signals ready to switch to a mode where it actively controls the elevator. However, in some extreme cases the actuator may be disengaged, i.e., it is loading but not shadowing. Thus, each PFCU has to decide for two actuators whether an actuator is disengaged, shadowing, or controlling and whether a feedback or a direct link controller is used for shadowing or controlling. These decisions depend on the mode of the other actuator, the state of the other PFCU and the detected failures. The best possible consistent state configuration of both PFCUs for a given failure situation is achieved by a complex iterative interaction of both PFCUs.

The hydraulic actuator design and the controller parameters may influence the overall behaviour of the aircraft significantly. Therefore, all contributing parts and phenomena of the aircraft such as aerodynamics, gravity, engines, etc. have to be considered in order to assess the design of the elevator control system. Because of the immense complexity and the intricate redundancy management model-based validation is required.

Formal verification techniques are widely used for pure discrete-event systems and much research has been carried out recently on the verification of hybrid systems. However, at present, the complexity of systems amenable to hybrid systems verification techniques is restricted to a low order continuous dynamics (typically not more than three continuous state variables) (Benedetto and Sangiovanni-Vincentelli, 2001, Lynch and Krogh, 2000, Vaandrager and van Schuppen, 1999). Consequently, formal methods are applied to the discrete-event part only, e.g., a so-called Failure Mode Effect (FME) analysis is employed to verify certain safety and reliability properties of the redundancy management system. However, its interaction with the continuous parts as well as the design of the position controllers and the hydraulic actuators can not be evaluated with formal verification techniques. Therefore, the only practical model-based approach for this task is to perform extensive simulation studies.

1.2 Model Design

In this contribution, we concentrate on the modelling and simulation of the elevator control system and the aircraft. The model formulation is driven by the assumption that the simulation studies have the purpose to assess whether the design of the evaluator control system meets the requirements with respect to the overall behaviour of the aircraft (e.g., lateral and longitudinal aircraft velocity and flight path angle). In particular, different sets of parameters of the controllers and of the hydraulic actuators have to be tested in combination with certain failure scenarios.

As a consequence, the simulation model has to incorporate a realistic model of the aircraft dynamics, including all essential effects and components such as aerodynamics, gravity, engines, and hydraulic oil supply. In order to automatically generate the correct Boolean input signals of the feedback controllers and the actuators depending on the sequence of failure events it is convenient to include at least the input-output behaviour of the redundancy management components. Since the sampling times of the PFCUs are very fast in comparison to the bandwidth of the actuators, the hardware aspects of the PFCUs can be neglected, i.e., the redundancy management model reacts instantaneously on failures and the controllers are modelled as ideal continuous controllers. Another idealisation is introduced for the hydraulic actuators. There are many small physical effects such as oil elasticity, viscosity, and fluid inertia which do not influence the overall dynamics significantly, but considerably increase the modelling effort, so that these effects are not considered in the corresponding models.

These basic model design decisions cause several difficulties for the modelling and the simulation. With respect to modelling, the complexity of the systems and their heterogeneous nature mandates the use of dedicated formalisms. These formalisms differ greatly in their visual representation and require the interoperation of specific and powerful modelling environments.

Present day simulation technology, on the one hand, can handle large systems of differential and algebraic equations (DAE), possibly extended by some discontinuous equations (ABACUSS, 1995). On the other hand, discrete-event simulators apply an event driven approach to manage the huge number of state changes in discrete-event models (Group, 1999). The combination of discrete and continuous behaviour requires the integration of a numerical integrator with some sort of discrete-event simulation. Especially, the detection and location of discrete events during continuous integration has to be supported. Furthermore, at event times discontinuities in continuous state variables may occur. For the aircraft model, this phenomenon emerges because the abstractions in the hydraulic actuator models result in a DAE with dynamically changing index. This requires a special simulation engine that switches the active equations and automatically reinitialises the state variables according to physical conservation laws, when the index changes. This contribution presents techniques that cope with all these problems.

1.3 The Modelling and Simulation Approach

For the components of the physical system we use object-oriented modelling. In this context, the term object-oriented modelling means that every physical object is modelled independently without making assumptions about its environment and preserving the physical connection structure of the object. The connections of a model component have to correspond to physical interactions the computational causality of which is not fixed a priori, i.e., the variables involved in an interaction are not a priori defined as inputs or outputs. Furthermore, the behaviour of the component should be defined in a declarative way where a set of (possibly implicit) equations is regarded as a set of behavioural constraints rather than as a calculation formula. To illustrate this, let us consider a hydraulic line. The component model of the line would have two connections which each incorporate a pressure and a flow variable. These variables represent neither inputs nor outputs, since depending on the structure of the environment the pressure drop causes the flow or the flow causes the pressure drop. In some cases the causality can even change dynamically so that a quantity that would be regarded as an input in signal flow diagrams becomes an output and vice versa. This is why the equation-based behavioural description is inherent to object-oriented modelling. Using equations (which may be written in an implicit style) for the description of the behaviour does not impose a specific calculation scheme. From the modelling perspective the equations of all model components and all connections of the overall aircraft model simply form a global set of differential and algebraic equations (DAE) so that simulation is the task to find a solution to these equations, i.e., functions over time that satisfy the equations. To generate efficient simulation code, the model equations must be processed by a symbolic engine and compiled into executable code.

The existing DAE based modelling languages such as gPROMS (Barton, 1992), VHDL-AMS (Heinkel, 2000, Christen, 1997), and MODELICA (Modelica Design Group, 2000) differ in many aspects. This work utilises an aircraft library (Moormann et al., 1999, Moormann, 2001) developed using MODELICA which allows to build domain-specific graphical component libraries and supports many features known from object-oriented programming such as inheritance, packages, etc. For the modelling and the symbolic processing task DYMOLA (Dymola, 2002) was used. It provides a graphical user interface for model composition. The symbolic engine of DYMOLA generates C-code from a MODELICA model. Then a standard C-compiler generates the executable simulation code.

This configuration is already powerful enough to model most parts of the aircraft and to simulate the resulting complex DAE system including certain discontinuities (a so-called 'hybrid DAE'). However, for simulating DAEs with dynamically changing index, the current symbolic engine of DYMOLA (version 4.1d) is too limited. Therefore, a specially developed environment, HYBRSIM (Mosterman and Biswas, 1999), which is based on hybrid bond graphs (Mosterman and Biswas, 1995) was used to model the components with variable index, i.e., the hydraulic actuators. The C-code generated by the two environments, DYMOLA and HYBRSIM was then merged manually and simulated using a general purpose hybrid dynamic system simulator MAsim (Mosterman, 2001).

The purely discrete-event parts of the elevator control, i.e., the redundancy management, are modelled by a domain-specific formalism including statecharts. It will be shown that the syntax and the semantics of the object-oriented modelling paradigm is not well suited to represent the objects of such a formalism. Instead, a separate modelling environment has been implemented which supports this formalism and generates a monolithic Modelica component that can be integrated into the Modelica aircraft model .The behaviour of this component is defined by an algorithm that is interpreted by Modelica as an additional model constraint, i.e., it is equivalent to one equation with multiple input and output variables. In contrast to the other parts of the model the causality of the interface variables and the calculation scheme of the resulting object are predetermined. While the actuator model is integrated on the data level, the redundancy management model is integrated on the model level.

Section 2 presents a system level view of the elevator redundancy control. Section 3 discusses the different parts in detail and presents the respective models. Simulation results are given in Sect. 4. Finally some conclusions are drawn in Sect. 5.

2 Aircraft Elevator Control System

The aircraft elevator control system includes several forms of redundancy (Seebeck, 1998). The system itself consists of two elevators, the control surfaces. Each of these are controlled by one of two hydraulic actuators while the other one is operating as a passive load. The four actuators take their power from three hydraulic subsystems as depicted in Fig. 2. Two primary flight control units are available to compute actuator control signals and modes.

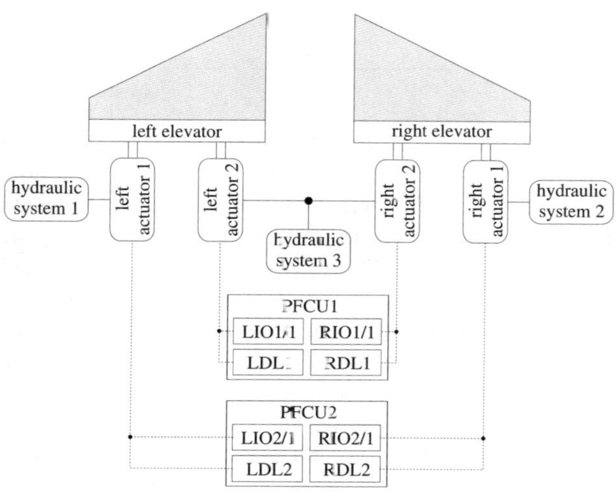

Fig. 2. Elevator system

The functionality of each actuator is specified in textual form in terms of a number of module actuator control modes (MACM) all with their specific behaviour characteristics. These are defined in Table 1. Note that the MACM definitions include behavioural information along with structural information about the particular mode of operation of the actuator components.

Table 1. Module actuator control modes

MACM	Description	Actuator Servo valve	Actuator Spool valve
active	The module controls the servo valve in closed loop mode. The corresponding actuator is active and controls the elevator movement.	controlled	open
hot and standby	The module controls the servo valve in closed loop mode. The corresponding actuator is not active and operates as a load.	controlled	closed
passive	The module is waiting and does not generate actuator control signals. *It can change its mode at any time to take on control of the corresponding actuator.*	not controlled	closed
off	The module is turned off temporarily because of an intermittent failure and does not generate actuator control signals. *As long as the failure has not been fixed, it cannot change to a mode where it controls the corresponding actuator.*	not controlled	closed
isolated	The module is turned off indefinitely. *A persistent fault in the control loop of the corresponding system isolates the module and it cannot change to a mode where it controls the corresponding actuator.*	not controlled	closed

The discrete outputs of the redundancy management system are transformed into physical behaviour by means of a spool valve and a servo valve in the hydraulic actuator. Power is supplied by one of the hydraulic systems and delivered to the actuator cylinder that positions the elevator. This flow of energy is modulated by the servo valve, the modulation is computed by a PID feedback control law. The control signals for the actuators are generated by two *primary flight control units* (PFCU) that can operate as *input-output* modules (IOM) or as *direct-link* modules (DLM) controlled by a switch in the control law. The IOMs calculate setpoint values for the actuators based on a PID control algorithm and monitor a number of critical system variables and change between the modes in response. The DLMs allow limited but direct control of the actuators in case the IOMs are not available. The control modules can be in different modes for each of the actuators separately. Moreover, they may control other aircraft actuators as well. In addition, the servo valve may not be controlled and its piston then is in a default position. Also, the spool valve can be turned on and off to switch between active control and passive loading. Continuous feedback control drives the elevator to its desired setpoint, while higher level redundancy management selects the active actuator and the control law to be used.

Interaction between the actuator and the aircraft model consist of forces and moments acting on the elevator that is stiffly connected to the actuator positioning cylinder as well as the pressure generated by the hydraulic systems. Three hydraulic systems supply the oil for the actuators shown in Fig. 1. When a failure occurs, the redundancy management switches between actuators and oil supply systems to achieve maximum control.

The behavioural redundancy requirements may be formalised by a set of rules for the redundancy management to switch between module actuator control modes as follows (Seebeck, 1998):

1. Mode changes only occur when
   - a system failure is detected, or
   - control of an uncontrolled elevator is requested, or
   - one module requests control of both elevators which are controlled by separate modules.
2. One module should be simultaneously in either *active*, *hot*, or *standby* for both elevators as long as possible.
3. If not overruled by the previous specification, the module priority is such that the switching sequence is IOM2/1 $\rightarrow$ IOM1/1 $\rightarrow$ DLM2 $\rightarrow$ DLM1.
4. There is always one and only one module that controls one elevator, i.e., that is *active*.
5. In case of a failure of the controlling module, control is assumed by a module that is *hot* or *standby*. If no module is in this mode, the one with highest priority that is *passive* assumes control.
6. A module switches to *hot* when the other module that controls the same elevator, and, therefore, is *active*, belongs to another PFCU and both elevators are controlled by IOMs.
7. A module switches to *standby* when the other module that controls the same elevator, and, therefore, is *active*, belongs to another PFCU and one of the elevators is controlled by a DLM.
8. In case of pressure failure, the 'low pressure' signal only serves for fault classification. It does not cause a direct mode change.
9. In case of 'low pressure' and if a sensor detects an elevator positioning system failure, the module switches *off*. The module switches back to *passive* only when no system failure is reported and the 'low pressure' condition does not hold anymore.
10. If 'low pressure' is not reported and the elevator positioning system is reported to fail then the module switches to *isolated*.

To prevent nondeterministic switching, priorities are assigned to the possible transitions. Because of the critical nature of switching to the *isolated* mode to prevent damage to the system, this transition has the highest priority. In addition this causes another module to immediately assume control. This is also desired when, e.g., a pressure loss is detected and the module switches *off*. Therefore, the corresponding transition has second highest priority. Another decision criterion is to allow modules to take over control as quickly as possible. As a result, modes that implement as much control as possible should have highest priority. So, when a module can be switched *active* this should be immediately executed rather than first switching to *standby* if this transition is also enabled. This yields the following priorities:

1. Transition to *isolated*
2. Transition to *off*
3. Transition to *active*
4. Transition to *hot*, *standby*, and *passive*.

Sensors in the elevator control system provide the PFCUs with information about the functioning of the system. In case of abnormal readings, the entire set of measurements is used to infer a particular failure mode. Details of this inference mechanism are beyond the scope of this paper. To test the redundancy management, failure mode effect (FME) analysis investigates the availability of the system for several test cases that embody a set of sensor readings:

- Pressure decrease in the hydraulic system (H1, H2, H3)
- Predefined set of failures (F)
  - IO module failure (1, 2)
  - DL module failure (1, 2)
  - Actuator failure (left inner/outer, right inner/outer).

These failures represent abstractions of actual physical phenomena underlying the failure detection. FME is still the most important step in verifying system safety and reliability of discrete-event control (Mai and Schröder, 1999, Osder, 1999).

The combined discrete redundancy management for two of the four actuators on each of the four modules results in eight redundancy modules. This adds up to a considerable discrete behavioural complexity. Each module consists of six possible local modes and there are eight such modules. Thus, the total number of modes of the redundancy management control is 48. There is always one and only one active state in each of the discrete-event models. But, because of the redundancy specification, each of the models needs to have information about the mode of each of the other ones. This interaction is based on the MACMs and causes logic connections between each of the actuator control modules. Finally, an additional discrete-event model is used to model possible fault scenarios by activating states that correspond to particular failure modes. This model has eleven states.

3 Modelling the Parts of the System

The elevator control system described in Sect. 2 contains a number of parts that are best captured by different modelling approaches: (i) the aircraft dynamics, (ii) the redundancy control, including control law switching, and (iii) the actuator switching behaviour.

3.1 Aircraft Dynamics

To investigate the effect of actuator switching on the overall flight characteristics such as nick rate (q) and angle of attack (α), an aircraft model is required. The more realistic this model, the higher is the probability that the analysis results also hold for the actual implementation on the aircraft.

Object-Oriented Modelling. The design of a realistic aircraft model is a tremendous task that combines several domains within aircraft design such as (i) aerodynamics, (ii) gravity, atmospheric, and wind models, (iii) engine/thrust models, (iv) rigid body models including the effects of fuel consumption, and (v) systems models for primary (attitude) control.

Traditionally, such complex aircraft models are written in a computer processable format such as, e.g., FORTRAN, and they are completely integrated with facilities for behaviour generation, e.g., the numerical solver. This, however, renders the models unwieldy, error-prone, and rather costly to implement and update.

Recently, a more structured approach to aircraft modelling has been developed based on object-oriented modelling techniques and the use of libraries of the domain specific components mentioned before (Moormann et al., 1999, Moormann, 2001). Object-oriented modelling techniques rely on the notion of *encapsulation* to hide the details of physical component models and to increase maintainability. Furthermore, the models are organised hierarchically which allows successive refinement of behaviours at increasing levels of detail.

Graphical Syntax. Figure 3 shows a top-level view of the aircraft model with the engine objects (left), the systems component (top) and the aerodynamics model (right), the rigid body model, and the gravity/atmosphere/wind models (bottom-right). These components can be decomposed hierarchically in similar object diagrams.

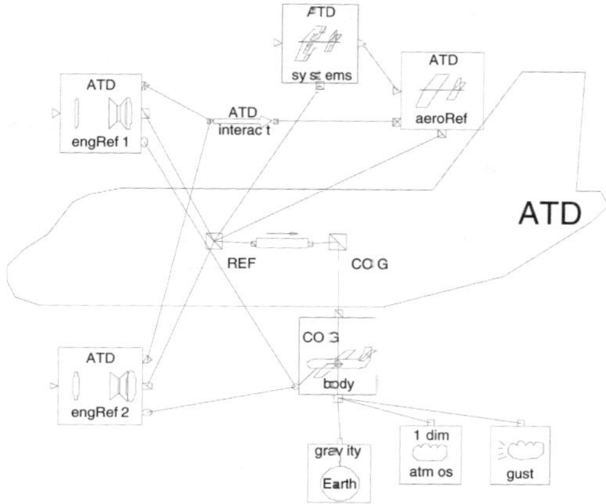

Fig. 3. Top level object diagram of the aircraft model

Communication between objects is realised through ports that also constitute the interface to the next level in the hierarchical decomposition. For a set of connected variables, v_i, these ports use two different connection semantics, (i) $\forall i(i \neq 0 | v_i = v_0)$, i.e., all connected variables are set equal, and (ii) $\Sigma_i v_i = 0$, i.e., the connected variables are summed to 0. This allows for a convenient implementation of energy flows across ports where the different semantics correspond to the across and through variables, respectively, the product of which constitutes power.

Execution Model. The behaviour of each of the primitive model objects is described in terms of algebraic and differential equations. These are treated as noncausal, i.e., no computational direction of the variables is assigned (it is not determined which variable is to be computed from an equation), which is a convenient way of modelling physical systems in terms of declarative constraint specification. Furthermore, it enhances model reuse.

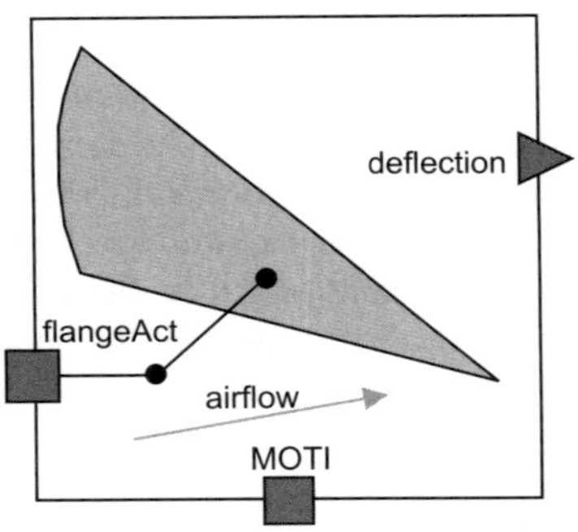

Fig. 4. Elevator control surface library component

To illustrate, consider the elevator control surface library component in Fig. 4. This surface consists of one or more movable parts to adjust the aerodynamic force acting on the aircraft. The library component is connected to the remainder of the aircraft model by three ports: (i) a mechanical port *flangeAct*, that contains the elevator deflection, δ, (ii) an aerodynamic port, *deflection*, that carries the forces because of the airflow around the elevator, and (iii) a mechanical port, *MOTI*, that contains the force acting on the aircraft. Table 2 itemises the most important interface variables of the ControlSurface class. The elevator computes the force F_{act} from

$$0 = f(F_{act}, V_a^2, \delta, \rho, S_{cs}, \ldots, m_{cs}, g, \ldots), \quad (1)$$

the deflection δ from

$$\delta = k_{kin} x_{act}, \quad (2)$$

and, finally, its rate of change $\dot{\delta}$ from

$$\dot{\delta} = k_{kin} v_{act} \ . \quad (3)$$

Note that k_{kin} is a parameter internal to the object that represents the kinematics of the mechanism.

Table 2. ControlSurface class interface variables

Interface Variables		
across	x_{act}	displacement of actuator flange
	v_{act}	displacement velocity of actuator flange
	V_a	airspeed velocity
	ρ	air density
	g	gravitational acceleration
through	F_{act}	force acting on actuator flange

To enable the execution, the primitive object equations and the connection constraints for across and through variables are accumulated by a global model interpretation scheme. It sorts and solves the overall system of differential and algebraic (DAE) equations by assigning causality so that the unknowns can be computed from the equations and input and state variables. Algebraic manipulations are performed to reduce the system of equations, e.g., (Andersson, 1994).

To represent switching, equations may be conditionally active. When the conditions change their truth value, this causes *events*. When events occur, variables may undergo discontinuous changes. In addition to the differential and algebraic equations, a 'pre' operator is defined to allow access to the value of a variable immediately before a discontinuous change. Because this introduces discrete state behaviour, an iteration is required to converge to a consistent state before the continuous simulation is resumed. Though this mechanism can be used for implementing discrete-event behaviour, it is difficult to mimic state transition diagrams using object diagrams and even more so to describe the state transition behaviour by local equations of the primitive states and transitions. The graphical syntax of object diagrams does not allow annotation of component connections, thus it is not possible to write conditions, events, and actions alongside a transition. Furthermore, transitions are not objects in object diagrams. Therefore, the transition behaviour requires a specific transition object to be inserted. Execution has to be described in terms of local algebraic constraints that communicate between states and transitions to evaluate whether a state is active and a transition is enabled (Mosterman et al., 1998).

The result of collecting the local equations, adding the connection constraints, and sorting and solving these leads to a global system of equations of the form

$$\begin{aligned} \dot{x} &= f_\alpha(x, u, t) \\ 0 &= g_\alpha(x, u, t) \\ \alpha^+ &= \phi_\alpha(x, u, t) \end{aligned} \tag{4}$$

where f_α specifies the dynamics in mode α, g_α the event generation functions ('zero crossings'), and ϕ_α the next mode function. Before continuous simulation can start or be resumed after an event occurred, a consistent mode α, i.e., $\alpha^+ = \alpha$, has to be found. Typically, this is performed by a fixed point iteration scheme.

3.2 Redundancy and Position Control

The main purpose of the two primary flight control units is the generation of appropriate continuous and discrete control signals for the four elevator actuators. Each PFCU contains a failure monitoring function, a specific discrete-event part for the redundancy management as well as one feedback and direct link controller per elevator. When a failure is detected, the redundancy management parts of both PFCUs interact tightly in order to achieve a consistent decision on the appropriate reaction, before switching the operating control laws. This is because each PFCU is responsible for different actuators and has to take the discrete state of the other PFCU into account in order to guarantee that each elevator is controlled by one actuator only. Therefore a simple failure may trigger a sequence of transitions in both PFCUs, where a discrete mode transition in one PFCU may lead to a state which forces another transition in the other PFCU and so on.

Graphics. Since hardware aspects are beyond the scope of this paper, the redundancy management parts of both PFCUs are unified in one discrete-event model component neglecting the distributed architecture of the system. As a consequence, the aircraft model contains only one elevator control component. This is divided into three parts: (i) a failure injection module that replaces the failure monitoring functions so that specific failure scenarios can be studied, (ii) the combined redundancy management parts of both PFCUs that react on changes of the failure configuration and (iii) the switched position controllers of both PFCUs the transfer functions of which depend on the actual modes of the redundancy management component (Fig. 5).

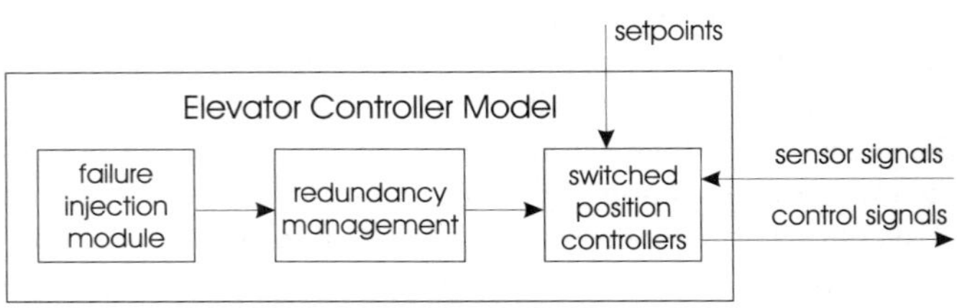

Fig. 5. The structure of the elevator controller model

The requirements for the redundancy management which were formulated informally in Section 2 state that each redundancy module contains 6 possible local modes. Since a redundancy module switches from one mode to another under certain conditions, the modules should be modelled by a kind of state transition diagram, where the modes are represented by discrete states and the transition arrows represent the possible mode switchings. In order to take the transition priorities into consideration, hierarchical states as known from the statechart formalism (Harel, 1987) are used. Additional states are introduced that do not correspond to a mode, but represent the priorities of the transitions: The higher a state in the hierarchy the higher is the priority of its outgoing transitions, e.g., the transition ToOff in Fig. 6 has a higher priority than ToAct and a lower priority than ToIso. The statechart model in Fig. 6 reflects the state transition aspects of a redundancy module declared in the informal description of the requirements.

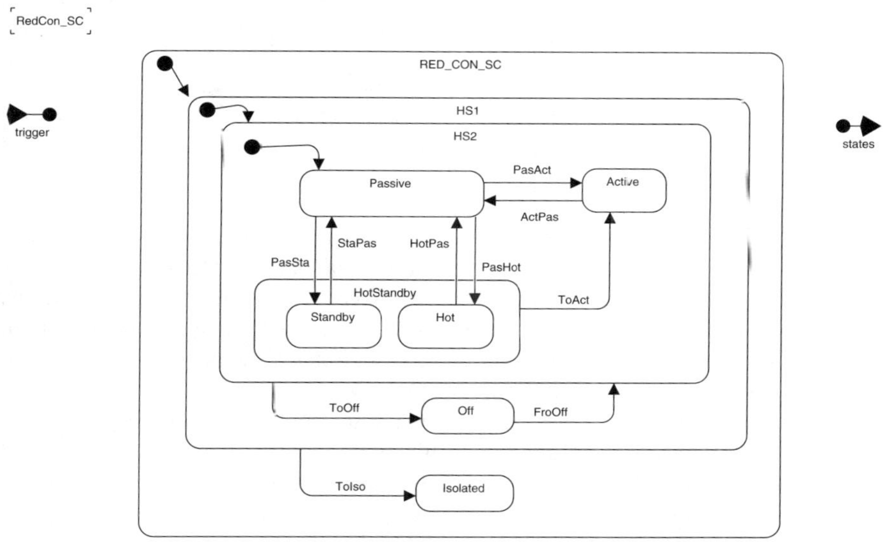

Fig. 6. The redundancy module statechart

The transition conditions can be derived from the switching rules of the requirements and differ for each statechart. In order to keep the statechart model generic and take architectural aspects into consideration, a specific hierarchical block diagram formalism is used (Fig. 7). Two blocks on the top level represent the two primary flight control units. The input port contains the failure values that originate from the failure injection module whereas the output ports transmit the actual module modes to the switched controllers. Each PFCU block contains four control modules (LIO, RIO, LDL and RDL) as subblocks. Their behaviour is defined by the statechart in Fig. 6. The transition conditions are calculated outside the statecharts in a special block (PFCU1_Logic and PFCU2_Logic) with no state behaviour.

Execution. The intended behaviour of the elevator control model is as follows: The failure injection module generates Boolean signals that indicate the presence of specific failures. When a failure signal changes, the transition conditions of both PFCUs are evaluated and their values are transmitted to the statechart blocks that

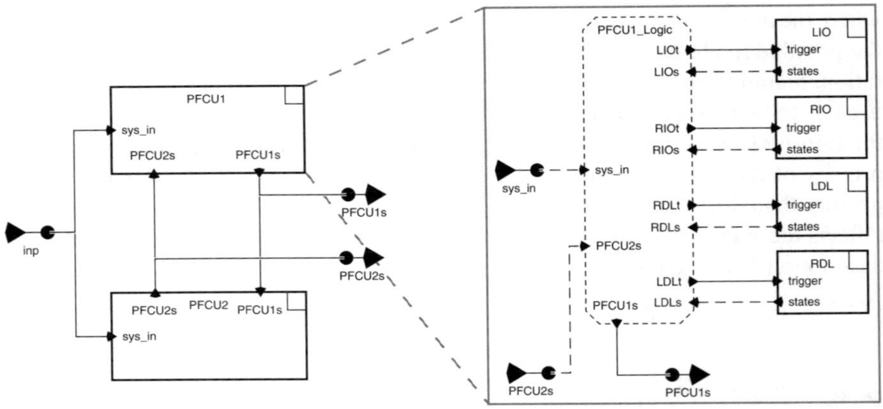

Fig. 7. Two block diagrams of the discrete-event part

perform their transitions independently. After all statecharts have converged to a persistent discrete state, i.e. no further transitions happen, the transition conditions are calculated again taking the new states into account and a new set of transitions may be performed in the modules again. When the overall discrete-event system reaches a stable state, this local event iteration is stopped, the output values are set, and the position controllers may change their mode.

In order to analyse the behaviour of the elevator control and the overall aircraft for different failure scenarios, the failure injection module generates predetermined sequences of failures. These scenarios can be modelled by equations containing logical expressions and inequalities over the independent variable time and parameters as shown in the following example where IO2failure is present from time t_1 to t_2:

$$\text{IO2failure} = (t > t_1) \wedge (t < t_2). \tag{5}$$

The output of the redundancy management part switches the position controllers that are easily described using equations. The following example shows the controller equations of PFCU1 for the left elevator:

$$e_{act,l1} = w_{act} - x_{act,l} \tag{6}$$

$$u_{act} = \begin{cases} 0 & \text{if PFCU1states.LIO is Off or Isolated} \\ w_{act} & \text{else if PFCU1states.LDL is Active,} \\ k_p e_{act,l1} + k_d v_{act,l} & \text{else} \end{cases} \tag{7}$$

$$u_{spool,l1} = \text{PFCU1states.LIO.Active} \tag{8}$$

3.3 Actuator Dynamics

The hydraulic actuators are the interface between the discrete-event domain of redundancy control and the continuous domain of the aircraft dynamics. The actuator here is not modelled with all details as this would lead to steep gradients in the behaviour that are difficult to handle and slow down simulation of the aircraft behaviour, even if efficient numerical solvers such as DASSL (Petzold, 1982) are used.

Higher Index DAE. The decision to remove small physical effects such as fluid storage in lines and oil elasticity and viscosity leads to DAEs with a higher complexity because state variables are then directly coupled instead of interacting through additional states with small time constants. These DAEs can be transformed by differentiation before simulation run, but the switching effects of the actuators may also cause such algebraic constraints to emerge during simulation, requiring two phenomena to be handled: (i) the state variables that become algebraically coupled are constrained to a subspace of reduced dimension and the values before the constraint becomes active have to be projected into this subspace, and (ii) the future dynamic behaviour of these state variables must be in this reduced subspace.

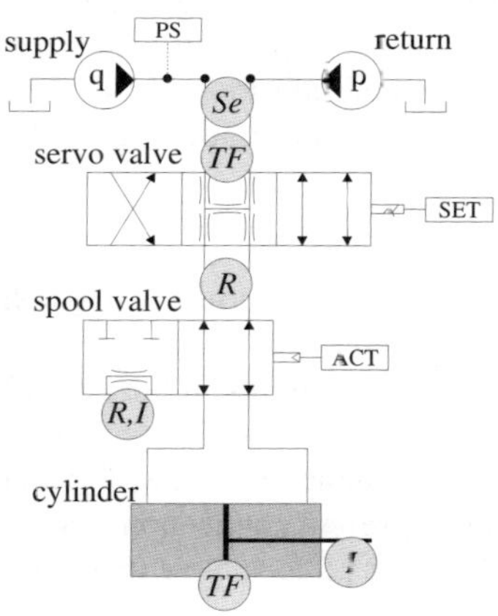

Fig. 8. Schematic of hydraulic actuator

To illustrate these effects, consider the actuator model in Fig. 8. When initially the actuator is *active*, the supply path is open, i.e., control signals generated by the servo valve are supplied to the positioning cylinder, causing the piston to accelerate. When, at a given point in time, the actuator is switched to be *off*, the loading path becomes active. Because of the inertial effects in the loading pathway, there is dependency between the piston and this fluid inertia and an algebraic constraint between these two variables ($v_{piston} = -A_p f_{load}$) restricts the state space in which the system evolves. This is illustrated in Fig. 9(a), where the double arrow heads on the dashed field lines indicate the direction of the discontinuous change. This algebraic dependency would be eliminated by introducing small parasitic storage effects for the piping and some oil elasticity and viscosity, but this adds very steep gradients to overall system behaviour as illustrated by Fig. 9(b) that complicate simulation and are not relevant for the overall behaviour of the aircraft.

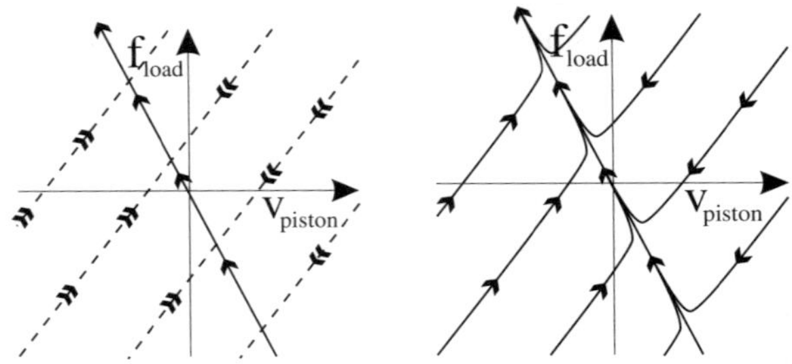

(a) Discontinuous changes (b) Steep gradients

Fig. 9. Phase space for v_p and f_{load}

The implicit jumps in the state variable values have to be computed during simulation. At present, commercially available simulation tools cannot handle such abrupt changes in DAE models. Therefore the experimental modelling and simulation environment HyBrSim (Mosterman and Biswas, 1999) was used which has been realised for the purpose of testing algorithms for the reinitialisation of switched systems with index changes. HyBrSim is based on bond graph modelling of the physical system.

Bond Graph Model of the Actuators. Figure 10 shows the hybrid bond graph model of the two left hydraulic actuators. The two Se elements[1] are sources (inputs) of a bond graph model which are connected to the hydraulic circuits in the aircraft model that provide the input pressure. The servo valve modulation is applied by the *TF* elements, where the *setL1* and *setL2* elements are connected to the setpoint generated by the aircraft control model. The I elements represent connections (equal flow points) and the attached R element captures dissipative effects. Note that these are modelled as linear phenomena. The *loadL1* (*loadL1*) connection also has some inertia associated with it, embodied by the *IloadL1* (*IloadL2*) element. The cylinder chamber is modelled by a 0 element, an equal pressure point. Both cylinders connect through a piston with area modelled by a *TF* element to one equal velocity point for the elevator control surface movement. This velocity, as well as the displacement and force are inputs to the aircraft model.

The switching behaviour is modelled by two *controlled junctions* (Mosterman and Biswas, 1995) in each actuator, in the left actuator these are *supplyL1* and *loadL1*. The local finite state machines that control their states are given in Fig. 11. The control event *actL1* is generated by the redundancy control in the enclosing part of the model. When the *supplyL1* junction is ON and *loadL1* is OFF, the actuator is active. When *supplyL1* is OFF and *loadL1* is ON, it is loading (either hot, standby, passive, or isolated). Note that the mutual switching constraints allow no other configurations.

[1] The element type is listed on the left of each element rectangle.

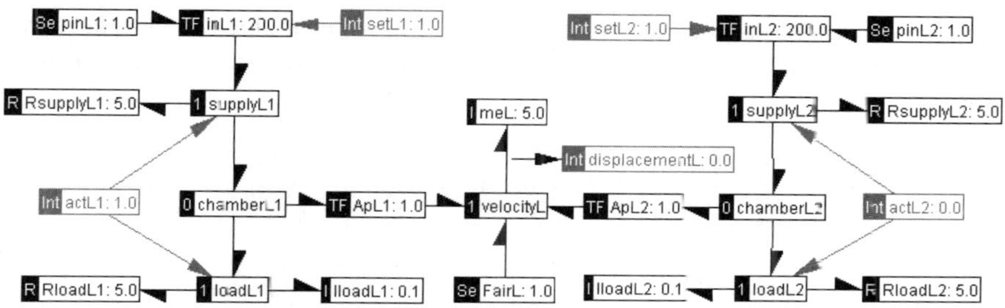

Fig. 10. Hybrid bond graph of the two left hydraulic actuators

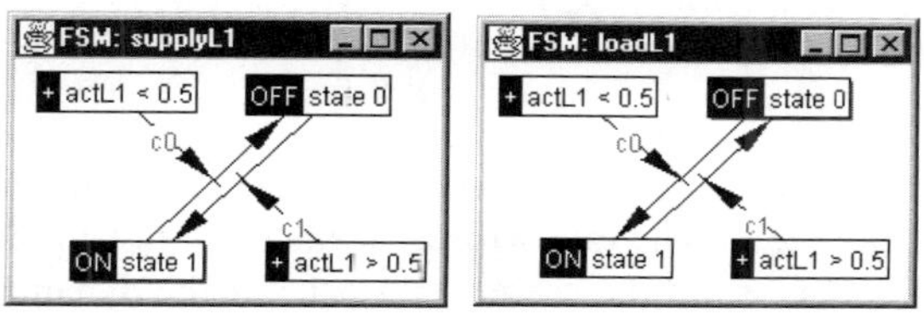

(a) Supply path (b) Load path

Fig. 11. Finite state machines of actuator 1 in the hybrid bond graph

Equations. The equations generated from the hybrid bond graph by HyBrSim incorporate the switching effect as guarded equations. This prevents the need for pre-enumeration which would cause an exponential growth of the number of modes.[2] For example, for the loading pathway, *loadL1*, the equation generated is

$$0 = (-chamberL1.p + IloadL1.p + RloadL1.p)\alpha_i + (loadL1.f) \cdot (1 - \alpha_i) \quad (9)$$

where α_i is the ith entry in the mode vector α. This ensures that in a mode where this connection is active, $\alpha_i = 1$, the pressure drops of the connected elements are balanced. When the connector is not active, $\alpha_i = 0$, the fluid flow through *loadL1* becomes 0. This models ideal switching but may lead to higher index DAEs (e.g., because *IloadL1* and *mpL* become algebraically related). A numerical solver such as DASSL can handle systems up to index 1 directly and up to index 2 with some provisions, e.g., the step-size control of index 2 variables needs to be switched off (Bujakiewicz, 1994). Another prerequisite is that DASSL should be given a set of consistent initial conditions, i.e., those that are in the correct subspace of continuous behaviours. This is achieved by applying a projection mechanism which is consistent with physical conservation laws (Griepentrog and März, 1986, van der Schaft and Schumacher, 1996, Verghese et al., 1981).

[2] For the hybrid bond graph in Fig. 10 there are already $2^4 = 16$ possible modes, but only two occur during normal operation.

The discontinuous changes are computed by first linearising the system with a finite difference method. Then a pseudo Weierstrass normal form is derived (up till index 2)

$$0 = \begin{bmatrix} \bar{E}_{11} & 0 & 0 \\ 0 & 0 & \bar{E}_{22,12} \\ 0 & 0 & 0 \end{bmatrix} \begin{bmatrix} \dot{\bar{x}}_1 \\ \dot{\bar{x}}_{2,1} \\ \dot{\bar{x}}_{2,2} \end{bmatrix} + \begin{bmatrix} \bar{A}_{11} & \bar{A}_{12,1} & \bar{A}_{12,2} \\ 0 & \bar{A}_{22,11} & \bar{A}_{22,12} \\ 0 & 0 & \bar{A}_{22,22} \end{bmatrix} \begin{bmatrix} \bar{x}_1 \\ \bar{x}_{2,1} \\ \bar{x}_{2,2} \end{bmatrix} + \begin{bmatrix} \bar{B}_1 \\ \bar{B}_{2,1} \\ \bar{B}_{2,2} \end{bmatrix} \begin{bmatrix} u \end{bmatrix} , \tag{10}$$

where $\bar{E}_{11,11}$, $\bar{A}_{22,11}$, and $\bar{A}_{22,22}$ are of full rank. This allows computation of the initial conditions as (Mosterman, 2000b)

$$\begin{aligned} \bar{x}_1 &= \bar{x}_1^0 + \bar{E}_{11}^{-1} \bar{A}_{12,1} \bar{A}_{22,11}^{-1} \bar{E}_{22,12} (\bar{x}_{2,2} - \bar{x}_{2,2}^0) \\ \bar{x}_{2,1} &= -\bar{A}_{22,11}^{-1} (\bar{B}_{2,1} u + \bar{E}_{22,12} \dot{\bar{x}}_{2,2} + \bar{A}_{22,12} \bar{x}_{2,2}) \\ \bar{x}_{2,2} &= -\bar{A}_{22,22}^{-1} \bar{B}_{2,2} u , \end{aligned} \tag{11}$$

where $\bar{x}_0$ are the user-provided initial values after the coordinate transformation to achieve the desired normal form, $\bar{x}_0 = Zx_0$. The values for $\bar{x}$ can then be transformed back to obtain initial values for x that are in the correct subspace of the dynamic behaviour, and in this manner the implicit jump is determined.

4 Simulation of the Overall System

The aircraft model, the redundancy control system, and the actuator feedback and discrete event control were modelled using different modelling formalisms and tools (DYMOLA, HYBRSIM, DOME). Each of these is best suited for the respective task. To enable a comprehensive analysis, however, the parts have to be integrated into a coherent model.

4.1 Integrating the Components

Since the descriptions of the failure injection module and the redundancy management system laws are based on equations, they can be incorporated easily into the object-oriented and equation-based aircraft model. This also holds for the hydraulic actuators, in principle, because the bond graph models correspond to a set of hybrid differential and algebraic equations. But due to present restrictions of the simulation software available for object-oriented modelling languages, specific simulation code is generated from the bond graphs of the actuators and merged with the C-code that results from the aircraft model.

For the redundancy management component, the modelling environment generates a simulation algorithm that defines the input-output behaviour of the discrete-event component. This automatically generated algorithm is designed in a way that is compatible to the MODELICA language so that it can be embedded directly into the aircraft model. In MODELICA such an algorithm is regarded simply as an additional model constraint that corresponds to an equation that contains a function with a fixed set of input and output variables.

To simulate the resulting hybrid model, MODELICA's hybrid DAE semantics is exploited. The temporal inequality expressions in the failure injection module are transformed into time events for the numerical integrator so that the continuous integration stops exactly when a switching time has elapsed. Then the whole set of equations is re-evaluated with the new values of the inequality expressions. Thereby, the algorithm of the redundancy management is also re-evaluated resulting possibly in a new state which may switch the feedback control laws.

4.2 Simulation Results

The phugoid in Fig. 12 is the result of two interacting phenomena: When the aircraft pitch angle increases, it gains altitude and at the same time loses airspeed. Because of this loss of airspeed, there is less upward thrust, which causes the aircraft to lose altitude in return. However, as it starts losing altitude, it picks up speed again and the airspeed rises. This results in a slightly damped oscillatory behaviour which is required to be stable in commercial aircrafts.

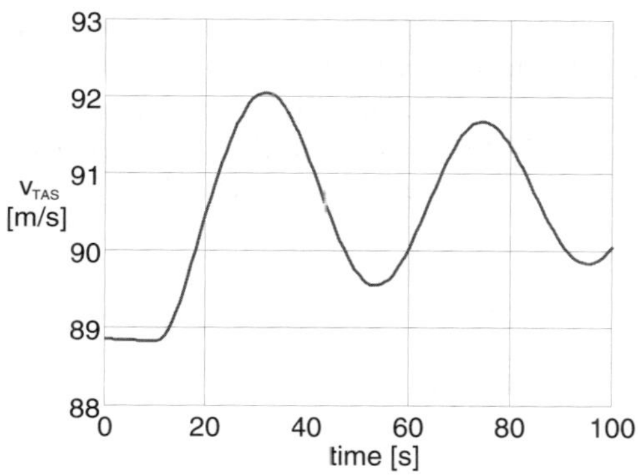

Fig. 12. Simulation shows a phugoid typical for aircraft

To investigate the effect of the redundancy control on the aircraft's behaviour, an actuator failure is introduced during a setpoint change. The setpoint change occurs at $t = 0.05$ $[s]$ and the actuator failure at $t = 0.08$ $[s]$. Figure 13 shows that the failure leads to an immediate change of the active actuators and the switching transients in the hydraulics cause a sharp drop in elevator velocity. Because small effects such as oil elasticity and viscosity are neglected in the simulation, this results in a discontinuous change that occurs because of the algebraic dependency between elevator inertia and fluid inertia of the new loading path.

During a short period of time, the PID control causes the elevator velocity to ramp up to the value which it would have assumed without the failure. Note the short delay that is possible because the actuator that switches to active was *hot* and *shadowing* the PID control.

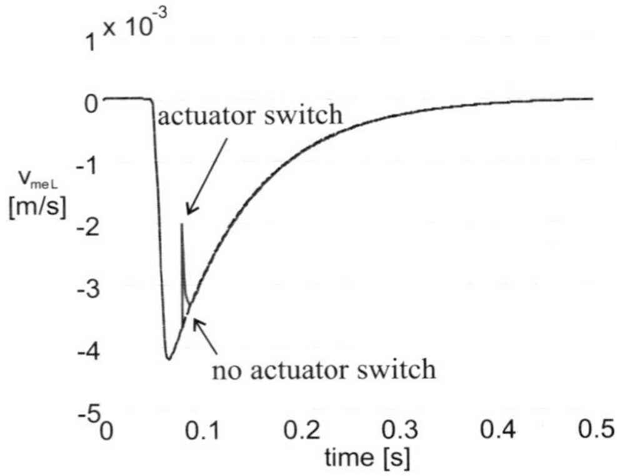

Fig. 13. Elevator velocity when a failure occurs at t = 0.08 shortly after a setpoint change at t = 0.05

The aircraft redundancy control is designed such that an actuator failure should not have a noticeable effect on the behaviour of the aircraft. Using the comprehensive model with switching logic and transients, and an extensive model of the aircraft dynamics, this effect can be studied as well. Figure 14(b) shows the effect of the actuator switch on the aircraft pitch angle, and Fig. 15(b) shows the effect on the pitch angle velocity. This verifies that the actuator switch has almost no effect on the overall aircraft behaviour which, because of the realistic aircraft model, provides much confidence for the real implementation. Note that the small effect of the actuator switching on the global behaviour manifests itself after a significant delay.

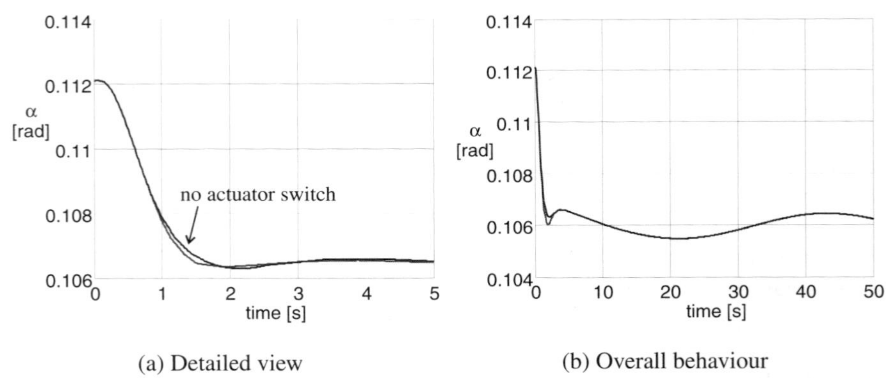

Fig. 14. Pitch angle for normal behaviour and for an actuator switch at $t = 0.08$

Table 3 illustrates how the redundancy management reacts, when the IO module failure occurs in PFCU2. In this case, all resulting state transitions are symmetrical, i.e., the modules of the right elevator have always the same state as the corresponding modules of the left elevator. Therefore the given states refer to both sides. In the first

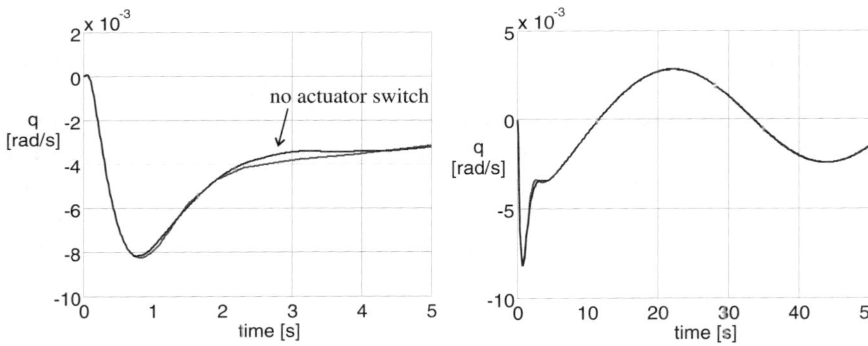

(a) Detailed view (b) Overall behaviour

Fig. 15. Pitch angle velocity for normal behaviour and for an actuator switch at $t = 0.08$

local transition the statecharts of LIO and RIO (Left / Right IO) of PFCU2 switch from *Active* to *Isolated*, since these modules should not be activated again (see rules 1 and 10 in Section 2). Then PFCU1 takes over the actuators by activating its LIO and RIO modules (rules 1, 3, 5). In the last local transition, the LDL and RDL (Left / Right DL) statecharts of PFCU2 switch into the *Hot* mode preparing the system for a possible second failure (rule 6). Since state 2 would violate rule 4 and the transition from state 3 to state 4 would violate rule 1, the internal iterations have to be hidden from the outer system in order to prevent inconsistent outputs. This is why only the global transition from state 1 to state 4 is made observable to the outside.

Table 3. State transitions of the redundancy management system

components		local steps			
		1	2	3	4
PFCU2	RIO/LIO	Active	Isolated	Isolated	Isolated
	RDL/LDL	Passive	Passive	Passive	Hot
PFCU1	RIO/LIO	Hot	Hot	Active	Active
	RDL/LDL	Passive	Passive	Passive	Passive
outer actuators		control	–	–	shadow
inner actuators		shadow	–	–	control
global visibility		yes	no	no	yes

5 Conclusions

The comprehensive model of the aircraft developed here incorporates the redundancy management system, the switched positioning controllers, the actuator models as well as a complex model of the general dynamics of the aircraft. Hence, it is possible to assess the design of the elevator control system with respect to the overall behaviour of the aircraft in the case of failures. Since the less important physical effects of the hydraulic actuators were neglected, the simulation is fast enough to be used also in the context of a multi-objective parameter optimisation (MOPS) (Joos, 1999). Such an optimisation may, e.g., reduce the elevator surface or the actuator power such that the switching transients still do not affect the level of aircraft handling.

The abstractions used in the actuator models, i.e. neglecting small physical effects such as oil elasticity and viscosity, result in a DAE that may change its index during simulation. A standard DAE solver, such as DASSL, can be applied for this model, if the re-initialisation at event times results in a consistent state. For a correct behavioural simulation, this re-initialisation has to satisfy the physical conservation laws. For the purpose of this feasibility study the actuators were modelled in HyBrSim, a modelling environment based on hybrid bond graphs that supports the necessary re-initialisation procedure. The C-code generated by this environment was manually combined with the C-code generated by Dymola which includes the rest of the aircraft model. The hybrid system simulator MAsim was used to generate behaviors. MAsim has facilities to compute discontinuous changes of generalized state variables as algebraic constraints between them become active. The discrete-event parts of the aircraft are modelled using a visual specification language and are translated into a Modelica algorithm that can be integrated into the aircraft model on the model level (Mosterman et al., 2002).

The presented modelling and simulation approach that combines an object-oriented modelling language such as Modelica, domain-specific model libraries, discrete-event modelling formalisms and powerful simulation methods including correct state re-initialisation, was successfully applied to the aircraft elevator control system and seems to be promising for general complex technological systems.

Development of Hybrid Component Models for Online Monitoring of Complex Dynamic Systems

Susanne Manz and Peter Göhner

University of Stuttgart
Institute of Industrial Automation and Software Engineering (IAS)
Pfaffenwaldring 47
70550 Stuttgart, Germany
ias@ias.uni-stuttgart.de
http://www.ias.uni-stuttgart.de

Abstract. Up to now model-based online monitoring is rarely applied in process automation and chemical industries. The main reason is the big effort, which is necessary to develop a comprehensive model for a technical system under various circumstances. However the growing complexity of plants and facilities requires increasingly the use of formal methods to analyze and to monitor the system behavior.

In this paper an online monitoring method based on qualitative models and combined with dynamic models is proposed. The qualitative description is very flexible in representing just as much information as it is actually needed and available. In that way also complex systems can be modeled. The dynamic description is only necessary for time dependent components. As a result dynamic systems can also be modeled.

The component-oriented approach is a basic feature of the method. Its ability to build automatically clusters of qualitative and dynamic components, which can be reused as single components, is very important.

An application example of a three-tank-system shows that such kind of models, the so-called hybrid models, are capable of solving monitoring problems.

1 Introduction

> "The world is infinitely complex. Our knowledge of the world is finite, and therefore always incomplete. The marvel is that we function quite well in the world in spite of never fully understanding it."
> *Benjamin Kuipers in* (Kuipers, 1994)

For industrial automation of plants the development of monitoring functions for online failure detection and diagnosis is as important as the realization of control functions. The aim of failure detection and diagnosis is to protect human beings and environment from danger and to avoid hazards as far as possible. Very often engineers use model-based solutions for failure detection and diagnosis. These models usually contain a detailed mathematical description of the plant. In this case the temporal changes of dynamic systems are described in the model. But the building of complete mathematical models for monitoring is very costly and difficult, in particular for complex dynamic systems. Therefore it is useful to build qualitative models instead

S. Engell, G. Frehse, E. Schnieder (Eds.): Modelling, Analysis and Design of Hybrid Systems, LNCIS 279, pp. 391–417, 2002.

of mathematical descriptions. The advantage of qualitative models is that the internal physical relations do not have to be represented exactly, so the qualitative models describe only situations in which something "happens" (Brown and de Kleer, 1990, Kuipers, 1994).

1.1 State of the Art

There are several model based methods to describe the dynamic behavior of plants. Normally modeling and simulation is performed in a quantitative way by application of differential equations (Brack, 1974, Buchholz, 1999). However most of the plants contain time-driven and event-driven system components[1], so that mathematical methods need additional ways to describe discrete system components. The hybrid system behavior is characterized by the interaction between continuous and discrete event system parts. Therefore a lot of new modeling methods for modeling and simulation of hybrid systems are developed (Chouikha and Krebs, 1998). A few of them were important for this research project and will be introduced in a short way:

Net-State-Models: Net-state-models consist of an extended state space model and an interpreted Petri Net, which describes the interaction of time-driven and event-driven system components. Such kind of models can be used to model and simulate hybrid dynamical systems and is applied to reachability analysis (Nenninger et al., 2001, Nenninger et al., 1999, Schlegl et al., 2000).

Hybrid automata: A hybrid automaton is a formal model for mixed discrete-continuous systems (Engell, 1997, Henzinger, 1996). Hybrid automata are used in (Buss et al., 1997) to simulate hybrid dynamical systems with numerical integration methods.

Hybrid Petri Nets: This approach of hybrid Petri Nets is similar to the net-state-models and expresses also the integration of Petri Nets and differential equations. The hybrid Petri Nets have the same structure as hybrid automata and can be used for analysis and synthesis of hybrid systems and also for reachability analysis using evolution graphs (Chouikha et al., 2000, Chouikha and Schnieder, 1998a, Chouikha et al., 2001).

Condition/Event systems: Condition/Event systems offer the possibility to approximate continuous systems by real-time discrete event models (Kowalewski et al., 1998). They are based on a block-diagram and signal-flow presentation and permit the separation of hybrid systems in discrete and continuous parts. In (Kowalewski et al., 2001a) condition/event systems are combined with timed automata to formally verify the correctness of hybrid systems.

Nondeterministic automata: In stochastic automata the abstraction of continuous dynamic systems is done in a qualitative way. In (Lichtenberg et al., 1999b, Lunze, 1998b) nondeterministic automata are either defined via qualitative input and output signals of the continuous systems or by direct partitioning of the continuous state space.

---

[1] Such kind of plants are called "hybrid systems"

All the above mentioned methods describe the behavior of the whole dynamic system. The modeling of complex dynamic systems however needs component oriented methods (Panreck, 1999), where not only the behavior of the system but also the structure can be described. In this case it is necessary to idealize the model in such a way, that on the one hand the model contains partly the dynamic behavior and on the other hand the model can be applied to describe complex systems. A further need beside the offline simulation and analysis is the use of this kind of model for *online* monitoring of complex dynamic systems.

Considering these requirements and the above mentioned methods only the use of nondeterministic automata with its qualitative approximation is important for the presented research. Qualitative methods aim to capture the fundamental behavior of a system in a computer model, while suppressing much of the detail. In these models imprecise, vague and incomplete expressions are used in order to simplify the approach for online monitoring and diagnosis of complex systems. Examples are found in (Lichtenberg et al., 1999c, Lunze and Schiller, 1997, Schiller, 1997) using nondeterministic automata or in (Console et al., 1992, de Kleer and Weld, 1990, Fröhlich, 1996) using *Qualitative Reasoning* methods. Qualitative Reasoning is an area of Artificial Intelligence (AI), which creates representations for continuous aspects, such as space, time and quantity, which support reasoning with very little information (Forbus, 1990). This method is an approach towards to human beings' thinking and reasoning (Brown and de Kleer, 1990, de Kleer and Weld, 1990, Kuipers, 1994) and offers therefore promising skills for research activities in online monitoring and diagnosis of complex dynamic systems.

1.2 Scope and Differentiation

Within the scope of the research area "Development of hybrid component models for online monitoring of complex dynamic systems" the use of already existing mathematical models in combination with situation based qualitative models (Fröhlich, 1996, Laufenberg, 1997) has been examined for online failure detection and hazard prediction of complex systems. In this context the hybrid modeling method SQMD (Situation based Qualitative Monitoring and Diagnosis) has been developed (Manz, 1999, Manz, 2000, Manz, 2001a, Manz, 2001b).

The component-oriented approach using Qualitative Reasoning methods will be introduced within the next sections. In the beginning a short introduction to model-based online monitoring is given. Then the SQMD concept, using qualitative and dynamic component descriptions, is introduced in more detail. Therefore the main focus is put on the building of the qualitative component models and its system structure. With the information out of the qualitative component models and the system structure conclusions can be drawn to the system behavior in order to recognize faulty behavior. The dynamic descriptions are used only for partly mapping the dynamic behavior into the qualitative models. The analysis of the complete dynamic system behavior (e.g. oscillation) is not the object of this research project. After introducing the concept it is applied to the well-known three-tank-system.

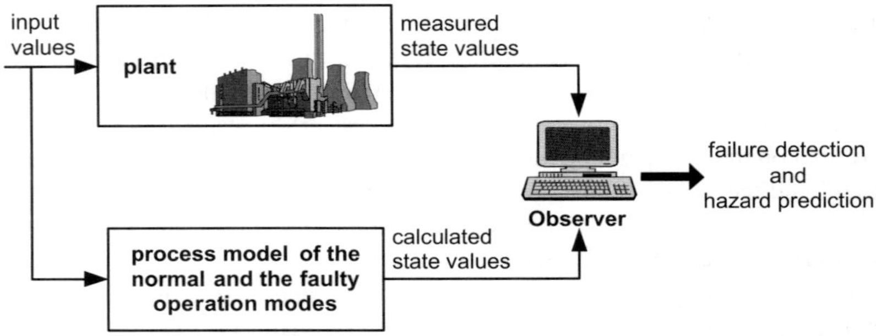

Fig. 1. Model-based online monitoring

2 Model-Based Online Monitoring

The basic concept of the model-based online monitoring is described in Fig. 1. The starting point is a process model of the normal and the faulty operation modes of the plant. The process model is executed online, that means in parallel to the activities in the plant and is feeded with the same input values.

The task of the observer is to compare the currently measured process states with the model states. If the model of the normal operation mode calculates the measured process states, then the technical process is in a fault-free operation mode. If the measured states are not in the calculated states of the normal operation mode model then it has to be assumed that the technical process is in a faulty mode.

Depending on the type of the technical process various models are used for process monitoring (Göhner and Lauber, 1999, Isermann, 1996a, Isermann, 1996b). For dynamic systems quantitative dynamic models such as systems of differential equations are usually applied. In these models a dynamic process is described in such a way that a deterministic behavior is guaranteed. The disadvantage however lies in the complexity of these models (Frank, 1998, Gilles et al., 1986, Liggesmeyer and Mäckel, 2000, Panreck, 1999). The more components the system contains, the more complex the model is. For complex models the calculations are too runtime consuming to be used in an online monitoring system. In opposite to that, qualitative models are less runtime consuming. Therefore qualitative models can be adopted for online monitoring of complex systems.

In general qualitative models can be built quite easy and fast, because of the simplified description of the technical process. But the qualitative models describe only the static behavior of a system and can therefore not be used for analyzing the dynamic behavior of systems. Another disadvantage is the non-deterministic behavior of these models because of the imprecise description of the process.

Based on the quantitative and qualitative modeling methods the SQMD (Situation based Qualitative Monitoring and Diagnosis) concept has been developed. SQMD uses hybrid models for online monitoring. Hybrid models consist of qualitative and dynamic components and combine the advantages of both methods. So it is possible

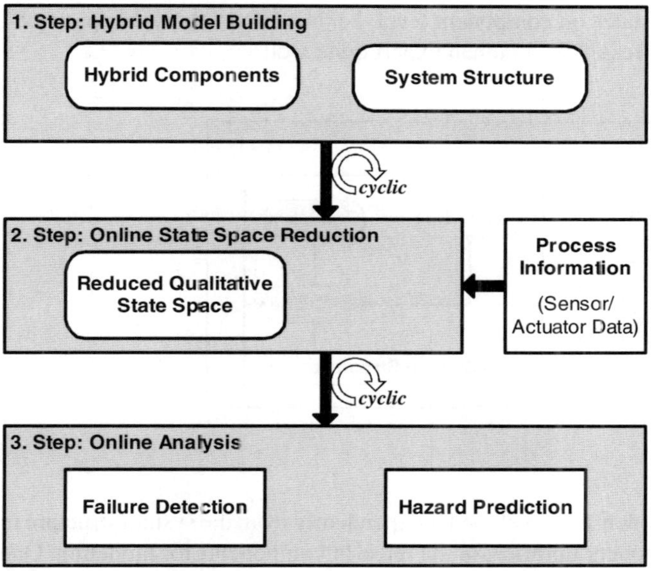

Fig. 2. SQMD concept

to realize online monitoring for the detection of failures and prediction of hazards in complex dynamic systems.

3 SQMD Concept

The SQMD concept is subdivided into three different steps, as shown in Fig. 2. The first step consists of the hybrid model building of the components and the specification of the system structure. The second step includes the main process of the SQMD concept: the online state space reduction based on the hybrid components, the system structure and the online information (sensor- and actuator data) of the real system. The reduced state space can be analyzed in the third step in order to detect possible failures and to predict their effects, i.e. hazardous system states.

In the following sections the concept with its three steps will be explained in more detail.

3.1 Hybrid Model Building

The problem of complex and dynamic systems is the building of suitable models, which describe the complete system behavior and can be used for online simulation. In this context the decomposition of the system offers a solution (Panreck, 1999). One of the major benefits of the SQMD concept is the easy component-oriented modeling. This includes a systematical determination and description of the most important effects and states on component level, i.e. local and surveyable parts of a system, which can be treated by a human expert quite well.

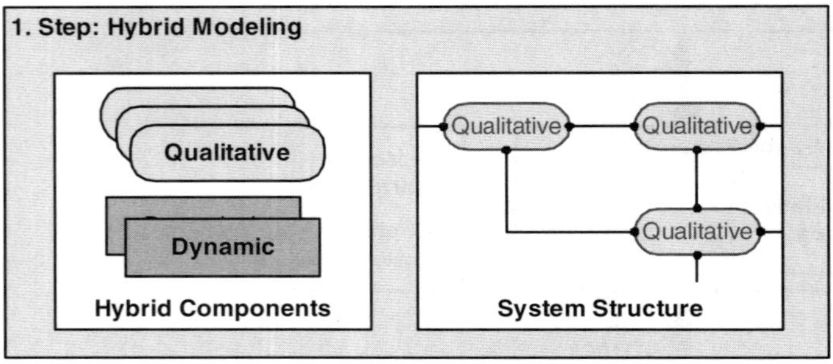

Fig. 3. Hybrid modeling

The component models are developed independently from the system structure in order to build up component libraries for the reuse of components for modeling. One of the basic ideas of SQMD is, that all effects, states and hazards, which are locally identified and specified on component level by human experts, can be completely taken into account within a complex system. By incorporating possible component failures, any possible hazardous states resulting from malfunctioning parts can also be detected.

So the first step of the SQMD concept consists of the "model building" of the hybrid model containing qualitative and dynamic components and the system structure. This is shown in Fig. 3.

For further understanding it is necessary to describe the definition of a *component* in more detail. Therefore the idea of the decomposition of a system (Panreck, 1999) plays an important role especially for complex systems. Such a component can be:

- a single component of a system, like a tank, a valve or a pipe.
- a bundle of components – called subsystem, like a valve and a pipe.
- a whole system, like a three-tank-system.

The degree of decomposition depends on the complexity of a system. Normally it is meaningful to describe the qualitative part at the single component level (or small subsystem level) and the dynamic part at the subsystem level (or for small systems at system level).

Dynamic model parts include the mathematical description of the components. The static behavior of the components is described by qualitative model parts, which are expressed in interval arithmetic. The graphical layout or specification of the system in form of piping- and instrument-diagrams (pi-diagram) yields the starting point for modeling. That means an engineer assigns to every physical quantity of a component different intervals, which describe qualitatively the normal and faulty behavior of this component.

All physically possible interval combinations of all quantities concerned are described by situations and their transitions and are stored in a situation table and a transition graph. This is similar to the concept of states in automata

The system structure contains only the connections between the qualitative components of the system and the outside terminals.

Figure 4 shows the hybrid specification of the system components in more detail. With regard to the handling of complex dynamic systems it is necessary to abstract the dynamic behavior in an easy way.

All components are described qualitatively by thresholds and rules. Furthermore components with temporal behavior, i.e. state based quantities are described additional dynamically by differential equations. All other components are called coupling components and are idealized, so only the qualitative description is enough (Brack, 1974).

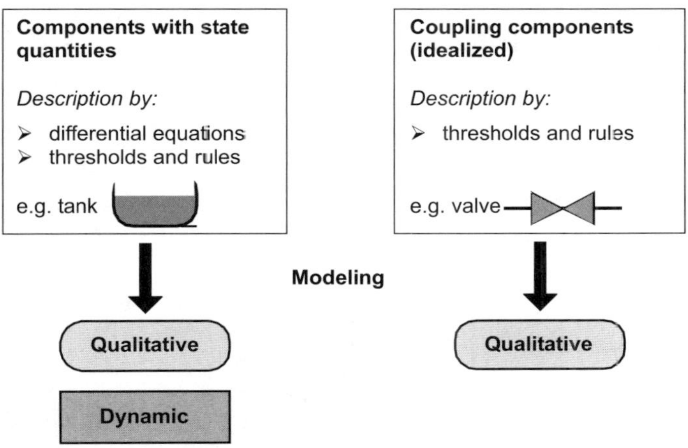

Fig. 4. Hybrid specification of components

Within the scope of the next three sections the qualitative modeling, the dynamic modeling and also the specification of the system structure is described in detail.

Qualitative Modeling. In qualitative modeling the essential behavior of a component is described by means of essential system variables and characteristic threshold values under normal and failure conditions. Beside the normal intended operation, non-standard operating modes are considered.

A qualitative component will be described as shown in Fig. 5 by the following elements:

- Terminals
- Qualitative Quantities
- Interval Arithmetic

- Situation Rules
- Comment Rules
- Transition Rules
- Situation Table
- Transition Graph

Terminals are interfaces to other components or to the outside world. *Qualitative quantities* are subdivided into intervals. Quantities can be physical quantities (flow/-pressure, current/voltage, heat/temperature, etc.) or information quantities (e.g. pressure variation). With these quantities, different states of matter, kinds of substance, etc. can be modeled.

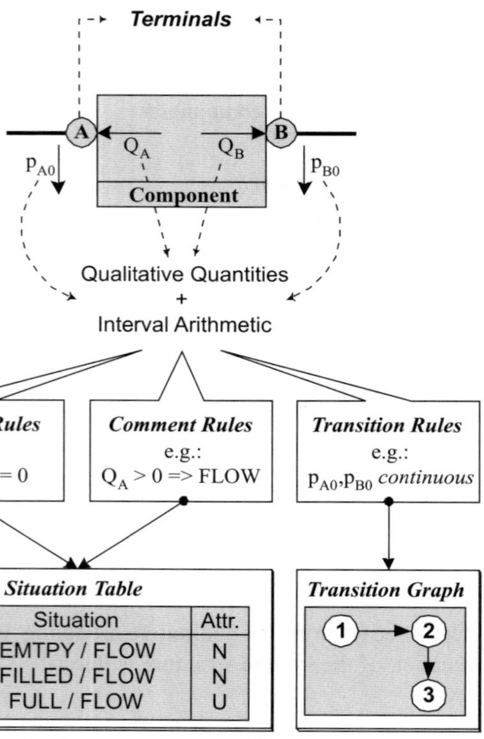

Fig. 5. Qualitative description of a component

Intervals are combined together using *interval arithmetic*. The description of the static component's behavior is based on *situation-* and *comment rules*. These are simplified "if-then" rules. The description of the quasi-dynamic[2] components' behavior is based on *transition rules*. These rules specify continuous quantities and the dependency of one quantity from others.

[2] The dynamic behavior is described only by following the sequence of the states.

The next step is the generation of the situations and transitions Each situation represents a physically possible set of qualitative values, which are characteristic for the behavior of the system. Situations can be marked with different attributes (see Fig. 7) and summarized as states. The entirety of all possible situations, which describe a component's behavior, is called situation space. It can be represented in tabular form, the *situation table*. The *transition graph* describes possible transitions from all situations in the situation table. These transitions are not explicitly time dependent, but they include for each situation only the information about the antecedent and the descendant situation.

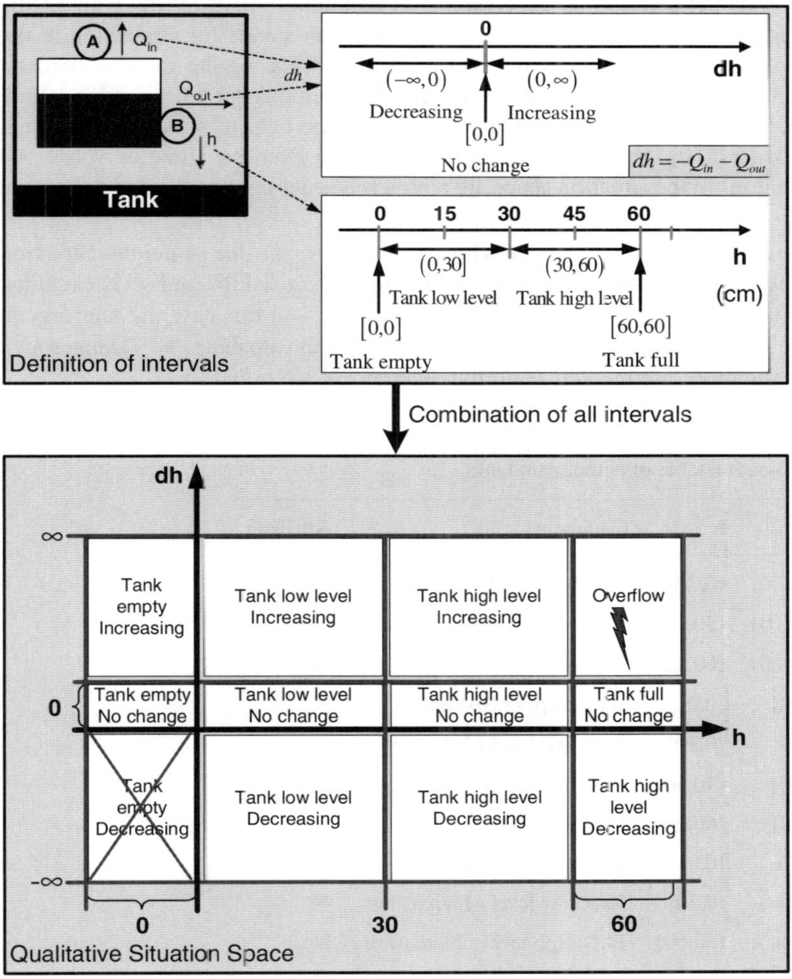

Fig. 6. Interval and situation definitions of the example tank

Figure 6 shows the specification of the intervals and the resulting situations at the example of the component tank. The tank has the terminals “A” with the inflow Q_{in} and “B” with the outflow Q_{out} and the filling level h. The physical flow quantities inflow and outflow can be replaced by the information quantity $dh = -Q_{in} - Q_{out}$ This information quantity describes the alteration of the filling level h.

The information quantity *dh* contains the simplest form of an interval definition, it behaves like a Signum-Variable: $dh < 0$ (Decreasing), $dh = 0$ (No change) and $dh > 0$ (Increasing). It is also possible to specify more intervals containing special thresholds like the definition of the filling level: $h = 0$ (Tank empty), $h = (0, 30]$ (Tank low level), $h = (30, 60)$ (Tank high level) and the threshold value $h = [60, 60]$ (Tank full).

The complete situation space contains all combinations of all intervals of the tank. These are $3 \cdot 4 = 12$ situations. Every situation stands for a partition in the situation space. The borders of the partitions are similar to the interval borders. But not all situations are possible in reality. For example the left situation “Tank empty/Decreasing” is physically impossible. With the help of situation rules (e.g. $h = 0 \rightarrow dh \geq 0$) these situations can be excluded from the situation space. All other situations in the situation space are stored for easier computation in a situation table.

Table 1 shows the situation table with all physically possible situations. Situation 1 stands for the impossible situation “Tank empty/Decreasing” and so is excluded from the table. The last situation is a dangerous one – in this case the tank has an overflow. This situation is marked as dangerous with the attribute “D” (Dangerous). All other situations are marked as normal with an ”N” (Normal).

Table 1. Situation table of component tank

	dh	h	Comment	Attribute
2	(-∞,0)	(0,30]	Tank low level / Decreasing	N
3	(-∞,0)	(30,60)	Tank high level / Decreasing	N
4	(-∞,0)	[60,60]	Tank full / Decreasing	N
5	[0,0]	[0,0]	Tank empty / No change	N
6	[0,0]	(0,30]	Tank low level / No change	N
7	[0,0]	(30,60)	Tank high level / No change	N
8	[0,0]	[60,60]	Tank full / No change	N
9	(0,∞)	[0,0]	Tank empty / Increasing	N
10	(0,∞)	(0,30]	Tank low level / Increasing	N
11	(0,∞)	(30,60)	Tank high level / Increasing	N
12	**(0,∞)**	**[60,60]**	**Overflow**	**D**

SQMD employs three different attributes to classify all situations and states into normal and abnormal operation modes. This classification is necessary for the monitoring task. Figure 7 illustrates these attributes.

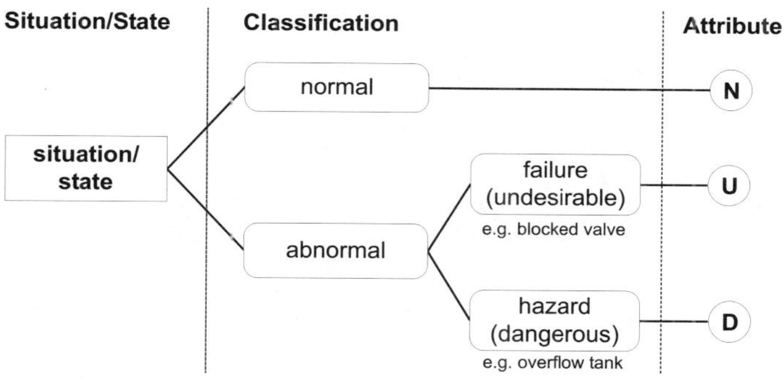

Fig. 7. Classification of situations and states

The first classification distinguishes between normal and abnormal. The "N"-marked situation represents the normal behavior of the component. Abnormal situations are subdivided into two classes. The first one regards the causes of abnormal behavior (i.e. failures) and marks them as "U" for undesirable situation. The second one regards from the effects of these faults and present normally hazards with dangerous consequences. So they are marked with a "D". The separation of the components' behavior into causes and effects is very important regarding the analyzing of the reduced qualitative state space.

The situation table describes qualitatively the static behavior of the tank. The qualitative description of the quasi-dynamic behavior is presented in Fig. 8. The transitions of all 11 situations are shown by a transition graph, which includes some interesting information, so for example:

- The dangerous situation 12 can only be reached from situation 8 and 11.
- Situation 4 and 9 are unstable (because of the point intervals[3] "Tank full" or "Tank empty" and the simultaneously de- or increasing filling level).

Up to now the components are described qualitatively by intervals and their combinations, the so-called situations. The qualitative description is stored in the situation table and the transition graph. The next step includes the quantitative dynamic description of all components including time behavior and will be presented within the next section.

Dynamic Modeling. Figure 9 shows several methods to describe a dynamic component (Buchholz, 1999). The first level includes the description of linear and non-linear

[3] Point intervals have on the left and on the right side identical borders

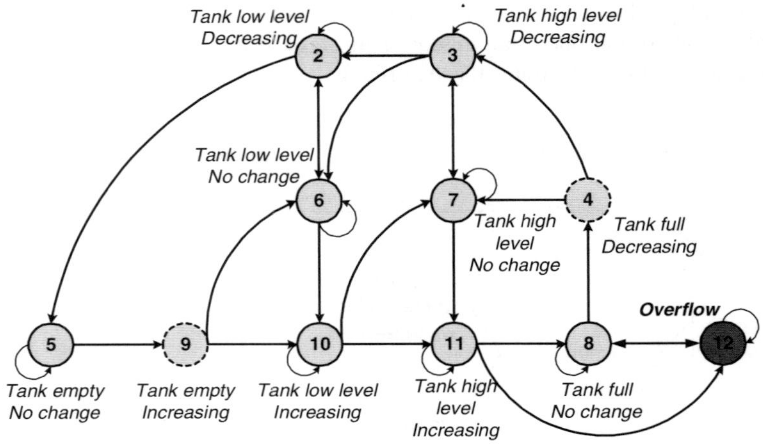

Fig. 8. Transition graph of component tank

systems. The SQMD concept takes only the linear branch into account. If there is a component with non-linear behavior, it can be linearized at a given operation point.

The second step in the linear branch includes continuous and discrete classes. A commercial simulation tool like Matlab/Simulink can easily perform the numerical calculation of these linear descriptions (and also the linearization of non-linear components), like continuous differential equations and the given state space. However the discrete methods are not taken into account. Meaningful for further calculations within this contribution is the continuous differential equation and the state space.

The numerical solutions of the dynamic descriptions will be used for the calculation of quantitative trajectories for a defined time slot. The calculated trajectories are necessary for reducing the qualitative state space, which will be described later.

Figure 10 shows the specification of the dynamic behavior of the tank expressed by a differential equation. The lower part of the figure illustrates the geometric solution of the differential equation in form of a vector field.

The arrows of the calculated vector field are tangential at the trajectories and represent the dynamic behavior. The vector field contains the alterations of the filling level (x-axis) depending from the inflow (y-axis). Each point in the x-y-phase is described by its position, direction and velocity and represents an instantaneous state of the tank.

All possible trajectories can now be calculated at every point of the field. The start state marks the beginning of the trajectory and is defined by a filling level value and an inflow value. The end state marks the end of the trajectory after a defined time and is also represented by a filling level value and an inflow value.

In Fig. 10 one possible trajectory is included. The start state contains the filling level of 30 cm and the inflow of 50 cm^3/sec. Within the time slot of 100 sec the trajectory follows the direction of the horizontal vectors until the end state of the

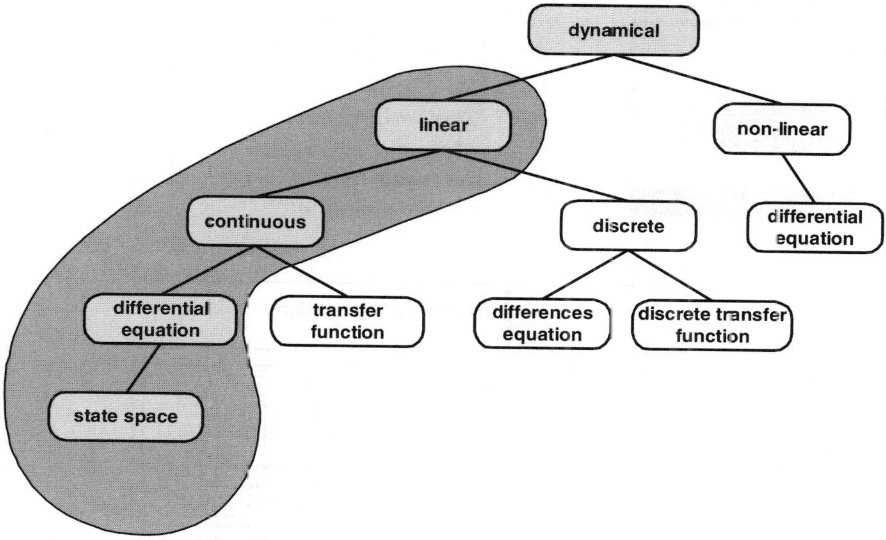

Fig. 9. Dynamic modeling methods

filling level at 8 cm. The alteration of the filling level is always in horizontal direction because of the one-dimension differential equation.

The so calculated start and end states of the trajectories for a given time slot are important for the following reduction of the qualitative state space.

System Structuring. Modeling the behavior of a component is not really part of modeling the system. In this case the *no-function-in-structure principle* is regarded (Brown and de Kleer, 1990). Rather, the components are modeled separately and stored in libraries. Then the system structure is modeled by describing the connections between the qualitative components of the systems. From the connections, rules can be derived based on physical and chemical laws, e.g. laws of conservation. So the structure is modeled by mesh and node rules analogous to the laws of Kirchhoff.

Therefore the behavior of the system is described by means of the functionality of the components and the mesh and node rules representing the interconnections between them. This principle is illustrated in Fig. 11.

Applying the system rules, the observer checks automatically all general possible combinations of component situations. This calculation is a part of the state space reduction.

3.2 State Space Reduction

The reduction of the qualitative state space is an essential part of the SQMD concept. It offers the analyzing of *only* a certain range of the state space and not of the complete one, so the online monitoring of complex systems can be done with low

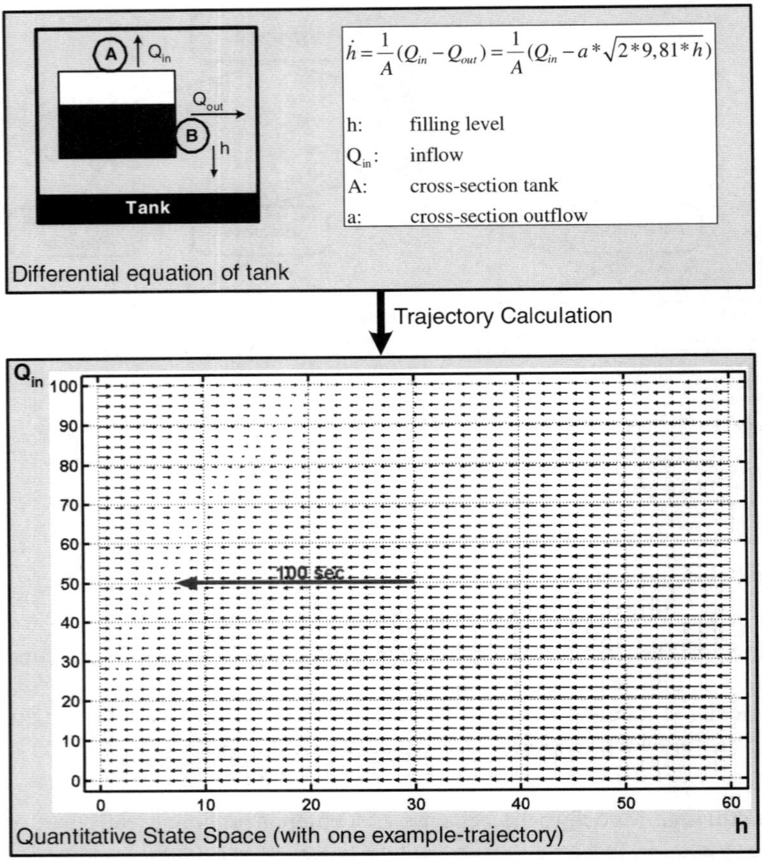

Fig. 10. Dynamic behavior of a tank expressed by a differential equation

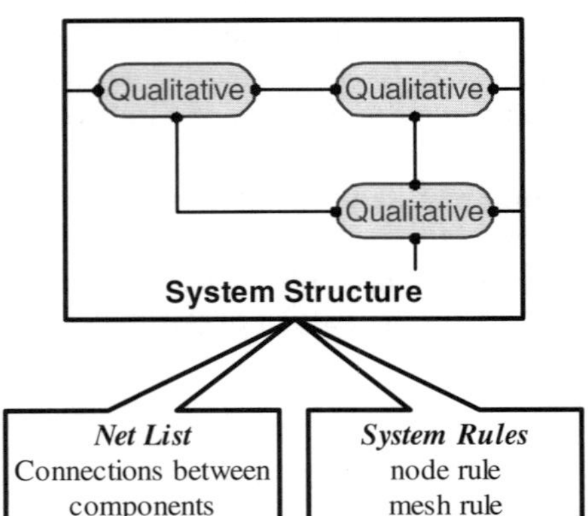

Fig. 11. Specification of system structure

computation power. This concept will be explained in some detail. The principle of the state space reduction is shown in Fig. 12. The appropriate calculation is based on the hybrid components, the system structure and the data of sensors and actuators of the technical process. With this input information the observer calculates periodically all possible *quantitative trajectories* for a defined time slot $[t_a, t_b]$, where t_a stands for the present time and t_b for a future time. This calculation is based on the dynamic description and the input data of the real system.

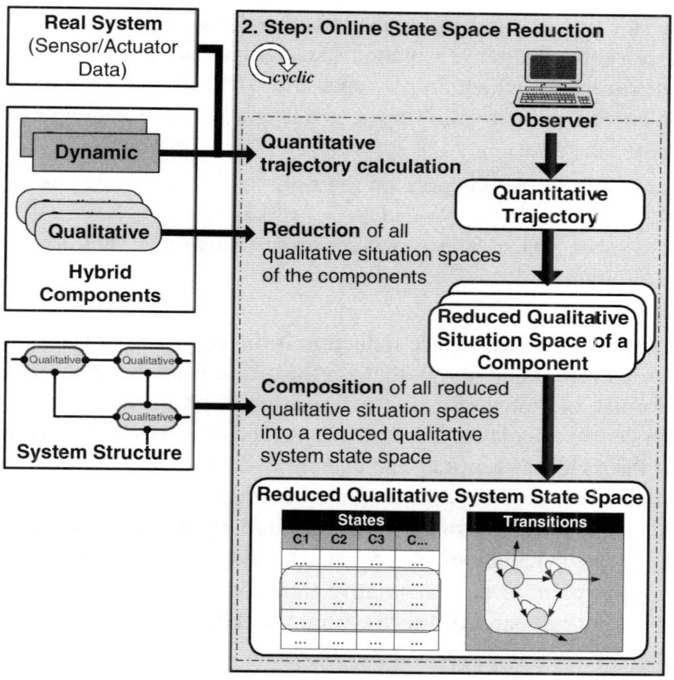

Fig. 12. Online state space reduction

Based on the quantitative trajectories and on the qualitative components all situation spaces of all qualitative components, i.e. their situation tables can be reduced. Then the *reduced situations* are connected to each other following the rules in the system structure. The result is a *reduced qualitative state space* of the system, which describes the static and quasi-dynamic qualitative system behavior for the given time slot.

The reduced qualitative state space contains all states and transitions of the system for the predefined time slot and can be screened for possible process deviations and process faults. This is the task of the online analysis including the failure detection and hazard prediction.

Figure 13 shows the situation space reduction of the component tank in more detail. On the left hand side the calculated quantitative trajectory is drawn into the quantitative state space, which is the vector field of the tank. The considered trajectory begins at time $t = 0$ sec and ends at $t = 20$ sec. So the trajectory is valid for the time slot $[0, 20]$. At the beginning of the time slot the start state is defined by the filling level of 40 cm and the small inflow of 10 cm^3/sec. That is the start state. Beginning at this point the trajectory follows now the horizontal direction and velocity of the given vector field. After 20 sec the trajectory ends at the filling level of 25 cm, which marks the end state.

Within this time slot and on the base of the calculated start and end point of the quantitative trajectory the qualitative situation space of the tank can be reduced. In this connection all situations, which do not touch the start and end state and its qualitative transition states[4], are not valid for the given time slot and subsequently excluded from the situation space.

In Fig. 13 the qualitative situation space on the right upper side marks the two partitions $[h = (0, 30), dh = (-\infty, 0)]$ and $[h = (30, 60), dh = (-\infty, 0)]$ as the reduced qualitative state space. These two resulting situations "Tank low level/Decreasing" and "Tank high level/Decreasing" can be stored for better computation in a situation table.

Important for the qualitative state space reduction is the further use of the start and end state of the calculated trajectory. With the information from the quantitative trajectory and the information from the qualitative transition graph all valid situations for the given time slot can be calculated. The transition graph is necessary to find out the situations passed through between start and end point. So the mapping from the quantitative state space to the qualitative situation space depends only on the start and end point of the calculated trajectories. This is an important fact based on the qualitative idea. For further calculations it is not necessary to know what happens between the start and end point of the quantitative trajectories. So for example an oscillating behavior cannot be mapped directly from the dynamic behavior to the qualitative behavior. Depending on the size of the time slot however it can be partly recognized.

That means that not the whole dynamic behavior of the system components is regarded in the qualitative model. The qualitative model itself has still the disadvantage of all other qualitative modeling methods: the nondeterministic behavior. The SQMD method offers however the possibility to reduce the nondeterministic behavior via the computation of the quantitative trajectories and the sensor and actuator data of the real process. The reduction of the situation space is done for all qualitative components, yielding several situation tables, which are valid for the given time slot. The next step includes the composition of the reduced qualitative situation tables of all components to a reduced qualitative state space of the system. This approach of the composition is presented in Fig. 14.

[4] The qualitative transition states between start and end state can be calculated from the given transition graph.

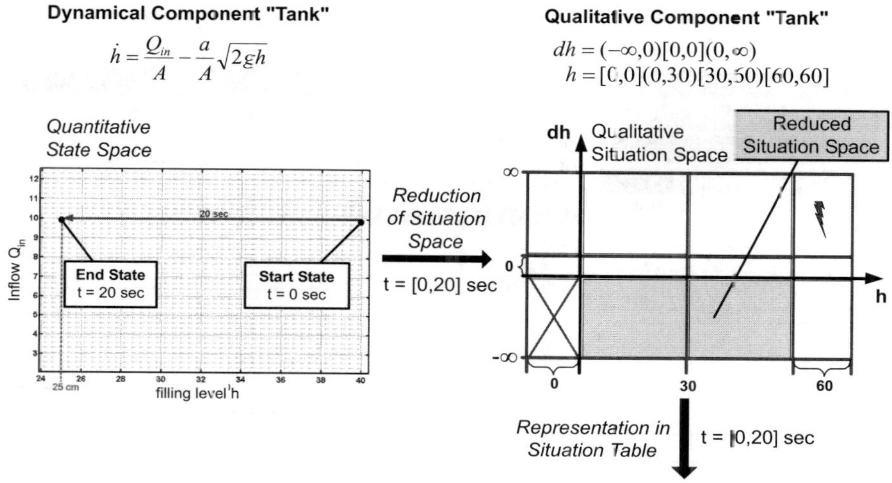

h	dh	State	Attribut e
(0,30]	(-∞,0)	Tank low level / Decreasing	B
(30,60)	(-∞,0)	Tank high level / Decreasing	B
~~[60,60]~~	~~(-∞,0)~~	~~Tank full / Decreasing~~	~~B~~
~~[0,0]~~	~~[0,0]~~	~~Tank empty / No change~~	~~B~~
~~(0,30]~~	~~[0,0]~~	~~Tank low level / No change~~	~~B~~
~~(30,60)~~	~~[0,0]~~	~~Tank high level / No change~~	~~B~~
~~[60,60]~~	~~[0,0]~~	~~Tank full / No change~~	~~B~~
~~[0,0]~~	~~(0, ∞)~~	~~Tank empyt / Increasing~~	~~B~~
~~(0,30]~~	~~(0, ∞)~~	~~Tank low level / Increasing~~	~~B~~
~~(30,60)~~	~~(0, ∞)~~	~~Tank high level / Increasing~~	~~B~~
~~[60,60]~~	~~(0, ∞)~~	~~Overflow~~	~~G~~

Fig. 13. Reduction of situation space at component level

The composition is done based on the system rules. All possible combinations of the reduced component situations are checked. During this checking senseless combinations are eliminated. The remaining set of states is the reduced qualitative state space of the system. It describes the static behavior of the whole system for the given time slot. The quasi-dynamic behavior of the whole system can be calculated by checking all system transitions in the reduced state space. The system transition is valid, if the concerned component transitions are also valid. The result is a transition graph of the reduced state space, which is necessary for the online analysis in order to examine the cause and effect paths to predict hazards in time.

For the next time slot, the reduced qualitative system state space and its transition graph has to be calculated again, beginning with the reduction of the components situation tables and ending with the composition of them. The resulting system situations allow a better understanding of the system as well as the identification of critical parts. Now, a closer look can be taken to the critical items of the system.

During the analyzing task the computer can search the solution space and detect and valuate all possibly undesirable and dangerous states as shown in the next section.

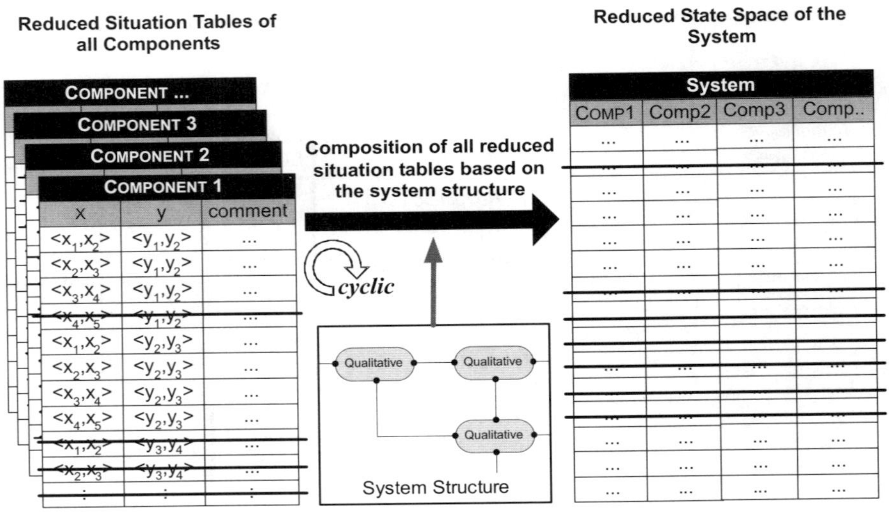

Fig. 14. Composition of the reduced qualitative situation tables of all components

3.3 Online Analysis

The reduced qualitative state space for the time slot $[t_a, t_b]$ is standing first for the actual system behavior $(t = t_a)$ and second for a prediction of the system behavior $(t_a < t \leq t_b)$. The next step consists of the online analysis of the reduced qualitative state space in order to detect undesirable and predict dangerous states. Figure 15 gives an idea about the monitoring tasks.

The first monitoring task is the failure detection and includes the screening of the reduced state space for special states marked with the attribute "U" for undesirable and "D" for danger. If the screening is successful and one or more *abnormal* states are detected, then the observer will store them in the fault-set.

The next monitoring task is the analyzing of the fault-set in order to predict possible hazards. A normal system behavior can be assumed, if the fault-set is empty. But if the fault-set is filled then the further investigation is necessary. In this context all undesirable states, which could be component failures and all dangerous states, which could effect to future hazards will be investigated in more detail according to the following principles:

- *Principle of optimism*: If there are *no* states in the fault-set with the attribute "D" then it can be assumed, that within the given time slot system hazards are not possible.
- *Principle of pessimism*: If there is at least one state in the fault-set with the attribute "D" then it has to be assumed, that the described damage in the system is possible, but because of the qualitative approach not 100% probable. When regarding the proportion between the dangerous states and the other states the observer has to react in time by an alarm in order to prevent hazards. Furthermore

a localization of the hazards is also possible by analyzing the cause-effect-graph, which can be generated from the system transitions.

In order to prevent wrong decisions of the observer regarding the prediction of an abnormal operation mode, it is necessary to give a highly probable information about the dangerous state. This can be done by the use of fuzzy intervals, which allows soft interval transitions and includes a probability factor into the qualitative quantity. This probability analysis may be a part of an ongoing research project (Rebolledo, 2002).

The analyzing of the states with the attribute "U" is necessary to find the causes of possible hazards in the system. For example the cause of a tank overflow could be a blocked valve, which stops the outflow.

Finally it can be summarized that it is the task of the online analysis to notice and react in time to faulty states in order to prevent danger. The analysis was the last calculation performed in the SQMD concept. The concept is now demonstrated with the application example "three-tank-system" in the next section.

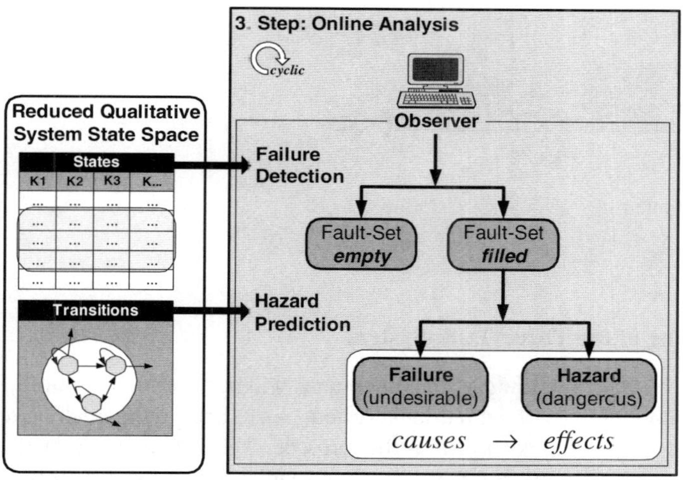

Fig. 15. Online Analysis

4 Application Example "Three-Tank-System"

In the following the SQMD concept will be described in more detail for the application example of a three-tank-system. Figure 16 shows the simple system of three coupled tanks. The system consists of eight components: three tanks, three pipes (with valves to simulate occlusions) between them and behind the last tank (flow out) and two pumps. The two pumps fill the first and last tank with a liquid. The pipes between the tanks have a limited flow rate.

The three-tank-system and the two-tank-system as well are found in the literature as a well-know application example in the qualitative modeling and monitoring area (Lichtenberg et al., 1999b, Lunze, 1998b), and also in modeling, simulation and analyzing of hybrid systems (Engell et al., 1997, Nenninger et al., 2001, Nenninger et al., 1999, Chouikha et al., 2001).

For the three-tank-system it is not very difficult to build a mathematical model using differential equations for the online simulation. Using the mathematical model in combination with qualitative component models allows not only an online simulation. It also allows a fast online detection of all possible failures and a prediction of all possible system hazards in time.

By means of the hybrid model it is possible to simulate the behavior under various conditions. The fluid in the tanks can oscillate between different maximum filling rates or can be in balance.

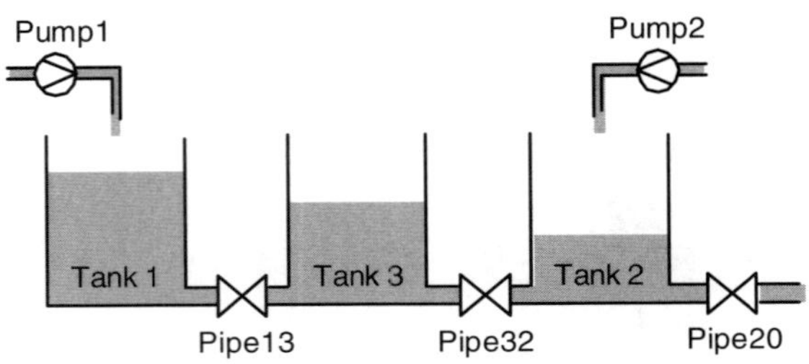

Fig. 16. Three-Tank-System

4.1 Model Building of the Three-Tank-System

In the next sections the model building of all components will be described. Figure 17 illustrates the model building for the three-tank-system. An extract of the qualitative model of the component $tank1$ is shown on the left side. This model contains the qualitative static part as a situation table with all possible situations (e.g. full or empty) and the qualitative quasi-dynamic part as a transition graph. The tank is modeled almost similar to the tank example in Fig. 6, but with three defined intervals for the filling level. The component pump is simply modeled and contains only the situations "on" or "off". The component pipe (including the valve) should be modeled according to the same procedure as the component tank. The physical quantity flow Q is subdivided into the intervals $Q > 0$ (flow is positive), $Q = 0$ (no flow) and $Q < 0$ (flow is negative). We also consider the faulty situation "blocked", which can be simulated by a closed valve.

All components of a technical system are modeled qualitatively using this method independent of their usage in the system. Furthermore the whole system can be modeled dynamically because of the low complexity. On the left side an extract of the differential equation is presented.

Beside the component models, individual system equations are needed. Those are derived from the system structure using the energy conversation laws (first and second Kirchhoff's laws). As an example consider the interface between the first tank and the $pipe13$ in Fig. 17. Using *Kirchoff's Laws*, two equations can be built:

$$tank1.h = pipe13.p_{in} \qquad \text{(\textit{First Kirchhoff's Law}: mesh rule)}$$
$$tank1.Q_{out} = pipe13.Q_{in} \qquad \text{(\textit{Second Kirchhoff's Law}: node rule)}$$

Based on all system equations the hybrid components are connected to each other and analyzed online in predefined time intervals. This is a part of the state space reduction.

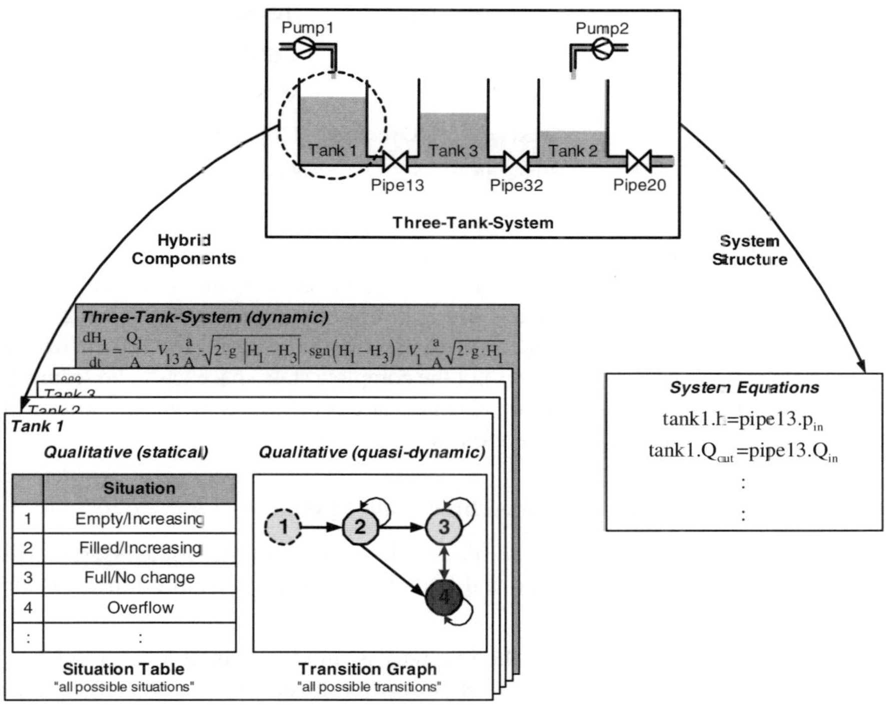

Fig. 17. Model building for the example of the three-tank-system

Table 2 gives a better understanding about the complete qualitative state space and their complexity. The table takes the included abnormal situations in the components into account. Furthermore number values are also presented in the table, like the number of states in the complete state space subdivided into the number of failures and hazards of every component.

From Table 2 some important information can be extracted. One remarkable information in the table indicates that $tank3$ does not have an overflow although it is

Table 2. Number values concerning the model of the three-tank-system

Component	Failures ("U")	Hazards ("D")	Number of states
System with faulty behavior	All	All	22704
System with normal behavior	None	None	244
Component Pump1	Pump defect	-	5912
Component Tank1	Leakage	Overflow	11596/2496
Component Pipe13	Blocked	-	10678
Component Tank3	Leakage	Overflow	10496/0
Component Pipe32	Blocked	-	10500
Component Pump2	Pump defect	-	5548
Component Tank2	Leakage	Overflow	11082/2820
Component Pipe20	Blocked	-	8540

regarded in the component model. This is such a kind of state, which is physically not possible and is therefore excluded from the state table based on the system equations.

The complete state space with all failures and hazards contains more than 20 000 states. That is a lot for such a small system and is the main reason for the online state space reduction. Because of this not the complete state space has to be scanned. Only a small part of the state space, which is valid for the next given time slot, will be regarded. This is described in the following section

4.2 State Space Reduction of the Three-Tank-System

In this section the results of the state space reduction for the three-tank-system will be discussed. The following initial and boundary conditions are presumed for the simulation:

- At the beginning ($t = 0$ sec) the first tank is filled with high level (level $h = 40$ cm), the second tank is filled with low level (level $h = 10$ cm) and the third tank is almost empty.
- The inflow in the first tank is high ($pump1$ on)
- No inflow in the last tank ($pump2$ off)
- All pipes are opened up to $t = 20$ sec, later on the first pipe ($pipe13$) will be blocked. A closed valve can simulate this faulty behavior.

Figure 18 presents the first two resulting state spaces for the given initial and boundary conditions, beginning at $t = 0$ sec. In this figure the reduced state space includes all possible states of the first time slot $t = [0, 20]$ secand the following time slot $t = [20, 40]$ sec. The next step consists of the analysis of the reduced qualitative state space in order to find faulty or dangerous states.

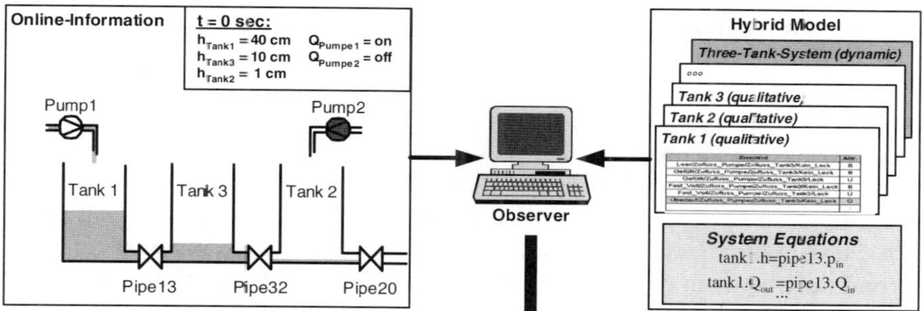

Trajectory Calculation
for time slot [0,20] sec

t = 20 sec:
h_{Tank1} = 49 cm
h_{Tank3} = 14 cm
h_{Tank2} = 2 cm

Q_{Pump1} = on
Q_{Pump2} = off

Reduced State Space for [0,20] sec

Pump1	Tank1	Pipe13	Tank3	Pipe32	Tank2	Pump2	Pipe20
On	Filled/Increasing	Positive	Filled/Increasing	Positive	Empty/Increasing	Off	Empty
On	Filled/Increasing	**Blocked**	Filled/Decreasing	Positive	Empty/Increasing	Off	Empty
On	Filled/Increasing	Positive	Filled/Increasing	**Blocked**	Empty/Unchanged	Off	Empty
On	Filled/Increasing	**Blocked**	Filled/Unchanged	**Blocked**	Empty/Unchanged	Off	Empty
On	Filled/Increasing	Positive	Filled/Increasing	Positive	Filled/Increasing	Off	Positive
On	Filled/Increasing	**Blocked**	Filled/Decreasing	Positive	Filled/Increasing	Off	Positive
On	Filled/Increasing	Positive	Filled/Increasing	**Blocked**	Filled/Decreasing	Off	Positive
On	Filled/Increasing	**Blocked**	Filled/Unchanged	**Blocked**	Filled/Decreasing	Off	Positive

Sudden appearance of a fault in the system: **Pipe13 totally blocked** !

Trajectory Calculation
for time slot [20,40] sec

t = 40 sec:
h_{Tank1} = Overflow
h_{Tank3} = 11 cm
h_{Tank2} = 3 cm

Q_{Pump1} = on
Q_{Pump2} = off

Reduced State Space for [20,40] sec

Pump1	Tank1	Pipe13	Tank3	Pipe32	Tank2	Pump2	Pipe20
On	Filled/Increasing	Positive	Filled/Increasing	**Blocked**	Empty/Unchanged	Off	Empty
On	Filled/Increasing	**Blocked**	Filled/Unchanged	**Blocked**	Empty/Unchanged	Off	Empty
On	Filled/Increasing	Positive	Filled/Increasing	Positive	Filled/Increasing	Off	Positive
On	Filled/Increasing	**Blocked**	Filled/Decreasing	Positive	Filled/Increasing	Off	Positive
On	Filled/Increasing	Positive	Filled/Increasing	**Blocked**	Filled/Decreasing	Off	Positive
On	Filled/Increasing	**Blocked**	Filled/Unchanged	**Blocked**	Filled/Decreasing	Off	Positive
On	Overflow	**Blocked**	Filled/Decreasing	Positive	Filled/Increasing	Off	Positive
On	Overflow	**Blocked**	Filled/Unchanged	**Blocked**	Filled/Decreasing	Off	Positive
On	Overflow	**Blocked**	Filled/Unchanged	**Blocked**	Empty/Unchanged	Off	Empty

Trajectory Calculation
for time slot [40,60] sec

...

Fig. 18. State space reduction for the three-tank-system

4.3 Online Analysis of the Three-Tank-System

The online calculation in Fig. 18 starts at t = 0 sec with the above described initial conditions. Then the observer calculates all possible trajectories for the first predefined time slot of $t = [0, 20]$ sec as a prediction for all possible states. The calculation of the trajectories is performed within the given dynamic model. The result shows, that the level h of the first tank ranges between $h = 40$ cm and $h = 49$ cm, of the second tank between $h = 10$ cm and $h = 14$ cm and of the third tank between $h = 1$ cm and $h = 2$ cm. With this information all impossible situations are excluded from the situation table and the result is a reduced qualitative state space for the given time slot. It is now very easy to survey the table. For example in the first state of the reduced table the $pump1$ is "On", the first and second tanks are in the situation "Filled/Increasing", the two pipes are in the state "Positive" and the $pump2$ is "Off". That means that there is a flow from the first tank into the second tank and then in the third tank. The fault-set of the reduced state space contains only possible faulty situations such as a blocked pipe and according to the described *principle of optimism* no further action is necessary.

The second reduced state space in Fig. 18 is calculated for the following time slot $t = [20, 40]$ sec. Considering the boundary conditions the first pipe is totally blocked, that means no outflow from $tank1$ in $tank3$. In this case tank1 will overflow within the time slot. This dangerous state is included in the fault set and according to the *principle of pessimism* a hazard of the system is probable. So the possible overflow of the tank can be recognized in time and the further action is a system alarm. After regarding the cause-effect-graph in Fig. 19 the cause of the hazard can be localized. In this case the blocked pipe between $tank1$ and $tank3$ is localized as the component fault, which has the dangerous effect of a tank overflow.

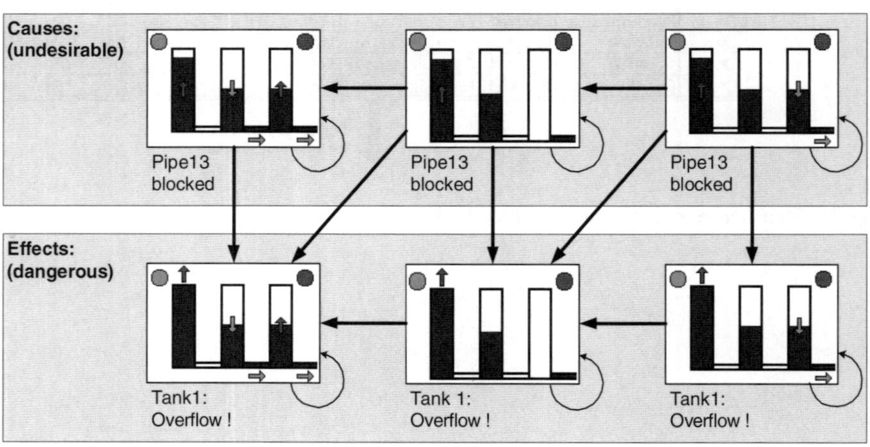

Fig. 19. Cause-effect-graph of the second time slot

The online state space reduction is part of the online monitoring and visualization realized within a Java application (Frank, 2001), performing the task of open and closed loop control of the real three-tank-system. The next section presents the results of the performed online analysis.

4.4 Monitoring Application for the Three-Tank-System

The calculation shown in Fig. 18 is performed periodically for every given time slot and yields different parts of the reduced qualitative state space. For the presentation of the monitoring application we will consider almost the same scenario described in Fig. 18. Only the time slot is set to 10 sec. In this context the monitoring application should find the answer to the question: "What are the effects of a blocked valve between the first and third tank?"

Figure 20 shows a part of the fault-set in a visualized form. This is done by another application written in Visual C++ (Laudwein, 1999, Ruhl, 1999), which supports besides the process visualization also the safety analysis of the three-tank-system based on the whole qualitative state space.

The figure illustrates an extract of all states, where the first tank is in the situation overflow. The task of the SQMD monitoring application is to detect this overflow in time. Figure 21 shows the visualization and monitoring user interface.

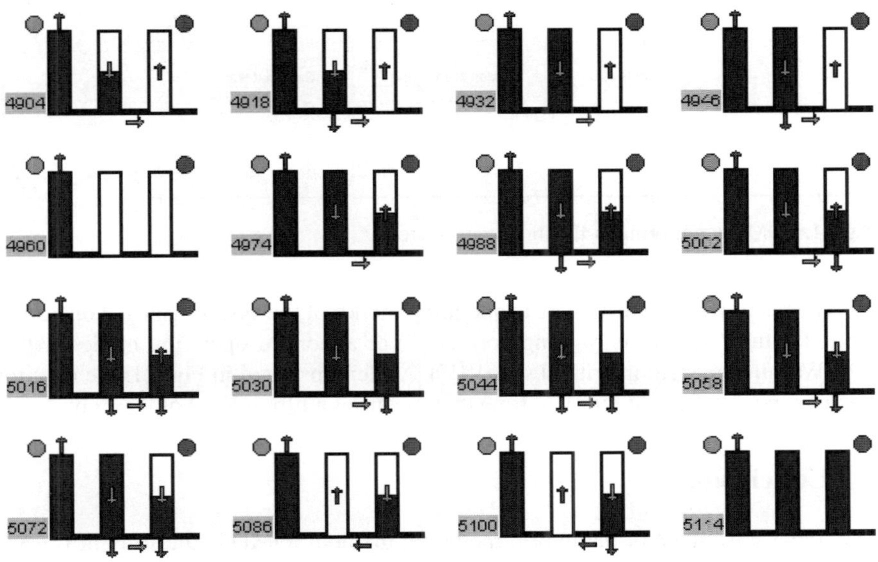

Fig. 20. Possible fault-set with an overflow of the first tank

The upper part of the user interface in Fig. 21 visualizes the sensor and actuator data of the real process. The lower part presents the output-window of the observer with the results of the analysis.

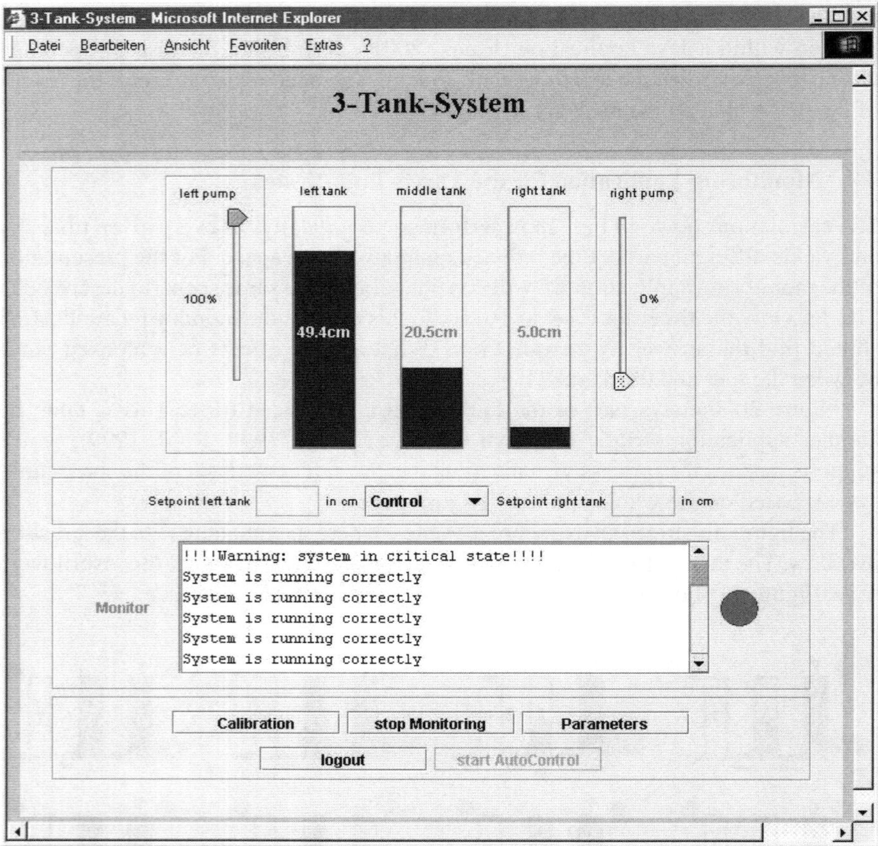

Fig. 21. SQMD monitoring of the three-tank-system

The result window displays the operating modes of the system: normal operation mode (status: "system is running correctly") or abnormal operation mode (status: "!!!! Warning: system in critical state!!!!"). As demonstrated in Fig. 21, the monitor predicts the overflow of tank1 in time that means at a filling level of 49.4 cm.

5 Conclusion

The major advantage of the hybrid modeling method SQMD is the straightforward modeling of complex dynamic systems. There are two important aspects. On the one hand already existing mathematical models are combined with qualitative models so that complex systems can be modeled and simulated. Therefore the described component oriented approach using qualitative reasoning methods plays an important role. This is the main difference to other modeling methods using for example Hybrid Automata or Petri Nets for discretization of the system behavior.

On the other hand only a certain reduced state space is analyzed online, that means only the relevant and possible states for a given time slot are observed online with low computation power.

The main part of the SQMD concept includes the online state space reduction, which realizes the mapping of the quantitative dynamic behavior to the qualitative behavior. Because of the qualitative idea however not the complete dynamic behavior is regarded. Only the start and end states of the calculated trajectories for every time slot are considered for further qualitative calculations, because it is not necessary to know what happens between the start and end states.

The online state space reduction is essential to handle complex systems and to reduce the nondeterministic behavior of the qualitative model.

A part of the model building and analysis is realized in a special SQMD toolbox in Matlab/Simulink. If the complete functionality will be integrated into the toolbox, the engineer is supported in a comfortable way and most tasks are performed automatically.

The use of the SQMD concept for model-based online monitoring has been successfully applied to the example of a three-tank-system. A new future task will be to apply the proposed concept for online monitoring of the real complex dynamic system of the air and gas concentration supervision in coal mines (Komarow and Skotschinski, 1956). This task arise from a collaboration with the university institute of computer science (LRT) in Donezk in the Ukraine, specialized in constructing mathematical models of the gas dynamics in coal mines. Within the collaboration it is intended to construct suitable models, which can be used for online failure detection and hazard prediction. The aim is to prevent methane explosions by reaction in time to a high methane concentration.

Modelling and Simulation of Controlled Road Traffic

Olaf Czogalla, Robert Hoyer, and Ulrich Jumar

Steinfeldstraße 3, 39179 Barleben, Germany
ocz@ifak.fhg.de, rho@ifak.fhg.de, ju@ifak.fhg.de

Abstract. A new combined approach to modelling of controlled urban road traffic was developed within the framework of the hybrid systems project KONDISK. The discrete event mesoscopic model and the cellular model have the benefit of faster simulation speed with respect to microscopic simulation especially for large scale models. The use of microscopic simulation instead is indispensable in the area of traffic actuated control and urban variable message signalling. Only the combination of modelling approaches and concurrent simulation allows the sound examination of telematic strategies. The research outlined in the paper is aimed at finding a suitable method on how to efficiently simulate overall urban road traffic systems by combining different modelling approaches.

1 Introduction

Processes of urban road traffic are characterised by the combination of continuous vehicle movement and discrete driver decisions. The hybrid driver-vehicle elements should be controlled in such a way that a minimal travel time and maximal traffic through-put result. Today, traffic modelling and simulation is increasingly used for the planning of guidance strategies for traffic management and individual route guidance. However, for various aspects of application, several approaches to model road traffic with different degrees of detail exist in literature. The research outlined in the paper is aimed at finding a suitable method on how to efficiently simulate overall urban road traffic systems by combining different modelling approaches.

At first, the most relevant modelling approaches are briefly introduced. Thereafter, a procedure is described which allows the transformation of an individual-continuous model into an individual-discrete description, where the time needed for simulation is drastically reduced and essential aspects of the model performance are maintained. A decisive premise to achieve a suitable overall model is a sufficiently precise approximation of vehicle actuated traffic control explained in the following paragraph. The traffic control algorithms are substituted by a generic fuzzy control module in order to avoid the requirement of re-implementation of the real traffic light control code. The final paragraph of the paper describes an approach to a concurrent simulation of different traffic models, the implementation of the outlined method and an example case.

S. Engell, G. Frehse, E. Schnieder (Eds.): Modelling, Analysis and Design of Hybrid Systems, LNCIS 279, pp. 419–435, 2002.

2 Approaches to Traffic Modelling

2.1 Motivation

Models for the simulation of road traffic serve different purposes. For example, they shall support the off-line planning of road networks and the programming of traffic actuated control of traffic lights, but also the models will become the basis of the on-line control of complex traffic guidance systems. Obviously it seems to be practical to increase the degree of abstraction in order to obtain a high computational performance, which is especially required in cases of on-line application. It would be helpful if it is possible to adapt the existing models, usually generated for the validation of traffic light control during the planning process, to the on-line application. For this a transformation procedure from very detailed models into models with less complexity is needed, however decisive effects of the road traffic behaviour must be preserved.

2.2 Collective-Continuous Road Traffic Flow Models

Collective-continuous road traffic flow models are also called macroscopic models, and describe the behaviour of a large amount of vehicles by common parameters such as the average speed v, the traffic density k and the traffic flow q. Here the behaviour of single vehicles is not taken into account. In the frame of collective-continuous traffic flow approaches, the motion of the total amount of single vehicles is assumed as a continuous flow process described by the continuity equation of traffic (Lighthill and Whitham, 1955):

$$\frac{\partial k(x,t)}{\partial t} + \frac{\partial q(x,t)}{\partial x} = 0 \text{ with } k = \left.\frac{dN}{dX}\right|_t \text{ and } q = \left.\frac{dN}{dT}\right|_x \tag{1}$$

The traffic flow q represents the number of vehicles N which pass a measurement line within the time interval T. The traffic density k denotes the number of vehicles N, which are present in a path of the length X at the time t. An empirical relation between traffic speed, traffic density and traffic flow can be observed, which is outlined in Fig. 1.

In particular, models for the determination of traffic demand and traffic assignment models are based on collective-continuous approaches. Due to the principle of continuity road traffic, flows can be calculated analogously normal in transportation networks for gasses or liquids. Since the collective-continuous approach disregards single vehicles, and since traffic actuated control of traffic lights reacts on single vehicles (especially on vehicles of public transport), this approach is not sufficient for the modelling of inner city road traffic. However, collective-continuous approaches require only little computational time for simulation. For that reason, the road traffic conditions on very large networks can be efficiently calculated.

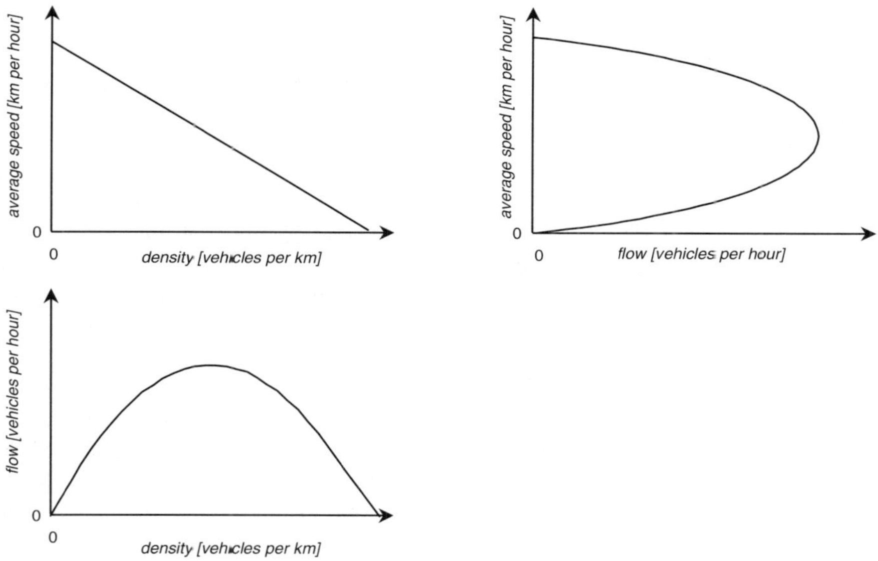

Fig. 1. Example of fundamental diagrams (idealised)

2.3 Individual-Continuous Traffic Flow Models

Increasing computer performance has made it possible to take into account single vehicles for road traffic simulation. These approaches are called individual-continuous or microscopic. Here the state of motion of every single vehicle is calculated for equidistant moments of time. Therefore, individual-continuous approaches are called time step based simulation models. The motion of vehicles is described through difference equations which distinguish the processes of

- free drive
- following a leading vehicle
- braking to a complete stop in front of a queue or red light
- when the right of way has to be given.

The simulation of every process is not homogeneously outlined in literature. For the process of following a leading vehicle, the reciprocal headway model (Gazis et al., 1959) is often used in various modifications. Here the acceleration of vehicles is calculated according to the following equation, where the acceleration depends on their own speed, the difference between the speeds of leading and following vehicles, and the headway to the leading vehicle:

$$\ddot{x}_{n+1}\big|_{t+T} = a \left. \frac{\dot{x}_{n+1}^{m}\left(\dot{x}_n - \dot{x}_{n+1}\right)}{\left(x_n - x_{n+1}\right)^l} \right|_t \tag{2}$$

where m, l are weighing exponents and a is a sensitivity parameter.

A further microscopic approach is based on (Wiedemann, 1974, Wiedemann, 1991). The model contains a psycho-physical car following model for longitudinal vehicle motion and a rule-based algorithm for lateral motions. It is implemented in a software package called $VISSIM^{TM}$ (Fellendorf, 1994), which was among others used in the KONDISK-project.

Compared with collective-continuous road traffic approaches, individual-continuous models have the advantage that simulation and visual representation of moving single vehicles achieve very detailed and plausible findings about phenomena of inner city road traffic considering the influence of traffic actuated traffic lights and traffic guidance systems. Only the taking into account of single vehicles in the model enables the simulation of interactions between the electronic transport system infrastructure and the driver vehicle unit. This approach makes the comprehensive study of the effects of collective and individual measures of traffic control and regulation possible. This advantage is accompanied with an essential shortcoming. The calculation of difference equations and the computing of more or less sophisticated algorithms for every vehicle in the road network (at least once per second) consumes a lot of simulation time. The on-line operation of this kind of microscopic models for the decision-support and network-wide optimisation of traffic responsive signalling is currently ruled out to a great extent.

2.4 Traffic Density Dependent Discrete Models

A way of avoiding a high consumption of simulation time is by eliminating the calculation of motion state of every single vehicle in equidistant time steps. This is done by a traffic density dependent discrete approach, which is also called mesoscopic. Here the motion state parameters of single vehicles are event driven computed. When a vehicle passes a new road section it is assigned a new speed which depends on the current number of vehicles in the section according to the speed - density relation of the fundamental diagram, see Fig. 1. This approach accelerates the simulation enormously as the arrival time of vehicles at the end of the section can be determined in only one simulation step. The execution of a simulation step is induced by an event such as the entry of a vehicle into a new section.

The event driven approach is only satisfactory for the simulation of road traffic. Relatively exact results can only be achieved for the traffic process at single intersections if the real length of traffic jams (not only the number of waiting vehicles) is of less interest. Compared with this application, the simulation of entire traffic networks requires the exact computing of the height of congestion. Here the KONDISK-project took remedial measures (see Sect. 3).

2.5 Cellular Automata

Another group of traffic simulation models is based on so called cellular automata which was originally developed for the flow simulation of granular materials. Following the principle of cellular automata, Schütt presented a bit-oriented simulation model for the highly efficient computing of road traffic (Schütt, 1990). The approach divides the road network into sections of the same length. The sections with a length of 7.5 m (average vehicle length with minimal headway) can be occupied by either one vehicle or by none. In a time interval of one second, vehicles move through one or more sections if the sections are not taken. Different speeds are simulated by the number of sections jumped. The essential benefit of the cellular automata is the ability to simulate very quickly, however the simulation result is inexact. This circumstance becomes quite clear if one considers that only a few speed steps can be simulated due to spatial and time discretisation. The cellular automata has been further developed in (Nagel and Schreckenberg, 1992).

3 Model Transformation

3.1 General Procedure

The general objective is to drastically reduce the consumption of simulation time by transforming an individual-continuous model into an individual-discrete description retaining an acceptable quality of simulation results such as plausibility of data and animation accuracy. For this, the following procedure is proposed:

- building up and calibration of an individual-continuous reference model
- definition of validation parameters taking into account the objectives of simulation
- one simulation run for measuring the validation parameters
- transformation of model structure by the partitioning of road sections
- definition of equivalent parameters
- adjustment and adaptation of equivalent parameters (e.g. calibration of the objective model by means of validation parameters)
- finally, the use of the objective model in order to efficiently achieve further simulation results.

Figure 2 shows the vehicle trajectories in the reference and objective model of one lane in front of a signalised intersection. The vehicles periodically stopped due to the traffic light. It can be seen that the speed changes discontinuously (infinite acceleration) in the individual-discrete model. The discontinuity is caused by events, where new speeds are assigned to the vehicles. This occurs when vehicles enter into a new section or reach the car set up distance at the end of congestion. Furthermore, it is shown that vehicles instantly follow the predecessor, which have to be corrected by an equivalent car set up distance and a lag for starting the move. The determination of these parameters among others will be done by the calibration process.

3.2 Calibration of the Objective Model

The calibration algorithm is formulated as an optimisation problem. The cost function applied to search for the objective model with the best validity is given by

$$c(\boldsymbol{x}^i) = \left[\max_i \frac{x^i_{\text{objective}} - x^i_{\text{reference}}}{x^i_{\text{reference}}} \right] \rightarrow \min \tag{3}$$

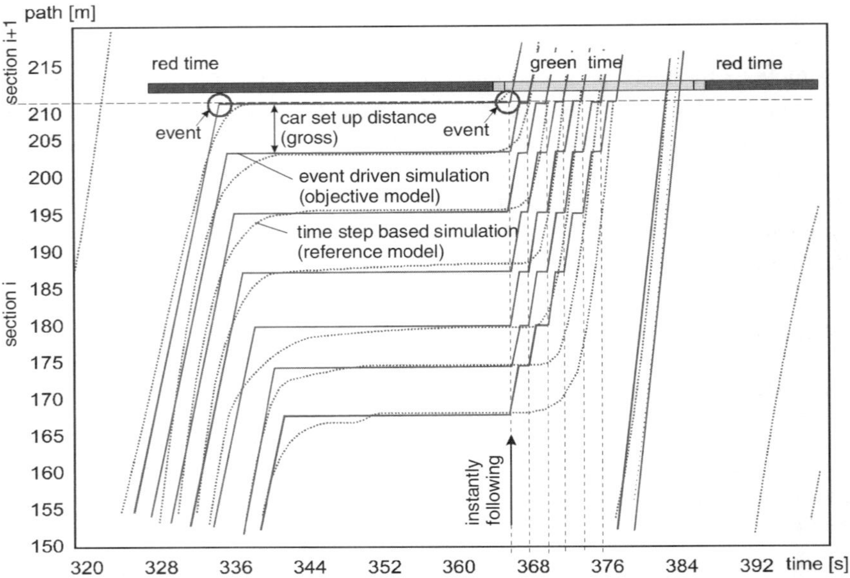

Fig. 2. Vehicle trajectories in reference and objective model

Setting the equivalent parameters by Min-Max criterion corresponds to a pessimistic view which means that the validity of the objective model could be better than the result of the calibration procedure. In contrast to this evaluation, an average criterion could lead to an optimistic assessment of model validity.

The optimisation task is solved by a selective algorithm. The method *Simulated Annealing* (Kirkpatrick et al., 1983, Vidal, 1993), an extension of the Monte Carlo method with random jumps of decreasing probability, was used by the authors. Selective algorithms are based on a simple search for local minima. One after the other, the cost function $c(\boldsymbol{x}^i)$ is calculated according to the parameter sets denoted by vector $\boldsymbol{x}^i$. Here the parameters are randomly varied within a given range. If the new value of $c(\boldsymbol{x}^i)$ is less than the old value of $c(\boldsymbol{x}^i)$, then the old parameter set is rejected. This rigid condition allows one to find local minima only. The way out of this dilemma is the introduction of a probability of acceptance, which avoids the hard rejection of less suitable parameters. This strategy is similar to tolerating a temporary risk in order to gain the best position in a game.

Figure 3 depicts the *Simulated Annealing* algorithm in general form used for experiments. Here z_s denotes a vector of random numbers $[-1, 1]$, a influences the step width, and ε denotes the termination bound. The probability of acceptance is introduced by a monotonous decreasing sequence a of control parameter $\{T_k\}_{k=0}^{l}$. Decreasing values of $\{T_k\}_{k=0}^{l}$ reduce the acceptance probability of parameter sets with unfavourable values of a suitable normalised cost function. With increasing k the chance, that the term $\exp(\Delta c(\boldsymbol{x}^i, \boldsymbol{x}^j)/T_k)$ is greater than a random number $p \in [0, 1)$, decreases, whereas the number p varies during execution of the algorithm.

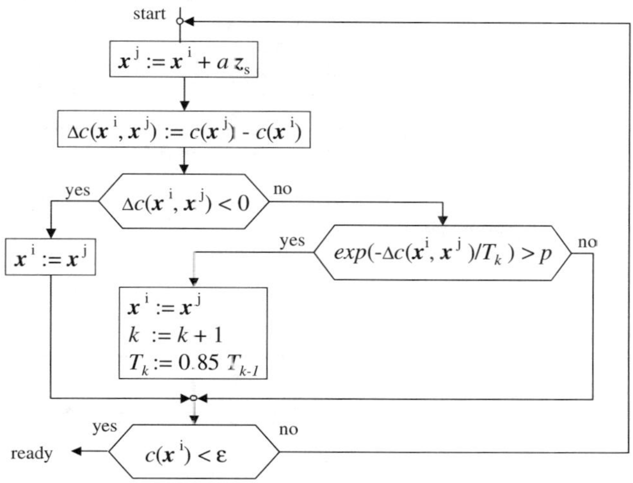

Fig. 3. Algorithm of calibration (Simulated Annealing)

The calibration process may be outlined as follows (see Fig. 4):

1. The simulation run of the reference model provides measurable validation parameters such as arrival times, travel times, traffic flows, lane utilisation rates, discharge rates and traffic lights.
2. The equivalent motion parameters of the objective model are set up as standard values from previous runs, and the simulation run of this model is started.
3. A set of validation quantities gained by simulation of the reference model is compared with a set of quantities determined by simulation of the objective model.
4. Based on the differences of the validation quantities of both models, the equivalent motion parameter set of the objective model is altered automatically considering also previous simulation runs. The aim of the additional simulation run is to lower the deviation between the quantities of both models.
5. As soon as threshold criteria are reached, the algorithm of calibration terminates.

3.3 Modelling of Traffic Adaptive Control

The decisive premise to achieve a satisfying model accuracy is a sufficiently accurate approximation of vehicle actuated traffic control algorithms. To solve this task, a method was devised to obtain a substitute controller which has a simplified structure, however a behaviour similar to the original algorithm. Basic steps to this approach are outlined in the following section.

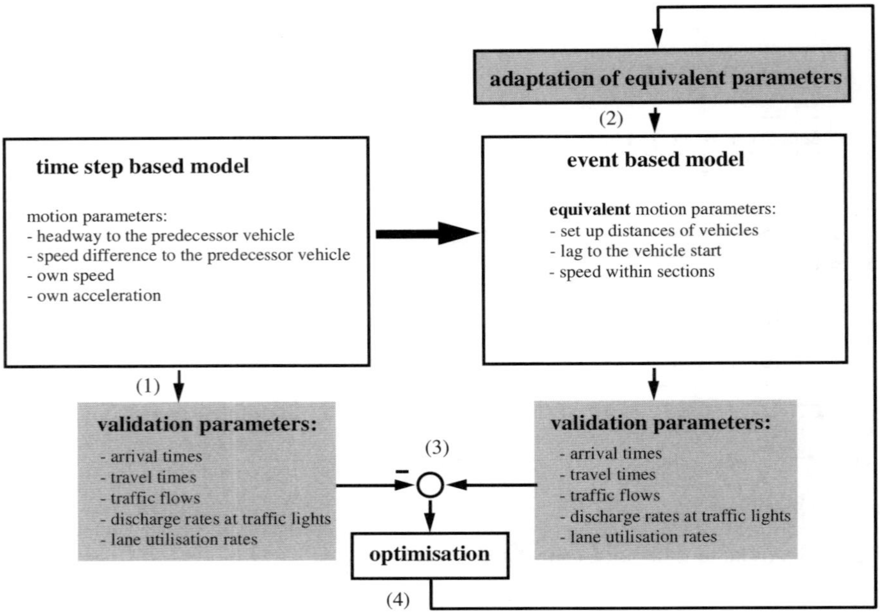

Fig. 4. Calibration procedure of the objective model

The original microscopic traffic controller is used to derive a substitute controller for the mesoscopic model by observing input and output values of the microscopic controller. Selected parts of the decomposed microscopic controller are used in the mesoscopic approximation as shown in Fig. 5. The substitute controller is based on a rule set for control decisions and undergoes a stepwise refinement of structure and parameters until a satisfying control behaviour is achieved. The measurements of simulated loop detectors and phases of green time periods of selected controller states are traced during simulation time. Standard traffic quantities such as traffic density and travel time were compared between both models to determine the performance. Additional experiments with the mesoscopic model and the substituted controller were conducted to validate the modelling process.

Decomposition of the microscopic traffic controller. In order to achieve a sufficient accuracy of the approximation, the microscopic traffic controller has to be decomposed into the following general components, see Fig. 5.

1. The *State Scheme* describes the combination of signals for each single phase as fixed signalling states and the set of transitions that allow the state to change from one phase to the following phase. The state scheme can be interpreted as a condition/event network with the output vector $\boldsymbol{S}_i$, representing the switched state of traffic lights at the places (phase) and attributed transitions $\boldsymbol{T}_{ij}$ representing the time dependent altering of signals from state $\boldsymbol{S}_i$ to $\boldsymbol{S}_j$.

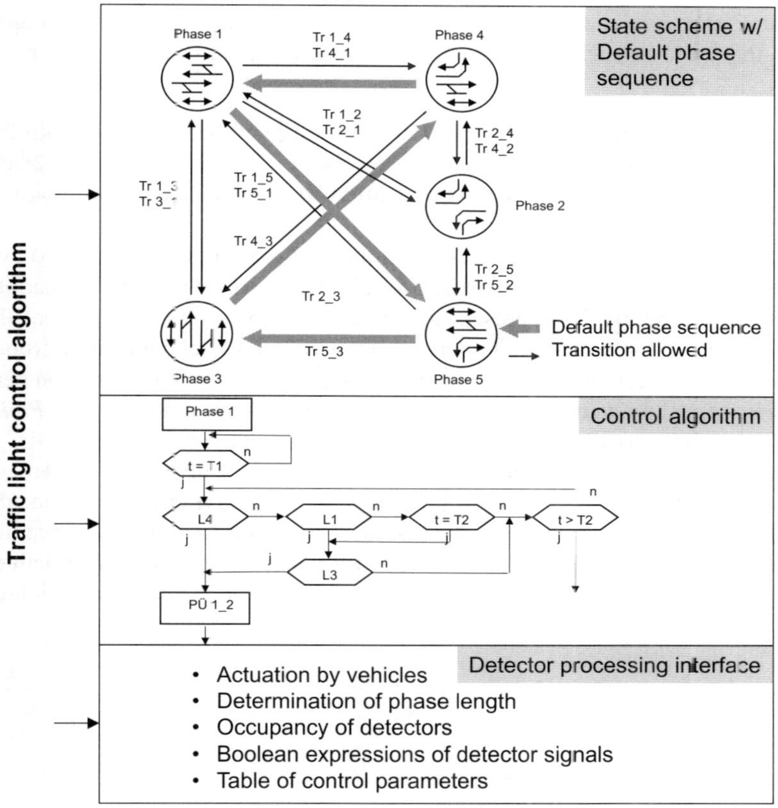

Fig. 5. Decomposition of traffic light control

2. The *Default Phase Sequence* is the preferred order of successive phases being processed in a loop until demands of higher prioritised actuations such as from public traffic are submitted from corresponding detectors.
3. The *Control Algorithm* describes the logical flow of conditions and events to be checked for the firing of each single transition. The duration of the relevant phase is calculated by checking the conditions for its termination. All conditions and events depend on results of the detector processing module. For parameter driven control methods, the control algorithm is pre-implemented within a procedure kernel of the control device. The system is configured by a set of parameters determining the desired control behaviour.
4. The *Detector Processing interface* includes algorithms and procedures for processing physical raw detector data. The raw data is converted into determined traffic quantities such as traffic volume, time gap (headway), occupancy and presence of vehicles.

This scheme allows to analyse various existing controller types of different origins and to fit them into a generic scheme of the corresponding substitute controller.

Design of the substitute controller. The basic structure of the substitute controller matches the scheme of decomposition described above. The components (1), (2) and (4) have to be re-implemented to ensure a control behaviour as adequate as possible to the original controller.

The *State Scheme* is modelled by a condition/event network to be initialised by a single set of parameters. The parameter set includes the matrix of admitted transitions $\boldsymbol{T}_a$. Rows of the matrix $\boldsymbol{T}_a$ represent the start phases $\boldsymbol{S}_i$, columns represent the target phases $\boldsymbol{S}_j$. The elements of $\boldsymbol{T}_a$ represent the availability of a transition from a start phase to the target phase. Zero values indicate a forbidden transition, non-zero values indicate an admitted transition. If the transition belongs to the *Default Phase Sequence*, the corresponding matrix element $\boldsymbol{S}_{ij}$ is set to a predefined value.

The time dependent alteration of the signal vector from state $\boldsymbol{S}_i$ to $\boldsymbol{S}_j$ is defined in a structure to be initialised for a formal description of phase transitions once the system is started. After initialisation, the system processes the default sequence in a loop until a dedicated actuation by public or pedestrian traffic demands a predefined service phase. After returning from the service phase, the system re-enters the default loop.

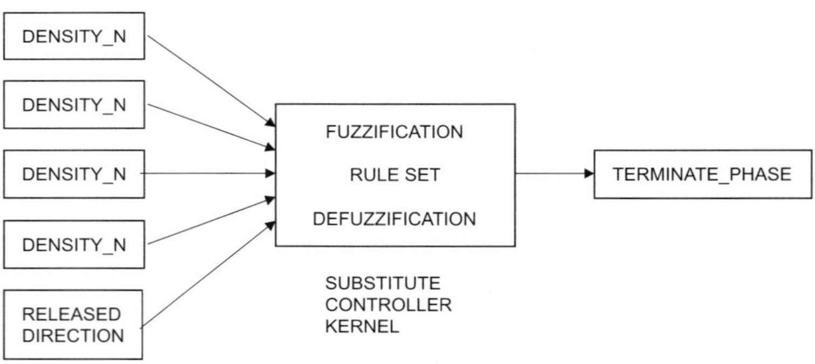

Fig. 6. Input and output values of the rule set

The *Control Algorithm* is substituted by a generic fuzzy control module used to determine an appropriate termination time between a period of minimal and maximal phase duration. The approach aims to avoid the requirement of re-implementation of large amounts of control code. In order to substitute microscopic traffic quantities like detector occupancy, another input quantity must be selected. The controller operates with 5 fuzzy input variables and one output variable. Since the traffic density is the only measurable quantity in the mesoscopic model, it was used at all approaching links of the intersection as input values for fuzzification, see Fig. 6. The state dependent values of traffic densities are assigned to the blocks DENSITY_N -

DENSITY_W. Extra information on the released direction of traffic flow has to be put into the rule set. Figure 7 depicts an example of membership functions used to

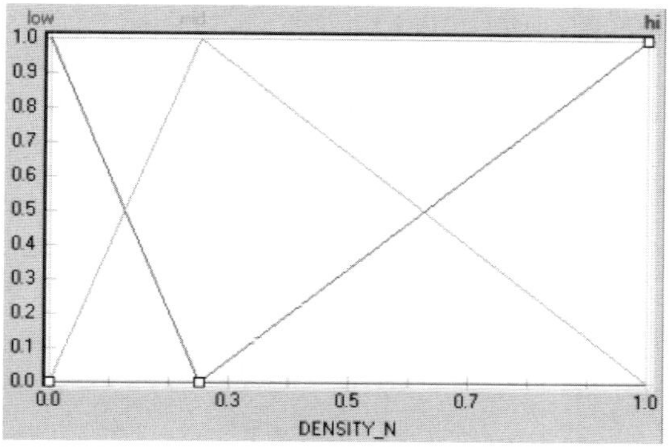

Fig. 7. Membership functions of normalised traffic density

define the linguistic terms *high, mid* and *low* of density by linear and triangular segments. Experiences have shown that the number of three linguistic terms is sufficient to describe the characteristics of a quantity as well as to keep the transparency of the designed rule set. To fit the membership functions exactly to introduced terms, technical knowledge was required from expert consultations. In this case, a density value of 0.5 is computed as "medium (mid)" by a membership degree of 0.7 and "high" by a membership degree of 0.3. The traffic density $\boldsymbol{k}$ was normalised to the maximum $\boldsymbol{k}_{\mathrm{max,norm}} = 1$ based on a stop distance of 7 m.

The binary output TERMINATE_PHASE is connected to the phase scheme, which takes over control after termination of the current phase and determines the next transition. The rule block contains a total of 30 fuzzy rules. The control strategy was developed following observations that a decreasing traffic density at a green signalised link leads to the termination of the current phase if the traffic density at blocked approaches is high. Information on released traffic flow directions is processed as a fuzzy input value. A multitude of rules extend the rule set to refine its control behaviour. Table 1 shows a selection of rules to illustrate the process of inference.

The rules are read as follows: If *dens_E* = *mid* and *dens_N* = *low* and *dens_S* = *mid* and *dens_W* = *hi* and *released_dir* = *East_West then terminate(phase)* = *false*. If instead *dens_E* = *med* and *dens_W* = *low* and *dens_N* = *hi* and *dens_S* = *hi* then *terminate(phase)* = *true*. Symmetries of the input term distribution in the rule set are intended to process all signals equally without preference. The developed rule set is the premise for a general purpose fuzzy control module containing no application specific variables and rules. This allows its application in a variety of intersection controllers without changes to the structure and rule set.

Simulation experiments were carried out to improve and refine the control behaviour of the substitute controller using the microscopic model as well as the mesoscopic for reference. The experiments aimed at the following objectives:

Table 1. Part of the Rule Set

IF					THEN
dens_E	dens_N	dens_S	dens_W	released_dir	terminate
mid	low	mid	hi	East_West	false
mid	mid	low	hi	East_West	false
hi	low	low	hi	East_West	false
mid	hi	hi	low	East_West	true
low	mid	hi	mid	North_South	false
mid	mid	hi	low	North_South	false
low	hi	hi	low	North_South	false
hi	mid	low	hi	North_South	true
⋮	⋮	⋮	⋮	⋮	⋮

- The microscopic traffic simulation incorporates the real traffic control unit and provides quantities for measurement of quality as distribution of green times.
- The approximated traffic controller as a part of the mesoscopic model has to be adjusted by means of fuzzification and rule adaptation. Density and time distribution can also be measured as in the microscopic model.
- To evaluate the measured quality of approximation, the density and green time distribution have been compared.

4 Concurrent Simulation

In recent years, the existing application fields of traffic simulation technology as traffic planning and control were extended by scopes such as urban traffic management and optimisation. Today, traffic simulation is increasingly used for the planning of guidance strategies for traffic management and individual route guidance systems. The utilisation of microscopic simulation in urban traffic management and control is discussed in more detail in (Czogalla and Hoyer, 1997).These application scopes require comprehensive traffic models of wide urban regions with sufficient accuracy as well as high performance simulation environments. Currently these requirements cannot be met only by means of microscopic traffic simulation. The trend in micro-simulation technology to refine the model accuracy is contrary to the impossibility to keep all traffic conditions and traffic signalling states in a model like in reality. Remarkable drops in simulation speed have still to be accepted, especially for models with a large spatial extent. Higher simulation speeds can be achieved by mesoscopic traffic models and cellular automata based simulation approaches due to simplified car following dynamics and event driven simulation results in reduced computing time.

The approximation of traffic control is an appropriate contribution to more efficient simulation as described in (Czogalla and Hoyer, 1999). The use of features of different simulation systems within an integrated model has a considerable advantage over a single model with regard to modelling effort and quality of results. Decisively in this context is the choice of spatial boundaries between the model parts, i.e. the definition of simulation tasks for each part of the complete model.

4.1 Model Characteristics and Limitations of Modelling Approaches

The primary aspect is the definition of criteria for the choice of modelling depth and the choice of an appropriate modelling approach. The main criteria have been identified as follows:

1. Requirements for traffic signalling and special telematics control elements as variable message signs, parking guidance systems, etc.
2. Demanded level of detail of the traffic network
3. Modelling of vehicle classes and the characteristics
4. Computing effort with regard to model features
5. Amount of data for animation of simulation results

By application of this set of criteria onto model characteristics, *microscopic models* have proved to be suitable for urban regions with mainly signalised intersections. Less applicable is this modelling approach for motorways and urban expressways. *Mesoscopic traffic simulation* is used for segments of interurban roads and minor city streets without signalling. The *Cellular approach* using a road segmentation of 7.5 m and a relatively coarse discretisation of space, time, velocity and acceleration can not be used for the detailed modelling of urban traffic phenomena but is appropriate for segments of motorways and expressways.

The separation of model parts can be achieved through the spatial structure of the traffic network. Digital maps with attributed road segments provide the required road classification and the topological network structure with nodes and links. Additional data such as the number of lanes and turning manoeuvres at intersections are obtained from traffic databases hosted by the civil engineering departments of local authorities. The topological and geographic data is converted by means of a *Geographic Information System* through the integrated programming language.

4.2 Substitution and Coupling of Model Parts

The coupling of model parts includes vehicle related aspects such as the passage of objects between models and network related topics, in order to regard interactions of models such as traffic congestion at boundaries. A control system was developed to allow a concurrent and synchronised simulation of all model components (Fig. 9). For the implementation of the microscopic model, the simulator *VISSIM*^{TM} (Fellendorf, 1994) was chosen. In addition, a method to control the simulation run and automatic generation of vehicles with defined type, length, velocity and routes was developed. This separate *Simulation Control Module* allows the measurement of vehicles in the microscopic model at predefined detection positions in the traffic network. The mesoscopic simulation system *VMAN* was developed by the authors of this paper and

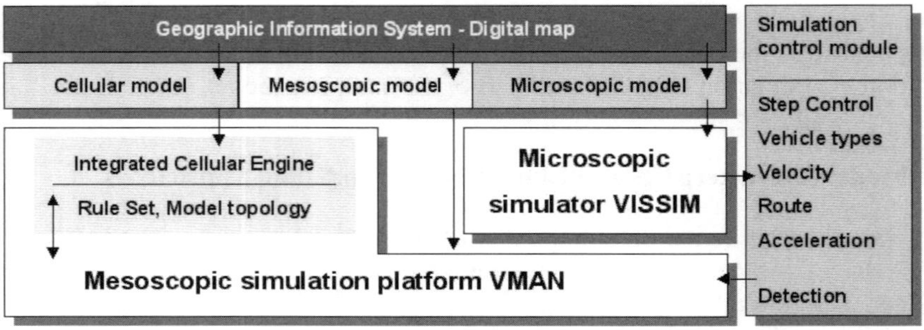

Fig. 8. Functional view on the structure of coupled model components

can be directly connected via the control module. The cellular engine was completely implemented within *VMAN* which allows easy access to cellular rule set and simulation control without an additional interface.

In order to synchronise the simulation progress in all model parts, a time management concept was devised using HLA (High level architecture, (HLA TMD Document, 1996)) which allows the regulation of models with different simulation step rates and autonomous clocks as well as event driven mesoscopic simulation. The microscopic model could be simulated with a variable step rate, whilst the cellular model allows only steps of one second. In the mesoscopic discrete event model, no equidistant time step rate is used because the calculation of the velocity and the position of vehicles are determined at the end points of road segments depending on traffic density along the considered stretch. In order to take these aspects into account, a synchronisation concept was introduced with the following premises. The time management should co-ordinate the interchange of events such as when vehicles cross model borders or information on congestion state at model borders, i.e. the simulation progress depends on time progress in each of the interconnected model parts. A model having an impact on the time progress of another model is called a "regulating model", the affected model is called "constrained model".

The synchronisation concept is shown in Fig. 9 exemplarily for the microscopic and mesoscopic model part. In this case the microscopic model is "regulating", because it generates *time stamp ordered events* (TSO), e.g. entry and exit of vehicles as simulation objects into and out of the model respectively. Although the simulation step rate is constant, more than one event can occur per time step at distinct model interfaces. The "lookahead" value of the regulating model is the time interval between current time t_{current1} and occurrence of the earliest TSO-event, i.e. all further events must occur after $t_{\text{current1}} + t_{\text{lookahead}}$. The constraint models are able to receive TSO-events and hold a *lower bound time stamp* (LBTS). This LBTS marks the time of the earliest possible TSO event to be received and is shifted onwards in the same quantity as the regulating model progresses in time. The lower bound time stamp is sent to the mesoscopic simulator and cellular automata via a control interface. Reversibly it is also the case that the mesoscopic model regulates the microscopic model by vehicles exiting the mesomodel and generating TSO-events which impact the LBTS-value of the micromodel. As a result, a mutual synchronisation of all model parts is given which prevents irregularities in simulation runtime control in spite of different clock rates and non-equidistant simulation steps. For a more detailed description of the synchronisation concept refer to HLA (High level architecture, (HLA TMD Document, 1996)).

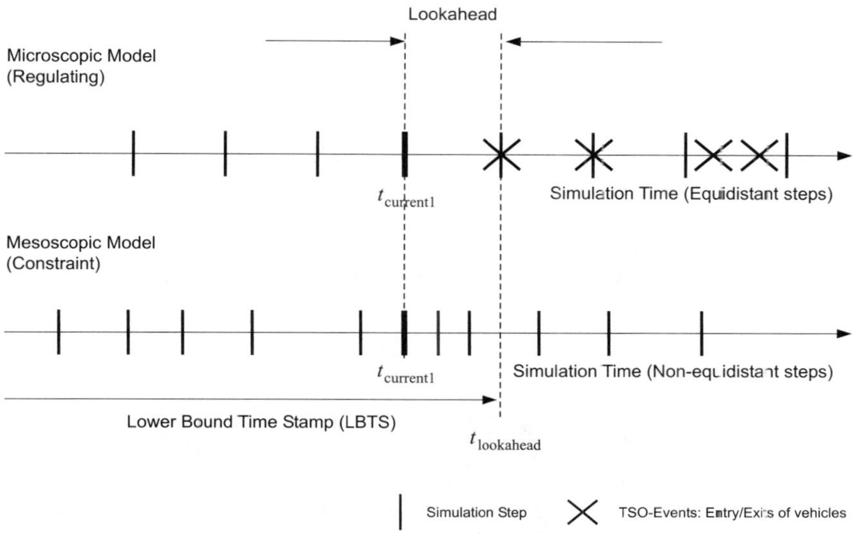

Fig. 9. Synchronisation of model parts

4.3 Simulation of a Sample Case

In the framework of the research work, a precise and realistic traffic model of a wide urban road network was created to be able to replicate typical traffic scenarios and interactions between different traffic modes, such as public and individual traffic. The geographic area of the traffic network was selected to represent multiple types of intersections and elements, such as

- single and multi-lane segments,
- simple T-junctions and
- crossings both signalised and non-signalised,
- lane restrictions,
- 2-lane expressway and
- traffic actuated control with public transport priority at intersections.

The reference model was developed as the first step in the form of a microscopic model. For selected segments of the road network, the model was replaced and extended by mesoscopic and cellular model parts.

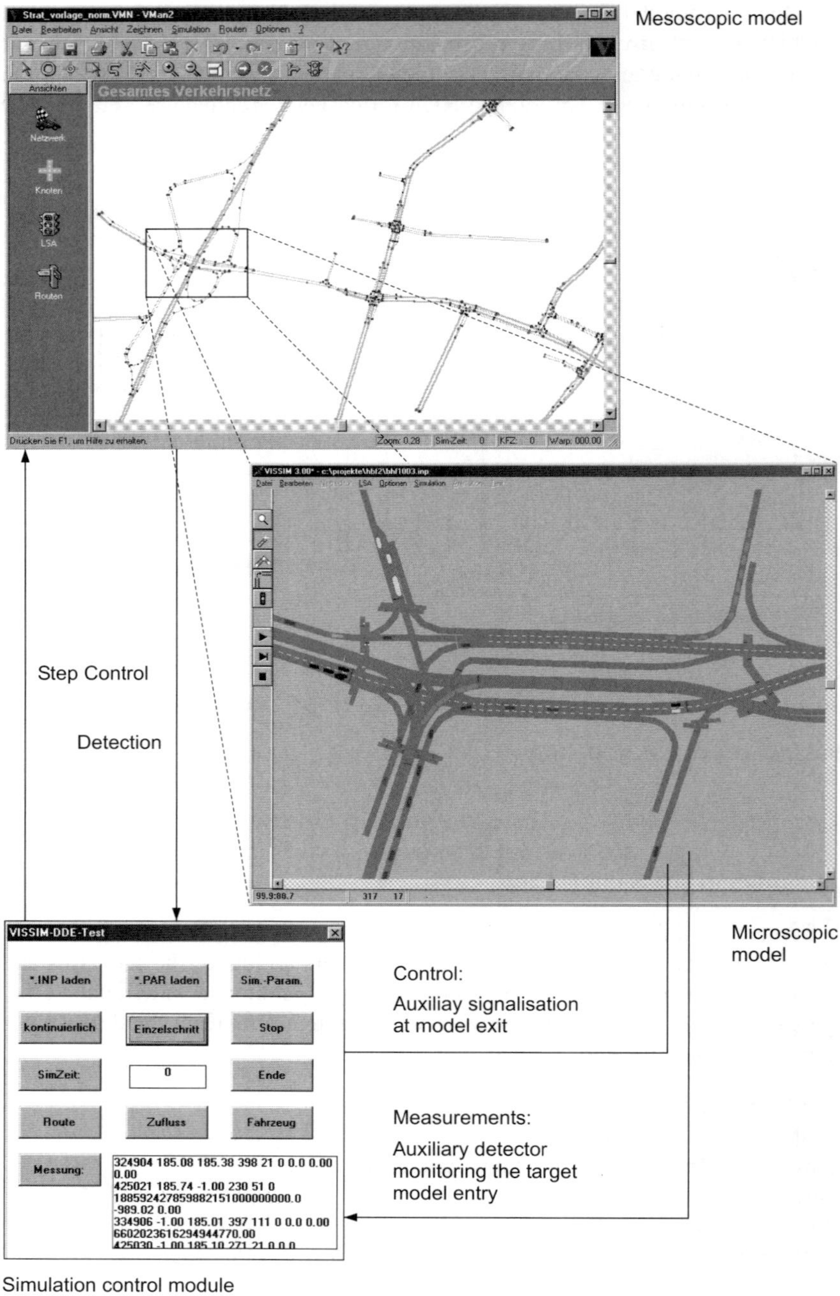

Fig. 10. Simulation of the coupled reference model

The mesoscopic model comprises an extended area of the city and regards street segments including lane number and intersections of major arterial roads which are identified by the functional class of available digital map data. The traffic load for each route is gained by traffic counts and original destination matrices as a result of the macroscopic equilibrium model. The multi-lane city expressway was modelled within the mesoscopic simulator as a cellular model. Ramps and exits of the expressway likewise represent the interfaces between mesoscopic and cellular model. For the concurrent simulation, the mesoscopic and microscopic have to be connected through the implemented control interface to allow simulation objects to pass over model borders.

Figure 10 illustrates the concurrent simulation of mesoscopic, microscopic and cellular model controlled by simulation control module (see also Fig. 8). The simulation control module communicates autonomously with both simulator applications and synchronises the simulation runs as described above.

Special interactions and simulation states such as congestion at model borders are regarded by auxiliary detection at model entries and auxiliary signalling at model exits in order to avoid the disappearance of simulation objects in this area. The auxiliary signalling at the model exit is switched to red by the simulation control module if the entry segment of the target model is occupied. When the entry segment is cleared, the auxiliary signal is turned to green and allows the vehicles to cross the model border. This concept of using semaphores for information exchange can also be applied and adapted at a more general level in distributed simulation technology.

5 Conclusions

For new application cases of traffic simulation, both precise and spatially wide traffic models of urban areas with various functional classes of streets are required. The utilisation of characteristics of distinct simulation methods in one integrated model is from the point of view of modelling effort and its exactness a significant advantage over the application of a single simulation method to the entire model space. The suitable decomposition of the model space and assignment to appropriate modelling techniques is important mainly to achieve adequate model quality. A new combined approach to modelling of controlled urban road traffic was developed within the framework of hybrid systems project KONDISK. The event discrete mesoscopic model and the cellular model have the benefit of faster simulation speed with respect to microscopic simulation especially for large scale models. The use of microscopic simulation instead is indispensable in the area of traffic actuated control and urban variable message signalling. Only the combination of modelling approaches and concurrent simulation allows the sound examination of telematic strategies.

Acknowledgements. The authors would like to express their acknowledgement to the German Research Council (Deutsche Forschungsgemeinschaft) for granting this project within the framework of KONDISK, the results of which contributed to this paper.

Hybrid Control of Multi-fingered Dextrous Robotic Hands

Thomas Schlegl[1], Martin Buss[2], and Günther Schmidt[3]

[1] Siemens VDO Automotive AG, Im Gewerbepark C25, D-93059 Regensburg, Germany*
[2] Control Systems Group, Technische Universität Berlin, D-10587 Berlin, Germany
[3] Institute of Automatic Control Engineering, Technische Universität München (TUM), D-80290 München, Germany

Abstract. This article presents a hybrid discrete-continuous dynamical systems approach to control of multi-fingered dextrous manipulations by robotic hands. A discrete-continuous modeling framework allows to derive a hybrid state model integrating time-driven dynamic features of manipulation systems as well as discrete event aspects resulting from varying contact situations between the robotic hand and manipulated objects. The combination of continuous and discrete dynamic aspects is typical for multi-fingered manipulation. A hybrid control architecture comprising a hybrid planning scheme for grasping and regrasping, impedance control algorithms based on 6D contact force sensor information, and a formal compensation method for discrete contact state errors are proposed. Results of dynamical simulations and experiments with a four-fingered hand grasping and manipulating objects demonstrate the robustness improvement in grasping control as achieved by use of the proposed hybrid control approach.

1 Introduction

Dextrous manipulation of objects by multi-fingered hands combines features from two interacting types of dynamical systems. On one hand, complex multi-body mechanics is modelled by a set of nonlinear differential equations subject to unilateral contact force constraints and/or kinematic (algebraic) constraint equations. These aspects belong to the domain of continuous variable dynamical systems – CVDS. On the other hand, aspects from discrete-event dynamical systems – DEDS arise due to varying contact situations between fingers and a grasped object. A finger can have stable contact on an object, it can slide on its surface, or there can be no contact at all. In the following, these situations are referred to as *discrete grasp states* of a finger. Especially, during *active manipulation* of objects by *regrasping* or *finger sliding*, the contact situation (discrete state) between fingers and grasped object varies dynamically, see Fig. 1(a). Discrete and continuous aspects are strongly coupled and cannot be treated independently without neglecting important effects. Therefore, an appropriate control architecture for dextrous manipulation needs to consider the *hybrid discrete-continuous nature* of its dynamics.

* The work reported here was performed while the first and second authors stayed with the Institute of Automatic Control Engineering, Technische Universität München, Germany. Experiments were conducted at the Tokyo Institute of Technology, Japan.

S. Engell, G. Frehse, E. Schnieder (Eds.): Modelling, Analysis and Design of Hybrid Systems, LNCIS 279, pp. 437–465, 2002.

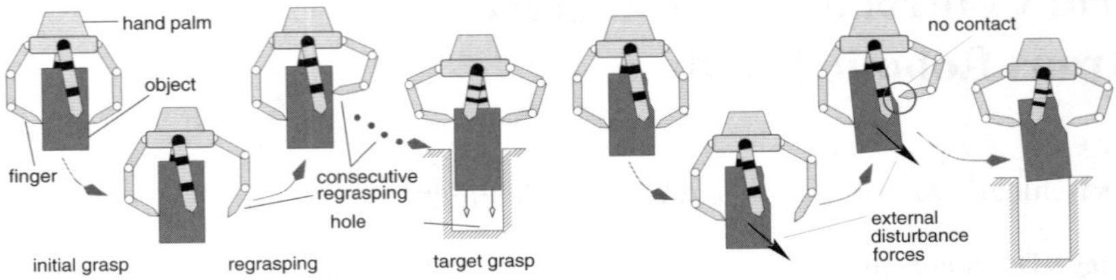

(a) Typical object manipulation task (b) Imprecise models and external forces

Fig. 1. Continuous dynamical and discrete event aspects of dextrous grasping

In recent years such hybrid dynamical systems have received increased attention by many researchers with different backgrounds (Antsaklis and Koutsoukos, 1998, Engell, 1997, Labinaz et al., 1996, Grossman et al., 1993, Grossman et al., 1993, Alur et al., 1996, Antsaklis et al., 1997, Antsaklis et al., 1999, Maler, 1997, Pnueli and Sifakis, 1995, Schumacher et al., 1999, Valavanis, 1997, Antsaklis and Nerode, 1998b). Various modeling and control approaches for hybrid systems have been proposed. In our work we have mainly concentrated on mechatronic multicontact systems such as multi-fingered robotic hands, or multi-legged walking machines, a particularly challenging subclass of hyrid systems (Buss, 2000, Buss and Schmidt, 1996, Buss, 1998). Differing from conventional research in the area of multicontact mechatronic systems a hybrid control approach is well suited to systematize and formalize modeling, trajectory planning, control, simulation, and implementation issues as will be shown in this article by the example of multi-fingered manipulation.

Control of multi-fingered manipulation is a manifold task because of the hybrid character of most control tasks involved. Indiviual problems, like *planning of grasps*, *grasp force generation*, and *synthesis of closed-loop control laws*, have to be solved for different grasp structures which originate from: i) the discrete contact state of the fingers, and ii) the (continuous) position of the fingertips on the surface of a grasped object. Various approaches to contact point and grasp force planning as well as closed-loop control laws have been suggested. Some of these propose controllers which are applicable merely for *static grasping* without manipulative aspects (Kleinmann, 1996). Others regard high-level planning of regrasping tasks without considering the problem of appropriate *low-level closed-loop control* and *grasp force planning* (Ricker et al., 1996, Cherif and Gupta, 1997). Especially the impact of modeling errors and external disturbances on the performance of manipulation are rarely considered in other works. A comprehensive approach to dextrous manipulation control from a hybrid systems viewpoint has not been reported otherwise to the best of our knowledge.

A major contribution of this article is the development of a robust control architecture for dextrous manipulation within a hybrid dynamical systems framework. An introducing example task will illustrate the features of multi-fingered manipulation that have to be considered during control synthesis.

1.1 Shaping the Planning and Control Problem

Figure 1(a) shows a typical manipulation task of a multi-fingered hand acting on an object. The goal of the task is to insert the object into the hole. During task execution the object is to be *fixed* by contacting fingertips stably and to be *moved* into the hole by coordinated finger motion. In order to avoid collisions between finger joints and the environment which is shaded in the figure, new contact points of the fingers closer to the upper boundary of the object have to be chosen. To achieve this change of contact configuration a sequence of regrasping tasks of single fingers needs to be performed.

This basic yet illustrative example demonstrates the high manipulative abilities of multi-fingered hands compared to other simpler manipulation systems. A multi-fingered hand is potentially capable to *freely* change a given grasp configuraion *during task execution* without having to release the object and grasp it again. This is amongst others one of the advantages of dextrous grasping which is supported by the control approach presented in this article.

Planning of hybrid trajectories: An important issue in this context is planning of hybrid reference trajectories for dextrous manipulation tasks. Eventually, as hybrid optimal control matures – see also a companion article in this volume (Buss et al., 2002) – it may become possible to derive optimal hybrid trajectories for dextrous manipulations. As the number of degrees-of-freedom in typical robotic hands currently prohibits solving of corresponding hybrid optimal control problems this remains an open problem for future research. Within this article only planning of *single* regrasping tasks is considered. The concatenation of regrasping primitives defines a *global* planning problem of how to transfer an initial grasp situation into a target grasp situation optimally with respect to a discrete-continuous optimization objective. Currently, only non-optimal desired hybrid reference trajectories are implemented in our hybrid reference generator for local planning of regrasping.

Competing control objectives: Manipulation of objects demands to change the position and orientation of a grasped object with respect to the hand palm, see the right part of Fig. 1(a). If the contact points of the fingertips on the object are known, this can be achieved by commanding appropriate trajectories for the fingers. In order to grasp the object stably it is necessary to compensate external forces, like the gravitational force or discturbance forces which may act on the object. This situation is depicted in the left part of Fig. 1(b). It is obvious that multi-fingered manipulation comprises aspects of position control and of force control simultaneously.

Effect of disturbances: During compensation of disturbing forces unilateral contact constraints between fingertips and object surface have to be considered. Fingertips can only push but not pull on the object. In addition nonlinear friction constraints – the violation of which may cause undesired slip of fingers – must be taken into account. Furthermore, multi-fingered hands often feature more mechanical degrees of freedom than the object. As a consequence, a variety of feasible contact forces exists which satisfies unilaterality and remains within friction limits. To choose the contact force vector from this manifold which produces a grasp as tight as necessary and as soft as possible is one of the key issues of *dextrous* manipulation.

As disturbing forces often change rapidly, an optimal high-speed contact force adaptation, i.e. within a few milliseconds, can increase the robustness of multi-fingered grasping dramatically.

Modeling errors: Model uncertainties have a major impact on the quality of manipulation task execution. During sequential regrasping these can cause *deviations* of the actual discrete contact state of the system from the desired one. Situations may arise in which not enough fingers are contacting the object in order to maintain a stable grasp. For instance, a finger may not be able to establish contact with the object at the end of a regrasping task, because the new contact point was planned using an *imprecise object model*, see the right part of Fig. 1(b). The regrasping finger reaches the target point of its regrasping trajectory without contacting the object surface, because the different shape of the upper-right object corner has not been considered in the object model. If in this example situation any other finger will be lifted off next, grasp stability is lost.

1.2 Focus and Outline of the Article

In this article the key components of a hybrid control architecture for multi-fingered grasping and regrasping of objects are presented. The hybrid controller enables robotic hands

- to robustly grasp objects
- to freely position and orientate objects
- to change a grasp by recontacting single fingers
- to robustify grasps against disturbance forces
- to cope with imprecise models by active compensation of discrete contact state errors and grasp stabilization.

The basic architecture of the proposed discrete-continuous controller is shown in Fig. 2. Its main components are a hybrid reference generator – HRG, a hybrid controller – HC and a set of continuous control laws – {C,D}IC, i.e. {Centralized, Decentralized} Impedance Control, respectively[1]. The continuous control laws are combined with a real-time grasping force optimization scheme – GFO, which ensures that all desired joint torques and velocities generate contact forces which satisfy friction conditions at all contact points. Although a *centralized* (CIC) and a *decentralized* concept (DIC) have been developed, only the latter is presented in this article due to space limitations. At this point of the discussion Fig. 2 is supposed to outline the basic structure of the controller. The exact meaning of the different blocks and symbols will be explained in the following sections.

The hybrid approach presented allows for the specification of discrete-continuous reference trajectories for the different phases of a manipulation task. Desired trajectories are generated by use of hybrid reference automata (Alur et al., 1993). A hybrid reference comprises the desired contact state $\boldsymbol{q}$ which is characterized by the number of contacting fingers, the type of contact and desired fingertip positions

[1] The structure {C,D}IC denotes an abbreviation and means *either* CIC *or* DIC.

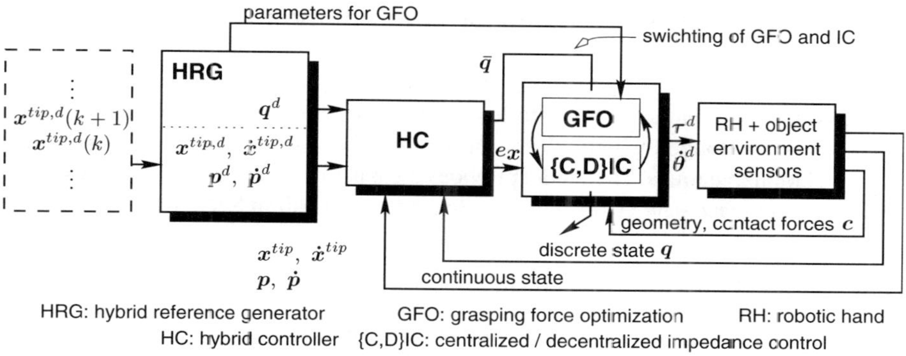

Fig. 2. Hybrid control architecture for dextrous manipulation

$x^{tip,d}$ as well as the corresponding velocities $\dot{x}^{tip,d}$. Because much work was done with respect to non-optimal *global* planning of multi-fingered manipulation, a simple task planner which generates a discrete sequence of fingertip contact points on the object $\ldots, x^{tip,d}(k), x^{tip,d}(k+1), \ldots$ is assumed to be given. The main focus here is on specification of single, elementary regrasping primitives into which each multi-fingered manipulation task can be decomposed.

The inner closed hybrid control loop guides the manipulation system along the commanded reference trajectory. The impedance control laws in combination with a grasping force optimization algorithm solve the aforementioned concurring control problems. They simultaneously allow for *active positioning* of fingers and for *generating contact forces* which guarantee a stable grasp. An object grasped in such a way will robustly follow coordinated motions of the fingers. Additional feedback of the discrete contact state to the hybrid controller enables compensation of contact state errors which typically arise during a sequence of regrasping tasks. A hybrid process model provides the necessary information to detect and correct wrong contact states. If the discrete system state is undesirable, an active contact search is initiated. As soon as the contact state is corrected, grasp stability is increased by locally correcting fingertip positions to enable further stable regrasping tasks. A Combination of the different controller components yields a control architecture for the robust manipulation of objects by multi-fingered hands, even in case of disturbance forces on the object or imprecise modeling. Continuous as well as discrete errors are reactively compensated on a lower control level thus avoiding burdening of a high-level task planner.

This rest of this article is organized as follows: Sect. 2 proposes a comprehensive analytical formulation of discrete and continuous aspects of multi-fingered manipulation. Based on the derived model the core components of the hybrid controller, as depicted in Fig. 2, are presented in Sect. 3. Section 4 discusses a model-based method to compensate discrete contact state errors and to stabilize a faulty grasp which robustifies object manipulation against model uncertainties. Results from dynamical simulations and experiments are presented in Sect. 5.

2 General Hybrid Modeling of Manipulation

The dynamic behavior of multi-fingered manipulation systems shows a hybrid discrete-continuous character. In the following section a comprehensive model for such systems is developped which allows for a comprehensive analytical description of manipulation dynamics. The model provides the basis of a strategy for compensation of discrete contact state errors which is discussed in detail in Sect. 4.

2.1 Basics of the HSM Formalism

An analytical formulation of the multi-fingered hand's discrete-continuous dynamics is given in the framework of the Hybrid State Model – HSM. The HSM builds a formal basis to modeling of a general class of hybrid dynamical systems – HDS. More details about the HSM are given in the companion article (Buss et al., 2002); a comparison with and survey of other hybrid modeling paradigms can be found in (Buss, 2000). The general structure of a system in HSM notation is illustrated in Fig. 3.

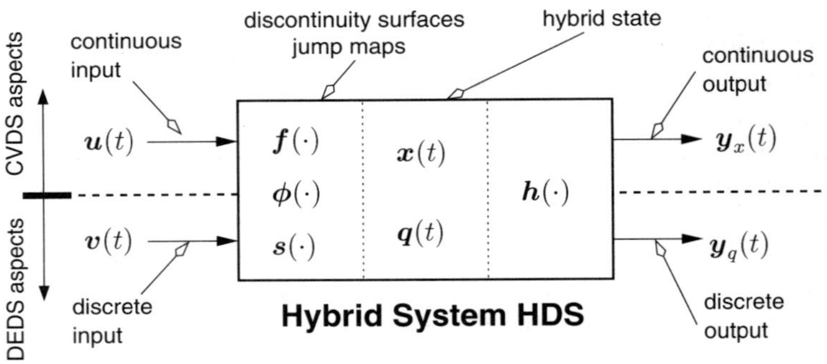

Fig. 3. Structure of a general hybrid system based on the Hybrid State Model

The HSM is defined as follows

$$\dot{\boldsymbol{x}} = \boldsymbol{f}(\boldsymbol{x}, \boldsymbol{u}, \boldsymbol{q}, t) \qquad \text{if} \quad s_i(\boldsymbol{x}, \boldsymbol{u}, \boldsymbol{q}, \boldsymbol{v}, t) \neq 0 \tag{1}$$

$$\left[\boldsymbol{x}(t^+), \boldsymbol{q}(t^+)\right] = \boldsymbol{\phi}_i(\boldsymbol{x}, \boldsymbol{u}, \boldsymbol{q}, \boldsymbol{v}, t) \qquad \text{if} \quad s_i(\boldsymbol{x}, \boldsymbol{u}, \boldsymbol{q}, \boldsymbol{v}, t) = 0 \tag{2}$$

$$\boldsymbol{y} = \boldsymbol{h}(\boldsymbol{x}, \boldsymbol{u}, \boldsymbol{q}, \boldsymbol{v}, t) \, , \tag{3}$$

where (1) describes the continuous dynamics, (2) the discrete dynamics and (3) is the output equation. The continuous state vector $\boldsymbol{x}(t) \in \mathcal{X} \subset \mathbb{R}^n$ and the discrete state vector $\boldsymbol{q}(t) \in \mathcal{Q} \subset \mathbb{N}^l$ form the *hybrid state* $\boldsymbol{\zeta}(t) = [\boldsymbol{x}(t), \boldsymbol{q}(t)] \in \mathcal{X} \times \mathcal{Q}$ of the HDS. The *continuous* control input vector is $\boldsymbol{u}(t) \in \mathcal{U} \subset \mathbb{R}^m$ and the set $\mathcal{U}$ denotes all permissible controls. The vector $\boldsymbol{v}(t) \in \mathcal{V} \subset \mathbb{N}^k$ is the *discrete* (symbolic) control input. Output of the system is the hybrid output vector $\boldsymbol{y}(t) = [\boldsymbol{y}_x(t), \boldsymbol{y}_q(t)] \in \mathcal{Y} \subset \mathbb{R}^p \times \mathbb{N}^r$ comprising a p-dimensional continuous output $\boldsymbol{y}_x(t)$ and a r-dimensional discrete output $\boldsymbol{y}_q(t)$; the hybrid output $\boldsymbol{y}(t)$ is generated by the output function

$$\boldsymbol{h} : \mathcal{X} \times \mathcal{U} \times \mathcal{Q} \times \mathcal{V} \times \mathbb{R} \to \mathbb{R}^p \times \mathbb{N}^r \,. \tag{4}$$

The continuous behavior of the HDS is defined by the vector field

$$\boldsymbol{f} : \mathcal{X} \times \mathcal{U} \times \mathcal{Q} \times \mathbb{R} \to \mathbb{R}^n \tag{5}$$

while the maps

$$s_i : \mathcal{X} \times \mathcal{U} \times \mathcal{Q} \times \mathcal{V} \times \mathbb{R} \to \mathbb{R} \,, \tag{6}$$

describe discontinuity surfaces in hybrid state space depending also on the continuous and discrete controls $\boldsymbol{u}(t)$, $\boldsymbol{v}(t)$. When the system trajectory intersects a discontinuity surface at $t = t_1$, i.e. $s_i(\boldsymbol{x}(t_1), \boldsymbol{u}(t_1), \boldsymbol{q}(t_1), \boldsymbol{v}(t_1), t_1) = 0$, the hybrid state $\boldsymbol{\zeta}(t_1)$ may instantaneously jump to the new state $\boldsymbol{\zeta}(t_1^+) = [\boldsymbol{x}(t_1^-), \boldsymbol{q}(t_1^+)]$ given by the (jump) map

$$\boldsymbol{\phi}_i : \mathcal{X} \times \mathcal{U} \times \mathcal{Q} \times \mathcal{V} \times \mathbb{R} \to \mathcal{X} \times \mathcal{Q} \,. \tag{7}$$

In the case that all $s_i(\boldsymbol{x}, \boldsymbol{u}, \boldsymbol{q}, \boldsymbol{v}, t_1) \neq 0$, the (continuous) system trajectory evolves according to (1).

Modeling of the continuous (1) and discrete aspects (2) of spatial multi-fingered manipulation using the HSM approach will be discussed next.

A typical manipulation system consists of a four- or five-fingered hand together with a grasped object, see Fig. 4. Links of the fingers and the object are assumed to be rigid bodies and the finger joints to be free of static friction. Fingertips and grasped object interact by point contacts with friction. As actuator dynamics are not a dominant factor in the control problem discussed here, they are not considered for simplicity.

2.2 Continuous Dynamic Aspects

The robotic hand is considered as a compound of N independently acting fingers. By applying the Euler-Lagrange method the equations of motion for a hand with freely moving fingers follow as

$$\underbrace{\begin{bmatrix} \boldsymbol{M}_{f_1} & & \boldsymbol{0} \\ & \ddots & \\ \boldsymbol{0} & & \boldsymbol{M}_{f_N} \end{bmatrix}}_{\boldsymbol{M}_h(\boldsymbol{\theta})} \ddot{\boldsymbol{\theta}} + \underbrace{\begin{bmatrix} \boldsymbol{N}_{f_1} \\ \vdots \\ \boldsymbol{N}_{f_N} \end{bmatrix}}_{\boldsymbol{N}_h(\boldsymbol{\theta}, \dot{\boldsymbol{\theta}})} = \underbrace{\begin{bmatrix} \boldsymbol{\tau}_1 \\ \vdots \\ \boldsymbol{\tau}_N \end{bmatrix}}_{\boldsymbol{\tau}} \,. \tag{8}$$

with $\boldsymbol{\theta}_i$ as joint angles, $\boldsymbol{\tau}_i$ as joint torques, $\boldsymbol{M}_{f_i}$ as inertia matrix and $\boldsymbol{N}_{f_i}$ as sum of gravitational, centripetal and Coriolis forces for finger i.

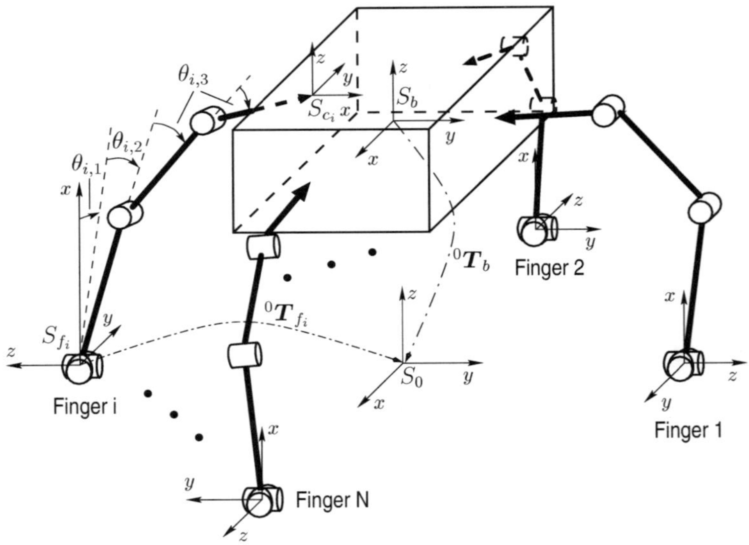

Fig. 4. Example of a multi-fingered manipulation system

The equations of motion of a body moving in free space can be formulated in local coordinates $\boldsymbol{p} = [{}^0\boldsymbol{r}_b,\ \boldsymbol{\psi}]$ as

$$\boldsymbol{M}_b(\boldsymbol{p})\,\ddot{\boldsymbol{p}} + \boldsymbol{N}_b(\boldsymbol{p}, \dot{\boldsymbol{p}}) = {}^b\boldsymbol{f} \tag{9}$$

with ${}^0\boldsymbol{r}_b$ as the position of the body frame origin and $\boldsymbol{\psi} = [\alpha, \beta, \gamma]$ as its RPY angles with respect to the $x-$ (α), $y-$ (β), and $z-$axis (γ) at the hand palm frame S_0; $\boldsymbol{M}_b$ denotes the mass matrix of the object and $\boldsymbol{N}_b$ the nonlinear force vector acting on it.

2.3 Discrete Event Dynamics

The actual contact situation between fingertips and grasped object is reflected in the *discrete grasp state* $\boldsymbol{q}$, with $q_i \in \{1, 2, 4\}$, $i = 1, \ldots N$, $\mathcal{Q} = \{1, 2, 4\}^N$. For each finger i we define

$$q_i = \begin{cases} 1, & \text{if finger } i \text{ contacts stably} \\ 2, & \text{if finger } i \text{ does not contact} \\ 4, & \text{if finger } i \text{ slides on object .} \end{cases} \tag{10}$$

The hybrid contact dynamics for one finger can be represented as a hybrid automaton (Alur et al., 1993), see Fig. 5. Events triggering a transition from one contact

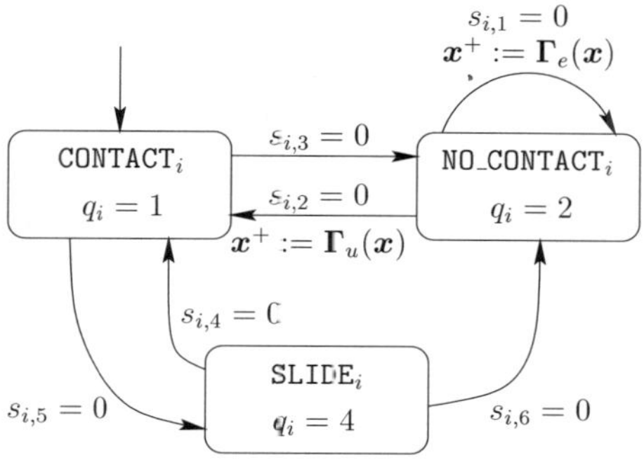

Fig. 5. Hybrid contact dynamics of finger i.

state of finger i to another are modeled by the following discontinuity surfaces

$$\begin{aligned} s_{i,1} &= 1 + \frac{1}{2}(q_i - 1)(q_i - 4) + \tilde{s}_i(\boldsymbol{x}_i^{tip}) \\ &\quad + 1 - \sigma((\{\dot{\boldsymbol{x}}_i^{tip} - \boldsymbol{v}_{o,i}\}_\perp)^2 - \epsilon^2) \end{aligned} \tag{11}$$

$$\begin{aligned} s_{i,2} &= 1 + \frac{1}{2}(q_i - 1)(q_i - 4) + \tilde{s}_i(\boldsymbol{x}_i^{tip}) \\ &\quad + \sigma((\{\dot{\boldsymbol{x}}_i^{tip} - \boldsymbol{v}_{o,i}\}_\perp)^2 - \epsilon^2) \end{aligned} \tag{12}$$

$$s_{i,3} = 1 - \frac{1}{3}(q_i - 2)(q_i - 4) + c_{i,1} \tag{13}$$

$$s_{i,4} = 1 - \frac{1}{6}(q_i - 1)(q_i - 2) + (\{\dot{\boldsymbol{x}}_i^{tip} - \boldsymbol{v}_{o,i}\}_\parallel)^2 \tag{14}$$

$$s_{i,5} = 1 - \frac{1}{3}(q_i - 2)(q_i - 4) + c_{i,1} - \frac{1}{\mu_i}\sqrt{c_{i,2}^2 + c_{i,3}^2} \tag{15}$$

$$s_{i,6} = 1 - \frac{1}{6}(q_i - 1)(q_i - 2) + c_{i,1} \tag{16}$$

where $0 < \mu_i < 1$ denotes the static friction coefficient, $\boldsymbol{c}_{i,1}$ the normal component of the contact force at the contact point of finger i, $\{\cdot\}_\perp$ and $\{\cdot\}_\parallel$ the normal and tangential component of the argument vector with respect to the object surface $\tilde{s}_i(\cdot) = 0$; $\sigma(\cdot)$ is the unit step function and $\epsilon > 0$ a small threshold value for the relative velocity between object and fingertip. The fingertip position $\boldsymbol{x}_i^{tip}$ can be calculated from the joint angles $\boldsymbol{\theta}_i$ using the forward kinematics map of finger i. Although the contact force $\boldsymbol{c}_i$ is not a system state from a dynamics viewpoint, it can be calculated from state variables of the system as shown in the next section. The discontinuity surfaces depend on both the discrete state q_i and the continuous state variables. For example, $s_{i,3}$ equals zero, iff $q_i = 1$ and $c_{i,1} = 0$, otherwise $s_{i,3} > 0$. If any

$s_{i,1..6} = 0$ the respective jump map

$$\phi_{i,1} = \begin{pmatrix} \boldsymbol{\Gamma}_{\boldsymbol{e}}(\boldsymbol{x}) \\ \boldsymbol{q} \end{pmatrix} \quad , \phi_{i,2} = \begin{pmatrix} \boldsymbol{\Gamma}_{\boldsymbol{u}}(\boldsymbol{x}) \\ (q_1..1..q_N)^T \end{pmatrix}, \phi_{i,3} = \begin{pmatrix} \boldsymbol{x} \\ (q_1..2..q_N)^T \end{pmatrix},$$
$$\phi_{i,4} = \begin{pmatrix} \boldsymbol{x} \\ (q_1..1..q_N)^T \end{pmatrix}, \phi_{i,5} = \begin{pmatrix} \boldsymbol{x} \\ (q_1..4..q_N)^T \end{pmatrix}, \phi_{i,6} = \begin{pmatrix} \boldsymbol{x} \\ (q_1..2..q_N)^T \end{pmatrix} \tag{17}$$

defines the new hybrid state

$$\boldsymbol{\zeta}^+ = [\boldsymbol{x}^{tip}, \boldsymbol{p}, (\dot{\boldsymbol{x}}^{tip})^+, .., \dot{\boldsymbol{p}}^+, q_1, .., q_i^+, .., q_n] ,$$

i.e. the contact state and the velocity of finger i as well as the body velocity are set to new values by $\phi_{i,1}, \phi_{i,2}$. In case of $\phi_{i,j}, j = 3, \ldots, 6$, the continuous state will not jump, i.e. $\boldsymbol{x}^+ = \boldsymbol{x}^-$. The maps $\boldsymbol{\Gamma}_e$ and $\boldsymbol{\Gamma}_u$ in ϕ_1 and ϕ_2 denote resets of the finger and object velocities in case of elastic and plastic impacts. The elastic impact can be calculated by use of Poisson's restitution equations (Glocker, 1995) depending on the restitution coefficient $0 < \rho < 1$. The computation of 3D impacts in real-time was discussed in (Ruspini and Khatib, 2000).

2.4 Overall Hybrid Model

Up to this point the formulated continuous dynamics of hand and object do not consider the possible contacts between fingertips and the surface of the object. Depending on the current contact state time-varying constraints on the velocity of the respective fingertip and the contact point on the grasped object have to be included into the equations of motion, see (Schlegl et al., 2002a) for further details.

The differential-algebraic system (8),(9) subject to contact constraints has to be transformed into a set of ordinary differential equations within the HSM. Therefore, the kinematic velocity constraints for the different contact points are formulated as

$$\boldsymbol{A}(\boldsymbol{q}) \left[\dot{\boldsymbol{\theta}}_1^T \ldots \dot{\boldsymbol{\theta}}_N^T \, \dot{\boldsymbol{p}}^T\right] = \boldsymbol{0} \, . \tag{18}$$

where $\boldsymbol{A}(\boldsymbol{q})$ denotes a matrix of coefficients which features an explicit multilinear dependency on the discrete contact state $\boldsymbol{q}$.

The constraints (18) are included into the continuous dynamics of the manipulation system by means of the Langrange-multiplier method. The result is the ordinary differential equation system

$$\underbrace{\begin{bmatrix} \boldsymbol{M}_h & \boldsymbol{0} \\ \boldsymbol{0} & \boldsymbol{M}_b \end{bmatrix}}_{\boldsymbol{M}} \begin{bmatrix} \ddot{\boldsymbol{\theta}} \\ \ddot{\boldsymbol{p}} \end{bmatrix} + \underbrace{\begin{bmatrix} \boldsymbol{N}_h \\ \boldsymbol{N}_b \end{bmatrix}}_{\boldsymbol{N}} + \boldsymbol{A}^T \boldsymbol{c} = \underbrace{\begin{bmatrix} \boldsymbol{\tau} \\ {}^b\boldsymbol{f} \end{bmatrix}}_{\boldsymbol{F}} \tag{19}$$

which expresses the equations of motion taking into account the discrete contact state $\boldsymbol{q}$. The term $\boldsymbol{A}^T\boldsymbol{c}$ generates forces which counteract the violation of velocity constraints (18). Hence, the Lagrange multipliers $\boldsymbol{c}$ set up contact forces which depend on state variables of the manipulation system. They follow to

$$\boldsymbol{c} = \left[\boldsymbol{A}\boldsymbol{M}^{-1}\boldsymbol{A}^T\right]^{-1}\left[\dot{\boldsymbol{A}}\begin{pmatrix}\dot{\boldsymbol{\theta}}\\ \dot{\boldsymbol{p}}\end{pmatrix} + \boldsymbol{A}\boldsymbol{M}^{-1}(\boldsymbol{F}-\boldsymbol{N})\right] .$$

Based on these considerations a hybrid state model of a multi-fingered manipulation system using HSM-notation can be formulated. Joint angles of all fingers, the local coordinates of an object along with the corresponding velocities form the continuous system state $\boldsymbol{x}$. The contact states of the fingertips build the discrete system state $\boldsymbol{q}$ and the finger joint torques are the system input, i.e. $\boldsymbol{u} = \boldsymbol{\tau}$. Performing some calculations on (19) yields

$$\begin{aligned}\cdot\; \dot{\boldsymbol{x}} &= \boldsymbol{f}(\boldsymbol{x},\boldsymbol{q},\boldsymbol{u}) = \begin{bmatrix}\dot{\boldsymbol{\theta}}\\ \dot{\boldsymbol{p}}\\ \boldsymbol{M}^{-1}\left(\boldsymbol{F}-\boldsymbol{N}-\boldsymbol{A}^T\boldsymbol{c}\right)\end{bmatrix}\\ \cdot\; [\boldsymbol{x}^+,\boldsymbol{q}^+] &= \boldsymbol{\phi}_{i,j}(\boldsymbol{x},\boldsymbol{u},\boldsymbol{q}) \quad \text{if} \quad s_{i,j}(\boldsymbol{x},\boldsymbol{q}) = 0\\ &\quad \text{with} \quad i = 1,\ldots,N\,, \quad j = 1,\ldots,6\end{aligned} \tag{20}$$

as hybrid system dynamics, where the variable structure piecewise continuous vector field $\boldsymbol{f}(\cdot)$ depends on the discrete state $\boldsymbol{q}$ via $\boldsymbol{A}(\boldsymbol{q})$. Assuming joint angles, angular velocities and discrete contact states to be measurable defines the hybrid output function of the system as $[\boldsymbol{y}_x,\ \boldsymbol{y}_q] = [\boldsymbol{x},\ \dot{\boldsymbol{x}},\ \boldsymbol{q}]$.

3 Discrete-Continuous Manipulation Control

The basic structure of the hybrid controller for multi-fingered grasping and regrasping has already been presented in Sect. 1. In this section the hybrid reference generation HRG, the hybrid controller HC, and the decentralized impedance control DIC are discussed. Furthermore, the beneficial usage of contact force information delivered by 6D intrinsic tactile sensors in the hybrid control loop is shown. An alternative centralized impedance control concept CIC and the real-time grasping force optimization algorithm GFO will *not* be presented in this article; for details on these topics see (Schlegl, 2002, Buss et al., 1996).

3.1 Reference Planning

After having established a model of the discrete-continuous behavior of dextrous grasping with variable contacts the problem of reference planning is addressed next. For this purpose a multi-fingered grasp configuration as shown in Fig. 4 is considered. One possibility to transfer a given grasp into a desired new one is by lifting off, moving and recontacting fingers. However, the transitions of a finger from stable contact to moving in free space and vice versa mean structural changes both in the mechanical system and in the grasp controller. Switching from an N-fingered grasp to a reduced

$(N-1)$-fingered grasp will cause non-smooth force trajectories which may degrade the performance of the manipulation system. This may, even if grasping forces are adapted by online grasping force optimization (GFO), lead to a complete loss of grasp stability. Therefore, it is necessary to *prepare* the system for the structural changes connected to regrasping. The basic regrasping operation of lifting off, moving, and recontacting *one finger* is planned as a sequence of four phases which are illustrated in Fig. 6(a):

1. `REDUCE` contact force;
2. `MOVE` the finger to the new contact point;
3. `INCREASE` contact force;
4. reenter a `STABLE` grasp configuration.

At the beginning of the regrasping task all fingers are assumed to be in stable contact with the grasped object. Consequently, the discrete state of the manipulation system is given by $\boldsymbol{q} = [1 \ \ldots \ 1]^T$ according to (10). In the initial state the finger to regrasp next imposes a contact force $\boldsymbol{c}_i^*$ onto the object. Before lifting the finger off the object its contact force is to be decreased over time and distributed to other fingers remaining in contact in the `REDUCE`–phase. At the instant when the regrasping finger starts moving its contact force shall almost be zero. On the other hand, as soon as the moving finger contacts the object again, its contact force is smoothly increased to a new final value $\boldsymbol{c}_i^{**}$.

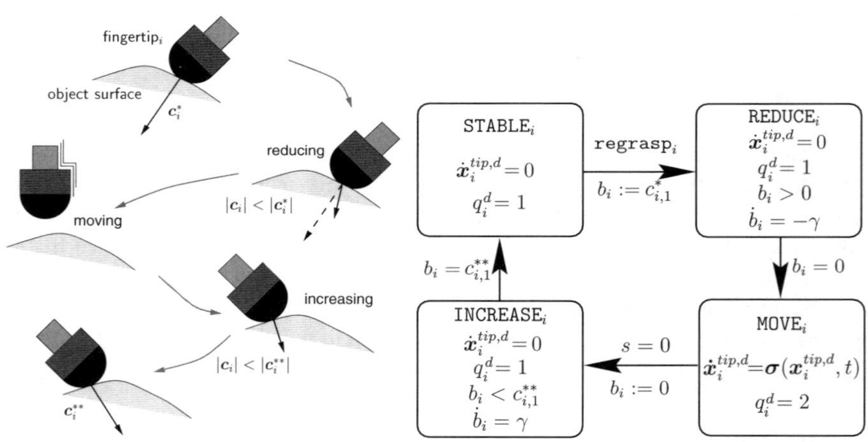

(a) Finger repositioning (b) Hybrid reference planning

Fig. 6. Planning of a single regrasping task

To each phase a corresponding hybrid grasp state is associated, as visualized in the hybrid automaton shown in Fig. 6(b). In this framework the formal reference specification for regrasping is disussed now. In the `STABLE`-state the condition $\dot{\boldsymbol{x}}_i^{tip,d} = \boldsymbol{0}$ commands a constant contact point position of finger i with respect to the object surface. Finger i is in stable contact with the object and is supposed to regrasp next. The corresponding desired discrete contact state is $q_i^d = 1$. Regrasping is initiated by the synchronization signal `regrasp` which triggers the transition of the HRG from the `STABLE` into the `REDUCE`-state. The current value of the finger normal grasp force component is stored in the variable b_i. The desired discrete contact state remains $q_i^d = 1$. Now the regrasping finger is to be unloaded to avoid non-smooth forces during its lift-off. This can be achieved by choosing

$$t \to t_{off} : \; |\boldsymbol{c}_i(t)| \to 0 \, , \quad \boldsymbol{c} \in \mathcal{C}^0$$

where $\mathcal{C}^0$ represents the class of smooth functions. A smooth reduction of the contact force is achieved by reducing the normal component $b_i(t)$ over time starting with the initial value $c^*_{i,1}$ at a rate $\gamma > 0$. Tangential contact force components are not considered, because the GFO as a core module of the control architecture implicitly ensures that tangential forces are reduced, too, while satisfying friction limits and balancing external forces acting on the grasped object. At the end of the `REDUCE`-phase the finger still contacts the object but imposes nearly no contact force onto it, i.e. $b_i = 0$.

After finger i is unloaded it lifts off without disturbing the grasp of the $(N-1)$ fingers remaining in contact with the object. Its discrete contact state changes to $q_i = 2$ and its tip moves along the specified trajectory $\dot{\boldsymbol{x}}_i^{tip,d} = \dot{\boldsymbol{x}}_i^d = \boldsymbol{\sigma}(\boldsymbol{x}_i^{tip,d}, t)$ within time T into the `MOVE`-state. This trajectory is calculated heuristically using standard interpolation methods.

The transition to the following `INCREASE`-phase is triggered by a condition $s = 0$. This signals the reestablishment of contact of the regrasping finger and can be detected by use of a contact sensor or, alternatively, by evaluation of the discontinuity surface $s_{i,2}$ (11) which is part of the hybrid model of the manipulation system.

Then, the finger is reintegrated into the grasp during the `INCREASE`-state over time as

$$t = t_{on} : \; |\boldsymbol{c}_i(t)| = 0 \, , \qquad t \to t_e : \; c_{i,1}(t) \to c^{**}_{i,1} \, ,$$

where the final contact force c^{**} is calculated on the basis of the new grasp situation and the current external object force. At the end of the contact force increase, which is indicated by $b_i = c^{**}_{i,1}$, the previously regrasping finger is fully integrated in the grasp again. The `INCREASE`-state is followed again by the `STABLE`-state, i.e. the regrasping task is completed.

3.2 Sensor-Based Closed-Loop Impedance Control

Multi-fingered manipulation problems usually require a sequence of regrasping operations as discussed above. The hybrid control architecture which was outlined in the Introduction enables a multi-fingered hand to follow specified hybrid reference trajectories and, consequently, to perform complex manipulation tasks. In the following, we refer to the scheme of the controller shown in Fig. 2 and concentrate on

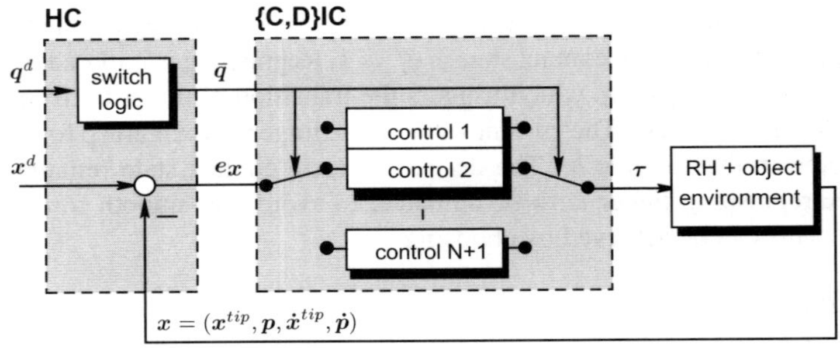

Fig. 7. Switching of impedance control laws {C,D}IC by the hybrid controller HC

the function of the hybrid control component HC and the continuous control laws {C,D}IC, see Fig. 7.

During a manipulation task the discrete grasp state undergoes frequent changes. A switching logic in the HC calculates a switching variable $\bar{q}$ from the discrete system state $\boldsymbol{q}$. This variable controls the selection of *one* continuous control law out of a set of $N+1$ different control laws. The blocks control 1 to control N+1 denote an individual impedance control law associated with a special mode of the GFO matching the current contact state in the system. The total number of $N+1$ control laws results from N controllers for grasps which have one finger lifted off plus 1 additional controller for a grasp where all fingers are contacting the object. Free moving fingers are trajectory controlled in case of the centralized impedance controller CIC. In addition to the discrete switching variable the continuous control error $\boldsymbol{e}_x = \boldsymbol{x}^d - \boldsymbol{x}$ is forwarded to the currently active impedance control law.

To achieve the desired behavior during regrasping as specified in Sect. 3.1 the HC commands different desired discrete contact states $\boldsymbol{q}^d$ of the fingers to the GFO. Each finger can be assumed to have either stable contact with the surface of the grasped object or to be lifted off and moving in free space. If the desired contact state of a finger changes the GFO switches to a transition mode of duration $t_{r/i}$. In case the respective finger is to be lifted off next, its contact force will be reduced to zero and optimally distributed to fingers remaining in contact during this transition phase. The GFO is switched to a reduced grasp mode where only the still contacting fingers are considered during force optimization. If the reference generator commands the recontaction of a moving finger the GFO is switched to another transition mode of duration $t_{r/i}$ during which the formerly moving finger is loaded and smoothly reintegrated in the grasp. When this process is completed, a full grasp mode is activated. Currently, modes for four and three contating fingers and for the transition $4 \leftrightarrow 3$ fingers are implemented.

The decentralized impedance control loops of the fingers are not directly affected by changes in the desired contact state. Rather, in case a certain finger is selected to move in free space a zero contact force is commanded to the impedance model of the respective controller.

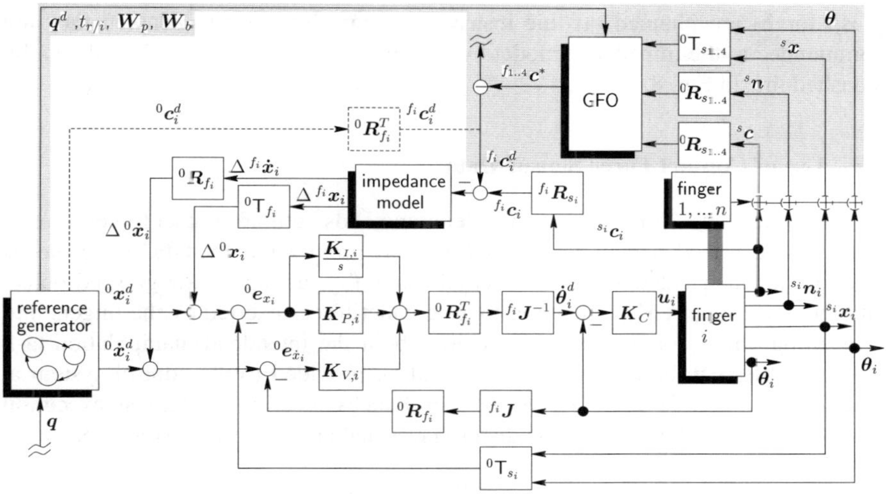

Fig. 8. Fingertip impedance control for dextrous manipulation

In the DIC case each finger is controlled by an impedance controller individually. Figure 8 shows this control loop for *one finger*. Its basis is formed by a trajectory controller for the position of the ith fingertip

$$\dot{\boldsymbol{\theta}}_i^d = ({}^{f_i}\boldsymbol{J}^{-1})\,{}^0\boldsymbol{R}_{f_i}^T \left\{ \boldsymbol{K}_{I,i} \int {}^0\boldsymbol{e}_{x_i} dt + \boldsymbol{K}_{P,i}\,{}^0\boldsymbol{e}_{x_i} + \boldsymbol{K}_{V,i}\,{}^0\boldsymbol{e}_{\dot{x}_i} \right\} \tag{21}$$

where ${}^{f_i}\boldsymbol{J}^{-1}$ is the inverse Jacobian of the fingertip position with respect to a finger base frame and ${}^0\boldsymbol{R}_{f_i}$ is the rotation matrix from finger base to hand palm coordinates; $\dot{\boldsymbol{\theta}}_i^d$ is the desired joint velocity commanded to an analog finger drive control loop. A linear impedance model

$$\boldsymbol{M}_i\,\Delta^{f_i}\ddot{\boldsymbol{x}}(t) + \boldsymbol{D}_i\,\Delta^{f_i}\dot{\boldsymbol{x}}(t) + \boldsymbol{K}_i\,\Delta^{f_i}\boldsymbol{x}(t) = {}^{f_i}\boldsymbol{c}_i - {}^{f_i}\boldsymbol{c}_i^d \tag{22}$$

with $\boldsymbol{M}_i$, $\boldsymbol{D}_i$, and $\boldsymbol{K}_i$ as mass, damping, and spring parameters generates position and velocity offsets $\Delta^{f_i}\boldsymbol{x}$ and $\Delta^{f_i}\dot{\boldsymbol{x}}$. These are used in addition to adapt the desired values ${}^0\boldsymbol{x}_i^d$ and ${}^0\dot{\boldsymbol{x}}_i^d$ of the fingertip position and velocity according to the difference between the desired and actual value of the contact force. Thus, a compliant behavior of the finger around the desired tip position ${}^0\boldsymbol{x}_i^d$ and velocity ${}^0\dot{\boldsymbol{x}}_i^d$ as a reaction to forces sensed by the force sensor is achieved.

For optimization of grasping forces measurements for all fingers are concatenated (denoted by $\oplus$ in Fig. 8) and mapped from sensor coordinates $S_{s_{1..4}}$ into palm coordinates S_0 using the forward kinematics maps ${}^0\mathsf{T}_{s_{1..4}}$ and rotation matrices ${}^0\boldsymbol{R}_{s_{1..4}}$ of the corresponding finger. The optimized forces ${}^{f_{1..4}}\boldsymbol{c}^*$ are then separated into subvectors containing the contact forces of the different fingers which is denoted by $\ominus$ in Fig. 8. They are used as desired contact forces whose difference to the actual contact forces builds the input of the impedance model described above. If desired grasp forces are planned off-line instead of being dynamically adapted they are commanded to the impedance model by the reference generator which is shown by a dashed line in Fig. 8.

3.3 Use of Contact Force Sensor Data

As mentioned before the hybrid controller commands optimal contact forces as reference into its impedance model. These forces are provided by the GFO which needs to know for their calculation the external force $\boldsymbol{f}_{ext}$ acting on the grasped object and the grasp matrix $\boldsymbol{W}$. The latter maps contact forces exerted by the fingers into a resulting object force. With information about the intendend manipulation task both variables can be calulated off-line and be provided to the control system at run-time by a task database. However, disturbances occuring during task execution may disturb this scheme. Possible disturbances and affected grasp parameters are

- unknown forces acting on the object ($\leadsto \boldsymbol{f}_{ext}$);
- imprecisely modeled object geometry ($\leadsto \boldsymbol{W}$);
- need for regrasping or rolling of fingertips ($\leadsto \boldsymbol{f}_{ext}$, $\boldsymbol{W}$).

On-line calculation of the necessary parameters from known sensor data is, in general, superior to off-line planning and increases flexibilty of the control architecture.

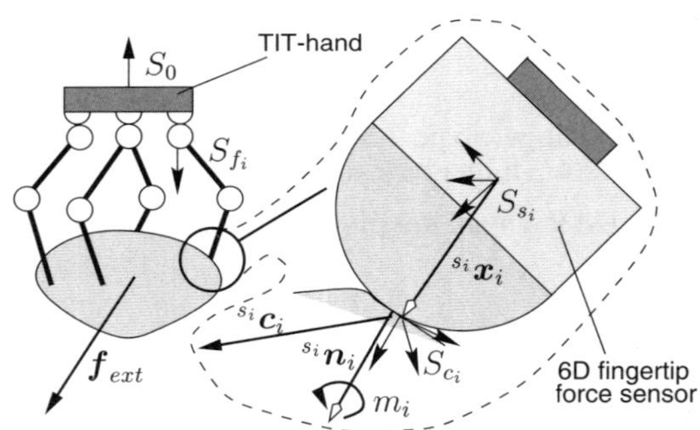

Fig. 9. Contact force sensor information

A 6D contact force sensor at each fingertip of the TIT-hand (Omata and Farooqi, 1996), see Fig. 9, provides contact point locations ${}^{s_i}\boldsymbol{x}_i$, contact forces ${}^{s_i}\boldsymbol{c}_i$, the contact normal vectors ${}^{s_i}\boldsymbol{n}_i$ and the moment m_i around ${}^{s_i}\boldsymbol{n}_i$, where ${}^{s_i}\bullet$ denotes the quantity $\bullet$ to be specified with respect to the sensor coordinate system S_{s_i}. As a first step all measurements are related to a common coordinate system, in this case the hand palm frame S_0. The external force acting on the object then follows as

$$ {}^{0}\boldsymbol{f}_{ext} = \begin{bmatrix} \sum_{i=1}^{n} {}^{0}\boldsymbol{c}_i \\ \sum_{i=1}^{n} \left({}^{0}\boldsymbol{x}_i \times {}^{0}\boldsymbol{c}_i + {}^{0}\boldsymbol{n}_i\, m_i \right) \end{bmatrix} \tag{23} $$

and the grasp matrix as

$$W = \begin{bmatrix} I_3 & \cdots & I_3 \\ {}^0\widehat{x}_1 & \cdots & {}^0\widehat{x}_n \end{bmatrix} \operatorname{diag}({}^0R_{c_i}({}^0n_i)) \tag{24}$$

with $\widehat{\bullet}$ as the skew-symmetric of $\bullet$ and $\widehat{a}b \equiv a \times b$, $a, b \in \mathbb{R}^3$.

As W and f_{ext} are calculated at run-time rolling motion of the contact points is compensated as well as a change of contact points due to regrasping.

4 Discrete Error Compensation

By application of the hybrid control architecture presented in the previous section a multi-fingered hand is capable of fixing objects and manipulating them by finger regrasping. External forces acting on the grasped object can robustly be compensated by the proposed combination of impedance control laws together with a grasping force optimization algorithm. An imprecise geometric model of the manipulation system, however, may cause deviations between the actual discrete contact state of the system and the desired reference. In the following, a novel strategy for compensation of discrete control errors is presented along with its implementation in the hybrid manipulation control structure, as shown in Fig. 2.

4.1 Model-Based Compensation of Discrete Contact State Errors

A discrete error arises when a discrete state transition occurs at a time sooner or later than expected, according to the desired discrete reference trajectory, see Fig. 10(a). In the latter case, a trigger surface $s_{i,1/2}$ is not hit at the desired time and, as a consequence, the corresponding state transition does occur.

In such a case the discrete states will differ from the desired ones, i.e. $q^d \neq q$. The comparison operation between desired and actual discrete state is denoted as

$$e_q = q^d \uplus q, \quad \text{with } \uplus : \mathbb{N}^N \times \mathbb{N}^N \to \mathbb{R}^k \tag{25}$$

where k is the dimension of the possibly continuously weighted deviation between q^d and q. Several possibilities exist for defining the operator $\uplus$, see (Buss, 2000). In

Table 1. Possible discrete error values at at the ith contact point

error $e_{q,i}$	interpretation
0	no error
1	finger shall move freely, but contacts the object stably
-1	finger shall contact the object stably, but moves freely
2*	finger shall slide on the object surface, but moves freely
-2	finger shall move freely, but slides on the object surface
3*	finger shall slide on the object surface, but contacts stably
-3	finger shall contact the object stably, but slides on its surface

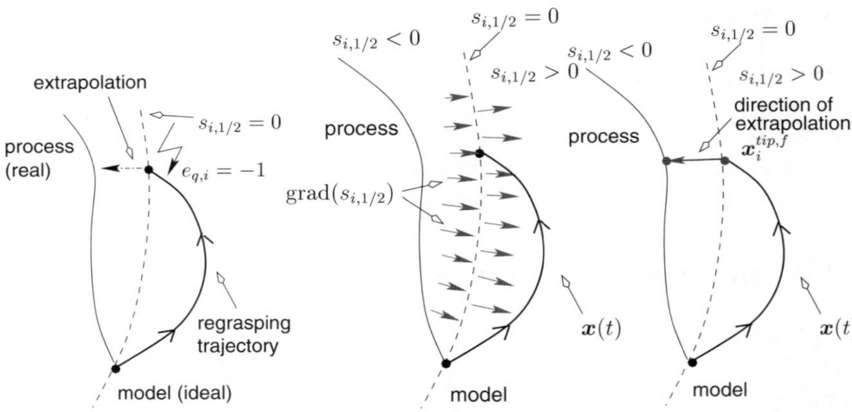

(a) Discrete error (b) Gradient field (c) Extrapolation

Fig. 10. Model-based compensation of discrete control errors

the following, ⊎ is interpreted as an arithmetic operator that calculates the discrete error $\boldsymbol{e}_q$ as the arithmetic integer difference between the desired and the actual output vector, resulting in an integer valued discrete error

$$\boldsymbol{e}_q = \boldsymbol{q}^d - \boldsymbol{q}, \quad \text{with} \uplus : \mathbb{N}^N \times \mathbb{N}^N \to \mathbb{Z}^N \,. \tag{26}$$

Based on definition (10) of the discrete contact state the type of discrete deviation can be identified uniquely by application of the simple arithmetic difference (26). Table 1 shows all possible cases for each component i $(1 < i < N)$ of the discrete error $\boldsymbol{e}_q$. The special values $e_{q,i} = 2$ and $e_{q,i} = 3$ are marked with $*$ to denote that they are not going to be discussed in the following. In this article robotic regrasping is performed by lifting off and recontacting fingers and not by desired sliding of fingers. Hence, the desired discrete state $q_i = 4$ is never commanded to the manipulation systems so that the marked error values cannot occur. The effects of the remaining error values are briefly discussed next.

- $e_{q,i} = 1$ means an undesired contact of a regrasping finger. Moving fingers are position controlled in the CIC case or impedance controlled in the DIC, see Sect. 3.2. A position controlled finger which has established contact will exert a large contact force. This force generates a major disturbance force onto the grasped object which can be compensated by the proposed controllers only within certain bounds. A loss of grasp stability may result.
- If $e_{q,i} = -2$, a finger slides on the grasped object, although it is supposed to move freely. A possibly big disturbance of the grasp as in case $e_{q,i} = 1$ is the consequence.
- $e_{q,i} = -3$ denotes undesired slip of a finger which should stably contact the object surface. This special error is prevented by the GFO as far as possible within the capabilites of the manipulation system.

- In the error case $e_{q,i} = -1$ a finger that is supposed to contact the object remains in free space without establishing contact. This situation occurs especially at the end of a regrasping motion, if the planned target contact point is not located on the object surface due to an imprecise model which was used for planning the regrasping trajectory, see Fig. 1(b). Of course, the grasp is not disturbed immediately, but during further regrasping tasks grasp stability may be lost at any time. Hence, before another finger is going to regrasp, the discrete error of the currently regrasping finger has to be compensated first to ensure a stable grasp even after lifting off another finger.

The last error case resembles the situation that the hybrid trajectory $[\boldsymbol{x}^{tip}, \boldsymbol{q}]$ does not intersect the discontinuity surface $s_{i,1/2}$, cf. (11) and (12), which would trigger the transition to a new contact state STABLE. The continuous trajectory reaches the specified target point which is located on the *modeled* discontinuity surface depicted by a dashed line in Fig. 10(a). However, the modeled surface is not identical with the *real* surface in the process also shown in the figure. A discrete error $e_{q,i} = -1$ results. In order to compensate this error the continuous trajectory $\boldsymbol{x}(t)$ has to be guided towards the real discontinuity surface of the process by extrapolation until the desired transition takes place. For this purpose, the direction in which the continuous trajectory has to evolve is approximated by the gradient field of the trigger surface $s_{i,j}$ of interest with respect to the continuous state variables of the system

$$\mathrm{grad}\big(s_{i,j}(\boldsymbol{x})\big) = \bigtriangledown(s_{i,j}) = \begin{pmatrix} \partial s_{i,j}/\partial x^{tip}_{1,x} \\ \vdots \\ \partial s_{i,j}/\partial x^{tip}_{N,z} \end{pmatrix} , \quad j = 1, 2 \tag{27}$$

which is depicted in Fig. 10(b). In order to determine the new desired trajectory, only the subspace of continuous state variables on which the respective trigger surface $s_{i,j}$ explicitly depends is regarded, i.e $\boldsymbol{x}_i^{tip}$. Then, on the assumption that the deviation between calculated and real gradient is bounded, a corrected desired trajectory resulting from

$$\dot{\boldsymbol{x}}_i^{tip,f} = -v\,\mathrm{sgn}(s_{i,j})\,\bigtriangledown_i\,(s_{i,j})\,, \quad v \in \mathbb{R}^+\,, \tag{28}$$

$$\boldsymbol{x}_i^{tip,f} = \boldsymbol{x}_i^{tip}(0) + \int_0^{t'} \dot{\boldsymbol{x}}_i^{tip,f}\, dt' \tag{29}$$

with $\bigtriangledown_i$ as projected gradient, t' as time index of the compensation process and parameter v sufficiently large, will guide the system towards s_i and trigger the desired state transition after some time, as shown in Fig. 10(c).

4.2 Extended Hybrid Control for Discrete Error Compensation

To integrate the principle of discrete error compensation into the hybrid control architecture depicted in Fig. 2 two approaches are proposed.

Prefilter: A straightforward implementation is given by prefiltering desired continuous values depending on the detected discrete error $\boldsymbol{e}_q$. As soon as an error $\boldsymbol{e}_q \neq \mathbf{0}$ is detected, the continuous reference values $\boldsymbol{x}^{tip,d}$ are modified using the extrapolation method presented in the previous subsection. The continuous trajectory is guided towards the real discontinuity surface in order to reach the desired discrete state. The discrete part of the reference trajectory and thus the mode of the continuous controller is left unchanged.

Reference Adaptation: Mere filtering of continuous references may not be sufficient to compensate discrete errors. A more general approach is the adaptation of the hybrid reference generator to compute new desired values for both, the continuous references $\boldsymbol{x}^{tip,d}$ and the discrete references $\boldsymbol{q}^d$. In contrast to the prefiltering method, here, an adequate continuous control law in combination with a corresponding mode of the GFO can be selected by the hybrid controller based on the new desired discrete state. Furthermore, the reference adaptation method allows to formulate sequences of discrete states for compensation of deviations. This is required in case of multi-fingered manipulation as will be shown in the following section.

4.3 Grasp Stabilization by Local Correction of Contact Points

The hybrid reference planning for regrasping and its representation as a hybrid automaton has been discussed in Sect. 3.1. In the nominal case the system evolves along the discrete state sequence $\texttt{STABLE}_i$, $\texttt{REDUCE}_i$, $\texttt{MOVE}_i$ and $\texttt{INCREASE}_i$. This is visualized in the upper part of the extended reference automaton depicted in Fig. 11. For reference adaptation additional states and transitions are introduced which branch off from the state $\texttt{MOVE}_i$. A state transition to the phase $\texttt{INCREASE}_i$ is triggered by detection of recontaction ($q_i = 1$ or $q_i = 4$). The reference planning of the regrasping trajectory is based on a duration T. If the recontaction of the regrasping finger takes place *before* reaching the target point at $t = T$, the system encounters the error $e_{q_i} = 1$, if a stable contact is established, or $e_{q_i} = -2$, if slip occurs. Both errors can be compensated in a straightforward manner, if the desired hybrid state of the regarded finger is set to the actual values, i.e $[\boldsymbol{x}_i^{tip,d}, q_i^d] = [\boldsymbol{x}_i^{tip}, 1]$, and the $\texttt{INCREASE}_i$-phase is initiated. If the error $e_{q,i} = -2$ occured, the reintegration of the finger into the optimal contact force distribution establishes a stable contact. If the regrasping finger cannot contact the object at the end of the regrasping motion ($t = T$), the discrete error will be $e_{q,i} = -1$. For its compensation a discrete sequence comprising the states $\texttt{M\_CONTACT}_i$, $\texttt{PARAMETERIZE}_i$ and $\texttt{STABILIZE}_i$ is initiated. The effect of the reference adaptation is visualized in the right part of Fig. 11. During $\texttt{M\_CONTACT}_i$ first the active contact search procedure of the regrasping finger as proposed in the previous section is activated. After contact could be established, surface parameters are estimated in phase $\texttt{PARAMETERIZE}_i$. Therefore, either tactile sensor information is evaluated or the surface normal vector at the contact point is calculated by repeatedly contacting the grasped object in the vicinity of the initial contact point after the contact search. Finally, in the $\texttt{STABILIZE}_i$-phase the stability of the grasp is evaluated. The solution of a scalar nonlinear optimization problem allows to determine the *unilateral* force closure of the grasp under consideration of

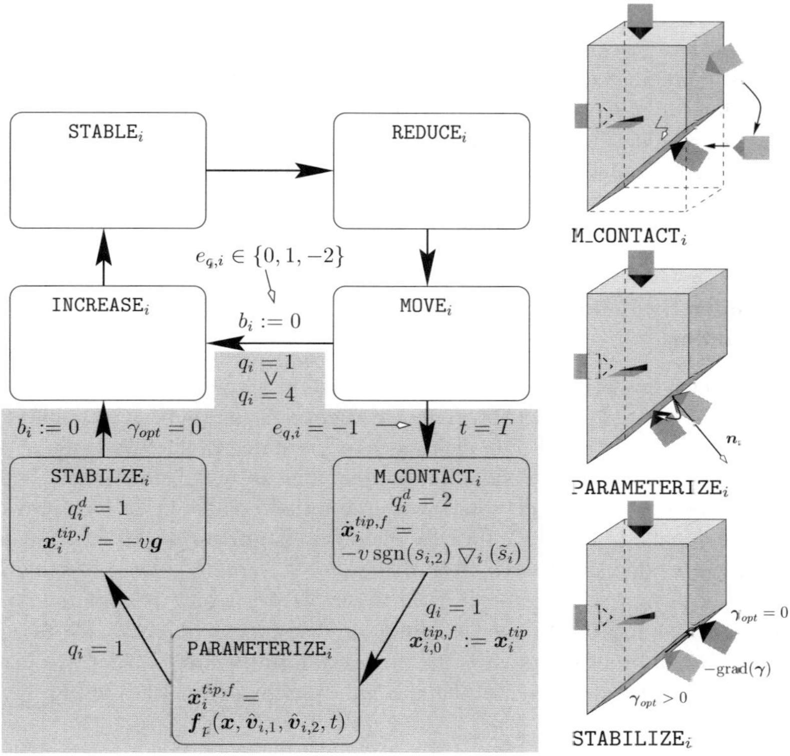

Fig. 11. Extended reference generator

the *next* planned regrasping task (Haidacher et al., 1999). If this analysis yields an unstable grasp in case another finger will be removed, i.e. the optimal solution of problem is $\gamma^* > 0$, the contact point of the currently regrasping finger is moved along the steepest gradient descent $-\boldsymbol{g}$ to reach a stable configuration, i.e. $\gamma^* = 0$.

5 Results from Dynamic Simulations and Experiments

The benefits of the proposed hybrid manipulation control are demonstrated in this section by results of dynamic simulations of a multi-fingered manipulation systems and experiments with the four-fingered hand of the TIT (Omata and Farooqi, 1996).

The hybrid modeling paradigm outlined in Sect. 2 yields the formal framework for the implementation of the multi-fingered simulation environment DYHASIM (Schlegl, 2002) within MATLAB. Based on the kinematic model of the four-fingered hand of the Technische Universität München (Woelfl, 1995), DYHASIM simulates multi-fingered manipulation accurately and efficiently. An animated graphical interface embedded within MAVERICK/OPENGL visualizes computed results and allows for their intuitive interpretation.

C-Code of the continuous finger dynamics was generated automatically using the tool AUTOLEV (Reckdahl and Mitiguy, 1996). Additionally, several example objects were implemented. A modified version of MATLAB's Runge-Kutta solver `ode45.m` with *event detection* activated in order to detect zero crossings of the discontinuity surfacces $s_{i,j}$ and a variable stepsize was used for numerical integration of the HSM based simulation model. Additional simulation results can be viewed at `http://www.rs.tu-berlin.de/~roha`. and are discussed in detail in (Schlegl et al., 2002a, Schlegl, 2002).

5.1 Simulative Evaluation of Discrete Error Compensation and Grasp Stabilization

The hybrid control law proposed above in combination with a simple heuristic approach to choose desired contact forces inside friction limits (instead of the GFO) controls the grasping and regrasping operations as presented in the following.

A sample object with a skew side-face is manipulated by the simulated robotic hand. The geometry of the object along with the initial contact point locations of the fingers are depicted in Fig. 12. Initially the object is grasped by all four fingers in one plane perpendicular to the z-axis of the object frame S_b. The objective of manipulation is to first regrasp finger 1 on the trajectory which is depicted by a dashed line and then to lift off finger 4. Although, the object with skew side-face is modeled for *simulation* of the process the model of a fullly cubic object was used for the *hybrid reference planning*. Its shape is indicated by dotted lines in Fig. 12.

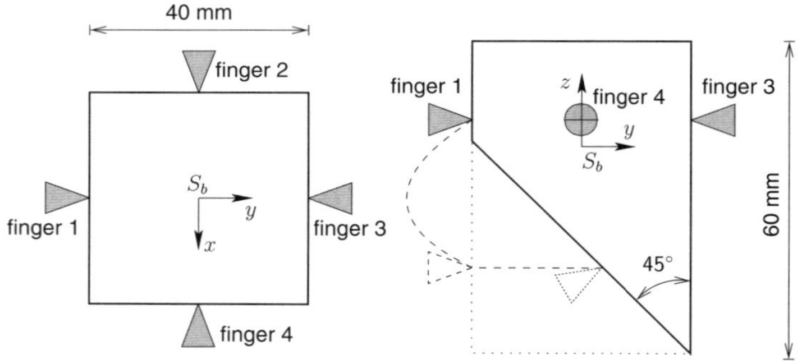

Fig. 12. Simulated example body with skew side-face

This intentional modeling error leads to a discrete contact error $e_{q,i} = -1$ of finger 1 at the end of its regrasping trajectory. Although the grasp set up by finger 2 to 4 is stable, the hand will obviously drop the object, if finger 4 is to lift off without previously having compensated the discrete error of finger 1. For compensation the extended reference planner as introduced in Sect. 4.3 is applied. After having detected the discrete error $e_{q,i} = -1$ by use of a simulated fingertip force sensor,

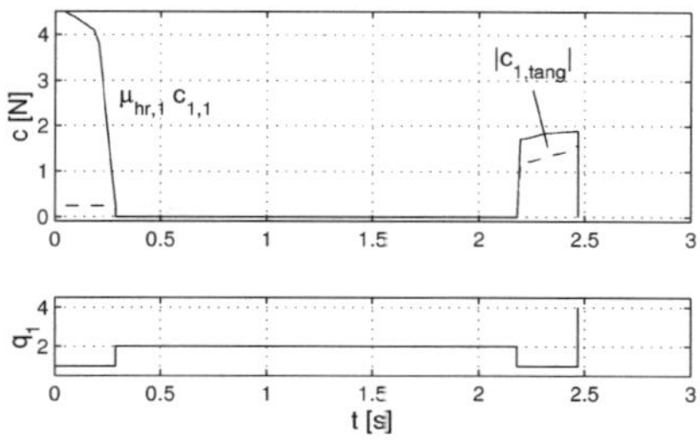

(a) Contact force and state finger 1

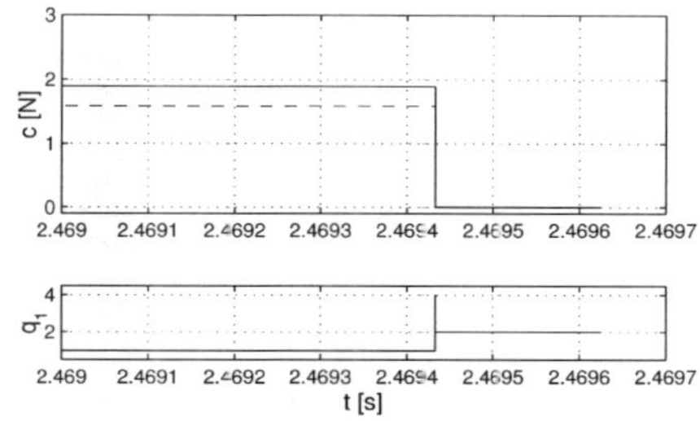

(b) Zoom out of (a)

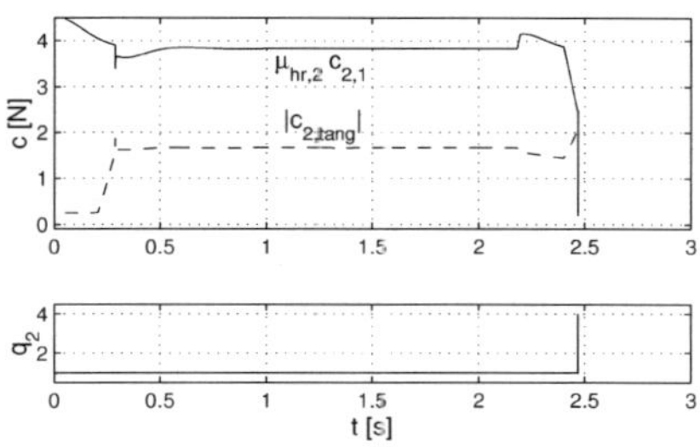

(c) Contact force and state finger 2

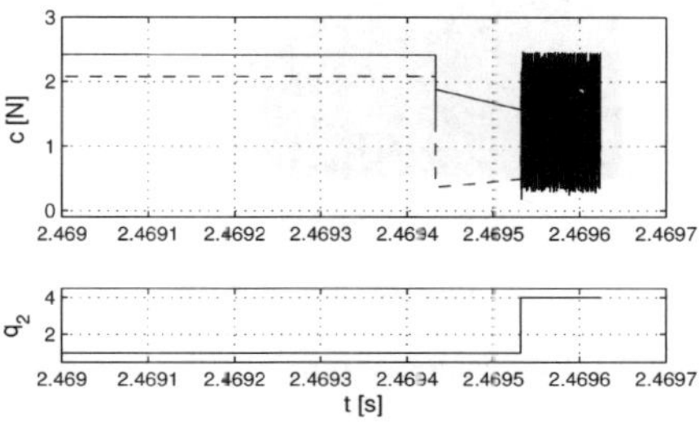

(d) Zoom out of (c)

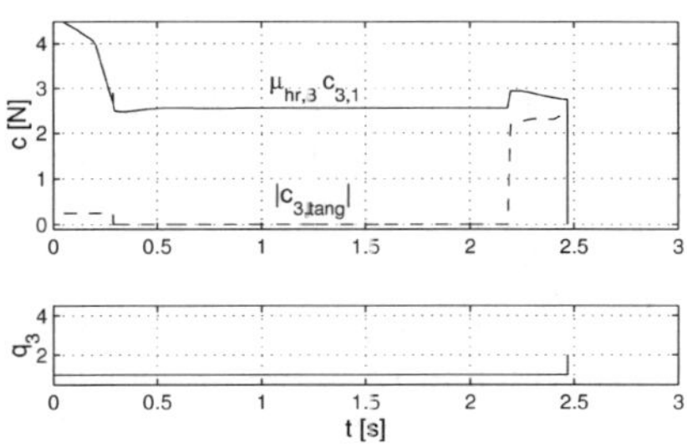

(e) Contact force and state finger 3

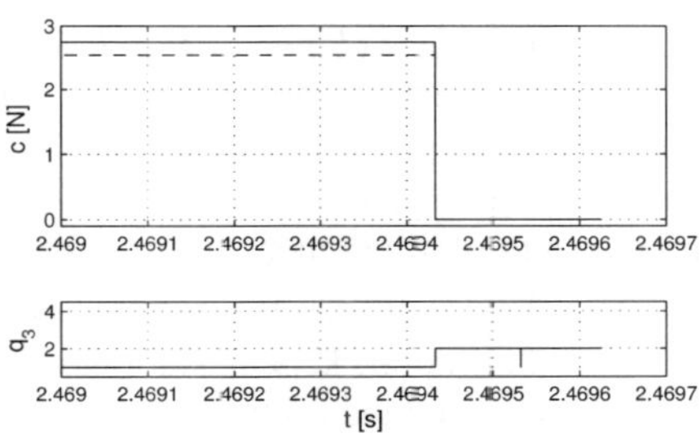

(f) Zoom out of (e)

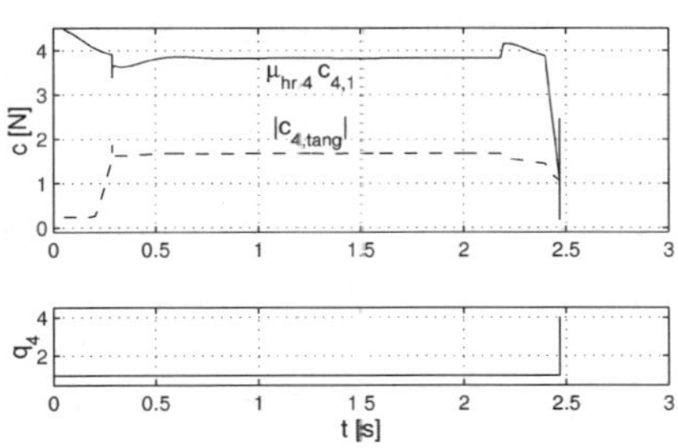

(g) Contact force and state finger 4

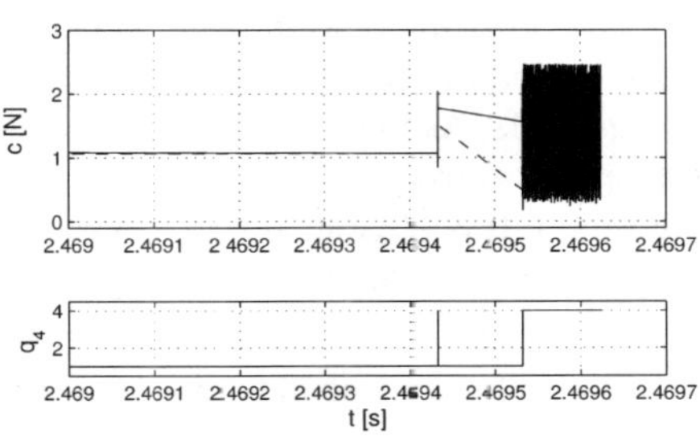

(h) Zoom out of (g)

Fig. 13. Simulation results for *unstable* regrasping of object with skew side-face

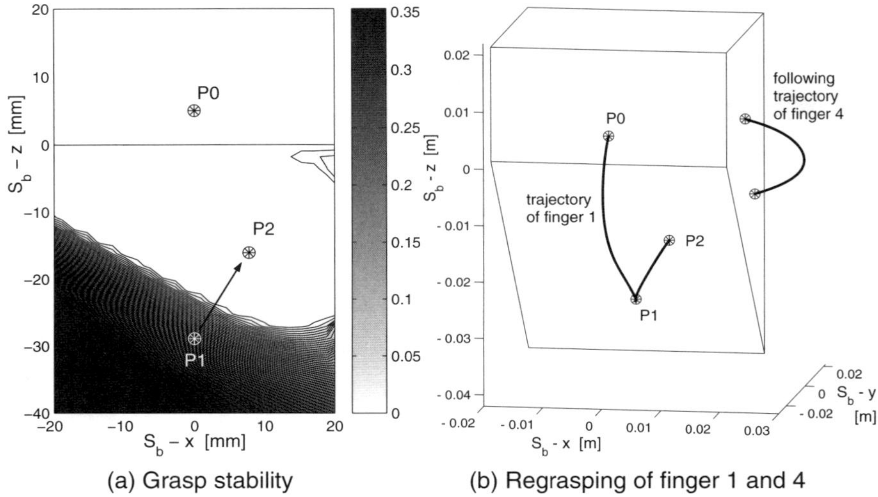

Fig. 14. Regrasping at object with skew side-face

the regrasping trajectory of finger 1 is extrapolated, i.e. the $\mathtt{M\_CONTACT}_1$ state of the extended reference planner in Fig. 11 is entered. As shown in Fig. 12 the extrapolated part of the trajectory evolves normal to the surface of the modeled *cubic* object. In a first simulation the $\mathtt{PARAMETERIZE}_1$ and $\mathtt{STABILIZE}_1$ states are *not entered*. As soon as finger 1 has established contact with the object it seems that a following regrasping task if another finger could be performed stably.

After having lifted off finger 4, however, the reduced grasp estabilished by fingers 1 to 3 cannot be kept up stably by the robotic hand. Fig. 13 shows the tangential force $|c_{i,tang}|$and the friction reserve $\mu_{hr,i}c_{i,1}$ where $\mu_{hr,i}$ denotes the static friction coefficient and $c_{i,1}$ the normal contact force at the ith contact. Additionally, the discrete contact state q_i is plotted. During the moving phase of finger 1 from $t = 0.287s$ to $t = 2.179s$ the other fingers remain in stable contact with the object. Soon after recontaction of finger 1 finger 4 is unloaded until it unintenionally starts sliding at $t = 2.4694s$. As the magnified plots on the right hand side show a sequence of further events is triggered which are detected by the event scheduler of the integration algorithm. It is obvious that a reduced grasp with only finger 1, 2 and 3 in contact with the object cannot be maintained, even though the discrete error of finger 1 was compensated.

In a following second simulation the error compensation method by extended reference planning is evaluated. Up to the time index at which $e_{q,i} = -1$ is compensated, the simulation runs similar to the one described previously. But now, before finger 4 starts regrasping reference adaptation as shown in Fig. 11 is active. Figure 14(a) shows the grasp stability measure γ^* mapped on the skew side-face of the example body for a *varying contact point location of finger 1*. The contact point

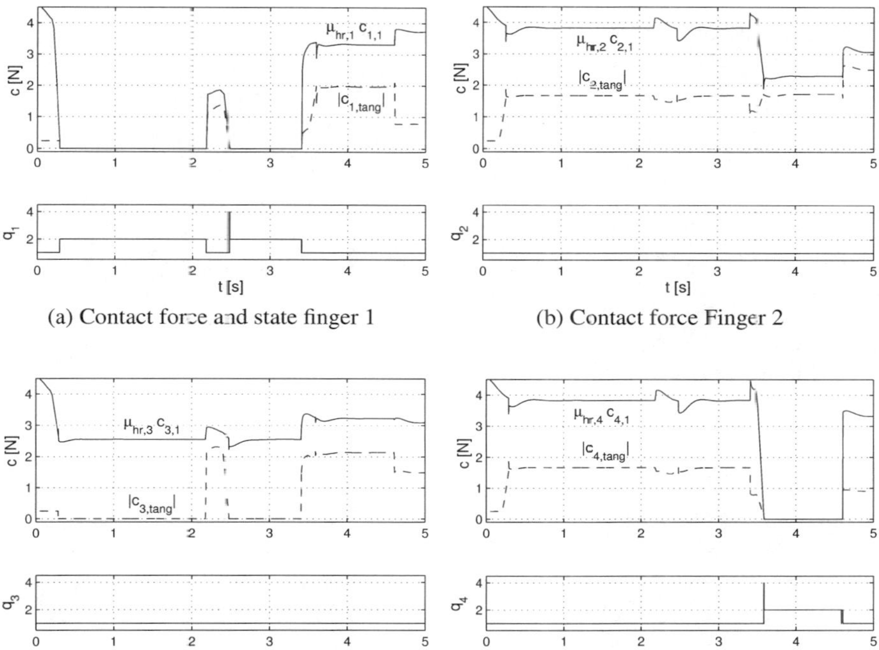

(a) Contact force and state finger 1

(b) Contact force Finger 2

(c) Contact force and state finger 3

(d) Contact force and state finger 4

Fig. 15. Simulation results for *stable* regrasping of object with skew side-face

locations of finger 2 and finger 3 are kept constant while finger 4 *was not considered* in the calculation as it is to regrasp next, see Fig. 14(b). If finger 1 contacts within the white areas, i.e. $\gamma^* = 0$, finger 4 can regrasp without losing grasp stability. The contact point location P1, at which finger 1 contacts after the $\texttt{M\_CONTACT}_1$ phase, is located in an unstable area on the object indicated by $\gamma^* > 0$. Local computation of the grasp stability measure at the initial contact point location P1 allows to specify the negative gradient $-\boldsymbol{g}$ of γ^*. The necessary geometric information therefore is derived in the $\texttt{PARAMETERIZE}_1$ phase by a simulated tactile sensor as described in Sect. 3.3. Thus, active exploration of the object surface close to the initial contact point is not necessary. Finally, to avoid instability of the grasp after finger 4 having started to regrasp, finger 1 is lifted off again and contacts at location P2 which is located within a stable area and allows for further regrasping. The evolution of contact forces and the discrete contact state for the second simulation run is depicted in Fig. 15. Stills taken from an animation of the sucessful double regrasping tasks are aggregated in Fig. 16.

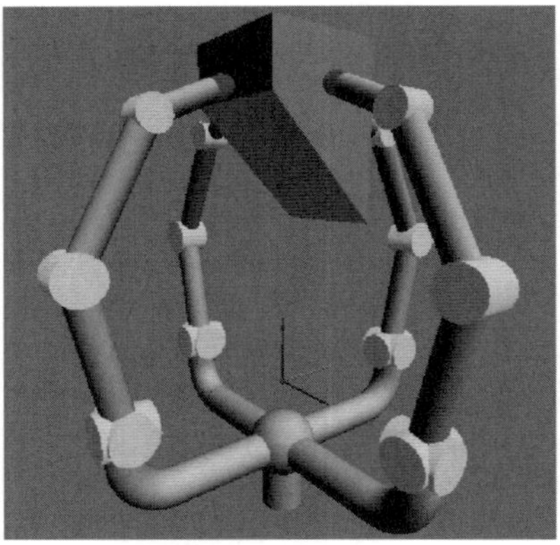

(a) Initial configuration

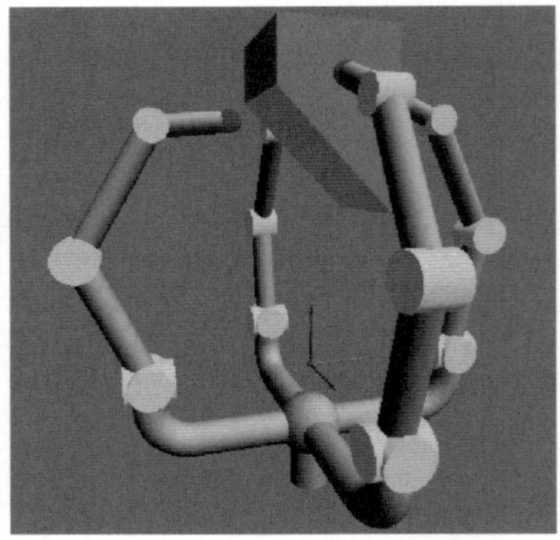

(b) Regrasping of finger 1

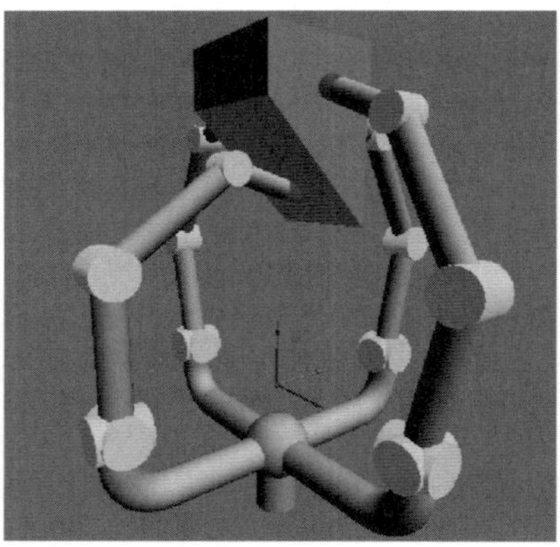

(c) Finger 1 at unstable location P1

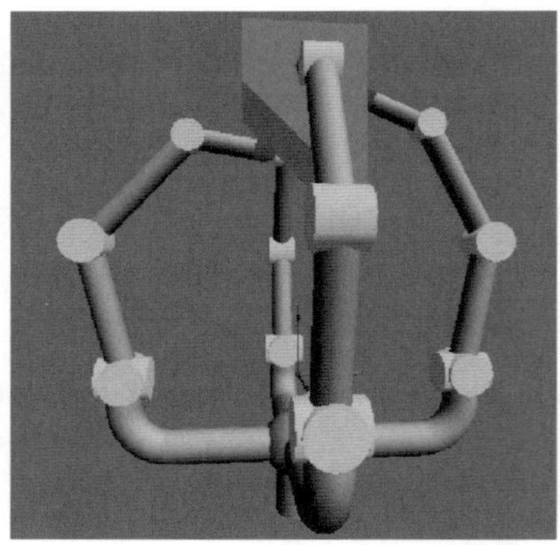

(d) Correcting the location of finger 1

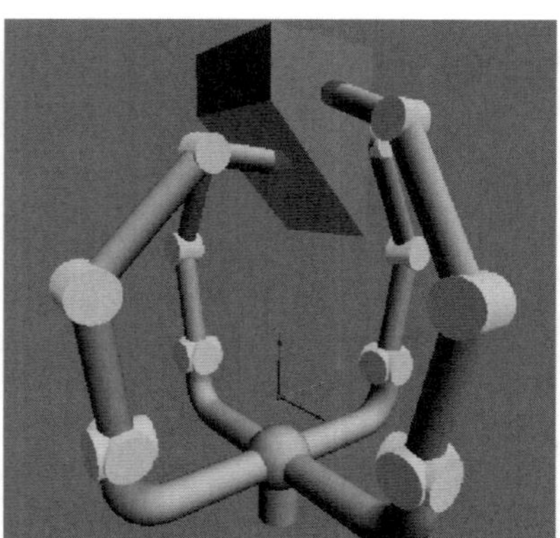

(e) Finger 1 at stable location P2

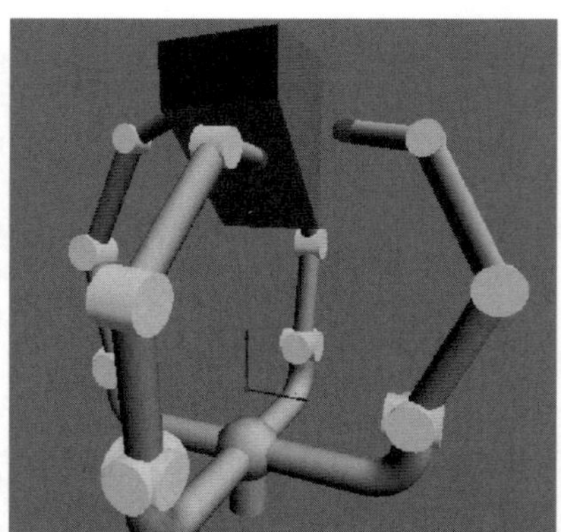

(f) Stable regrasping of finger 4

Fig. 16. Visualization of discrete error compensation and grasp stabilization

5.2 Experimental Results Using the Hybrid Control Approach

During experiments the hybrid control architecture as presented in Sect. 3 was applied to perform complex manipulation tasks one of which is presented in the following. A significant difference to the control applied in the simulations is the use of the real-time GFO in the experiments. The discussion of the considered experiment is kept on a qualitative level. A detailed quantitative analysis of numerical results can be found in (Schlegl, 2002, Schlegl et al., 2001). The main purpose of this section is to demonstrate the benefits of a combined application of hybrid control, discrete error compensation, and real-time GFO.

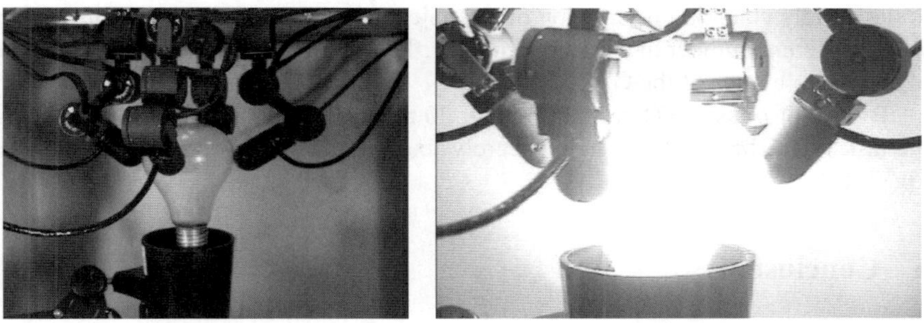

(a) Start (b) End

Fig. 17. Screwing an electric bulb into a socket

The robotic hand is supposed to screw an electric bulb with a mass of $38g$ and a diameter of $6.5cm$ into a socket, see Fig. 17(a). The manipulation task is separated into the following phases:

1. establish a stable four-fingered grasp on the bulb in the plane of maximal bulb diameter;
2. move bulb towards socket until contact is detected;
3. turn bulb about $0.63rad$ with respect to its vertical axis during $0.14s$;
4. consecutively regrasp the fingers during $0.75s$ to new contact points on the bulb;
5. repeat turning and regrasping until the bulb is screwed into the socket.

A central objective was to reduce the time needed for task completion as far as possible. Therefore, the duration of the $\texttt{REDUCE}_i$ and $\texttt{INCREASE}_i$ phases in the reference generator was chosen to a very short time $t_{r/i} = 0.0675s$.

During screwing the bulb moves slowly into the direction of the socket. This motion is *intentionally not* considered in contact point planning where the bulb is assumed to have a purely cylindrical shape. As a consequence, the diameter in the plane of contact point locations is decreasing over time resulting in discrete contact state errors $e_{q,i} = -1$. Without compensation of this error the task could not be completed successfully. Furthermore, the importance of fast optimization of contact forces was examined in the experiments. Table 2 shows the results from a series of

Table 2. Sucess of bulb experiment over sample time

time	4ms	20ms	50ms	75ms	125ms	0.25s	0.5s	1s
sucess	ok	ok	ok	fail	fail	fail	fail	fail

experiments to insert the bulb into the socket. The sample time of the GFO was varied between $4ms$ and $1s$. It could be observed that for sampling times less than 50ms the task could be successfully accomplished. For sample times greater than $75ms$ the object was pushed out of the grasp during regrasping of fingers. The reason for this behavior is that the grasp forces were not adjusted sufficiently fast with respect to the rapidly changing orientation of the object after rotation and new grasp states with a different number of contacting fingers. This renders the grasp unstable by use of low sample rates of the GFO.

By use of a fast GFO and after about 20 repetitions of rotating and regrasping the task is finished with the bulb securely fixed in the socket and operational, see Fig. 17(b).

6 Conclusions

The results from dynamic simulations and experiments with a real multi-fingered hand which were presented in the previous section demonstrate the effectiveness of hybrid manipulation control compared to conventional methods. Complex grasping and regrasping tasks can be performed robustly, even if system models are imprecise and external disturbances affect the manipulation process.

In general, this article pointed out the *inherent* mixed discrete-continuous character of multi-fingered manipulation. The classical problems of modeling, reference, and force planning as well as closed-loop control of robotic grasping have been reconsidered from a discrete-continuous systems viewpoint. Solutions to all these problems have been – at least – outlined in the framework of this article. For further details appropriate citations of own or other work have been made.

Specifically, *hybrid dynamic modeling* of multi-fingered manipulation as proposed Sect. 2 allows for the highest known level of *formalization* in modeling dextrous grasping. The *detailed planning* of regrasping tasks is also formulated in a hybrid framework. It allows for *comprehensive* simultaneous specification of desired continuous reference trajectories and discrete contact state transitions.

Hybrid closed-loop control as reported in Sect. 3 and Sect. 4 increases the *robustness* of performing robotic grasping and regrasping operations in two respects: First, *modeling errors* lead to erroneous task planning which can render unstable grasps during a regrasping sequence. In this case *model-based reference adaptation* as a key functionality of the hybrid controller allows for compensation of discrete errors and for grasp stabilization such that a desired task can be continued. Second, a *variable structure real-time algorithm for grasping force optimization* in combination with a set of *impedance control laws* robustifies a grasp against *external disturbance forces*.

Furthermore, it enables a high speed of multi-fingered manipulation of objects which is not reported for other approaches to grasping control.

Further research will focus on various objectives. Concerning multi-fingered grasping we will concentrate on solving the hybrid optimal control problem for *global* regrasping specification during manipulation tasks. Furthermore, hybrid planning and optimal control problems in the field of multi-legged walking machines, which like robotic hands also belong to the challenging class of mechatronic multicontact systems, will be investigated.

References

ABACUSS (1995). http://yoric.mit.edu/abacuss/abacuss.html. Massachussets Institute of Technology.

Abadi, M. and Cardelli, L. (1996). *A Theory of Objects*. Springer, New York.

Abel, D. (1990). *Petri-Netze für Ingenieure*. Springer, Berlin, Germany.

Adjiman, C., Schweiger, C., and Floudas, C. (1998). Mixed-integer nonlinear optimization in process synthesis. In Du, D.-Z. and Pardalos, P., editors, *Handbook of Combinatorial Optimization*, volume 1, pages 1–76. Kluwer Acadademic Publisher.

Albro, J. and Bobrow, J. (2001). Optimal motion primitives for a 5 DOF experimental hopper. In *Proceedings of the IEEE International Conference on Robotics and Automation (Seoul, Korea)*, pages 3630–3635.

Allgor, R. and Barton, P. (1997). Mixed integer dynamic optimization. *Computational Chemical Engineering*, 21:451–456.

Alur, R., Courcoubetis, C., Halbwachs, N., Henzinger, T., Ho, P.-H., Nicollin, X., Olivero, A., Sifakis, J., and Yovine, S. (1995). The algorithmic analysis of hybrid systems. *Theoretical Computer Science*, 138:3–34.

Alur, R., Courcoubetis, C., Henzinger, T. A., and Ho, P. H. (1993). Hybrid Automata: An Algorithmic Approach to the Specification and Verification of Hybrid Systems. In Grossmann, R. L., Nerode, A., Ravn, A. P., and Rischel, H., editors, *Hybrid Systems*, volume 736 of *Lecture Notes in Computer Science*, pages 209–229. Springer.

Alur, R., Dang, T., Esposito, J., Fierro, R., Hur, Y., Ivanĉić, F., Kumar, V., Lee, I., Mishra, P., Pappas, G., and Sokolsky, O. (2001). Hierarchical hybrid modeling of embedded systems. In Henzinger, T. and Kirsch, C., editors, *EMSOFT 2001: First International Workshop on Embedded Software, Tahoe City, CA, USA, October 8–10, 2001*, volume 2211 of *Lecture Notes in Computer Science*, pages 14–31. Springer.

Alur, R. and Dill, D. (1990). A theory of timed automata. *Theoretical Computer Science*, 126:183–235.

Alur, R., Grosu, R., Hur, Y., Kumar, V., and Lee, I. (2000a). Modular specification of hybrid systems in Charon. In *Proc. HSCC'00*, Springer LNCS 1790.

Alur, R., Henzinger, T., Lafferiere, G., and Pappas, G. (2000b). Discrete abstractions of hybrid systems. *Proceedings of the IEEE*, 88(7):971–984.

Alur, R. and Henzinger, T. A. (1999). Reactive modules. *Formal Methods in System Design: An International Journal*, 15(1):7–48.

Alur, R., Henzinger, T. A., and Sontag, E. D., editors (1996). *Hybrid Systems III: Verification and Control*, volume 1066 of *Lecture Notes in Computer Science*. Springer.

Andersson, M. (1994). *Object-Oriented Modeling and Simulation of Hybrid Systems*. PhD dissertation, Department of Automatic Control, Lund Institute of Technology, Lund, Sweden.

Antsaklis, P., editor (2000). *Special Issue on Hybrid Systems: Theory and Applications*, volume 88, no. 7 of *Proceedings of the IEEE*.

Antsaklis, P., Kohn, W., Lemmon, M., Nerode, A., and Sastry, S., editors (1999). *Hybrid Systems V*, volume 1567 of *Lecture Notes in Computer Science*. Springer.

Antsaklis, P. and Koutsoukos, X. D. (1998). On Hybrid Control of Complex Systems: A Survey. In *Proceedings Hybrid Dynamical Systems*, ADPM '98, pages 1–8, Reims, France.

Antsaklis, P. and Nerode, A., editors (1998a). *Special Issue on Hybrid Control Systems*, volume 43 of *IEEE Transactions on Automatic Control*.

Antsaklis, P., Nerode, A., Kohn, W., and Sastry, S., editors (1995). *Hybrid Systems II*, volume 999 of *Lecture Notes in Computer Science*. Springer.

Antsaklis, P., Nerode, A., Kohn, W., and Sastry, S., editors (1997). *Hybrid Systems IV*, volume 1273 of *Lecture Notes in Computer Science*. Springer.

Antsaklis, P. J. and Nerode, A. (1998b). Special issue on hybrid systems. *IEEE Transactions on Automatic Control*, 43.

Apt, K. R., Francez, N., and de Roever, W.-P. (1980). A proof system for communicating sequential processes. *ACM Transactions on Programming Languages and Systems*, 2(3):359–385.

Asarin, E., Bournez, O., Dang, T., and Maler, O. (2000a). Reachability analysis of piecewise-linear dynamical systems. In 3rd *Int. Workshop of Hybrid Systems: Comp. and Control*, volume 1790 of *LNCS*, pages 20–31. Springer.

Asarin, E., Bournez, O., Dang, T., Maler, O., and Pnueli, A. (2000b). Effective synthesis of switching controllers for linear systems. *Proceedings of the IEEE*, 88:1011–1025.

Automatica 35(3) (1999). A special issue on hybrid systems. *Automatica*, 35:347–519.

Back, A., Guckenheimer, J., and Myers, M. (1993). A dynamical simulation facility for hybrid systems. In Grossmann, R., Nerode, A., Ravn, A., and Rischel, H., editors, *Lecture Notes in Computer Science: Hybrid Systems*, volume 736, pages 255–267. Springer.

Balas, E. (1985). Disjunctive programming and a hierarchy of relaxations for discrete optimization problems. *SIAM Journal Alg. Disc. Meth.*, 6(3):466–486.

Barros, F. J. (1996). The dynamic structure discrete event system specification formalism. *Transactions of the SCS International*, 13(1):35–46.

Barton, P. I. (1992). *The Modelling and Simulation of Combined Discrete/Continuous Processes*. PhD dissertation, University of London.

Bastide, R. (1995). Approaches in unifying Petri nets and the Object-Oriented Approach. In *Object-Oriented Programming and Models of Concurrence 16th International Conference on Application and Theory of Petri Nets*, Italy.

Baumgarten, B. (1990). *Petri-Netze: Grundlagen und Anwendungen*. BI-Wissenschaftsverlag, Mannheim, Wien, Zürich.

Bellman, R. (1957). *Dynamic Programming*. Princeton University Press.

Bemporad, A., Borelli, F., and Morari, M. (2002). On the optimal control law for linear discrete time hybrid systems. In *Hybrid Systems: Computation and Control*, volume 2289 of *LNCS*, pages 105–119. Springer.

Bemporad, A., Mignone, D., and Morari, M. (1999). An efficient branch and bound algorithm for state estimation and control of hybrid systems. In *Proc.* 5th *European Control Conference*.

Bemporad, A. and Morari, M. (1999a). Control of systems integrating logic, dynamics, and constraints. *automatica*, 35(3):407–427.

Bemporad, A. and Morari, M. (1999b). Verification of hybrid systems using mathematical programming. In Vaandrager, F. W. and van Schuppen, J. H., editors, *Hybrid Systems: Computation and Control, Proc. 2nd Int. Workshop, HSCC'99, Berg en Dal, The Netherlands, March 1999*, Lecture Notes in Computer Science 1569, pages 31–45. Springer.

Bender, K. and Kaiser, O. (1995). Simultaneous Engineering durch Maschinenemulation. *CIM Management*, 11(4):14–18.

Benedetto, M. D. D. and Sangiovanni-Vincentelli, A. L., editors (2001). *Hybrid Systems: Computation and Control*, volume 2034 of *Lecture Notes in Computer Science*. Springer.

Bergstra, J. and Klop, J. (1984). Process algebra for synchronous communication. *Information and Control*, 60(1):109–137.

Betts, J. (1998). Survey of numerical methods for trajectory optimization. *AIAA Journal of Guidance, Control, and Dynamics*, 21(2):193–207.

Bhat, G., Cleaveland, R., and Grumberg, O. (1995). Efficient on-the-fly model checking for CTL*. In *LICS '95: 10th Annual IEEE Symposium on Logic in Computer Science, San Diego, California, USA, June 26–29, 1995*, pages 388–397. IEEE Computer Society Press.

Blanke, M., Frei, C., Kraus, F., Patton, R., and Staroswiecki, M. (2000a). Fault-tolerant control systems. In Isidori, A., Aström, K. J., Blanke, M., Schaufelberger, W., Albertos, P., and Sanz, R., editors, *Control of Complex Systems*, chapter 8, pages 165–189. Springer.

Blanke, M., Frei, C. W., Kraus, F., Patton, R. J., and Staroswiecki, M. (2000b). What is fault-tolerant control? In *Proceeding of SAFEPROCESS 2000: 4th Symposium on Fault Detection*, page 40. IFAC.

Bobbio, A., Garg, S., Gribaudo, M., Horváth, A., Sereno, M., and Telek, M. (1999). Modeling software systems with rejuvenation, restoration and checkpointing through fluid stochastic petri nets. In *Proc. Eighth International Workshop on Petri Nets and Performance Models - PNPM'99*, pages 82–91.

Bolognesi, T. and Brinksma, E. (1987). Introduction to the ISO specification language LOTOS. *Computer Networks*, 14:25–59.

Brack, G. (1974). *Dynamik technischer Systeme*. VEB Deutscher Verlag für Grundstoffindustrie, Leipzig.

Branicky, M. (1993). Topology of hybrid systems. In *Proceedings of the 32nd IEEE Conference on Decision and Control (San Antonio, TX)*, pages 2309–2314.

Branicky, M. (1994a). Analyzing continuous switching systems: Theory and examples. In *Proceedings of the American Control Conference (Baltimore, MD)*, pages 3110–3114.

Branicky, M. (1994b). Stability of switched and hybrid systems. In *Proceedings of the 33rd IEEE Conference on Decision and Control (Lake Buena Vista, FL)*, pages 3498–3503.

Branicky, M. (1994c). A unified framework for hybrid control. In *Proceedings of the 33rd IEEE Conference on Decision and Control (Lake Buena Vista, FL)*, pages 4228–4234.

Branicky, M. (1995). *Studies in Hybrid Systems: Modeling, Analysis and Control*. PhD thesis, Massachusetts Institute of Technology, Department of Electrical Engineering and Computer Science.

Branicky, M. (1996). General hybrid dynamical systems: Modeling, analysis, and control. In Alur, R., Henzinger, T., and Sontag, E., editors, *Lecture Notes in Computer Science: Hybrid Systems III*, volume 1066, pages 186–200. Springer.

Branicky, M. (1998). Multiple Lyapunov Functions and Other Analysis Tools for Switched and Hybrid Systems. *IEEE Trans. Aut. Control*, 43(4):475–482.

Branicky, M., Borkar, V., and Mitter, S. (1998). A unified framework for hybrid control: Model and optimal control theory. *IEEE Transactions on Automatic Control*, 43(1):31–45.

Branicky, M., Hebbar, R., and Zhang, G. (1999). A fast marching algorithm for hybrid systems. In *Proceedings of the 38th IEEE Conference on Decision and Control (Phoenix, AZ)*, pages 4897–4902.

Brenan, K. E. and Campbell, S. L. (1996). *Numerical Solution of Initial-Value Problems in Differential-Algebraic Equations*. siam.

Brockett, R. (1993). Hybrid models for motion control systems. In Trentelmann, H. and Willems, J., editors, *Essays on Control: Perspectives in the Theory and its Applications*, pages 29–53. Boston: Birkhäuser.

Broenink, J., Hilderink, G., and Bakkers, A. (1998). Conceptual design for controller software of mechatronic systems. In Bradshaw, A. and Counsel, J., editors, *Computer aided Conceptual Design '98*.

Bröhl, A. and Dröschel, W. (1995). *Das V-Modell*. Oldenburg.

Brooke, A., Kendrick, D., Meeraus, A., and Raman, R. (1998). *GAMS/CPLEX – A User's Guide*. GAMS Development Corporation.

Brookes, S., Hoare, C., and Roscoe, A. (1984). A theory of communicating sequential processes. *Communications of the ACM*, 31(3):560–599.

Broucke, M., Di Benedetto, M., Di Gennaro, S., and Sangiovanni-Vincentelli, A. (2000). Theory of optimal control using bisimulations. In *Proc.* 3rd *Int. Workshop of Hybrid Systems: Comp. and Control*, volume 1790 of *LNCS*, pages 89–102. Springer.

Brown, J. S. and de Kleer, J. (1990). A qualitative physics based on confluences. In *Qualitative Reasoning about Physical Systems*, pages 88–126. Morgan Kaufmann Publishers, San Mateo, CA.

Broy, M. (2001). Refinement of time. *Theoretical Computer Science*, 253(1):3–26.

Bryant, R. E. (1986). Graph-based algorithms for Boolean function manipulation. *IEEE Transactions on Computers*, C-35(8):677–691.

Bryant, R. E. (1992). Symbolic Boolean manipulation with ordered binary-decision diagrams. *ACM Computing Surveys*, 24(3):293–318. Preprint version published as CMU Technical Report CMU-CS-92-160.

Buchholz, J. J. (1999). Systemsimulation. Vorlesungsmanuskript.

Bühler, M. and Koditschek, D. (1993). From stable to chaotic juggling: Theory, simulation, and experiments. In Spong, M., Lewis, F., and Abdallah, C., editors, *Robot Control – Dynamics, Motion Planning, and Analysis*, pages 525–530. New York: IEEE Press.

Bujakiewicz, P. (1994). *Maximum weighted matching for high index differential algebraic equations*. PhD dissertation, TU Delft, Delft, Netherlands. ISBN 90-9007240-3.

Buss, M. (1998). Multi-fingered Regrasping using a Hybrid Systems Approach. In *Proceedings of the 2nd IMACS/IEEE International Multiconference on Computational Engineering in Systems Applications (CESA'98)*, pages 857–861, Hammamet, Tunisia.

Buss, M. (2000). *Control Methods for Hybrid Dynamical Systems – Models, Control Loops, Optimal Control, Computation Tools, and Mechatronic Applications – (in German)*. PhD thesis, Institute of Automatic Control Engineering, Technische Universität München.

Buss, M., Glocker, M., Hardt, M., von Stryk, O., Bulirsch, R., and Schmidt, G. (2002). Nonlinear hybrid dynamical systems: Modeling, optimal control, and applications. In Engell, S., Frehse, G., and Schnieder, E., editors, *Modelling, Analysis, and Design of Hybrid Systems*, Lecture Notes in Control and Information Science. Springer. (This volume).

Buss, M., Hashimoto, H., and Moore, J. (1996). Dextrous Hand Grasping Force Optimization. *IEEE Transactions on Robotics and Automation*, 12(3):406–418.

Buss, M., Schlegl, T., and Schmidt, G. (1997). Development of Numerical Integration Methods for Hybrid (Discrete-Continuous) Dynamical Systems. In *Advanced Intelligent Mechatronic AIM97*, Tokyo, Japan.

Buss, M. and Schmidt, G. (1996). Hybrid System Behavior Specification for Multiple Robotic Mechanisms. In *Proceedings of the IEEE/RSJ International Conference on Intelligent Robots and Systems IROS*, pages 140–147, Osaka, Japan.

Buss, M., von Stryk, O., Bulirsch, R., and Schmidt, G. (2000a). Towards hybrid optimal control. *at–Automatisierungstechnik*, 48:448–459.

Buss, M., von Stryk, O., Bulirsch, R., and Schmidt, G. (2000b). Towards hybrid optimal control. *Automatisierungstechnik*, 9:448–459.

Cellier, F., Elmqvist, H., and Otter, M. (1996). Modelling from physical principles. In Levine, W., editor, *The Control Handbook*, pages 99–107. CRC Press, Boca Raton, FL.

Champagnat, R., Esteban, P., Pingaud, H., and Valette, R. (1996). Petri Net Based Modeling of Hybrid Systems. In *Proc. of ASI'96*, pages 53–60, Toulouse, France. Advanced Summer Institute.

Champagnat, R., Esteban, P., Pingaud, H., and Valette, R. (1998). Modeling and Simulation of a Hybrid System Through PR/TR PN-DAE Model. In *Proc. of the 3rd Int. Conf. on Automation of Mixed Processes*, pages 131–137, Reims, France.

Chase, C., Serrano, L., and Ramadge, P. J. (1993). Periodicity and chaos from switched flow systems: examples of discretely controlled continuous systems. *IEEE Trans. Automatic Control*.

Cherif, M. and Gupta, K. K. (1997). Practical Motion Planning for Dextrous Re-Orientation of Polyhedra. In *Proceedings of the IEEE/RSJ International Conference on Intelligent Robots and Systems IROS*, pages 291–297, Grenoble, Frankreich.

Chouikha, M. (1999). *Entwurf disktret-kontinuierlicher Steuerungssysteme - Modellbildung, Analyse und Synthese mit hybriden Petri-Netzen*. PhD thesis, TU Braunschweig.

Chouikha, M., Decknatel, G., Drath, R., Frey, G., Müller, C., Simon, C., Thieme, J., and Wolter, K. (2000). Petri net-based descriptions for discrete-continuous systems. *at - Automatisierungstechnik*, 48(9):415–425.

Chouikha, M. and Krebs, V. G. (1998). Beschreibungsmittel und Methoden für kontinuierlich-diskrete Systeme. In Abel, D. and Lemmer, K., editors, *Theorie ereignisdiskreter Systeme*, München, Wien. Oldenbourg.

Chouikha, M., Ober, B., and Schnieder, E. (2001). Automatisierter Steuerungsentwurf für diskrete und kontinuierlich-diskrete Systeme. *at - Automatisierungstechnik*, 49(6):280–289.

Chouikha, M. and Schnieder, E. (1998a). Beschreibung kontinuierlich-diskreter Systeme mit hybriden Petrinetzen. In *GMA-Kongress '98 Mess- und Automatisierungtechnik*, pages 365–372, Ludwigsburg. Institut für Regelungs- und Automatisierungstechnik, TU Braunschweig, VDI-Verlag. VDI-Bericht 1397.

Chouikha, M. and Schnieder, E. (1998b). Modelling of Continuous-discrete Systems with hybrid Petri Nets. In *IEEE: Computational Engineering in Systems Applications*, pages 606–612.

Chouikha, M. and Schnieder, E. (1999). Model-based control synthesis of continuous-discrete systems. In *Proc. IEEE Int. Conf. Systems, Man and Cybernetics*, pages 452–456.

Chow, A.-H. (1996). Parallel DEVS: A parallel, hierarchical, modular modeling formalism and its distributed simulator. *Transaction of the SCS International*, 13(2):55–67.

Christen, E. (1997). The VHDL 1076.1 Language for Mixed-Signal Design. http://-www.analogy.com/support/wp/vhdl_ern.htm.

Chutinan, A. and Krogh, B. H. (1999a). Computing approximating automata for a class of linear hybrid systems. In *Hybrid Systems V: Proc. Int. Workshop, Notre Dame, USA*, Lecture Notes in Computer Science 1567, pages 16–37. Springer.

Chutinan, A. and Krogh, B. H. (1999b). Verification of polyhedral-invariant hybrid automata using polygonal flow pipe approximation. In 2nd *Int. Workshop on Hybrid Systems: Computation and Control*, volume 1569 of *LNCS*, pages 76–90. Springer.

Ciardo, G., Nicol, D., and Trivedi, K. (1999). Discrete-event simulation of fluid stochastic petri nets. *IEEE Trans. Softw. Eng.*, 25(2):207–217.

Clarke, E. M. and Emerson, E. A. (1982). Design and synthesis of synchronization skeletons for branching time temporal logic. In Kozen, D., editor, *Logics of Programs Workshop, IBM Watson Research Center, Yorktown Heights, New York, May 1981*, volume 131 of *Lecture Notes in Computer Science*, pages 52–71. Springer.

Clarke, E. M., Grumberg, O., and Peled, D. A. (1999). *Model Checking*. MIT Press.

Clarke, E. M. and Kurshan, R. P. (1996). Computer-aided verification. *IEEE Spectrum*, pages 61–67.

Collins, D. (1995). *Designing Object-Oriented User Interfaces*. Benjamin/Cummings Publishing Company, Inc., Redwood City, CA.

Console, L., de Kleer, J., and Hamscher, W., editors (1992). *Readings in Model-based Diagnosis*, San Mateo, CA. Morgan Kaufmann Publishers.

Courcoubetis, C., Vardi, M. Y., Wolper, P., and Yannakakis, M. (1992). Memory-efficient algorithms for the verification of temporal properties. *Formal Methods in System Design*, 1(2/3):275–288.

Cury, J. E. R., Krogh, B. A., and Niinomi, T. (1998). Synthesis of supervisory controllers for hybrid systems based on approximating automata. *IEEE Transactions on Automatic Control, Special issue on hybrid systems*, 43:564–568.

Czogalla, O. and Hoyer, R. (1997). Simulation based design of control strategies for urban management and control. In *4th World Congress on Intelligent Transport Systems*, Berlin.

Czogalla, O. and Hoyer, R. (1999). Model based approximation of traffic actuated signal control for mesoscopic traffic simulation. In *6th World Congress on Intelligent Transport Systems*, Toronto.

Dang, T. and Maler, O. (1998). Reachability analysis via face lifting. In Henzinger, T. and Sastry, S., editors, *Hybrid Systems: Computation and Control, Proc. 1st Int. Workshop, HSCC'98, Berkeley, USA, March 1998*, Lecture Notes in Computer Science 1386, pages 96–109. Springer.

David, R. and Alla, H. (1987). Continuous Petri Nets. In *8th European Workshop on Applications and Theory of Petri Nets*, pages 275–294, Spain.

David, R. and Alla, H. (1992). *Petri nets and Grafcet - Tools for modelling discrete event systems*. Prentice Hall, New York, London.

David, R. and Alla, H. (1994). Petri Nets for Modeling of Dynamic Systems - A Survey. *Automatica*, 30(2):175–202.

David, R. and Alla, H. (1998). Continuous and hybrid Petri nets. *International Journal of Circuits and Systems*, 8(1):159–188.

Davoren, J. M. and Nerode, A. (2000). Logics for hybrid systems. *Proceedings of the IEEE*, 88:985–1010.

de Kleer, J. and Weld, D. S., editors (1990). *Readings in Qualitative Reasoning about Physical Systems*, San Mateo, CA. Morgan Kaufmann Publishers.

de Roever, W.-P. (1998). The need for compositional proof systems: A survey. In de Roever, W.-P., Langmaack, H., and Pnueli, A., editors, *Compositionality: The Significant Difference, Proceedings of the International Symposium COMPOS '97, Malente, Germany, September 7–12, 1997*, volume 1536 of *Lecture Notes in Computer Science*, pages 1–22. Springer.

de Roever, W.-P., de Boer, F., Hannemann, U., Hooman, J., Lakhnech, Y., Poel, M., and Zwiers, J. (2001). *Concurrency Verification: Introduction to Compositional and Noncompositional Methods*. Number 54 in Cambridge Tracts in Theoretical Computer Science. Cambridge University Press.

De Schutter, B. (1999). Optimal control of a class of linear hybrid systems with saturation. In *Proc.* 38th *IEEE Conf. Decision and Control*, pages 3978–3983.

Decknatel, G. and Schnieder, E. (1998). Hybrid petri nets as a new formalism for modelling railway systems. In *Computers in Railways VI*, pages 773–782. Computational Mechanics Publications/WIT Press.

DEDS'98 (1998). Special issue on hybrid systems. *Discrete Event Dynamic Systems: Theory and Application*, 8:99–222.

Denk, J. (1999). Online optimal control strategies for mechatronic systems under multiple contact configurations. Technical report, Institute of Automatic Control Engineering, Technische Universität München. Internal Report.

Deparade, A., Pereira Remelhe, M., and Engell, S. (2001). Eine Modellierungs- und Simulationsumgebung für hybride technische Systeme mit ereignis-diskreten Steuerungen. In *3. VDI/VDE-GMA Aussprachetag, Rechnergestützter Entwurf von Regelungssystemen, Dresden*, volume 36 of *GMA-Berichte*, Düsseldorf. GMA-Aussprachetag FA-6.23, VDI/VDA-GMA.

Design/CPN (2002). Design/CPN Version 4.0.1. http://www.daimi.au.dk/design-CPN. University of Aarhus, Department of Computer Science, CPN Group.

Dijkstra, E. W. (1969a). On understanding programs (EWD 264). Published in an extended version as (Dijkstra, 1969b).

Dijkstra, E. W. (1969b). Structured programming. In Buxton, J. and Randell, B., editors, *Software Engineering Techniques, Report on a conference sponsored by the NATO Science Committee*, pages 84–88. NATO Science Committee.

Dill, D. (1990). Timing assumptions and verification of finite-state concurrent systems. In Sifakis, J., editor, *International Workshop on Automatic Verification Methods for Finite State Systems, Grenoble, France, June 12–14, 1989*, volume 407 of *Lecture Notes in Computer Science*, pages 197–212. Springer.

Dimitriadis, V., Shah, N., and Pantelides, C. (1997). Modelling and safety verification of dicrete/continuous processing systems. *AIChE Journal*, 43(4):1041–1059.

Dimitriadis, V. D., Shah, N., and Pantelides, C. C. (1996a). A case study in hybrid process safety verification. *Computers and Chem. Eng.*, 20, Suppl.:S503–S508.

Dimitriadis, V. D., Shah, N., and Pantelides, C. C. (1996b). Optimal design of hybrid controllers for hybrid process systems. In *R. Alur, T.A. Henzinger, and E.D. Sontag, editors,* Lecture Notes in Computer Science 1066: Hybrid Systems III, volume 1066 of *LNCS*, pages 224–257. Springer.

Doğruel, M. and Özgüner, U. (1995). Modeling and stability issues in hybrid systems. In Antsaklis, P., Kohn, W., Nerode, A., and Sastry, S., editors, *Lecture Notes in Computer Science: Hybrid Systems II*, volume 999, pages 148–165. Springer.

Doğruel, M., Özgüner, U., and Drakunov, S. (1996). Sliding-mode control in discrete-state and hybrid systems. *IEEE Transactions on Automatic Control*, 41:414–419.

DoME (1999). DoME guide. http://www.htc.honeywell.com/dome/, Honeywell Technology Center, Honeywell. version 5.2.1.

Drath, R. (1999). *Modellierung hybrider Systeme auf der Basis modifizierter Petri-Netze*. PhD thesis, TU-Ilmenau, Fachgebiet Automatisierungsanlagen und Prozeßleittechnik, Ilmenau. ISBN-Nr.: 3-932633-40-7.

Drath, R., Engmann, U., and Schwuchow, S. (1999). Hybrid aspects of modelling manufacturing systems using modified petri nets. In *5th Workshop on Intelligent Manufacturing Systems*, Granado, Brasil.

Drath, R. and Schwuchow, S. (1997). Modellierung diskret-kontinuierlicher Systeme mit Petri-Netzen. In Schnieder, E., editor, *Entwurf komplexer Automatisierungsysteme 5. Fachtagung*, pages 265–283, Braunschweig.

Dymola (2002). Dymola version 4.2a. http://www.dynasim.se/.

Elmqvist, H., Cellier, F. E., and Otter, M. (1993). Object-oriented modeling of hybrid systems. In *ESS'93, European Simulation Symposium*, Delft.

Engell, S. (1997). Modellierung und Analyse hybrider dynamischer Systeme. *at–Automatisierungstechnik*, 45(4):152–162.

Engell, S., editor (2000). *Special Issue on Discrete Event Models of Continuous Systems*, volume 6, no. 1 of *Mathematical and Computer Modelling of Dynamical Systems*.

Engell, S., Hoffmann, I., and Sapronowa, L. (1997). Chaos in einfachen kontinuierlich-diskreten dynamischen Systemen. *at–Automatisierungstechnik*, 45(9):399–406.

Engell, S., Kowalewski, S., and Zaytoon, J., editors (2000). *4th Int. Conf. on Automation of Mixed Processes: Hybrid Dynamic Systems (ADPM 2000), Dortmund, Germany*. Shaker.

Enste, U. (2001). *VDI Fortschritt-Berichte, Reihe 8, Nr. 884, Generische Entwurfsmuster in der Funktionsbausteintechnik und deren Anwendung in der operativen Prozessführung*. VDI-Verlag.

Enste, U. and Epple, U. (1998). Standardisierte Prozessfuehrungsbausteine - die Basis fuer Applikationsmodelle zur operativen Fuehrung von verfahrenstechnischen Produktionsanlagen. In *VDI Bericht 1397*. VDI-Verlag.

Enste, U. and Epple, U. (2001). Technical application of hybrid modeling methods to specify function block systems. *Automatisierungstechnik - at*, 49(2):52–59.

Enste, U. and Fedai, M. (1998). Flexible process control structures in multi-product and redundant-routing-plants. In *9th IFAC Symposium on Automation in Mining, Mineral and Metal Processing*, pages 211–214.

Enste, U. and Kneissl, M. (2000). Modelling of software structures in process control systems - avoiding bugs by using graph grammars. In *IMACS Symposium on MATHEMATICAL MODELLING, ARGESIM Report No. 15: Proceedings Vol.1, Vienna*, pages 381–384.

Enste, U. and Uecker, F. (2000). Use of supervision information in process control. *IEE Computing & Control Engineering Journal*, pages 234–241.

Epple, U. (1994). Operational control of process plants. In *Process Control Engineering*. VCH-Verlagsgesellschaft, Weinheim.

Ernst, T., Jähnichen, S., and Klose, M. (1997). Object-oriented physical systems modeling, Modelica, and the SmileM simulation environment. In Sydow, A., editor, *Proceedings of the 15th IMACS World Congress on Scientific Computation, Modelling and Applied Mathematics*, volume 6, pages 653–658.

Ernst, T., Klein-Robbenhaar, C., Nordwig, A., and Schrag, T. (2000). Modellierung und Simulation hybrider Systeme mit Smile. *Informatik Forschung und Entwicklung*, 5.

Evans, R. and Savkin, A., editors (1999). *Systems and Control Letters, Special issue on Hybrid Control Systems*, volume 38(3).

Fábián, G., van Beek, D. A., and Rooda, J. E. (1998). Integration of the discrete and the continuous behaviour in the hybrid chi simulator. In *1998 European Simulation Multiconference, Manchester*, pages 207–257.

Fahrland, D. (1970). Combined discrete event continuous systems simulation. *Simulation*, 14(2):71–72.

Fellendorf, M. (1994). VISSIM: A microscopic Simulation Tool to evaluate Actuated Signal Control including Bus priority. In *64th ITE Annual Meeting*, Dallas.

Fieldbus DDLS (1996). Device description language specification. Technical report, Fieldbus Foundation, Austin Texas.

Floyd, R. W. (1967). Assigning meanings to programs. In Schwartz, J., editor, *Proceedings AMS Symposium Applied Mathematics*, volume 19, pages 19–31, Providence, RI. American Mathematical Society.

Föllinger, O. (1994). *Regelungstechnik. Einführung in die Methoden und ihre Anwendung*. Hüthig.

Forbus, K. D. (1990). Qualitative reasoning. Draft chapter.

Förstner, D. (2001). *Qualitative Modellierung für die Prozeßdiagnose und deren Anwendung auf Dieseleinspritzpumpen*. PhD thesis, TU Hamburg-Harburg.

Frank, P. M. (1998). Komplexe Systeme - Nichtlineare Rückkopplungssysteme jenseits der Stabilität. *at - Automatisierungstechnik*, 46(4):167–179.

Frank, R. (2001). Entwicklung einer Internetanbindung für den Modellprozess Drei-Tank-System. Diplomarbeit, Institut für Automatisierungs- und Softwaretechnik (IAS), Universität Stuttgart.

Franke, D., Moor, T., and Raisch, J. (2000). Discrete supervisory control of switched linear systems. *at-Automatisierungstechnik*, 48:9:461–467.

Frehse, G., Stursberg, O., Engell, S., Huuck, R., and Lukoschus, B. (2002). Modular analysis of discrete controllers for distributed hybrid systems. In *b '02: The XV. IFAC World Congress, Barcelona, Spain, July 21–26, 2002*. To appear.

Frehse, G. F., Stursberg, O., Engell, S., Huuck, R., and Lukoschus, B. (2001). Verification of hybrid controlled processing systems based on decomposition and deduction. In *ISIC 2001: 16th IEEE International Symposium on Intelligent Control, Mexico City, Mexico, September 5–7, 2001*, pages 150–155. IEEE Control Systems Society, IEEE Press.

Friesen, V. (1995). An exercise in hybrid system specification using an extension of Z. In Bouajjani, A. and Maler, O., editors, *Second European Workshop on Real-Time and Hybrid Systems*, pages 311–316.

Friesen, V. (1997). *Objektorientierte Spezifikation hybrider Systeme*. PhD thesis, Technical University of Berlin.

Friesen, V. (1998). A logic for the specification of continuous systems. LNCS 1386, Berlin, Germany. Springer.

Friesen, V., Nordwig, A., and Weber, M. (1998a). Object-oriented specification of hybrid systems using UML^{h} and ZimOO. In *Proc. 11th Int. Conf. on the Z Formal Method (ZUM)*, LNCS 1493. Springer.

Friesen, V., Nordwig, A., and Weber, M. (1998b). Toward an object-oriented design methodology for hybrid systems. Proceedings of the Colloquium on Object Technology and System Re-Engineering, Oxford.

Fröhlich, P. (1996). *Überwachung verfahrenstechnischer Prozesse unter Verwendung eines qualitativen Modellierungsverfahrens*. PhD thesis, Institut für Automatisierungs- und Softwaretechnik (IAS), Universität Stuttgart.

Gamma, E., Helm, R., Johnson, R., and Vlissides, J. (1995). *Design Patterns, Elements of Reusable Object-Oriented Software*. Addison-Wesley.

Gao, Z. and Antsaklis, P. J. (1991). Stability of the pseudo-inverse method for reconfigurable control systems. *International Journal of Control*, 53:717–729.

Gazis, D. C. et al. (1959). Car following theory of steady-state traffic flow. *Operns. Res.*, 7:499–505.

Geisler, R., Klar, M., and Pons, C. (1998). Dimensions and dichotomy in metamodeling. Technical Report 98-2, Technical University of Berlin.

Genrich, H. J. (1978). Ein Kalkül des Planens und Handelns. In *Ansätze zur Organsiationstheorie rechnergestützter Informationssysteme*, GMD Bericht 111, pages 77–92. Oldenbourg.

Genrich, H. J. (1987). Predicate/transition nets. *Advances in Petri nets 1986, part I. Lecture Notes in Computer Science*, 254:207–247.

Genrich, H. J. and Lautenbach, K. (1981). System modelling with high-level petri nets. *Theoretical Computer Science*, 13.

Ghezzi, C., Mandrioli, D., Morasca, S., and Pezzè, M. (1991). A unified high-level petri net formalism for time-critical systems. *IEEE Transactions On Software Engineering*, 17(2):160–172.

Gill, P., Murray, W., and Saunders, M. (1997). *User's guide for SNOPT 5.3: a fortran package for large-scale nonlinear programming*. Department of Mathematics, Univ. of California San Diego.

Gilles, E. D., Holl, P., and Marquardt, W. (1986). Dynamische Simulation komplexer chemischer Prozesse. *Chem.-Ing.-Tech*, 58(4):268–278.

Giua, A. and Piccaluga, A. (2002). Bibliography on hybrid petri nets. http://bode.-diee.unica.it/~ hpn/.

Giua, A. and Usai, E. (1996). High-level hybrid petri nets: a definition. In *35th Conference on Decision and Control*, pages 148–150, Kobe, Japan.

Giua, A. and Usai, E. (1998). Modeling hybrid systems by high-level petri nets. In *ADPM'98*, pages 316–323.

Glocker, C. (1995). *Dynamik von Starrkörpersystemen mit Reibung und Stößen*. PhD thesis, TU München, München.

Glover, F. (1975). Improved linear integer programming formulations of nonlinear integer problems. *Managem. Science*, 22(4):455–460.

Göhner, P. and Lauber, R. (1999). *Prozessautomatisierung 2*, volume 2. Springer, Berlin Heidelberg, 1 edition.

Gokbayrak, K. and Cassandras, C. G. (2000). Hybrid controllers for hierarchically decomposed systems. In *Proc. 3rd Int. Workshop of Hybrid Systems: Computations and Control*, volume 1790 of *LNCS*, pages 117–129. Springer.

Goldstein, H. H. and von Neumann, J. (1947). Planning and coding problems of an electronic computing instrument. In Taub, A., editor, *J. von Neumann—Collected Works*, pages 80–151. McMillan, New York.

gPROMS (2002). Homepage: http://www.psenterprise.com/.

Greenstreet, M. and Mitchell, I. (1999). Reachability analysis using polygonal projections. In Vaandrager, F. W. and van Schuppen, J. H., editors, *Hybrid Systems: Computation and Control, Proc. 2nd Int. Workshop, HSCC'99, Berg en Dal, The Netherlands, March 1999*, Lecture Notes in Computer Science 1569, pages 103–116. Springer.

Gribaudo, M., Sereno, M., and Bobbio, A. (1999). Fluid stochastic petri nets: An extended formalism to include non-markovian models. In *Proc. Eighth International Workshop on Petri Nets and Performance Models - PNPM'99*, pages 74–81, Zaragoza, Spain.

Griepentrog, E. and März, R. (1986). *Differential-Algebraic Equations and Their Numerical Treatment*. BSB Teubner, Leipzig. ISBN 3-322-00343-4.

Grossman, R. L., Nerode, A., Ravn, A. P., and Rischel, H., editors (1993). *Hybrid Systems*, volume 736 of *Lecture Notes in Computer Science*. Springer.

Grosu, R., Krüger, I., and Stauner, T. (2000). Hybrid Sequence Charts. In *Proc. of ISORC 2000*. IEEE.

Grosu, R., Stauner, T., and Broy, M. (1998). A modular visual model for hybrid systems. In *Proc. of FTRTFT'98*, LNCS 1486. Springer.

Group, I. . W. (1999). IEEE standard 1076.1-1999. http://www.vhdl.org.

Haidacher, S., Schlegl, T., and Buss, M. (1999). Grasp Evaluation Based on Unilateral Force Closure. In *Proceedings of the IEEE/RSJ International Conference on Intelligent Robots and Systems IROS*, pages 424–429, Kyongju, Korea.

Hamscher, W., deKleer, J., and Console, L., editors (1992). *Readings in Model-Based Diagnosis*. Morgan Kaufman.

Hanisch, H.-M. (1992). *Petri-Netze in der Verfahrenstechnik*. Oldenbourg, München, Wien.

Hanisch, H.-M., Lautenbach, K., Simon, C., and Thieme, J. (1998a). Timestamp nets in technical applications. In *IEEE International Workshop on Discrete Event Systems*, San Diego, CA, USA.

Hanisch, H.-M., Lautenbach, K., Simon, C., and Thieme, J. (1998b). Timestamp petri nets in technical applications. In Giua, A., Smedinga, R., and Spathopoulos, M. P., editors, *IEE International Workshop on Discrete Event Systems*, IEE Control, pages 321–326, Cagliari, Sardinia, Italy.

Hanisch, H.-M., Lautenbach, K., Simon, C., and Thieme, J. (1998c). Zeitstempelnetze in technischen Anwendungen. Fachberichte Informatik 2–98, Universität Koblenz-Landau, Institut für Informatik, Rheinau 1, D-56075 Koblenz.

Hardt, M., Helton, J., and Kreutz-Delgado, K. (2000). Numerical solution of nonlinear $\mathcal{H}_2$ and $\mathcal{H}_\infty$ control problems with application to jet engine compressors. *IEEE Transactions on Control Systems Technology*, 8(1):98–111.

Hardt, M. and von Stryk, O. (2000). Towards optimal hybrid control solutions for gait patterns of a quadruped. In *CLAWAR 2000 – 3rd International Conference on Climbing and Walking Robots, Madrid, 2–4 October, Professional Engineering Publishing, UK*, pages 385–392.

Harel, D. (1987). Statecharts: A Visual Formalism for Complex Systems. *Science of Computer Programming*, 8:231–274.

Harel, D. and Gery, E. (1996). Executable object modeling with Statecharts. In *Proceedings of the 18th International Conference of Software Engineering*, IEEE Press.

Harel, D., Pnueli, A., Schmidt, J., and Sherman, R. (1987). On the formal semantics of statecharts. In *2nd IEEE Symp. on Logic in Computer Science*, pages 54–64. IEEE Press.

He, K. X. and Lemmon, M. D. (1998). Lyapunov Stability of Continous-Valued Systems Under the Supervision of Discrete-Event Transition Systems. In Henzinger, T. A. and Sastry, S., editors, *Hybrid Systems: Computation and Control*, LNCS 1386, pages 175–189, Berlin, Germany. Springer.

Hedlund, S. and Rantzer, A. (1999). Optimal control of hybrid systems. In *Proceedings of the 38th IEEE Conference on Decision and Control (Phoenix, AZ)*, pages 3972–3977.

Heinkel, U. (2000). *The VHDL reference*. Wiley, Chichester.

Henzinger, A., Kopke, P., Puri, A., and Varaiya, P. (1995). What's decidable about hybrid automata. In *Proceedings of the 27th Annual ACM Symposium on Theory of Computing (STOC1995)*, pages 373–382.

Henzinger, T., Ho, P., and Wong-Toi, H. (1997). Hytech: A model checker for hybrid systems. *Software Tools for Technology Transfer*, 1(1,2):110–122.

Henzinger, T., Ho, P., and Wong-Toi, H. (1998a). Algorithmic analysis of nonlinear hybrid systems. *IEEE Transactions on Automatic Control*, 43(4):540–554.

Henzinger, T., Kopke, P., Puri, A., and P.Varaiya (1998b). What's decidable about hybrid automata. *J. Comp. Syst. Science*, 57:94–124.

Henzinger, T., Qadeer, S., Rajamani, S., and Tasiran, S. (1998c). You assume, we guarantee: Methodology and case studies. In *Proc. 10th Int. Conf. on Computer-Aided Verification*, volume 1427 of *Lecture Notes in Computer Science*, pages 440–451. Springer.

Henzinger, T. A. (1996). The theory of hybrid automata. In *Proc. of 11th Annual IEEE Symposium on Logic in Computer Science (LICS'96)*, pages 278–292. IEEE Computer Society Press.

Henzinger, T. A., Minea, M., and Prabhu, V. (2001). Assume-guarantee reasoning for hierarchical hybrid systems. In *HSCC '01: 4th International Workshop on Hybrid Systems: Computation and Control*, volume 2034 of *Lecture Notes in Computer Science*, pages 275–290. Springer.

Henzinger, T. A. and Sastry, S., editors (1998). *Hybrid Systems – Computation and Control (HSCC'98)*, volume 1386 of *Lecture Notes in Computer Science*. Springer.

HLA TMD Document (1996). HLA time management design document, version 1.0.

Hoare, C. (1969). An axiomatic basis for computer programming. *Communications of the ACM*, 12(10):576–580, 583.

Hoare, C. (1985). *Communicating Sequential Processes*. Prentice-Hall International, Engelwood Cliffs.

Holzmann, G. J. (1997). The model checker SPIN. *IEEE Transactions on Software Engineering*, 23(5):279–295.

Horton, G., Kulkarni, V. G., Nicol, D. M., and Trivedi, K. S. (1998). Fluid stochastic petri nets: Theory, applications and solution. *European Journal of Operations Research*, 105(1):184–201.

Huber, F., Schätz, B., and Einert, G. (1997). Consistent graphical specification of distributed systems. In *FME '97: 4th International Symposium of Formal Methods Europe, LNCS 1313*, pages 122 – 141.

Hubert, P., Jensen, K., and Shapiro, R. (1991). Hierarchies in couloured petri nets. *Lecture Notes in Computer Science*, 483.

Huuck, R., Engell, S., Kowalewski, S., Lakhnech, Y., Preußig, J., and Urbina, L. (1997). Comparing timed c/e systems with timed automata. In *International Workshop on Hybrid and Real-Time Systems (Hart '97)*, LNCS 1201, pages 81–86, Grenoble. Springer.

IEC 1131 (1993). International standard IEC 1131 programmable controllers, part 3, programming languages.

IEC 61131-3 (1992). Programming language for programmable controllers. Technical report, Committee IEC 61131-3.

IEC SC 65C WG7 (1999). Function blocks for process control. Technical report, Committee IEC 61804-1.

IEC TC65 WG6 (1999). Function blocks for industrial-process measurement and control systems. Technical report, Committee IEC 61499-1.

Ioannou, P. (1996). *Robust Adaptive Control*. Prentice-Hall Upper Saddle River NJ.

Isermann, R. (1996a). Modellgestützte Überwachung und Fehlerdiagnose Technischer Systeme (Teil 1). *atp*, 38(5):9–20.

Isermann, R. (1996b). Modellgestützte Überwachung und Fehlerdiagnose Technischer Systeme (Teil 2). *atp*, 38(6):48–57.

ITU (1999). ITU-T Recommendation Z.120: Message Sequence Charts (MSC).

Jähnichen, S. and Klein-Robbenhaar, C. (2000). Generic modeling and simulation of hybrid systems with adaptive modeling depth. Technical report, Technical University of Berlin. (in German).

Jensen, H. E. (1999). *Abstraction-Based Verification of Distributed Systems*. PhD thesis, Aalborg University.

Jensen, H. E., Larsen, K. G., and Skou, A. (2000). Scaling up Uppaal – automatic verification of real-time systems using compositionality and abstraction. In Joseph, M., editor, *FTRTFT 2000: 6th International Symposium on Formal Techniques in Real-Time and Fault-Tolerant Systems, September 20–22, 2000, Pune, India*, volume 1926 of *Lecture Notes in Computer Science*, pages 19–30. Springer.

Jensen, K. (1992). *Coloured Petri Nets: Basic Concepts, Analysis Methods and Practical Use*, volume 1. Springer.

Jensen, K. (1997). *Coloured Petri Nets: Basic Concepts, Analysis Methods and Practical Use*, volume 2. Springer.

Jensen, K. and Rozenberg, G., editors (1991). *High-level Petri Nets: theory and application*. Springer.

Jhala, R. and McMillan, K. L. (2001). Microarchitecture verification by compositional model checking. In Berry, G., Comon, H., and Finkel, A., editors, *CAV 2001: 13th International Conference on Computer Aided Verification, Paris, France, July 18–22, 2001*, volume 2102 of *Lecture Notes in Computer Science*, pages 396–410. Springer.

Jirstrand, M. (1998). *Constructive Methods for Inequality Constraints in Control*. PhD thesis, Department of Electrical Engineering, Linköping University, Linköping, Sweden.

Johannson, M. and Rantzer, A. (1998). Computation of Piecewise Quadratic Lyapunov Functions for Hybrid Systems. *IEEE Trans. Aut. Control*, 43(4):555–559.

John, S. (2001). Transition selection algorithms for Statecharts. *Proceedings of the GI/OCG annual congress*, 1:pp. 622–627.

Jones, C. B. (1981). *Development Methods for Computer Programs including a Notion of Interference*. PhD thesis, Oxford University Computing Laboratory. Printed as: Programming Research Group, Technical Monograph 25.

Jones, C. B. (1983). Tentative steps toward a development method for interfering programs. *ACM Transactions on Programming Languages and Systems*, 5(4):596–619.

Joos, H.-D. (1999). A methodology for multi-objective design assessment and flight control synthesis tuning. *Aerospace Science and Technology*, 3(3):161–176.

Kaiser, R. and Beaumariage, T. (1997). Conceptual design of an artificial intelligence architecture for decision making in manufacturing simulation. In Wallace, J. and Beaumariage, T., editors, *Object-Oriented Simulation Conf. OOS'97*, pages 11–15. SCS International, San Diego.

Kienle, A. (2000). Low-order models for ideal multicomponent distillation processes using nonlinear wave propagation theory. *Chemical Engineering Science*, 55:1817–1828.

Kirkpatrick, S., Gelatt, C. D., and Vecchi, M. P. (1983). Optimization by simulated annealing. *Science*, 220(4598):671–680.

Klar, M. and Mann, S. (1998). A Metamodel for Object-Oriented Statecharts. *The Second Workshop on Rigorous Object Oriented Methods, ROOM 2*.

Klein, E., Itigin, A., Raisch, J., and Kienle, A. (2000). Automatic generation of switching start-up schemes for chemical processes. *Proc. ESCAPE10 – 10th European Symposium on Computer Aided Process Engineering*, pages 619–624.

Klein, E., Kienle, A., and Raisch, J. (1998). Synthesizing a supervisory control scheme for the start-up procedure of a distillation column - an approach based on approximating continuous dynamics by des models. *Proc. LSS'98 - 8th IFAC Colloquium on Large Scale Systems*, pages 716–721.

Klein, E., Kienle, A., Raisch, J., and Wehlan, H. (1999). Synthese einer Anfahrregelung für eine Destillationskolonne auf der Grundlage einer ereignisdiskreten Approximation der kontinuierlichen Dynamik. *6. Fachtagung Entwicklung and Betrieb komplexer Automatisierungssysteme (EKA99)*, pages 447–464.

Klein, E. and Raisch, J. (1998). Safety enforcement in process control systems - a batch evaporator example. In *Proc. WODES'98 - International Workshop on Discrete Event Systems, Cagliari, Italy*, pages 327–333. IEE.

Kleinmann, K. P. (1996). *Lernende Regelung eines mehrfingrigen Robotergreifers.* PhD thesis, TU Darmstadt, Darmstadt.

Kloas, M., Friesen, V., and Simons, M. (1995). Smile — A simulation environment for energy systems. In Sydow, A., editor, *Proceedings of the 5th International IMACS-Symposium on Systems Analysis and Simulation (SAS'95)*, pages 503–506. Gordon and Breach Publishers.

Komarow, W. B. and Skotschinski, A. A. (1956). *Grubenbewetterung*. VEB Verlag Technik Berlin.

Kondak, K. and Hommel, G. (2001). Computation of time optimal movements for autonomous parking of non-holonomic mobile platforms. In *Proceedings of the IEEE International Conference on Robotics and Automation (Seoul, Korea)*, pages 2698–2703.

König, R. and Quäck, L. (1988). *Petri-Netze in der Steuerungs- und Digitaltechnik.* Oldenbourg, München, Wien.

Koutsoukos, X., Antsaklis, P. J., Stiver, J. A., and Lemmon, M. D. (2000). Supervisory control of hybrid systems. *Proceedings of the IEEE*, 88:1026–1049.

Kowalewski, S. (1996). *Modulare diskrete Modellierung verfahrenstechnischer Anlagen zum systematischen Steuerungsentwurf.* PhD thesis, Fachbereich Chemietechnik, Dortmund.

Kowalewski, S. (2002). Introduction to the analysis and verification of hybrid systems. In *this volume*.

Kowalewski, S., Engell, S., Preussig, J., and Stursberg, O. (1999). Verification of logic controllers for continuous plants using timed condition/event system models. *Automatica*, 35(3):505–518.

Kowalewski, S., Herrmann, P., Engell, S., Huuck, R., Krumm, H., Lakhnech, Y., and Lukoschus, B. (2001a). Approaches to the formal verification of hybrid systems. *at-Automatisierungstechnik*, 49(2):66–74.

Kowalewski, S. and Preußig, J. (1996). Timed condition/event systems: A framework for modular models of chemical plants and verification of their real-time discrete control. In Margaria, T. and Steffen, B., editors, *Tools and Algorithms for the Construction and Analysis of Systems, Proc. 2nd International Workshop TACAS'96*, Lecture Notes in Computer Science 1055, pages 225–240, Passau. Springer.

Kowalewski, S., Stursberg, O., and Bauer, N. (2001b). An experimental batch plant as a test case for the verification of hybrid systems. *European Journal of Control*, 7.

Kowalewski, S., Stursberg, O., and Treseler, H. (1998). Diskrete Modellierung verfahrenstechnischer Prozesse zur Steuerungsverifikation. *at - Automatisierungstechnik*, 4:180–187.

Kramer, D. (1997). *JDK 1.1.1 Documentation*. Sun Microsystems, Inc.

Krebs, V. G. and Schnieder, E., editors (2000). *Hybrid Systems I: Modeling and Control*, volume 48.

Kripke, S. A. (1963). Semantical considerations on modal logic. *Acta Philosophica Fennica*, 16:83–94.

Krogh, B. (1993). Condition/event signal interfaces for block diagram modeling and analysis of hybrid systems. In *8th Int. Symp. on Intelligent Control Systems*, pages 180–185.

Kuipers, B. (1986). Qualitative simulation. *Artificial Intelligence*, 29:289–338.

Kuipers, B. (1994). *Qualitative Reasoning*. MIT Press.

Kupferman, O., Vardi, M. Y., and Wolper, P. (2000). An automata-theoretic approach to branching-time model checking. *Journal of the ACM*, 47(2):312–360.

Kurz, H. (1990). Realisierung gehobener Methoden der Regelungstechnik auf Prozessleitsystemen - Ein Diskussionsbeitrag. *Automatisierungstechnische Praxis - atp*, 32(10):489–494.

Labinaz, G., Bayoumi, M. M., and Rudie, K. (1996). Modeling and Control of Hybrid Systems: A Survey. In *Proc. IFAC 13th Triennial World Congress*, pages 293–304, San Francisco, USA. IFAC.

Lafferiere, G., Pappas, G., and Yovine, S. (1999). A new class of decidable hybrid systems. In Vaandrager, F. W. and van Schuppen, J. H., editors, *Hybrid Systems: Computation and Control, Proc. 2nd Int. Workshop, HSCC'99, Berg en Dal, The Netherlands, March 1999*, volume 1569 of *Lecture Notes in Computer Science*, pages 137–151. Springer.

Lafferriere, G., Pappas, G., and Sastry, S. (2000). O-minimal hybrid systems. *Mathematics of Control, Signals, and Systems*, 13(3):1–21.

Larsen, K. G., Pettersson, P., and Yi, W. (1997). UPPAAL in a nutshell *International Journal on Software Tools for Technology Transfer*, 1(1–2):134–152.

Laudwein, A. (1999). Konzeption und Entwicklung einer Steuerungs- und Regelungssoftware für den Modellprozess "Drei-Tank-System". Diplomarbeit, Institut für Automatisierungs- und Softwaretechnik (IAS), Universität Stuttgart.

Laufenberg, X. (1997). *Ein modellbasiertes qualitatives Verfahren für die Gefahrenanalyse*. Dissertation, Institut für Automatisierungs- und Softwaretechnik (IAS), Universität Stuttgart.

Lautenbach, K. and Simon, C. (1999). Erweiterte Zeitstempelnetze. Fachberichte Informatik 03–99, Universität Koblenz-Landau, Institut für Informatik, Rheinau 1, D-56075 Koblenz.

Lautenbach, K. and Simon, C. (2000). Verification in a logic of actions. In *7. Workshop Algorithmen und Werkzeuge für Petrinetze*, Koblenz.

Lautenbach, K. and Simon, C. (2001). Modellierung der Dynamik einer Batch-Anlage. In Schnieder, E., editor, *Engineering komplexer Automatisierungssysteme, EKA 2001*, Braunschweig.

Le Bail, J., Alla, H., and David, R. (1991). Hybrid petri nets. In *European Control Conference*, pages 1472–1477.

Lee, J.-D. e. a. (2000). Analysis of moving and fixed autoblock systems for korean high speed railway. In *Computers in Railways VII*, pages 843–851. WIT Press, Bologna.

Lee, S. and Grossmann, I. (2000). New algorithms for nonlinear generalized disjunctive programming. *Comp. and Chemical. Eng.*, 4:2125–2141.

Lemmon, M., He, K., and Markovsky, I. (1999). Supervisory hybrid systems. *IEEE Control Systems Magazine*, 19:42–55.

Leue, S., Mehrmann, L., and Rezai, M. (1998). Synthesizing ROOM Models from MSC Specifications. Technical Report TR-98-06, University of Waterloo.

Levin, G. M. and Gries, D. (1981). A proof technique for communicating sequential processes. *Acta Informatica*, 15(3):281–302.

Li, Z., Soh, C., and Xu, X. (2000). Lyapunov stability of a class of hybrid dynamic systems. *Automatica*, 36:297–302.

Liberzon, D. and Morse, A. S. (1999). Basic problems in stability and design of switched systems. *IEEE Control Systems Magazine*, 19.

Lichtenberg, G. and Kamau, S. (2001). A classification of the input-output behaviour of hybrid systems. In *European Control Conference*.

Lichtenberg, G., Lunze, J., and Raisch, J. (1999a). Two approaches to modeling the qualitative behaviour of dynamic systems. *at-Automatisierungstechnik*, 47:187–198.

Lichtenberg, G., Lunze, J., and Raisch, J. (1999b). Zwei Wege zur Modellierung des qualitativen Verhaltens dynamischer Systeme. *at - Automatisierungstechnik*, 47(5):187–198.

Lichtenberg, G., Lunze, J., Scheuring, R., and Schröder, J. (1999c). Prozessdiagnose mittels qualitativer Modelle am Beispiel eines Wasserstoffverdichters. *at - Automatisierungstechnik*, 47(3):101–109.

Lichtenstein, O. and Pnueli, A. (1985). Checking that finite state concurrent programs satisfy their linear specifications. In *Twelfth ACM Symposium on the Priciples of Programming Languages*, pages 97– 105.

Liggesmeyer, P. and Mäckel, P. (2000). Automtisierung erweiterter Fehlerbaumanalysen für komplexe technische Systeme. *at - Automatisierungstechnik*, 48(2):67–76.

Lighthill, M. J. and Whitham, G. B. (1955). On kinematic waves. ii. a theory of traffic flow on long crowded roads. In *Roy. Society*, volume 229 A, pages 317–345, London.

Lincoln, B. and Rantzer, A. (2001). Optimizing linear system switching. In *Proc. 40th IEEE Conf. Decision and Control*, pages 2063–2068.

Lorch, O., Denk, J., Seara, J., Buss, M., Freyberger, F., and Schmidt, G. (2000). Vigwam — an emulation environment for a vision guided virtual walking machine. In *Proceedings of the First IEEE-RAS International Conference on Humanoid Robots HUMANOIDS 2000 (Cambridge, MA, USA)*.

Lötzbeyer, H. and Pretschner, A. (2000). AutoFocus on Constraint Logic Programming. In *Proc. (Constraint) Logic Programming and Software Engineering*.

Lunze, J. (1994). Qualitative modelling of linear dynamical systems with quantized state measurements. *Automatica*, 30:417–431.

Lunze, J. (1995). Stabilisation of nonlinear systems by qualitative feedback controllers. *Intern. J. Control*, 62:109–128.

Lunze, J. (1998a). On the Markov property of quantised state measurement sequences. *Automatica*, 34:1439–1444.

Lunze, J. (1998b). Qualitative Modellierung dynamischer Systeme durch stochastische Automaten. *at - Automatisierungstechnik*, 46(6):271–283.

Lunze, J. (1999). A timed discrete-event abstraction of continuous-variable systems. *Intern. J. Control*, 72:1147–1164.

Lunze, J. (2000). Process supervision by means of qualitative models. *Annual Reviews in Control*, 24:41–54.

Lunze, J. (2001). Control reconfiguration. In *Encyclopedia of Live Support Systems*. EOLSS Publishers. submitted.

Lunze, J. (2002). *Regelungstechnik, Band 2*. Springer.

Lunze, J., Heiming, B., and et. al., M. S. (2000). Three-tank control reconfiguration. In Aström, K., editor, *Control of Complex Systems*. Springer.

Lunze, J. and Nixdorf, B. (2002). Representation of hybrid systems by means of stochastic automata. *Mathematical Modelling of Systems*, 7:383–422.

Lunze, J. and Nixdorf, B. (2003). Discrete reachability of hybrid systems. *Intern. J. Control, submitted*.

Lunze, J., Nixdorf, B., and Richter, H. (1997). Hybrid modelling of continuous-variable systems with application to supervisory control. In *Proceedings of the European Control Conference 1997*.

Lunze, J. and Raisch, J. (2002). Discrete models for hybrid systems. In Engell, S., Frehse, G., and Schnieder, E., editors, *Modelling, Analysis, and Design of Hybrid Systems*, Lecture Notes in Control and Information Science. Springer. (This volume).

Lunze, J. and Schiller, F. (1997). Qualitative Prozessdiagnose auf wahrscheinlichkeitstheoretischer Grundlage. *at - Automatisierungstechnik*, 45(8):351–359.

Lunze, J. and Schröder, J. (1999). Process diagnosis based on a discrete-event description. *at – Automatisierungstechnik*, 47:358–365.

Lunze, J. and Steffen, T. (2000). Reconfigurable control of a quantised system. In *Proceeding of SAFEPROCESS 2000: 4th Symposium on Fault Detection*, pages 822–827. IFAC.

Lunze, J. and Steffen, T. (2002). Hybrid reconfigurable control. In *this volume*.

Lüth, T. (1998). *Technical Multiagent Systems*. Hanser Publisher. (in German).

Lygeros, J., Tomlin, C., and Sastry, S. (1999). Controllers for reachability specifications for hybrid systems. *Automatica*, 35:349–370.

Lynch, N. and Krogh, B. H., editors (2000). *Hybrid Systems – Computation and Control (HSCC 2000)*, volume 1790 of *Lecture Notes in Computer Science*. Springer.

Lynch, N., Segala, R., Vaandrager, F., and Weinberg, H. B. (1996). Hybrid I/O automata. In Alur, R., Henzinger, T. A., and Sontag, E. D., editors, *Hybrid Systems III*, LNCS 1066, pages 496–510. Springer.

Maciejowski, J. (2002). *Predictive control with constraints*. Prentice Hall.

Mai, G. and Schröder, M. (1999). Simulation of a Flight Control Systems' Redundancy Management System using Statemate. 7. User group meeting STATEMATE.

Maler, O., editor (1997). *Hybrid and Real-Time Systems (HART'97)*, volume 1201 of *Lecture Notes in Computer Science*. Springer.

Maler, O., editor (2001). *Special Issue on Verification of Hybrid Systems*, volume 7, issue 4 of *European Journal of Control*.

Manz, S. (1999). Qualitative Modeling of a Three-Tank-System. In *Interkama-ISA Tech Conference*, Düsseldorf.

Manz, S. (2000). On-line monitoring and diagnosis based on hybrid component models. In *13th International Conference on Software & Systems Engineering and Applications ICSSEA 2000*, Paris.

Manz, S. (2001a). Fuzzy based qualitative models in combination with dynamical models for online monitoring of technical systems. In *International Conference on Computational Intelligence for Modelling, Control and Aut. CIMCA 2001*, Las Vegas.

Manz, S. (2001b). Online fault detection and diagnosis of complex systems based on hybrid component models. In *14th International Congress on Condition Monitoring and Diagnostics Engineering Managem. COMADEM2001*, Manchester.

Mareczek, J., Buss, M., and Schmidt, G. (1998). Robust Global Stabilization of the Underactuated 2-DOF Manipulator R2D1. In *Proceedings of the IEEE International Conference on Robotics and Automation (Leuven, Belgium)*, pages 2640–2645.

Mareczek, J., Buss, M., and Schmidt, G. (1999). Robust Control of a Non-Holonomic Underactuated SCARA Robot. In Tzafestas, S. and Schmidt, G., editors, *Lecture Notes in Control and Information Sciences: Progress in System and Robot Analysis and Control Design*, volume 243, pages 381–396. Springer.

Marsan, M. A., Balbo, G., Chiola, G., Donatelli, S., and Francheschinis, G. (1995). *Modelling with Generalized Stochastic Petri Nets.* John Wiley & Sons.

Martin, B. and Bobrow, J. (1997). Minimum effort motions for open chain manipulators with task-dependent end-effector constraints. In *Proceedings of the IEEE International Conference on Robotics and Automation (Albuquerque, New Mexiko)*, pages 2044–2049.

Matlab (2002). Homepage: http://www.mathworks.com.

Matsuno, H. and Doi, A. (2000). Hybrid Petri Net Representation of Gene Regulatory Network. In *Pacific Symposium on BioComputing 2000*, pages 341–352, Hawaii.

Matsuno, H., Doi, A., Drath, R., and Miyano, S. (2000). Genomic object net: Object representation of biological systems. *Genome Informatics*, 11.

Matsuno, H., Doi, A., Drath, R., and Miyano, S. (2001). Genomic object net: Hybrid petri net for describing biological systems. In *Fifth Annual International Conference on Computational Molecular Biology*, Montreal, Canada.

Matsuno, H. and Miyano, S. (2000). A platform for virtual cells; simulation of gene regulatory control by hybrid object net. *bit*, 32:22–31. (in Japanese).

McMillan, K. L. (1992). *Symbolic Model Checking: An Approach to the State Explosion Problem.* PhD thesis, Carnegie Mellon University. CMU Technical Report CMU-CS-92-131.

McMillan, K. L. (1995). A Technique of a State Space Search Based on Unfolding. In *Formal Methods in System Design 6 (1)*, pages 45–65.

McMillan, K. L. (2000). *The SMV system.* Carnegie Mellon University. Manual for SMV version 2.5.4.

Merz, R. and Litz, L. (2000). Objektorientierte mathematische Modellierung. *Informatik Spektrum*, pages 90–99.

Merz, S. (2001). Model checking: A tutorial overview. In Cassez, F., Jard, C., Rozoy, B., and Ryan, M. D., editors, *Modeling and Verification of Parallel Processes*, volume 2067 of *Lecture Notes in Computer Science*, pages 3–38. Springer.

Meyer, B. (1992). *Eiffel: The Language*. Object-Oriented Series. Prentice Hall, New York, NY.

Meyer, B. (1997). *Object-Oriented Software Construction, Second Edition*. The Object-Oriented Series. Prentice-Hall, Englewood Cliffs (NJ), USA.

Michalewicz, Z. and Fogel, D. (2000). *How to solve it: Modern Heuristics*. Springer.

Millington, D. and Stapleton, J. (1995). Special report: Developing a RAD Standard. In *IEEE Software*, volume 12(5).

Milner, R. (1980). *A Calculus of Communicating Systems*, volume 92 of *Lecture Notes in Computer Science*. Springer.

Milner, R. (1989). *Communication and Concurrency*. Prentice-Hall International, Engelwood Cliffs.

Misra, J. and Chandy, K. M. (1981). Proofs of networks of processes. *IEEE Transactions on Software Engineering*, 7(4):417–426.

Modelica Design Group, T. (2000). Modelica – a unified object-oriented language for physical system modeling v1.4. http://www.modelica.org.

Moody, J. O. and Antsaklis, P. J. (1998). *Supervisory Control of Discrete Event Systems Using Petri Nets*. Kluwer Academic Publishers.

Moor, T. (1998). Event driven control of switched integrator systems. In *Proc. ADPM'98 (Automatisation des Processus Mixtes: les Systèmes Dynamiques Hybrides)*, pages 271–277, Reims, France.

Moor, T. (2000). *Approximationsbasierter Entwurf diskreter Steuerungen für gemischtwertige Regelstrecken*, volume 2 of *Forschungsberichte aus dem Max-Planck-Institut für Dynamik komplexer technischer Systeme*. Shaker, Aachen, Germany. Also PhD thesis, Fachbereich Elektrotechnik, Universität der Bundeswehr Hamburg.

Moor, T., Davoren, J. M., and Raisch, J. (2001a). Modular supervisory control of a class of hybrid systems in a behavioural framework. In *Proceedings of the European Control Conference 2001*, pages 870–875, Porto, Portugal.

Moor, T. and Raisch, J. (1999a). Discrete control of switched linear systems. In *Proceedings of the European Control Conference 1999*, Karlsruhe, Germany.

Moor, T. and Raisch, J. (1999b). Supervisory control of hybrid systems within a behavioural framework. *Systems and Control Letters*, 38:157–166.

Moor, T. and Raisch, J. (2000). Approximation of multiple switched flow systems for the purpose of control synthesis. In *Proc. of the 39th International Conference on Decision and Control, CDC'00*. IEEE Press.

Moor, T. and Raisch, J. (2002). Abstraction based supervisory controller synthesis for high order monotone continuous systems. In *this volume*.

Moor, T., Raisch, J., and Davoren, J. M. (2001b). Computational advantages of a two-level hybrid control architecture. In *Proc. of the 40th International Conference on Decision and Control, CDC'2001*, pages 358–362. IEEE Press.

Moor, T., Raisch, J., and O'Young, S. D. (1998). Supervisory control of hybrid systems via l-complete cpproximations. In *Proc. WODES'98 - International Workshop on Discrete Event Systems, Cagliari, Italy*, pages 426–431. IEE.

Moor, T., Raisch, J., and O'Young, S. D. (2002). Discrete supervisory control of hybrid systems based on l-complete approximations. *Journal of Discrete Event Dynamic Systems*, 12:83–107.

Moormann, D. (2001). *Automatisierte Modellbildung der Flugsystemdynamik*. PhD dissertation, Aachen Technical University (RWTH Aachen), Aachen, Germany. in German.

Moormann, D., Mosterman, P., and Looye, G.-J. (1999). Object-Oriented Computational Model Building of Aircraft Flight Dynamics and Systems. *Aerospace Science and Technology*, 3:115–126.

Mosterman, P. and Biswas, G. (1999). A Java implementation of an environment for hybrid modeling and simulation of physical systems. In *International Conference on Bond Graph Modeling (ICBGM '99)*, pages 157–162. San Francisco.

Mosterman, P., Otter, M., and Elmqvist, H. (1998). Modeling Petri Nets as Local Constraint Equations for Hybrid Systems Using Modelica. In *Proceedings of SCS Summer Simulation Conference*, pages 314–319, Reno, Nevada.

Mosterman, P., Remelhe, M. P., Engell, S., and Otter, M. (2002). Simulation for analysis of aircraft elevator feedback and redundancy control. In Engell, S., Frehse, G., and Schnieder, E., editors, *Modelling, Analysis, and Design of Hybrid Systems*. Springer.

Mosterman, P. J. (1999). An overview of hybrid simulation phenomena and their support by simulation packages. In *Hybrid Systems Computation and Control (HSCC'99)*, LNCS 1569. Springer.

Mosterman, P. J. (2000a). HYBRSIM - a modeling and simulation environment for hybrid bond graphs. *Journal of Systems and Control*.

Mosterman, P. J. (2000b). Implicit modeling and simulation of discontinuities in physical system models. In Engell, S., Kowalewski, S., and Zaytoon, J., editors, *The 4th International Conference on Automation of Mixed Processes: Hybrid Dynamic Systems*, pages 35–40.

Mosterman, P. J. (2001). MASIM. Technical Report DLR-IB, DLR Oberpfaffenhofen, Oberpfaffenhofen, Germany.

Mosterman, P. J. and Biswas, G. (1995). Modeling discontinuous behavior with hybrid bond graphs. In *1995 International Workshop on Qualitative Reasoning*, pages 139–147, Amsterdam. University of Amsterdam.

Müller, C. (2002). *Analyse und Synthese diskreter Steuerungen hybrider Systeme mit Petri-Netz-Zustandsraummodellen*, volume 930 of *Fortschritt-Berichte VDI Reihe 8*. VDI-Verlag, Düsseldorf, Germany.

Müller, C., Orth, P., and Rake, H. (2001). Analyse und Synthese diskreter Steuerungen hybrider Systeme mit einem Petri-Netz-Zustandsraummodell. In Schnieder, E., editor, *Engineering komplexer Automatisierungssysteme*, EKA 2001, pages 113–131, Braunschweig, Germany.

Müller, C. and Rake, H. (1999). Modellbildung und Analyse hybrider Systeme mit Petri-Netzen und geschalteten Differentialgleichungen. In Schnieder, E., editor, *Entwicklung und Betrieb komplexer Automatisierungssysteme*, EKA '99, pages 233–246, Braunschweig, Germany.

Müller, C. and Rake, H. (2000). A Petri Net-State-Model for the Analysis and the Control Synthesis of Hybrid Technical Systems. In *Proceedings Hybrid Dynamic Systems*, ADPM 2000.

Müller, K. (1996). *Entwurf robuster Regelungen*. B.G. Teubner Stuttgart.

Münnemann, A. and Enste, U. (2001). Systemtechnische Integration gehobener Regelungsverfahren. *atp - Automatisierungstechnische Praxis*, 43(7):40–48.

Nagel, K. and Schreckenberg, M. (1992). A cellular automaton model for free-way traffic. *Journal Phys.*, 2:2221.

Naur, P. (1966). Proof of algorithms by general snapshots. *BIT (Nordisk tidskrift for informationsbehandling)*, 6(4):310–316.

Nenninger, G. (2001). *Modellbildung und Analyse hybrider dynamischer Systeme als Grundlage für den Entwurf hybrider Steuerungen*, volume 902 of *Fortschritt-Berichte VDI Reihe 8*. VDI-Verlag.

Nenninger, G., Frehse, G., and Krebs, V. (2000). Hybrid regions of attraction of piecewise affine hybrid systems. In *4th Conference on Automation of Mixed Processes: Hybrid Dynamic Systems ADPM 2000*, pages 87–92.

Nenninger, G. and Krebs, V. (1998). Analysis of Hybrid Systems using Hybrid Dynamical Models. In *Hybrid Dynamical Systems: 3rd International Conference on Automation of Mixed Processes*, pages 428–431.

Nenninger, G., Schnabel, M., and Krebs, V. (1999). Modellierung, Simulation und Analyse hybrider dynamischer Systeme mit Netz-Zustands-Modellen. *Automatisierungstechnik*, 47(3):118–126.

Nenninger, G. M., Nixdorf, B., Krebs, V. G., and Lunze, J. (2001). Erreichbarkeitsanalyse hybrider Systeme. *at - Automatisierungstechnik*, 49(2):75–85.

Nerode, A. and Kohn, W. (1993). Models for hybrid systems: Automata, topologies, controllability, observability. In Grossmann, R., Nerode, A., Ravn, A., and Rischel, H., editors, *Lecture Notes in Computer Science: Hybrid Systems*, volume 736, pages 317–356. Springer.

Nicol, D. M. and Miner, A. S. (1995). The fluid stochastic petri net simulator. In *Proc. Sixth International Workshop on Petri Nets and Performance Models - PNPM'95*, pages 214–215, Durham, North Carolina, USA. IEEE-CS Press.

Nicollin, X., Olivero, A., Sifakis, J., and Yovine, S. (1992). An approach to the description and analysis of hybrid systems. In *Proceedings of Workshop on Theory of Hybrid Systems*, volume 736 of *Lecture Notes in Computer Science*, pages 149–178, Lyngby, Denmark. Springer.

Ning, B. (1998). Absolute braking and relative distance braking train operation control modes in moving block systems. In *Computers in Railways VI*, pages 991–1001. WIT Press, Lisbon.

Nixdorf, B. and Lunze, J. (2000a). Control of a manufacturing cell. Technical report, Arbeitsbereich Regelungstechnik, TU Hamburg-Harburg. Internal document.

Nixdorf, B. and Lunze, J. (2000b). KONDISK benchmark of an automated manufacturing cell. Technical report, Technical University of Hamburg-Harburg. (in German).

Nordwig, A. (2000). the zooed homepage. Technische Universität Berlin. ISTI. http://swt.cs.tu-berlin.de.

Nordwig, A. (2002). Formal integration of structural dynamics into the object-oriented modeling of hybrid systems. In *Proceedings of the European Simulation Multi-Conference '02*. to appear.

Nöth, G. (1998). Randbedingungen für den Einsatz von regelungstechnischen Methoden. In *GMA-Kongress'98 Meß- und Automatisierungstechnik, VDI Bericht 1397*. VDI-Verlag.

Nytsch-Geusen, C. (2001). *Berechnung und Verbesserung der Energieeffizienz von Gebäuden und ihren energietechnischen Anlagen in einer objektorientierten Simulationsumgebung*. PhD thesis, TU Berlin.

Olivero, A. and Yovine, S. (1993). *KRONOS: A Tool for Verifying Real-Time Systems. User's Guide and Reference Manual*. Verimag, Grenoble, France.

Omata, T. and Farooqi, M. A. (1996). Regrasps by a Multifingered Hand Based on Primitives. In *Proceedings of the IEEE International Conference on Robotics and Automation ICRA*, pages 2774–2780, Minneapolis, Minnesota, USA.

Osder, S. (1999). Practical view of redundancy management application and theory. *Journal of Guidance, Control, and Dynamics*, 22(1):12–21.

Otter, M., Elmqvist, H., and Mattson, S. (1999). Hybrid modeling in Modelica based on the synchronous data flow principle. In *CACSD'99*, Hawaii, USA.

Otter, M., Remelhe, M., Engell, S., and Mosterman, P. (2000). Hybrid models of physical systems and discrete controllers. *at-Automatisierungstechnik*, 48:426–437.

Owicki, S. S. and Gries, D. (1976). An axiomatic proof technique for parallel programs I. *Acta Informatica*, 6:319–340.

Pachl, J. (1999). *Systemtechnik des Schnieneverkehrs*. B. G. Teubner, Stuttgart.

Panreck, K. (1999). Systembeschreibungen zur Modellierung komplexer Systeme. *at - Automatisierungstechnik*, 47(4):157.

Park, T. and Barton, P. (1997). Implicit model checking of logic based control systems. *AIChE Journal*, 43(9):2246–2260.

Pawletta, T. and Lampe, B. (2001). KONDISK project report no. la $724/8-2$ — Modeling and simulation of modular-hierarchical systems with discret event oriented structure dynamics. Technical report, University of Rostock. (in German).

Pawletta, T., Lampe, B., Pawletta, S., and Drewelow, W. (1996). An object oriented framework for modeling and simulation of variable structure systems. In Ingalls, V., Cynamon, J., and Saylor, A., editors, *SCS Summer Simulation Conf., Portland, Oregon*, pages 8–13. SCS International.

Pawletta, T., Lampe, B., Pawletta, S., and Drewelow, W. (2002). A DEVS-based approach for modeling and simulation of structure dynamics in hybrid systems. In Engell, S., Frehse, G., and Schnieder, E., editors, *Modelling, Analysis, and Design of Hybrid Systems*, Lecture Notes in Control and Information Science. Springer. (This volume).

Pawletta, T., Lampe, B., Pawletta, S., Drewelow, W., and Schildmann, P. (2001). Modeling of temporal objects with self-dynamics in hybrid systems. In Panreck, K. and Dörrscheidt, F., editors, *15th Symp. of Simulation, Paderborn*, Frontiers in Simulation, pages 73–78, Ghent, Belgium. SCS Publishing House. (in German).

Pawletta, T., Pawletta, S., and Dimitrov, E. (1994). Modeling and simulation of structure variable systems. In Kampe, G. and Zeitz, M., editors, *Progress in Simulation*, pages 59–64. Vieweg Publisher. (in German).

Pawletta, T., Pawletta, S., Schildmann, P., and Drewelow, W. (1997). Interactive modeling and simulation of time-invariant system structures. In Kuhn, A. and Wenzel, S., editors, *Progress in Simulation*, pages 649–655. Vieweg Publisher. (in German).

Paynter, H. M. (1961). Analysis and design of engineering systems. *The M.I.T. Press, Cambridge, Massachusetts.*

Pearson, R. (1984). Modern control: Why don't we used it? *InTech*, 11:47–49.

Pereira Remelhe, M., Deparade, A., and Engell, S. (2001). Integration und Synchronisierung von diskreten Beschreibungsformen und kontinuierlichen Systemmodellen in Modelica. In Panreck, K. and Dörrscheidt, F., editors, *Simulationstechnik, ASIM 2001, 15. Symposium*, pages 95–100. ASIM, SCS.

Péter, I., Pretschner, A., and Stauner, T. (2000). Heterogeneous development of hybrid systems. In *Proc. GI workshop Rigorose Entwicklung software-intensiver Systeme*, pages 83–93.

Petri, C. (1962). Kommunikation mit automaten. Technical Report 2, Institut für Instrumentelle Mathematik, Bonn. Schriften des IIM.

Petterson, S. (1999). *Analysis and Design of Hybrid Systems*. PhD thesis, Chalmers University of Technology.

Petzold, L. R. (1982). A description of DASSL: A differential/algebraic system solver. Technical Report SAND82-8637, Sandia National Laboratories, Livermore, California.

Philips, P. (2001). *Modeling, Control and Fault Detection of Discretely Observed Systems*. PhD thesis, TU Eindhoven.

Philips, P., Weiss, M., and Preisig, H. A. (1999). Control based on diskrete-event models of continuous systems. In *Proceedings of the European Control Conference 1999*.

Plank, J. (1997). *State Events in Continous Modelling and Simulation*. PhD thesis, Technical University of Vienna.

PNO (1999). Profibus-pa profile for process control devices, revision 3.0. Technical report, PNO, Karlsruhe.

Pnueli, A. (1977). The temporal logic of programs. In *Proceedings of the 18th IEEE Symposium on Foundations of Computer Science (FOCS 1977)*, pages 46–57.

Pnueli, A. (1981). The temporal logic of concurrent programs. *Theoretical Computer Science*, 13:45–60.

Pnueli, A. (1984). In transition for global to modular temporal reasoning about programs. In *Logics and Models of Concurrent Systems*, volume 13 of *NATO ASI-F*. Springer.

Pnueli, A. and Sifakis, J. (1995). Special issue on hybrid systems. *Theoretical Computer Science*, 138:1–239.

Prähofer, H. (1991). *System Theoretic Foundations for Combined Discrete-Continuous System Simulation*. PhD thesis, Johannes Kepler University of Linz.

Prähofer, H. (1996). An environment for DEVS-based multi-formalism modeling und simulation in C++. In *6th Annual Conference on AI, Simulation and Planning in High Autonomy Systems*, page 8pp. SCS International, San Diego.

Prähofer, H. and Zeigler, B. (1992). Modelling and simulation. In Pichler, F. and Schwaertzel, H., editors, *CAST - Methods in Modelling*, pages 123–241. Springer Publisher, Berlin.

Pretschner, A. (2001). Classical search strategies for test case generation with Constraint Logic Programming. In *Proc. Formal Approaches to Testing of Software*, pages 47–60.

Pretschner, A., Lötzbeyer, H., and Philipps, J. (2001). Model Based Testing in Evolutionary Software Development. In *Proc. 11th IEEE Intl. Workshop on Rapid System Prototyping*, pages 155–160.

Pretschner, A., Slotosch, O., and Stauner, T. (2000). Developing Correct Safety Critical, Hybrid, Embedded Systems. In *Proc. New Information Processing Techniques for Military Systems, NATO Research.*

Preußig, J. (2000). *Formale Überprüfung der Korrektheit von Steuerungen mittels rektangulärer Automaten*. PhD thesis, Department of Chemical Engineering, University of Dortmund, Germany. (in German).

Preußig, J., Kowalewski, S., Henzinger, T., and Wong-Toi, H. (1998). An algorithm for the approximate analysis of simple rectangular automata. In *Proc. 5th Int. School and Symposium on Formal Techniques in Fault Tolerant and Real Time Systems, Lyngby, Denmark, 1998*, Lecture Notes in Computer Science 1486, pages 228–240. Springer.

Preußig, J., Stursberg, O., and Kowalewski, S. (1999). Reachability analysis of a class of switched continuous systems by integrating rectangular approximation and rectangular analysis. In Vaandrager, F. W. and van Schuppen, J. H., editors, *Hybrid Systems: Computation and Control, Proc. 2nd Int. Workshop, HSCC'99, Berg en Dal, The Netherlands, March 1999*, Lecture Notes in Computer Science 1569, pages 209–222. Springer.

Preußig, J. and Wong-Toi, H. (2000). An procedure for the reachability analysis of rectangular automata. In *Proc. American Control Conference*, pages 1674–1678.

Queille, J.-P. and Sifakis, J. (1982). Specification and verification of concurrent systems in CESAR. In Dezani-Ciancaglini, M. and Montanari, U., editors, *Proceedings of the 5th International Symposium on Programming, Turin, April 6–8, 1982*, pages 337–350. Springer.

Raisch, J. (1998). A hierarchy of discrete abstractions for a hybrid plant. *JESA - European Journal of Automation, Special Issue on Hybrid Dynamical Systems*, 32(9-10):1073–1095.

Raisch, J. (2000a). Complex systems – simple models? In *Proc. ADCHEM2000 - International Symposium on Advanced Control of Chemical Processes, Pisa*, pages 275–286.

Raisch, J. (2000b). Discrete abstractions of continuous systems - an input/output point of view. *Mathematical and Computer Modelling of Dynamical Systems*, 6(1):6–29.

Raisch, J., Iitgin, A., and Moor, T. (2001). Hierarchical strategies for hybrid process control problems. In *Proceedings of the European Control Conference 2001*, pages 2534–2539, Porto, Portugal.

Raisch, J. and Itigin, A. (2000). Synthesis of hierarchical process control systems based on sequential aggregation. In *Proc. 3rd Mathmod, Vienna*, pages 385–389.

Raisch, J., Itigin, A., and Moor, T. (2000). Hierarchical control of hybrid systems. In Engell, S., Kowalewski, S., and Zaytoon, J., editors, *Proc. 4th International Conference on Automation of Mixed Processes: Dynamic Hybrid Systems*, pages 67–72. Shaker.

Raisch, J., Klein, E., O'Young, S. D., Meder, C., and Itigin, A. (1998). Approximating automata and discrete control for continuous systems - two examples from process control. In Antsaklis, P., Kohn, W., Nerode, A., and Sastry, S., editors, *Hybrid Systems V*, LNCS 1567, pages 279–303. Springer.

Raisch, J. and O'Young, S. (1997). A totally ordered set of discrete abstractions for a given hybrid or continuous system. In Antsaklis, P., Kohn, W., Nerode, A., and Sastry, S., editors, *Hybrid Systems IV*, volume 1273 of *LNCS*, pages 342–360. Springer.

Raisch, J. and O'Young, S. D. (1998). Discrete approximation and supervisory control of continuous systems. *IEEE Transactions on Automatic Control, Special issue on hybrid systems*, 43:569–573.

Rakoto-Ravalontsalama, N. and Aguilar-Martin, J. (1998). Diagnosing uncertain parameters to improve hybrid process model. In *Hybrid Dynamical Systems. 3rd International Conference on Automation of Mixed Processes*, pages 49–53, Reims.

Ramadge, P. J. and Wonham, W. M. (1987). Supervisory control of a class of discrete event systems. *SIAM J. Control and Optimization*, 25:206–230.

Ramadge, P. J. and Wonham, W. M. (1989). The control of discrete event systems. *Proceedings of the IEEE*, 77:81–98.

Rational (1999). *Unified Modeling Language*. Rational Software Corporation. Version 1.3.

Rational UML (1997). Unified modeling language, version 1.1. Rational Software Corporation.

Rausch, M. and Hanisch, H.-M. (1995). Netz-Condition/Event-Systeme. In Schnieder, E., editor, *Entwurf komplexer Automatisierungssysteme - Methoden, Anwendungen und Tools auf der Basis von Petrinetzen und anderer formaler Beschreibungsmittel*, pages 55–71, Braunschweig.

Raymond, P., Weber, D., Nicollin, X., and Halbwachs, N. (1998). Automatic testing of reactive systems. In *Proc. 19th IEEE Real-Time Systems Symposium*.

Rebolledo, M. (2002). Development of a Concept for the Handling of Vagueness in the SQMA Modeling Approach. Diplomarbeit, Institut für Automatisierungs- und Softwaretechnik (IAS), Universität Stuttgart.

Reckdahl, K. J. and Mitiguy, P. C. (1996). *AUTOLEV 3 Tutorial*. OnLine Dynamics, Inc., Sunnyvale, USA.

Reisig, W. (1985). *Petri Nets, An Introduction*. EATCS, Monographs on Theoretical Computer Science. Springer, Berlin.

Ricker, S. L., Sarkar, N., and Rudie, K. (1996). A Discrete-Event Systems Approach to Modeling Dextrous Manipulation. *Robotica*, 14:515–525.

Royce, W. W. (1970). Managing the development of large software systems: Concepts and techniques. In *Proc. IEEE WESTCON*.

Ruhl, H. (1999). Konzeption und Implementierung einer Visualisierungssoftware für den Modellprozess "Drei-Tank-System". Diplomarbeit, Institut für Automatisierungs- und Softwaretechnik (IAS), Universität Stuttgart.

Rumbaugh, J. (1991). *Object-Oriented Modeling and Design*. Prentice-Hall Inc., New Jersey.

Ruspini, D. and Khatib, O. (2000). A Framework for Multi-Contact Multi-Body Dynamic Simulation and Haptic Display. In *Proceedings of the 2000 IEEE/RSJ International Conference on Intelligent Robots and Systems*, Takamatsu,Japan.

Schätz, B. and Pretschner, A. (2002). Model based development of embedded systems. Submitted to Model-Driven Approaches to Software Development, OOIS'02.

Schildmann, P. (2000). Benchmarks for the simulator prototype MATSIM-2. Technical report, University of Rostock. (in German).

Schiller, F. (1997). *Diagnose dynamischer Systeme auf der Grundlage einer qualitativen Prozessbeschreibung*. Dissertation, TU Hamburg-Harburg.

Schlegl, T. (2002). *Diskret-kontinuierliche Regelung mehrfingriger Roboterhände zur robusten Manipulation von Objekten*. Number 928 in Fortschrittsberichte VDI, Reihe 8: Meß-, Steuerungs- und Regelungstechnik. VDI-Verlag, Düsseldorf.

Schlegl, T., Buss, M., Omata, T., and Schmidt, G. (2001). Fast Dextrous Regrasping with Optimal Contact Forces and Contact Sensor Based Impedance Control. In *Proceedings of the IEEE International Conference on Robotics and Automation ICRA*, pages 103–107, Seoul, Korea.

Schlegl, T., Buss, M., and Schmidt, G. (1997). Development of numerical integration methods for hybrid (discrete-continuous) dynamical systems. In *Proceedings of the IEEE/ASME International Conference on Advanced Intelligent Mechatronics AIM'97 (Tokyo, Japan, Paper No. 154)*.

Schlegl, T., Buss, M., and Schmidt, G. (2002a). A Hybrid Systems Approach towards Modeling and Dynamical Simulation of Dextrous Manipulation. *IEEE Transactions on Mechatronics, under review*.

Schlegl, T., Buss, M., and Schmidt, G. (2002b). Hybrid control of multi-fingered dextrous hands. *This volume*.

Schlegl, T., Schnabel, M. K., Buss, M., and Krebs, V. G. (2000). State Reconstruction and Error Compenstation in Discrete-Continuous Control Systems. *at - Automatisierungstechnik*, 48(9):439–447.

Schnabel, M. (2001). *Diskret-kontinuierliche dynamische Systeme: Steuerung und Beobachtung*, volume 900 of *Fortschritt-Berichte VDI Reihe 8*. VDI-Verlag.

Schöneburg, E., Heinzmann, F., and Feddersen, S. (1996). *Genetische Algorithmen und Evolutionsstrategien*. Addison-Wesley.

Schröder, J. (2002). *Modelling, State Observation and Diagnosis of Quantised Systems*. Lecture Notes in Control and Information Sciences. Springer, Berlin.

Schuler, H. (1992). Was behindert den praktischen Einsatz moderner regelungstechnischer Methoden in der Prozessindustrie? *atp - Automatisierungstechnische Praxis*, 34(3):116–123.

Schumacher, J., Morse, A., Pantelides, C., and Sastry, S., editors (1999). *Special Issue on Hybrid Systems*, volume 35 of *Automatica*.

Schürr, A. (1994). Logic based structure rewriting systems. In *Lecture Notes in Computer Science*. Springer.

Schütt, H. (1990). *Entwicklung und Erprobung eines sehr schnellen, bitorientierten Verkehrssimulationssystems für Straßennetze*. PhD thesis, TU Hamburg-Harburg.

SDL92 (1992). Specification and Description Language SDL, blue book. CCITT Recommendation Z.100.

Seebeck, J. (1998). *Modellierung der Redundanzverwaltung von Flugzeugen am Beispiel des ATD durch Petrinetze und Umsetzung der Schaltlogik in C-Code zur Simulationssteuerung*. Diplomarbeit, Arbeitsbereich Flugzeugsystemtechnik, Technische Universität Hamburg-Harburg.

Seiche, W. (1991). *Analyse und Synthese diskret gesteuerter Systeme mit Petri-Netzen*, volume 269 of *Fortschritt-Berichte VDI Reihe 8*. VDI-Verlag, Düsseldorf, Germany.

Seiche, W. and Abel, D. (1993). Entwurf verklemmungsfreier Steuerungen auf der Grundlage einer graphentheoretischen Petri-Netz-Analyse. *Automatisierungstechnik*, 41:88–93.

Selic, B., Gullekson, G., and Ward, P. T. (1994). *Real-Time Object-Oriented Modeling*. John Wiley & Sons Ltd, Chichester.

Simon, C. (2001a). Developing software controllers with petri nets and a logic of actions. In *IEEE International Conference on Robotics and Automation, ICRA 2001*, Seoul, Korea.

Simon, C. (2001b). *A Logic of Actions and Its Application to the Development of Programmable Controllers*. PhD thesis, Universität Koblenz-Landau.

Simon, C., Ridder, H., and Marx, T. (1997). The petri net tools neptun and poseidon. Fachberichte Informatik 15–97, Universität Koblenz-Landau, Institut für Informatik, Rheinau 1, D-56075 Koblenz.

Simon, C. and Thieme, J. (1999). Transformation zeitbewerteter Netzmodelle. Fachberichte Fakultät Elektrotechnik 3–99, Otto-von-Guericke-Universität Magdeburg, Institut für Automatisierungstechnik, Postfach 4120, D-39016 Magdeburg.

Six, J. (1996). Abstandhaltung und Streckenleistungsfähigkeit. *Signal+Draht*.

Smith, H. (1995). *Monotone Dynamical Systems*. American Mathematical Society, Providence.

Sreenivas, R. S. and Krogh, B. H. (1991a). On condition/event systems with discrete state realizations. *Discrete Event Dynamic Systems: Theory and Application 1*, pages 209–236.

Sreenivas, R. S. and Krogh, B. H. (1991b). Petri net based models for condition/event systems. *Proceedings of 1991 American Control Conference*, 3:2899–2904.

Stahl, K. (1998). Comparing the expressiveness of different real-time models. Master's thesis, Christian-Albrechts-University of Kiel.

Stauner, T. (2001). *Systematic development of hybrid systems*. PhD thesis, Technische Universität München.

Stauner, T. (2002). Discrete-Time Refinement of Hybrid Automata. In *Proc. HSCC'02*. To be published.

Stauner, T., Pretschner, A., and Péter, I. (2001). Approaching a Discrete-Continuous UML: Tool Support and Formalization. In *Proc. UML'2001 workshop on Practical UML-Based Rigorous Development Methods*, pages 242–257.

Steffen, T. (2001). Rekonfiguration linearer Systeme durch eine Ergänzung des Reglers. Technical report, Ruhr University Bochum, Institute for Automation and Computer Control.

Strikwerda, J. C. (1989). *Finite Difference Schemes and Partial Differential Equations*. Wadsworth & Brooks/Cole.

Stursberg, O. (2000a). *Analyse gesteuerter verfahrenstechnischer Prozesse durch Diskretisierung*. PhD thesis, Department of Chemical Engineering, University of Dortmund, Germany. (in German).

Stursberg, O. (2000b). Analysis of switched continuous systems based on discrete approximation. In *Proc. 4th Int. Conf. on Automation of Mixed Processes*, pages 73–78.

Stursberg, O. and Engell, S. (2001). Optimized startup-procedures of processing systems. In *Proc.* 6th *IFAC Symp. Dynamics and Control of Process Sys.*, pages 231–236.

Stursberg, O. and Engell, S. (to appear in July 2002). Optimal control of switched continuous systems using mixed-integer programming. In *Proc.* 15th *IFAC World Congress on Automatic Control*.

Stursberg, O. and Kowalewski, S. (1999). Approximating switched continuous systems by rectangular automata. In *Proc. European Control Conference*. CD-ROM, file 1014–4.

Stursberg, O. and Kowalewski, S. (2000). Analysis of controlled hybrid processing systems based on approximation by timed automata using interval arithmetics. In *Proc. 8th IEEE Mediterranean Conference on Control and Automation*. CD-ROM, file TA1–3.

Stursberg, O., Kowalewski, S., and Engell, S. (2000). On the generation of timed discrete approximations for continuous systems. *Mathematical and Computer Modelling of Dynamical Systems*, 6(1):51–70. Special Issue on "Discrete Event Models of Continuous Systems".

Stursberg, O., Kowalewski, S., Hoffmann, I., and Preußig, J. (1997). Comparing timed and hybrid automata as approximations of continuous systems. In Antsaklis, P., Kohn, W., Nerode, A., and Sastry, S., editors, *Hybrid Systems IV*, volume 1273 of *LNCS*, pages 361–377. Springer.

Stursberg, O. and Panek, S. (to appear in 2002). Control of switched continuous systems based on disjunctive formulations. In 5th *Int. Workshop on Hybrid Systems: Computation and Control*, LNCS. Springer.

Sussmann, H. (1999). A maximum principle for hybrid optimal control problems. In *Proc.* 38th *IEEE Conf. Decision and Control*, pages 425–430.

Tavernini, L. (1987). Differential automata and their discrete simulators. *Nonlinear Analysis, Theory, Methods, and Applications*, 11:665–683.

Thieme, J. (2002). *Symbolische Erreichbarkeitsanalyse und automatische Implementierung struktureller, zeitbewerteter Steuerungsmodelle.* PhD thesis, Martin-Luther-Universität Halle-Wittenberg, Mathematisch-Naturwissenschaftlich-Technische Fakultät.

Thieme, J. and Lüder, A. (1999). Transformation von Netzmodellen zur Analyse technischer Systeme. Fachberichte Fakultät Elektrotechnik 2–99, Otto-von-Guericke-Universität Magdeburg, Institut für Automatisierungstechnik, Postfach 4120, D-39016 Magdeburg.

Thomas, C. (1996). *An Object Oriented Approach to Modeling and Simulation of Complex Systems.* VDI-Verlag. (in German).

Thomas, J. (1995). *Numerical Partial Differential Equations: Finite Difference Methods.* Springer.

Tittus, M., Egardt, B., and Lennartson, B. (1994). Hybrid systems in process control. In *3rd IEEE Conference on Decision and Control*, pages 3587–3595.

Tomlin, C. (1999). Towards efficient computation of solutions to hybrid systems. In *Proceedings of the 38th IEEE Conference on Decision and Control (Phoenix, AZ)*, pages 3532–3537.

Tomlin, C. and Greenstreet, M. R., editors (2002). *Hybrid Systems: Computation and Control, 5th International Workshop (HSCC'02)*, volume 2289 of *Lecture Notes in Computer Science*, Stanford, CA, USA. Springer.

Tomlin, C., Lygeros, J., and Sastry, S. (2000). A game theoretic approach to controller design for hybrid systems. *Proceedings of the IEEE*, 88(7):949–970.

Treseler, H. (2001). *Ein Rechnerwerkzeug zur formalen Verifikation diskret gesteuerter verfahrenstechnischer Prozesse.* PhD thesis, Department of Chemical Engineering, University of Dortmund, Germany. (in German).

Trontis, A. and Spathopoulos, M. (2001). Target control for hybrid systems with linear continuous dynamics. In *Proc.* 40th *IEEE Conf. on Decision and Control*, pages 1229–1234.

Turing, A. M. (1949). On checking a large routine. In *Report of a Conference on High Speed Automatic Calculating Machines*, pages 67–69, Cambridge. University Mathematics Laboratory.

Uebel, H. (2000). Durchsatz von Strecken und Stationen bei Bahnen. In *Gesamtverkehrsforum 2000*, number 1545 in VDI Berichte, pages 257–275. VDI-Verlag, Düsseldorf.

Uhrmacher, A. M. and Arnold, R. (1994). Distributing and maintaining knowledge: Agents in variable structure environment. In *5th Annual Conference on AI, Simulation and Planning in High Autonomy Systems*, pages 178–194.

Utkin, V. (1992). *Sliding Modes in Control Optimization.* Springer.

Vaandrager, F. and van Schuppen, J., editors (1999). *Hybrid Systems – Computation and Control, Proc. 2nd Int. Workshop HSCC'99, Berg en Dal, The Netherlands, March 1999*, volume 1569 of *Lecture Notes in Computer Science*. Springer.

Valavanis, K. (1997). Special issue on applications of discrete event and hybrid systems. *IEEE Robotics and Automation Magazine*, 4.

van der Schaft, A. and Schumacher, H. (2000). *An Introduction to Hybrid Systems*, volume 251 of *Lecture Notes in Control and Information Science*. Springer, London.

van der Schaft, A. J. and Schumacher, J. M. (1996). The complementary-slackness of hybrid systems. *Math. Contr. Signals Syst.*, 9:266–301.

Vardi, M. Y. and Wolper, P. (1994). Reasoning about infinite computations. *Information and Computation*, 115(1):1–37.

Vecchietti, A. and Grossmann, I. (1999). Logmip: A disjunctive 0-1 nonlinear optimizer for process system models. *Comp. and Chemical. Eng.*, 23:555–565.

Verghese, G. C., Lévy, B. C., and Kailath, T. (1981). A generalized state-space for singular systems. *IEEE Transactions on Automatic Control*, 26(4):811–831.

Vidal, R. (1993). *Applied Simulated Annealing*. Springer, Berlin.

Vidal, R., Schaffert, S., Shakernia, O., Pappas, G., and Sastry, S. (2001). Decidable and semi-decidable controller synthesis for classes of discrete-time hybrid systems. In *Proc.* 40th *IEEE Conf. Decision and Control*, pages 1243–1248.

Visual Object Net ++ (2000). http://www.r-drath.de/VON/von_e.htm.

von Stryk, O. (2000). Numerical hybrid optimal control and related topics. Habilitation Dissertation, Technische Universität München.

von Stryk, O. (2001). User's guide for DIRCOL version 2.1: A direct collocation method for the numerical solution of optimal control problems. Technical report, Simulation and Systems Optimization Group, Technische Universität Darmstadt. WWW: www.sim.informatik.tu-darmstadt.de/sw/.

von Stryk, O. and Bulirsch, R. (1992). Direct and indirect methods for trajectory optimization. *Annals of Operations Research*, 36:357–373.

von Stryk, O. and Glocker, M. (2000). Decomposition of mixed-integer optimal control problems using branch and bound and sparse direct collocation. In *ADPM – 4th Int'l Conf. on Automation of Mixed Processes: Hybrid Dynamic Systems*, pages 99–104.

von Stryk, O. and Glocker, M. (2001). Numerical mixed-integer optimal control and motorized traveling salesmen problems. *APII – JESA (Journal européen des systèmes automatisés – European Journal of Control)*, 35(4):519–533.

Vries, R. d., Tretmans, J., Belinfante, A., Feenstra, J., Feijs, L., Mauw, S., Goga, N., Heerink, L., and Heer, A. d. (2000). Côte de Resyste in Progress. In *Progress 2000 – Workshop on Embedded Systems*, pages 141–148.

W3C (1998). Extensible markup language XML. http://www.w3.org/TR/REC-xml.

Wiedemann, R. (1974). Simulation des Straßenverkehrsflusses. Technical Report 8, Instituts für Verkehrswesen der Universität, Karlsruhe, Germany.

Wiedemann, R. (1991). Modelling of rti-elements on multi-lane roads. In of the European Community, C., editor, *Advanced Telematics in Road Transport*, Brussels.

Wieting, R. (1996). Modeling and simulation of hybrid systems using hy-brid high-level nets. In *8th European Simulation Symposium ESS'96*, volume 1, pages 96–100.

Wieting, R. (1998). *Modellbildung und Simulation mit hybriden höheren Netzen*. PhD thesis, Carl von Ossietzky Universität, Oldenburg. ISBN 3-8265-3291-0.

Willems, J. C. (1989). Models for dynamics. *Dynamics Reported*, 2:172–269.

Willems, J. C. (1991). Paradigms and puzzles in the theory of dynamic systems. *IEEE Transactions on Automatic Control*, 36:258–294.

Williams, H. P. (1978). *Model Building in Mathematical Programming*. J. Wiley P., 1st edition.

Woelfl, K. (1995). *Planung von Manipulationsvorgängen einer Roboterhand*. Number 455 in Fortschrittsberichte VDI, Reihe 8: Meß-, Steuerungs- und Regelungstechnik. VDI-Verlag, Düsseldorf.

Wolf, A. (2001). *Components and Interfaces for Modeling and Simulation of Continuous-Discrete Systems*. PhD thesis, Technical University of Magdeburg. (in German).

Wöllhaf, K. (1995). *Object Oriented Modeling and Simulation of Multi-Product Batch Plants*. PhD thesis, University of Dortmund. (in German).

Wolter, K. (1999). *Performance and Dependability Modelling with Second Order Fluid Stochastic Petri Nets*. Shaker, Aachen.

Wolter, K. (2001). A performability model for a hybrid reactor system. In Djemame, K. and Kara, M., editors, *Proc. 17th annual UK Performance Engineering Workshop*, pages 13–22, Leeds, UK.

Wolter, K. and Zisowsky, A. (2001). Performance evaluation. *On Markov Reward Modelling with FSPNs*, 44:165–186.

Xu, X. and Antsaklis, P. (2001). An approach for solving general switched linear quadratic optimal control problems. In *Proc.* 40th *IEEE Conf. Decision and Control*, pages 2478–2483.

Yovine, S. (1997). Kronos: a verification tool for real-time systems. *Software Tools for Technology Transfer*, 1(1,2):123–133.

Zaytoon, J., editor (1998). *3rd Int. Conf. on Automation of Mixed Processes: Hybrid Dynamic Systems (ADPM'98), Reims, France*. Université de Reims.

Zeigler, B. (1976). *Theory of Modelling and Simulation*. John Wiley & Sons.

Zeigler, B. (1984). *Multifacetted Modelling and Discrete Event Simulation*. Academic Press, Inc.

Zeigler, B. (1990). *Object-Oriented Simulation with Hierarchical, Modular Models*. Academic Press, Inc.

Zeigler, B. and Prähofer, H. (2000). *Theory of Modelling and Simulation*. Academic Press, London, second edition.

Zhang, P. and Cassandras, C. (2001). An improved forward algorithm for optimal control of a class of hybrid systems. In *Proc.* 40th *IEEE Conf. Decision and Control*, pages 1235–1236.

Zhivoglyadov, P. and Middleton, R. (1999). A novel approach to systematic switching control design for a class of hybrid systems. In *Proc. of the 38th International Conference on Decision and Control, CDC'99*. IEEE Press.

Zhu, P. (2001). *Betriebliche Leistung von Bahnsystemen unter Störungsbedingungen*. VDI-Verlag, Düsseldorf.

Zimmermann, A., German, R., Freiheit, J., and Hommel, G. (2000). Petri net modelling and performability evaluation with timenet 3.0. In *Proc. 11th Int. Conf. on Computer Performance Evaluation; Modelling Techniques and Tools*, number 1786 in LNCS, pages 188–202, Schaumburg, IL, USA.

Zisowsky, A. (1998). Entwurf und Implementierung eines Verfahrens für die transiente Analyse fluider stochastischer Petri-Netze. Master's thesis, TU Berlin.

Index

Lecture Notes in Control and Information Sciences

Edited by M. Thoma and M. Morari
1998–2002 Published Titles:

Vol. 231: Emel'yanov, S.V.; Burovoi, I.A.; Levada, F.Yu.
Control of Indefinite Nonlinear Dynamit Systems
196 pp. 1998 [3-540-76245-0]

Vol. 232: Casals, A.; de Almeida, A.T. (Eds)
Experimental Robotics V: The Fifth International Symposium Barcelona, Catalonia, June 15–18, 1997
190 pp. 1998 [3-540-76218-3]

Vol. 233: Chiacchio, P.; Chiaverini, S. (Eds)
Complex Robotic Systems
189 pp. 1998 [3-540-76265-5]

Vol. 234: Arena, P.; Fortuna, L.; Muscato, G.; Xibilia, M.G.
Neurat Networks in Multidimensional Domains: Fundamentals and New Trends in Modelling and Control
179 pp. 1998 [1-85233-006-6]

Vol. 235: Chen, B.M.
H∞ Control and Its Applications
361 pp. 1998 [1-85233-026-0]

Vol. 236: de Almeida, A.T.; Khatib, O. (Eds)
Autonomous Robotic Systems
283 pp. 1998 [1-85233-036-8]

Vol. 237: Kreigman, D.J.; Hager, G.D.; Morse, A.S. (Eds)
The Confluence of Vision and Control
304 pp. 1998 [1-85233-025-2]

Vol. 238: Elia, N.; Dahleh, M.A.
Computational Methods for Controller Design
200 pp. 1998 [1-85233-075-9]

Vol. 239: Wang, Q.G.; Lee, T.H.; Tan, K.K.
Finite Spectrum Assignment for Time-Delay Systems
200 pp, 1998 [1-85233-065-1]

Vol. 240: Lin, Z.
Low Gain Feedback
376 pp. 1999 [1-85233-081-3]

Vol. 241: Yamamoto, Y.; Hara S.
Learning, Control and Hybrid Systems
472 pp. 1999 [1-85233-076-7]

Vol. 242: Conte, G.; Moog, C.H.; Perdon A.M.
Nonlinear Control Systems
192 pp. 1999 [1-85233-151-8]

Vol. 243: Tzafestas, S.G.; Schmidt, G. (Eds)
Progress in Systems and Robot Analysis and Control Design
624 pp. 1999 [1-85233-123-2]

Vol. 244: Nijmeijer, H.; Fossen, T.I. (Eds)
New Directions in Nonlinear Observer Design
552 pp. 1999 [1-85233-134-8]

Vol. 245: Garulli, A.; Tesi, A.; Vicino, A. (Eds)
Robustness in Identification and Control
448 pp. 1999 [1-85233-179-8]

Vol. 246: Aeyels, D.; Lamnabhi-Lagarrigue, F.; van der Schaft, A. (Eds)
Stability and Stabilization of Nonlinear Systems
408 pp. 1999 [1-85233-638-2]

Vol. 247: Young, K-D.; Özgüner, Ü. (Eds)
Variable Structure Systems, Sliding Mode and Nonlinear Control
400 pp. 1999 [1-85233-197-6]

Vol. 248: Chen, Y.; Wen C.
Iterative Learning Control
216 pp. 1999 [1-85233-190-9]

Vol. 249: Cooperman, G.; Jessen, E.; Michler, G. (Eds)
Workshop on Wide Area Networks and High Performance Computing
352 pp. 1999 [1-85233-642-0]

Vol. 250: Corke, P.; Trevelyan, J. (Eds)
Experimental Robotics VI
552 pp. 2000 [1-85233-210-7]

Vol. 251: van der Schaft, A.; Schumacher, J.
An Introduction to Hybrid Dynamical Systems
192 pp. 2000 [1-85233-233-6]

Vol. 252: Salapaka, M.V.; Dahleh, M.
Multiple Objective Control Synthesis
192 pp. 2000 [1-85233-256-5]

Vol. 253: Elzer, P.F.; Kluwe, R.H.; Boussoffara, B.
Human Error and System Design and Management
240 pp. 2000 [1-85233-234-4]

Vol. 254: Hammer, B.
Learning with Recurrent Neural Networks
160 pp. 2000 [1-85233-343-X]

Vol. 255: Leonessa, A.; Haddad, W.H.; Chellaboina V.
Hierarchical Nonlinear Switching Control Design with Applications to Propulsion Systems
152 pp. 2000 [1-85233-335-9]

Vol. 256: Zerz, E.
Topics in Multidimensional Linear Systems Theory
176 pp. 2000 [1-85233-336-7]

Vol. 257: Moallem, M.; Patel, R.V.; Khorasani, K.
Flexible-link Robot Manipulators
176 pp. 2001 [1-85233-333-2]

Vol. 258: Isidori, A.; Lamnabhi-Lagarrigue, F.; Respondek, W. (Eds)
Nonlinear Control in the Year 2000 Volume 1
616 pp. 2001 [1-85233-363-4]

Printing (Computer to Film): Saladruck Berlin
Binding: Stürtz AG, Würzburg